NEURAL CONTROL
OF LOCOMOTION

ADVANCES IN BEHAVIORAL BIOLOGY

Editorial Board:

Jan Bures	*Institute of Physiology, Prague, Czechoslovakia*
Irwin Kopin	*National Institute of Mental Health, Bethesda, Maryland*
Bruce McEwen	*Rockefeller University, New York, New York*
James McGaugh	*University of California, Irvine, California*
Karl Pribram	*Stanford University School of Medicine, Stanford, California*
Jay Rosenblatt	*Rutgers University, Newark, New Jersey*
Lawrence Weiskrantz	*University of Oxford, Oxford, England*

Recent Volumes in this Series

Volume 8 • DRUGS AND THE DEVELOPING BRAIN
Edited by Antonia Vernadakis and Norman Weiner • 1974

Volume 9 • PERSPECTIVES IN PRIMATE BIOLOGY
Edited by A. B. Chiarelli • 1974

Volume 10 • NEUROHUMORAL CODING OF BRAIN FUNCTION
Edited by R. D. Myers and René Raúl Drucker-Colín • 1974

Volume 11 • REPRODUCTIVE BEHAVIOR
Edited by William Montagna and William A. Sadler • 1974

Volume 12 • THE NEUROPSYCHOLOGY OF AGGRESSION
Edited by Richard E. Whalen • 1974

Volume 13 • ANEURAL ORGANISMS IN NEUROBIOLOGY
Edited by Edward M. Eisenstein • 1975

Volume 14 • NUTRITION AND MENTAL FUNCTIONS
Edited by George Serban • 1975

Volume 15 • SENSORY PHYSIOLOGY AND BEHAVIOR
Edited by Rachel Galun, Peter Hillman, Itzhak Parnas, and Robert Werman • 1975

Volume 16 • NEUROBIOLOGY OF AGING
Edited by J. M. Ordy and K. R. Brizzee • 1975

Volume 17 • ENVIRONMENTS AS THERAPY FOR BRAIN DYSFUNCTION
Edited by Roger N. Walsh and William T. Greenough • 1976

Volume 18 • NEURAL CONTROL OF LOCOMOTION
Edited by Richard M. Herman, Sten Grillner, Paul S. G. Stein, and Douglas G. Stuart • 1976

Volume 19 • THE BIOLOGY OF THE SCHIZOPHRENIC PROCESS
Edited by Stewart Wolf and Beatrice Bishop Berle • 1976

A Continuation Order Plan is available for this series. A continuation order will bring delivery of each new volume immediately upon publication. Volumes are billed only upon actual shipment. For further information please contact the publisher.

NEURAL CONTROL OF LOCOMOTION

Edited by

Richard M. Herman
Krusen Research Center, Temple University
and Moss Hospital
Philadelphia, Pennsylvania

Sten Grillner
Institute of Physiology, GIH
Stockholm, Sweden

Paul S. G. Stein
Washington University
St. Louis, Missouri

and

Douglas G. Stuart
University of Arizona
Tuscon, Arizona

PLENUM PRESS · NEW YORK AND LONDON

Library of Congress Cataloging in Publication Data

Main entry under title:

Neural control of locomotion.

(Advances in behavioral biology; v. 18)
"Proceedings of an international Conference on Neural Control of Locomotion held in Valley Forge, Pennsylvania, September 29–October 2, 1975."
Includes index.
1. Human locomotion—Congresses. 2. Animal locomotion—Congresses. 3. Neurophysiology—Congresses. I. Herman, Richard M.
QP301.N45   591.1'852   76-18949
ISBN 0-306-37918-X

Proceedings of an International Conference on Neural Control of
Locomotion held in Valley Forge, Pennsylvania, September 29–October 2, 1975

© 1976 Plenum Press, New York
A Division of Plenum Publishing Corporation
227 West 17th Street, New York, N.Y. 10011

All rights reserved

No part of this book may be reproduced, stored in a retrieval system, or transmitted,
in any form or by any means, electronic, mechanical, photocopying, microfilming,
recording, or otherwise, without written permission from the Publisher

Printed in the United States of America

SAN FRANCISCO
PUBLIC LIBRARY

3 1223 01365 8795

The manuscripts were prepared for publication by

Anna M. Tummillo

with the assistance of

Ruth D. Christaldi

Due to the length of these proceedings the discussion summaries could not be included in this volume. They may be obtained from

>Krusen Research Center
>Moss Hospital
>12th Street and Tabor Road
>Philadelphia, Pennsylvania 19141

Preface

In March, 1974, John Desmedt in Brussels and Sten Grillner in Göteborg encouraged Richard Herman to convene an international conference concerned with control of locomotion. It was the opinion of these scientists that the amount of data accumulated in the fields of invertebrate and vertebrate locomotion warranted such a meeting. Commonality of behavior across species was to be the conference theme. Grillner, Desmedt, and Herman concluded that the scientific investigations presented at the conference would stimulate discussion of the concepts which underlie certain characteristics and control properties.

A planning committee was formed, comprised of investigators from the fields of both invertebrate and vertebrate locomotion. These included Richard Herman (Philadelphia), Sten Grillner (Stockholm), H.J. Ralston (San Francisco), P.S.G. Stein (St. Louis), Douglas Stuart (Tucson), and Roy Wirta (Philadelphia). Ms. Ronnie Zuckerman (Philadelphia) acted as the coordinator throughout the entire project.

The goals and structure of the conference were:

1. To provide a comprehensive analysis of a stereotypic motor behavior, namely locomotion, and of the neural control factors related to the performance of this behavior;
2. To stress characteristics of bodily progression common to a number of invertebrate and vertebrate species;
3. To describe possible solutions or strategies developed by the nervous system in dealing with certain physical properties of the organism;
4. To present views pertaining to central and peripheral control of the developing animal and to compare these views with those derived from experimentation in the mature and developing animal;
5. To provide a unique forum for established investigators who utilize various species and experimental techniques to communicate with each other and with others who are interested in the analysis and the control of locomotion;
6. To emphasize certain aspects of human locomotion with respect to characteristics of performance, to adaptation to the environment

and disease and to neural implications as compared to the control of the step-cycle in other classes of animals;
7. To encourage collaboration among scientists to establish guidelines for new techniques of experimentation;
8. To offer a conceptual framework for studies of other stereotyped motor behaviors; and
9. To contribute to the neural implications of normal and abnormal motor function.

To achieve the stated objectives of this conference, the sessions were structured to describe the characteristics of locomotion across a number of species and the mechanisms of central and peripheral control, emphasizing the central view of control and modulation of these central mechanisms.

The speakers selected by the committee were among the leading contributors to the field. Each speaker was given a charge consistent with the theme of the assigned session (see Contents) and with the charges assigned to other speakers in the same category and subcategory.

Unfortunately, a number of the noted Russian scientists who were invited did not come and did not submit papers (see "Lost Opportunity," Camhi et al., Science, 190, 422, Oct. 31, 1975). The conference program was modified to maintain the structure and goals of the conference. A series of papers on Human Solutions to Locomotion (Herman, Craik, Cook) replaced, in part, the intended presentation by V.S. Gurfinkel; Grillner's paper entitled "Some Aspects on the Descending Control of the Spinal Circuits Generating Locomotor Movements" replaced M.L. Shik; Perret replaced G.M. Orlovsky with a presentation entitled "Neural Control of Locomotion in the Decorticate Cat" and characteristics of fish locomotion were detailed by Grillner, who had previously collaborated with S.M. Kashin.

In a conference of this magnitude, numerous individuals and groups must contribute substantially to the total effort. The editors acknowledge the contributions of the many members of the Krusen Research Center of Temple University and Moss Rehabilitation Hospital, the administration of the Faculty of Medicine of Temple University and the Department of Rehabilitation Medicine whose efforts helped make the conference a notable success. Further, the editors specifically acknowledge the participation of the grantors, i.e., The Alfred P. Sloan Foundation, the Rehabilitation Services Administration of HEW, the Department of Rehabilitation Medicine of Temple University, and the Faculty of Medicine of Temple University. Particular indebtedness for the successful outcome of the conference is due to Ms. Ronnie Zuckerman, Eileen Klaus, and Helen Bleil, who spent innumerable hours in the preparation and administration of the conference. Thanks goes to Sue Ritter for her design of the logo which appears on the dust cover.

PREFACE

The editors thank the chairmen of the sessions for their tireless efforts in controlling the flow of questions pertaining to each presentation, in contributing to scientific detail during each session, and for their review of the session manuscripts. The editors also wish to thank all the participants in the free communication sessions.

Richard M. Herman

Contents

I. CHARACTERISTICS OF LOCOMOTION
(Session Chairmen: T.D.M. Roberts, P.S.G. Stein, F. Todd)

Physical Principles of Locomotion 1
    H.D. Eberhart

Human Solutions for Locomotion: Single
        Limb Analysis 13
    R. Herman, R. Wirta, S. Bampton, and
    F.R. Finley

Human Solutions for Locomotion: Interlimb
        Coordination 51
    R. Craik, R. Herman, and F.R. Finley

Human Solutions for Locomotion: The
        Initiation of Gait 65
    T. Cook and B. Cozzens

Energetics of Human Walking 77
    H.J. Ralston

Movements of the Hindlimb During Locomotion
        of the Cat 99
    M.C. Wetzel, A.E. Atwater, and D.G. Stuart

Arthropod Walking 137
    G. Hoyle

On the Generation and Performance of Swimming
        in Fish 181
    S. Grillner and S. Kashin

Analysis of Tetrapod Gaits: General Considerations
        and Symmetrical Gaits 203
    M. Hildebrand

Robot Locomotion 237
    R.B. McGhee

II. CENTRAL - PERIPHERAL MECHANISMS

A. CENTRAL VIEW OF CONTROL
(Session Chairmen: W.J. Davis, G. Hoyle)

Organizational Concepts in the Central Motor
    Networks of Invertebrates 265
    W.J. Davis

Command Neurons

Command Interneurons and Locomotor Behavior
    in Crustaceans 293
    J.L. Larimer

Trigger Neurons in the Mollusk Tritonia 327
    A.O.D. Willows

Some Aspects on the Descending Control of the Spinal
    Circuits Generating Locomotor Movements 351
    S. Grillner

Pattern Generation: Single Limb

Neuronal Mechanisms for Rhythmic Motor Pattern
    Generation in a Simple System 377
    A.I. Selverston

Central Nervous Control of Cockroach Walking 401
    C.R. Fourtner

Neural Control of Flight in the Locust 419
    M. Burrows

Central Generation of Locomotion in Vertebrates 439
    V.R. Edgerton, S. Grillner, A. Sjöstrom,
    and P. Zangger

Pattern Generation: Interlimb

Mechanisms of Interlimb Phase Control 465
    P.S.G. Stein

Basic Programs for the Phasing of Flexion and
    Extension Movements of the Limbs
    During Locomotion 489
    J. Halbertsma, S. Miller, and F.G.A. van der Meché

B. MODULATION OF CENTRAL CONTROL
(Session Chairman: G. Wendler)

Function of Segmental Reflexes in the Control of
    Stepping in Cockroaches and Cats 519
  K.G. Pearson and J. Duysens

The Role of Vestibular and Neck Receptors
    in Locomotion 539
  T.D.M. Roberts

Non-Rhythmic Sensory Inputs: Influence on
    Locomotory Outputs in Arthropods 561
  J.M. Camhi

Neural Control of Locomotion in the
    Decorticate Cat 587
  C. Perret

Modulation of Proprioceptive Information
    in Crustacea 617
  W.H. Evoy

Phasic Control of Reflexes During Locomotion
    in Vertebrates 647
  H. Forssberg, S. Grillner, S. Rossignol,
  and P. Wallén

Motor Behavior Following Deafferentation in the
    Developing and Motorically Mature Monkey 675
  E. Taub

C. DEVELOPMENT
(Session Chairman: D. Stuart)

The Development of Neural Circuits in the Limb
    Moving Segments of the Spinal Cord 707
  L. Landmesser

Developmental Aspects of Locomotion 735
  G. Székely

III. FREE COMMUNICATIONS

A Kinetic Analysis of the Trot in Cats 759
  G. Ariel and R. Maulucci

Ipsilateral Extensor Reflexes and Cat Locomotion 763
  J. Duysens and K.G. Pearson

Long-Term Peripheral Nerve Activity During
      Behavior in the Rabbit 767
  J.A. Hoffer and W.B. Marks

Chemical Lesioning of the Spinal Noradrenaline
      Pathway: Effects on Locomotion in the Cat 769
  L.M. Jordan and J.D. Steeves

Metamorphosis of the Motor System During the
      Development of Moths 775
  A.E. Kammer and M.B. Rheuben

Cerebellar Neuronal Firing Patterns in the Intact
      and Unrestrained Cat During Walking 781
  J.G. McElligott

Proprioception from Nonspiking Sensory Cells in
      a Swimming Behavior of the Sand Crab,
      Emerita analoga 785
  D.H. Paul

Discharge Patterns of Motor Units During Cat
      Locomotion and Their Relation to
      Muscle Performance 789
  F.E. Zajac and J.L. Young

IV. CONCLUSION
(Session Chairman: K.G. Pearson)

Intrasegmental Mechanisms for Stepping 796
  K.G. Pearson

Coordination of Movements 798
  E. Bizzi

Central Activation of Movements 804
  J. Davis

The Interaction of Central Commands and
      Peripheral Feedback in Pyramidal
      Track Neurons (PTNs) of the Monkey 808
  E. Evarts

Index 819

PHYSICAL PRINCIPLES OF LOCOMOTION

H.D. Eberhart

Biomechanics Laboratory

University of California, Berkeley

Locomotion is the act of moving from place to place. Since motion is involved, Newton's laws governing the motion of a particle, as used in mechanics and in particular in dynamics, are recalled and used as a starting point. It is obvious that the application of a force to a stationary body produces motion by overcoming inertia and by overcoming such restraining forces as friction and viscosity of the surrounding environment. Without the application of a force or forces no voluntary movement, in this case locomotion, can take place.

It is equally obvious that there must be a source of energy for the performance of mechanical work. The source of both force and energy for locomotion in all species of animals is internal and derived from muscle.

The resulting movement may be creeping, walking by means of legs, paddling through water, or flying through air. In general, forward movement is obtained by pushing backward against surroundings, whether ground, water, or air. This pushing takes many forms.

The bipedal gait of man is used to illustrate the fundamental mechanical factors involved in locomotion.

Introduction

Locomotion, as a subject of interest, has been studied by various individuals for some hundreds of years. The desire of man to move about with greater ease, or speed, or through the air, or in or on the water, led to a study of naturally occurring solutions of locomotion problems and some attempts at imitation. About five hundred years ago Leonardo da Vinci formulated some rather sophisticated ideas of the mechanics of flight. But such study was sporadic until more recent times when equipment such as the motion picture camera was developed, thereby making it possible to analyze the gait of man and other animals. Whether continuing study of nature's solution to locomotion problems has contributed very much to man's development of machines for the purpose of assisting movement is doubtful, except after the fact, since the movement of any animal is very complex, requiring special equipment to determine what is happening and patience and ability on the part of the investigator to understand why and how it happened. Watching the flight of birds, both powered and soaring, created a desire in man to fly, but a detailed understanding of how a bird flies contributed little to the development of the airplane. Lift and drag as a function of shape and orientation were determined by intuition, experiment, and analysis. However, an understanding of basic principles contributed to explaining some of nature's solutions to flight, which in turn led to improvements in airplane design.

Whether or not a study of the locomotion of various animals that inhabit the earth has a utilitarian value is not of particular importance. Although there are indications in a number of fields that mechanical and electronic systems can benefit by a detailed understanding of some of the processes developed in nature, the knowledge of such processes is of interest for its own sake. But the possibility of valuable contributions does exist.

A major area of human disability involves various types of functional loss of arms and legs due to disease or accident resulting in paralysis or amputation. To restore function often requires various devices, braces or prostheses, that must be accurately fitted and that can be controlled to be useful. Much improvement in devices has taken place during the last thirty years through coordinated research activity of various organizations in several parts of the world. But the one area that has lagged behind and prevented greater progress, particularly in connection with arm function, has been the lack of adequate control of movement. It is recognized that miniature power sources exist, but satisfactory functional control of such external power has been lacking. Possibly some progress in this most difficult area might result from the presentations and discussions at this conference.

GENERAL SUMMARY OF LOCOMOTION

Locomotion involves movement from one place to another. Movement requires the application of a reaction force to change from a condition of rest to one of motion. A force sufficient to cause movement must overcome the resistance or restraint to such change of position.

Briefly, there must be:

(1) an internal source of energy, necessary for the performance of mechanical work, and

(2) an internal development of force, which can be applied to the environment to generate a reaction force.

It has been said that there are three basic mechanisms used in achieving motility: amoeboid, ciliary, and muscle action. The first involves changes in cell shape by flow of cytoplasm and pseudopodal activity. Cilia used for movement function in an aqueous medium and are found in cell surfaces covered by a fluid or film. Ciliary movement as a method of locomotion is only of use to very small creatures and is produced by wave-like motion of few or numerous very fine "hairs" known as cilia. The majority of animal movements, however, depend on the use of muscle. Although the fundamental chemistry involved in all types of contractility may be similar, only locomotion due to muscle action will be considered further.

Muscles contain within themselves a complex chemical system, capable of providing the energy for mechanical work. Contractile elements of the system convert chemical energy into mechanical energy, resulting in the generation of force. If such force is exerted against a rigid surface, such as the ground, and if there is sufficient friction so that slippage does not occur, the ground reaction force essentially pushes the body forward. In a water or air environment the reaction force is obtained by displacing a quantity of water or air, necessitating a large movement of the driving elements relative to forward progression.

Regardless of the type of organism, a source of energy and a means of generating a reaction force are prerequisites for locomotion.

In mechanics, work is defined as the product of force and the distance through which the force acts. By shortening, muscle can change the distance between its points of attachment, this being possible because of the presence of a movable joint in the system. For locomotion, as considered here, to take place, it is essential that there be relative motion between body segments in order to develop a reaction force that will produce motion of the body as a whole.

RESTRAINTS TO LOCOMOTION

Given an organism with muscles that contract and cause relative motion between parts of the system, which then produces a reaction force from the environment, what restraints must be overcome in order to achieve locomotion?

There must be a component of the reaction force that supports the body against the pull of gravity. Resistance of the surrounding medium, such as air or water, impedes movement due to viscosity of friction; the magnitude of this effect is a function of velocity. Internal friction within the system also may be a factor but generally is negligible. Nature of the terrain affects locomotion, i.e., roughness and slopes. Inertia of the body itself resists movement and requires a force proportional to the mass to produce a change in velocity. For locomotion generally, and for bipedal especially, the necessity to maintain balance or stability requires muscle action.

Mechanical analogies include a machine equipped with an engine that produces motion by turning wheels in contact with the ground, or the propellers of a ship or airplane, which push backward against the ground or surrounding water or air with sufficient force to overcome resistance to motion. For successful operation there must be ability to steer and to apply brakes. This involves the control of the amount and the rate of energy expenditure as well as control of direction and of stability. In living organisms the muscles are the engines and the bones are levers that function in place of wheels or propellers. Control is achieved by the action of the nervous system.

The basic physical principles mentioned above with examples from the field of human locomotion will be presented.

HUMAN WALKING

Movement is produced by angular displacement of body segments resulting from muscle contraction. Details of such motion depend on the functional characteristics of isolated skeletal muscle, compensatory skeletal arrangements of the muscles within the body, and the phasic action of all muscles involved in the action.

Isolated human muscle under completely normal voluntary control has been studied in amputees who have had cineplastic operations. The isometric length-tension diagram shows that maximal force under voluntary contraction is developed when the muscle is at rest length. Rest length is that which an inactive muscle assumes when stretched by a barely measurable force (Ralston et al.,

1947). At other lengths the force that can be developed is less and some arrangements have been built into the system that nearly compensate for the reduction. For example, flexion of the elbow joint involves an increase in the lever arm as the force in the muscle decreases with shortening, resulting in nearly constant movement over a considerable range. A muscle that passes over two or more joints can be maintained at or near its rest length for maximal force while simultaneous rotations take place at the joints with little total change in muscle length. The hamstring muscles acting over the hip and knee joints, are a good example.

During locomotion the major groups of muscles of the lower limb are active at or around the beginning and termination of the stance and swing phases of walking. These are periods of acceleration and deceleration of the leg and transference of body weight from one foot to the other. During midstance and midswing, most muscles are quiet. Muscles that are used to initiate movement are generally stretched beyond their rest lengths before being activated, and are not active when in a shortened state. On the other hand, muscles that decelerate moving skeletal parts or that resist the external force of gravity may be activated when short and may be elongated during activity.

It follows that muscles function in three ways:

(1) They cause acceleration of joint movement, with the muscle shortening and doing physical (positive) work.

(2) They decelerate joint movement, with the muscle being stretched while under active tension as physical (negative) work is done on it.

(3) They exert active tension with no change in length (isometric contraction); in this case there is no joint movement and no physical work.

All three types of action, however, require the expenditure of metabolic energy. A muscle also may undergo passive stretch, but this does not require energy expenditure.

Bipedal walking of a normal human being requires the interaction of certain forces. Standing quietly requires a modest amount of muscle activity to maintain balance or stability. There is a vertical reaction equal to the body weight divided between the two feet (Figure 1a). To initiate walking, one foot is raised and the leg is accelerated forward. Muscle action in the supporting leg causes the body center of gravity to move forward through the creation of a horizontal reaction force at the foot. The greater this reaction force, the greater the acceleration of the body since the force equals the mass times the acceleration (Figures 1a, b).

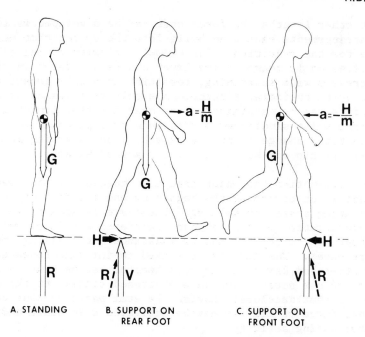

Figure 1. Floor reactions. $\underline{G}$: gravitational force. $\underline{R}$: reaction force at the floor. $\underline{H}$: horizontal component of $\underline{R}$. $\underline{V}$: vertical component of $\underline{R}$. $\underline{a}$: acceleration. $\underline{m}$: mass of body.

Every person has a characteristic way of walking. Each learns to integrate the numerous variables of the neuromusculoskeletal system for smooth performance. There are gross similarities between individuals. The legs oscillate; the body rises and falls with each step. Movements in the plane of progression are large and the individual variations in relation to the size of the total displacements are relatively small. Average values help to provide a general understanding of the basic relationships that exist between the major segments of the lower extremity.

Superimposed on these basic activities are numerous less obvious movements of individual parts of the body. Small movements occur in the frontal and transverse planes, and the individual variations are relatively large. It is these small magnitudes but relatively large individual differences that give each of us a distinctive manner of walking. Average values of these movements may be of little help in understanding some of the interrelationships that must exist between elements of the anatomy.

In order to understand the dissimilarities as well as the similarities in the details of the locomotor process between individuals it is desirable to look for a basic principle that may reveal the reasons for a basic pattern, modified with growth, which also allows for individual variations. The following principle seems to apply.

The human body integrates the motions of its various segments and controls the activity of the muscles in such a way that the energy required for each unit of distance traversed is minimized.

Walking is characterized by a pattern of bodily movement that is repeated over and over, step after step. There are variations between steps of the same individual as well as between individuals. More importantly, there are major differences in an individual. Speed, stride length, type of footwear, and probably motivation and psychological attitude alter the pattern of movement.

Human walking then, is a process of locomotion by which the erect moving body is supported ultimately by one leg and then the other. As the moving body passes over the supporting leg, the other leg swings forward in preparation for its next support phase. One foot or the other always contacts the ground. When the support of the body is transferred from the trailing to the leading leg, both feet are momentarily on the ground.

The two basic requisites of walking are continued ground reaction forces that support the body's center of gravity, and periodic movement of each foot from one position of support to the next in the direction of progression. These elements are necessary for any form of bipedal walking, even though affected by physical disability. As a consequence, certain displacements of body segments occur in walking. The reciprocal swinging of the legs is accompanied by rhythmical swinging of the arms. This tends to create a balance in the torques about the longitudinal axis of the body. As the body passes over the weight-bearing extremity it tends to be displaced toward the weight-bearing side, producing a certain amount of side-to-side movement (Figure 2a). The body rises and falls with each step (Figure 2b). It rises as it passes over the weight-bearing leg and falls to a minimum during the period of double weight-bearing. As the body accelerates upward, an increase in the vertical floor reaction to a value greater than body weight is required, and as it falls to a minimum the downward acceleration reduces the total vertical reaction to a value less than body weight (Figure 3). During normal walking there is a push-off phase when the resultant of the floor reaction on the supporting foot is inclined and passes through or close to the center of gravity. The horizontal component of this reaction causes an acceleration of the center of gravity in the direction of motion. In the same manner, when the support is transferred to the leading foot, the reaction is inclined backward, the horizontal component causing a deceleration in the direction of motion (Figures 1b, c, and 4). Forward

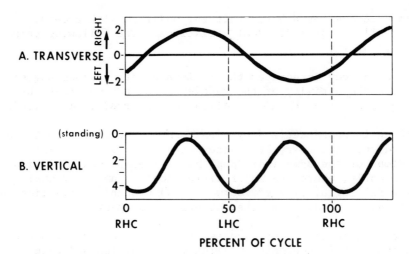

Figure 2. Pelvis displacement, cm, walking at 110 steps/min, in the transverse and vertical planes. RHC: right heel contact. LHC: left heel contact.

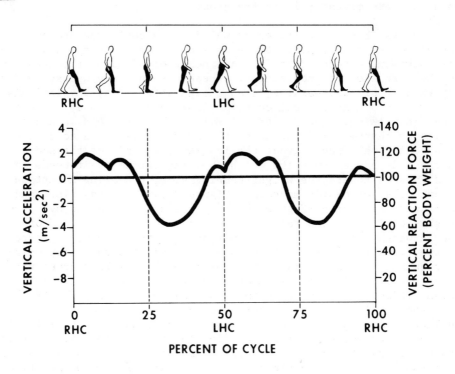

Figure 3. Vertical acceleration and reaction. RHC: right heel contact. LHC: left heel contact.

PHYSICAL PRINCIPLES OF LOCOMOTION

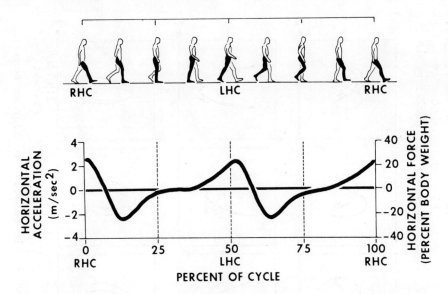

Figure 4. Horizontal acceleration and horizontal reaction, fore and aft. <u>RHC</u>: right heel contact. <u>LHC</u>: left heel contact.

progress, therefore, in bipedal locomotion is not at a constant velocity but oscillates above and below a certain intermediate value.

Magnitudes of the movements indicated above for any one individual change considerably relative to step frequency and step length. Measurements made without recording this information produce data that cannot be correlated with results of other studies of human locomotion. Figure 5 shows the variation of movement of the pelvis in the three orthogonal planes for six frequencies from 78 to 136 steps per minute as recorded for one individual walking on a treadmill. This illustrates the motion of the center of gravity of the body during walking. Of interest are the smoothness of the curves and the similarity to a lazy figure eight when viewed from the rear. Minimal energy requirement depends on the smoothness of the path and on the transfer between potential and kinetic energy. Smoothness is achieved by optimal joint rotations, which are in turn produced by muscle forces causing external reactions that result in accelerations and decelerations.

Normally there is approximate symmetry of movement; the pattern repeating itself with each successive stride. While the displacements of the entire body through space may be described as

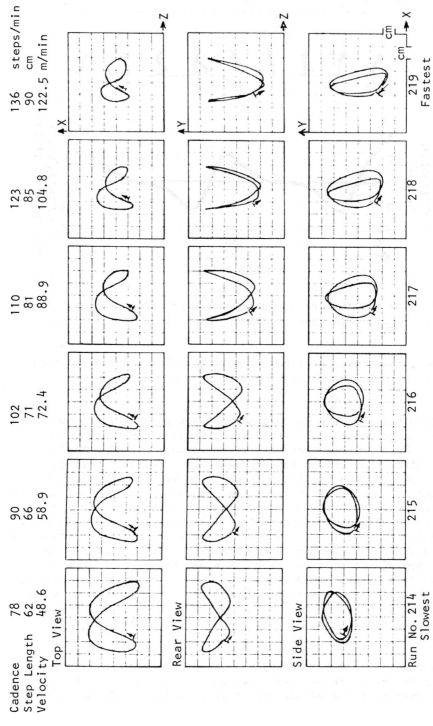

Figure 5. Pelvis displacements, cm. Linear displacements of pelvis relative to average position as a function of cadence and step length.

translatory, this translation is achieved by the angular rotations of various segments of the body about axes that lie in the proximity of joints. Such rotations are the result of muscle contractions, programmed and controlled for minimal energy expenditure by a built-in computer.

The attempt has been made to point out the basic requirements and the fundamental principles involved in locomotion. Measured values used to describe the process can be affected in many ways. For example, the gait of individual human beings varies depending on their aspirations and motivation. For some situations speed is the most important. Safe arrival may be paramount, or the expenditure of minimal energy. Some individuals prefer the most graceful walk, or at least the most unobtrusive appearance, and consequently may expend extra energy.

Normal or pathological locomotion in the human, or the movement of quadrupeds, birds, fish, or insects, depends on an internal source of energy, on the development of an external reaction force to produce motion, and also on the assimilation and utilization of feedback information to develop habit patterns or to provide signals calling for overrides that require conscious control of possibly numerous muscles to obtain the desired result.

REFERENCES

Ralston, H.J., Inman, V.T., Strait, L.A. and Shaffrath, M.D., (1947) Mechanics of human isolated voluntary muscle. J. Physiol. 151, 612-620.

Schmidt-Nielson, K., (1975) Animal Physiology. Cambridge University Press, Cambridge, (p. 499).

HUMAN SOLUTIONS FOR LOCOMOTION

I. SINGLE LIMB ANALYSIS

R. Herman, R. Wirta, S. Bampton, F.R. Finley

Department of Rehabilitation Medicine, Temple University, Health Sciences Center, and *

A volitional motor program for steady state locomotion is ascertained in normal human subjects as a function of an array of operational or sub-behavioral input to output levels. The tactics and strategies employed during a structured series of protocols are revealed by describing certain features or phenomena of gait during each <u>stride</u> cycle.

LEVEL I   - cyclical behavior, assessing velocity, stride frequency and stride length;
LEVEL II   - ground reaction measures, stressing temporal details of stride (swing and stance), and vertical and horizontal components of ground reaction forces during single and double support periods of stance;
LEVEL III   - joint motion, emphasizing time relations of intra-joint and inter-joint displacement and the nature of forces about selected joints associated with displacement;
LEVEL IV   - EMG activity, examining spatial, temporal, and quantitative features of various muscles; and
LEVEL V   - the relationship between mechanical and neural events (i.e., levels III to IV).

---

*Krusen Center for Research and Engineering, Moss Rehabilitation Hospital, Philadelphia, Pa.

Statistical treatment of data of all parameters permitted assessment of (1) reliability and reproducibility of the data, (2) variability of each parameter as a function of relations among trials and sets (task related), and inter-sets (inter-task related) in each subject and inter-sets (task related) between subjects, and (3) the presence and extent of correlation between parameters (utilizing both computer processed averaging techniques and cycle by cycle analysis).

The protocol was designed essentially as a 4-by-4 block design with each subject performing a series of tasks either in an ordered series or pseudo-random manner; in one series of experiments each subject was instructed to select a "brisk, natural or comfortable, slow, and extremely slow" rate during free walking; the rate and frequencies elicited by each subject were later used to control (by cue) (1) cadence, (2) velocity, and (3) both velocity and cadence. This scheme and the subsequent statistical analysis were utilized to test the limits, i.e., the dynamic range, and the adaptability of the stereotyped behavior. The protocol, as stated, provided for the examination of performance during extremely low rates (e.g., below 0.6 m/sec.) of locomotion because low rates are usually exhibited by patients with both musculoskeletal and neurosensory disorders.

Examination of the phenomena within each performance level leads to the conclusion that a rather uniform and rigid functional program exists within the dynamic range of stride frequencies. In this range, the synergistic construction of the motor program or patterns of mechanical-biological organization leads to an anticipated and delineated outcome of ground reaction measures and of cyclical behavior. Below this performance range, the synergistic behavior of gait is significantly altered.

Introduction

Historically, investigations of human locomotion have been guided by requirements for prosthetic and orthotic design and, hence, by biomechanical considerations (Morton, 1952; Steindler, 1953; Hellebrandt, 1964). Aside from the perspectives of the Moscow school of Bernstein, Gurfinkel, Tsetlin and others (Gurfinkel et al., 1971; and from the automata concepts of McGhee, see this volume), there has been virtually no identifiable programmatic approach to such questions as: what are the critical or essential parameters of gait to be controlled and how are these variables controlled?

The work of the Russian investigators implies that individual motor programs are directed with rather simple executive commands implemented by the so-called low-level "function generators" that constrain large sub-sets of free variables (i.e., multiple number of physical degrees of freedom during movement). Thus, a locomotor program which must control a considerable number of degrees of mechanical freedom operates efficiently with few economically controlled sets (Szentágothai and Arbib, 1974). According to these interpretations, "function generators" provide for synergic patterns of muscular firing. Usually disregarded in such discussions, however, is the role of the passive and active (contractile) states of muscle which can serve, in part, as autogenic regulators in the control of dynamic performance (Grillner, 1972; Herman et al., 1973).

Such definitions of synergy are generally not useful in providing a conceptual framework for behavioral interpretation of such stereotyped movements as locomotion. The operational problem of control of locomotion, as stated by Bernstein and his pupils relates to how the nervous system regulates, in a precise and orderly coordinated manner, a complex behavior involving multiple degrees of mechanical freedom. According to Kots et al., (1971), coordination is ensured by functional synergies which limit the number of degrees of freedom of the control system. In this manner, each characteristic form of motor activity has a correspondingly small number of connected groups (joints, muscles) of the sort that one degree of freedom of the control system is sufficient for the control of each of them. This concept has been tested by deafferentation experiments. In man, deafferentation leads to either diminished coordination of movements as a result of impaired functional synergies or control of different muscle groups (Kots et al., 1971). This is in marked contrast to the observation that deafferentation does not appear to lead to substantial reduction in control of an individual group of muscles governing control of one joint. In order to control a large number of mechanical degrees of freedom, interaction among joints must exist. A unitary concept of control of a multilink system is made more plausible by the observation that an increased number of movable limbs is not associated with an increase in regulation error (see Results, Gurfinkel et al., 1971b).

In order for higher levels of the central nervous system to solve effectively various tasks and sub-tasks commonly confronting man, it is essential that both the number of control parameters and afferentation requiring analysis be limited. The basic synergies, having the simplest physiological mechanism, then form an "orchestration of movement." Once the "feature extraction" determines the synergy to be selected, the synergies are able to simplify the processing of afferentation, having organized it according to the motor problem (Gurfinkel et al., 1971b; Szentágothai and Arbib, 1974).

The principle of "least interaction" seems to be pertinent to the features of the locomotor synergy. If a subsystem (presumably at the spinal level) is responsible for a particular goal related to locomotion, it must limit the influence from the periphery, other subsystems, and supraspinal centers in order to ensure a coordinated, stereotyped performance. Thus, a subsystem would tend to minimize external interactions that might disturb it. According to Gelfand et al., (1971), "The tendency to minimization of interaction leads to the coordination working of individual subsystems subordinating autonomous activity of each of them in the interest of the solution of the overall problem assigned by supraspinal influences."

These concepts have led to the development of automata theories, e.g., the development of robot processes (McGhee, this volume). They also underscore the basis of the remarkable motor synergic patterns observed for various upper limb movements (Hellebrandt, 1964; Finley, 1969). These observations indicated that consistent and predictable patterns of motor discharges can be identified for individual and specified motor tasks and are virtually unchanged following amputation of the upper limb (Finley, 1969). This later formed the basis for the design of a control system of an above-elbow artificial limb (i.e., an externally powered myoelectric prosthesis, Wirta and Taylor, 1970). A pattern recognition system sensing myoelectric activity from various muscle sites about the shoulder was developed for the amputee utilizing a kinetic myoelectric algorithm, permitting the subject to activate the mechanical replacements with repertoires of movement similar to those observed in normal subjects.

The relatively "tight" spatio-temporal organization of myoelectric potentials from proximal muscles (e.g., shoulder and periscapular muscles), observed in upper limb amputees attempting a complex shoulder-hand motion, is viewed as representing a central pattern of motor outflow signals concerned with performance of a particular act. Integrated amplitudes of EMG, developed at each of seven or more muscle sites at any given time, are presented to a pattern recognition network or circuit which serves to activate appropriate motors to move the joints of the prosthetic limb through variable positions and loads. Hence, a single neural program for a specific task can be recognized and predicted for an efficient coordination of a multi-link mechanical system of six movements about three axes.

Based upon some of the above principles, we have conducted our investigations to answer, in part, the following questions: (1) given a definable and regulated experimental regimen, how variable and adaptable is the performance of steady-state locomotion, i.e., what are the statistical limits to the dynamic performance with respect to plasticity of walking which would lead to an analysis of synergic behavior (cf. Barnes, 1975) and (2) to what extent does the synergic construction of gait change at rates observed in various pathological states (e.g., mean rates of hemiplegics and rheumatoid arthritics are 0.46 and 0.53 meters per second, respectively).

Synergic patterns related to locomotion have not been adequately described principally for the following reasons: (1) the lack of unique and innovative experimental protocols, (2) failure to approach the investigations systematically, i.e., from the neural input stage to the output stage of cyclical behavior, (3) the absence of statistical information regarding variability and adaptability of the behavior in a normal and pathological gait under various environmental conditions, and (4) the particular emphasis placed on lower limb amputees and biomechanical assessment.

The characteristics of steady-state locomotion to be presented here were derived from two different protocols: (1) a "forced" protocol in which subjects verbally were given instructions after each trial (a series of strides during a complete pass on the locomotion "walkway," Herman et al., 1974), which directed them to alter their rates of progression by small increments through an exceedingly wide range of velocities, and (2) a "preferred" protocol in which each subject was instructed to select a walking speed in accordance with each of the following four verbal commands: brisk, normal or comfortable, slow, and very or super slow. The latter protocol was tested under two conditions, namely, repeated trials at any one rate (a "grouped" condition) and a randomized order of velocities (a "pseudorandom" condition) in which the influence of repeated performance was obviated.

Parametric evaluation estimates and tests the hypothesis that populations of data are constant at each of our operational output levels, (i.e., cyclical behavior, ground reaction measures, joint motion, and myoelectric potentials), across a wide range of velocities and conditions. The statistical treatment measures: (1) repeatability of variables, (2) correlation or co-variance of variables, and (3) sources of error, i.e., systematic, biological or measurement errors which are due to instrumentation, computer processing, protocol design, etc.

RESULTS

I. CYCLICAL BEHAVIOR

A. <u>Kinematic Phenomena</u>

(1) <u>General Comments</u>. Whole body velocity* during steady-state conditions of walking varies cyclically as evidenced by the tachometer traces in Figure 1. The periods between subsequent peaks are utilized to depict step (e.g., time duration of heel strike of one limb to heel strike of the contralateral limb) frequency. The periods between alternate peaks (not illustrated) are used to indicate stride (i.e., time duration of heel strike of one limb to the following heel strike of the same limb) or single limb frequency. In Figure 1, the "hump" prior to heel strike demonstrates the effect of deceleration of the swinging limb upon whole body velocity. The magnitude of the "hump" is usually directly related to the magnitude of the mean velocity and, hence, is considered to reflect the transfer of kinetic energy from the decelerating limb to the body.

(2) <u>Waveform Analysis</u>. The manner in which velocity of the step cycle varies with respect to time is analyzed by using a Fourier technique. (Note that the step cycle rather than the stride cycle is employed for convenience of presentation.) This method yields an equation of the following general form:

$$V_t = V_o + A_1 \cos 2\pi f\, t + B_1 \sin 2\pi f\, t + A_2 \cos 4\pi f\, t + B_2 \sin 4\pi f\, t$$

where
V_t = velocity at time t

V_o = mean velocity

A = coefficient for the cosine component

B = coefficient for the sine component

f = step frequency

Subscripts 1 and 2 identify the first and second harmonics.

---

*The whole body velocity is measured with a tachometer-generator attached with a thin cord to the lumbar area of the subject in an attempt to register velocity near the center of gravity.

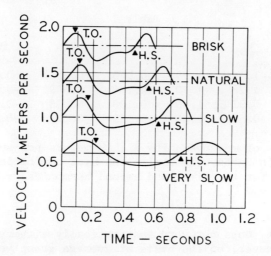

Figure 1. Typical tachometer traces showing velocity as a function of time for four different speeds of walking. The interval from one peak to the next represents one step cycle. The dash-dot lines represent mean velocities for each trace and the symbols H.S. and T.O. designate single limb occurrences of Heel Strike and Toe-Off, respectively.

Velocity data obtained are analyzed for the first and second harmonics only. Examination of the data reveals no contributions from higher harmonics which might have been reliably distinguished from measurement system error (approximately ±0.04 meters per second). The average values for a group of five subjects at each of the preferred rates, namely brisk, normal, slow and super slow, are given in Table 1. The first harmonic or fundamental frequency coefficient A_1 is independent of velocity while the second harmonic coefficient A_2 is proportional to velocity. The coefficient B_1 is proportional to velocity and reflects the asymmetry of the deceleration and acceleration portions of the velocity pattern. B_2 is too small to contribute significantly. The magnitude of the first harmonic coefficient A_1 (fundamental frequency) was found to be a function of both the height and mass of the subject in accordance with the following empirical expression:

$$A_1 = 0.21 H - 0.028 \sqrt{m}$$

where H is the height of the subject in meters and m is the mass in kilograms. The magnitude of the second harmonic coefficient A_2 is proportional to the mean velocity V_o, and a function of height H:

$$A_2 = 0.053 V_o (H - 1.1)$$

Table 1

	V_o	f	A_1	A_2	B_1	B_2
Brisk	1.86	1.98	0.140	0.059	+0.011	+0.013
Normal	1.28	1.70	0.136	0.042	−0.003	−0.004
Slow	0.92	1.40	0.138	0.033	−0.011	−0.003
Super Slow	0.55	1.06	0.139	0.029	−0.018	−0.001

Average values of mean velocity V_o, in meters per second, the step frequency f, in Hertz, the cosine and sine coefficients A and B, respectively, for the first and second harmonics, subscripts 1 and 2, respectively.

The value of B_1 does not vary systematically either within or among subjects. However, on the basis of average group performance, B_1 may be estimated by the following expression:

$$B_1 = 0.025 \, V_o - 0.034$$

where V_o is the mean velocity in meters per second.

The whole body velocity variation during a step cycle contains another feature which reflects movement control, namely, acceleration and deceleration. Since the cyclic velocity variation, ΔV, remains relatively independent of velocity, then the rates of change of velocity (i.e., accelerations) must correspond to the step frequency. The accelerations and decelerations may be calculated by differentiating the velocity expression with respect to time to yield the following expression:

$$a_t = 2\pi f \, [-A_1 \sin 2\pi f \, t + B_1 \cos 2\pi f \, t - 2A_2 \sin 4\pi f \, t + 2B_2 \cos 4\pi f \, t]$$

Experimental evidence shows that the relation of peak acceleration to step frequency is non-linear and that peak acceleration may be estimated by the following expression:

$$a = bf^{1.59}$$

where a = peak acceleration in meters/sec^2

f = step frequency in Hertz

b = coefficient 0.94 ± 0.20

If the step cycle is visualized as consisting of 360° from one peak of velocity to the next, the maximum deceleration and the maximum acceleration occur at about 60° and 300°, respectively. Heel strike occurs at or slightly before maximum acceleration.

(3) *Temporal and Spatial Phenomena*. The two phenomena associated with whole body velocity, namely, stride frequency and stride length, are derived from the tachometer by digital processing of rate information. The cycle-to-cycle data obtained from all subjects reveal an exceptionally strong direct correlation (note correlation coefficient or r values in Figure 2) between each of the two variables, i.e., stride frequency or stride length, and velocity (confidence levels are greater than 0.99). Sample linear regression lines (Figure 2) approximate the scatter values throughout the range of velocities; however, the linear expression of data points is particulary strong (e.g., r>0.95) at rates between 0.7 meters per second and 2.0 meters per second. Beyond this range, the direct linear relationship is altered slightly. Under these conditions, when the stride frequency and stride length are treated by applying the square root value for each data point, the relationship between the dependent and independent variables becomes strongly linear throughout the operational range of velocities. By this technique, linear correlation coefficients for data obtained from subjects with wide dynamic range of rates of movement, e.g., below 0.6 meters per second and above 2.0 meters per second, do not indicate a significant alteration in the correlation between parameters or variables when compared with similar coefficient values for data obtained from subjects transcending a more limited dynamic range of velocities.

a. *Stride Frequency*. As indicated above, stride frequency shows a strong direct linear relation with the rate of progression. Observation of the frequencies elicited at a preferred "comfortable or normal" rate of steady-state walking reveals an unusual narrow range of frequencies (e.g., 0.82 - 0.87 Hz) which occur over a relatively wide spectrum of velocities (e.g., 0.87 - 1.61 meters per second). The range of frequencies selected by the subjects are somewhat lower than the frequencies observed (0.946 ± 0.18, 2 standard deviations) in a study of a large population sample (n = 1106) of urban pedestrians, studied covertly under non-laboratory, non-instrumented and natural environmental conditions (e.g., at shopping centers, residential areas, etc., Finley and Cody, 1970). These differences are attributed to prevailing natural constraints within a laboratory setting (i.e., instrumentation) and environmental conditions which demand the employment of different tactics and strategies.

b. *Stride Length*. Similarly, stride length increases linearly with the speed of progression (Figure 2). The stride length may be considered as consisting of two components,

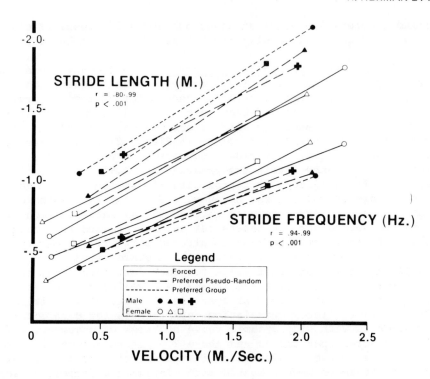

Figure 2. Sample regression lines for the relationship between stride length (meters) and stride frequency (Hz) and velocity (m/s) in a population of both males and females under two protocol conditions (see text); r values indicate the range of correlation coefficients and p values indicate the strong statistical relationship between the dependent and independent variables.

namely, the distance traversed during the swing phase and during the stance phase. As observed with the total distance traveled during the stride cycle (e.g., in Figure 2), there is a direct linear correlation between swing and stance distance and velocity (not illustrated). The slope values for both stance distance (range 0.24 - 0.33) and swing distance (range 0.25 - 0.30) are similar. All correlation coefficients (r values) exceed 0.90 and are statistically significant beyond a level of 0.99 (i.e., $p < .001$).

The virtual parallel nature of the sample regression lines for distance traveled during the swing phase and during the stance phase (note this latter phase includes a single support equivalent in time to the swing phase, and the double support period) suggests that the distance traversed during double support is invariant with velocity. This is not surprising, given the quadratic relationship between the percentage of time that double support occupies during the stride cycle and velocity or frequency (see below).

c. <u>The Relationship Between Frequency and Stride Length</u>.
When actual values for frequency and stride length are compared to rates of progression, normalized about their selected "normal or comfortable speed" (Figure 3), it becomes evident that a remarkably limited range of frequencies is accompanied by a considerable dispersion of the regression lines plotted for stride length. Similarly, in the urban pedestrian study by Finley and Cody (1970), a delimited spread of frequencies is associated with considerable dispersion of values for stride length. Nevertheless, the stride length to frequency relation is constant over a velocity range of 0.7 to 2.0 meters per second. The invariance of this ratio with the rate of progression has been implicated as the principal output feature of the stride cycle (Molen, 1973).

A statistical difference between males and females with respect to all three parameters, i.e., velocity, frequency, and stride length, has been reported (Murray, 1967; Finley and Cody, 1970). However, among these variables, the greatest statistical difference between males and females is stride length. Further, there is statistical evidence that, in women, the stride length is smaller and the stride frequency greater than values observed in males. These observations are supported by our findings. Molen (1973) also demonstrates a statistical difference between males and females by calculating the ratio between the amplitude (i.e., stride length) of movement (the spatial factor), and the frequency of movement (the temporal factor). The ratios for males and females, as specified by Molen, are also supported by data derived from our own laboratory investigations.

d. <u>Variability of the Three Parameters</u>. For each parameter, the variability noted at slow and very slow rates is statistically greater than that observed at normal and at brisk rates. At each preferred rate, variability for velocity is greater than that observed for either stride length or frequency and the variability for stride length is always greater than that of frequency (also, Finley and Cody, 1970). There is no systematic effect of the different protocol conditions on correlation and variability values.

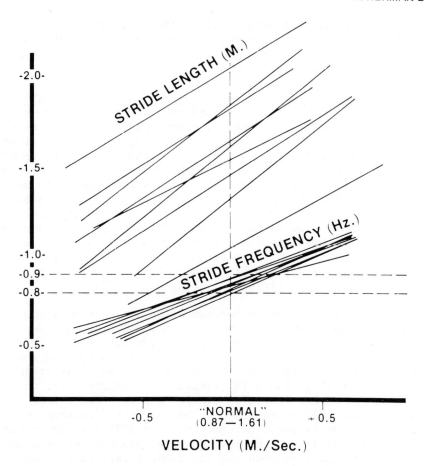

Figure 3. Sample regression lines for the relationship between the actual data points for the variables of stride length (meters) and stride frequency (Hz) and the "normalized" values of velocity (normalized about the preferred "normal" rate of progression). Please note the limited spread of frequencies and the wide dispersion of stride lengths and velocities at and about ($\pm$50%) the selected "normal" velocity for the entire group of subjects studied.

II. GROUND REACTION MEASURES

A. <u>Kinetic Phenomena</u>

(1) <u>Peak Longitudinal Shear Forces</u>. The longitudinal shear forces (see Figure 7) are developed to move the body from one point to another in the direction of progression. The peak values encountered, both accelerative and decelerative, are approximately

proportional to the <u>momentum</u> (mass times velocity) of the body. The decelerative component, acting in opposition to the movement of the body, is experienced at the time the limb accepts the weight of the body (see Figure 7). The accelerative component, acting in the direction of the movement of the body, is experienced near the end of the stance phase when the weight of the body is being transferred to the contralateral limb (see Figure 7). The relation between the peak accelerative and decelerative forces varies among individuals, but the accelerative peaks are approximately 20 to 30 percent greater than the decelerative peaks (Figure 4). (Asymmetries also may be noted when interlimb performance is considered.) Approximations of the mean peak shear forces may be derived from the following expressions:

$$\text{accelerative } (+) \; F_x = 1.6 \; mV$$

$$\text{decelerative } (-) \; F_x = 1.3 \; mV$$

where F_x = peak longitudinal shear force, newtons
  m = mass of the body, kilograms
  V = average velocity, meters per second

While the peak values of the longitudinal shear forces vary with velocity (see also Figure 7), the area under the <u>force-time</u> <u>curve tends to remain constant and independent of walking speed.</u> This observation is due to the combined effect of the direct relationship between peak values and velocity (Figure 4) and to the relationship between stance duration and velocity (hence, force duration). The area under the force-time curve represents impulse (product of force and time). The nominal value for impulse, both under the decelerative and accelerative portions of the longitudinal shear force curve, is approximately 20 newtons-seconds. This figure varies with the mass of the subject and with gait characteristics of individual subjects. Under conditions where the subject is observed to be walking at constant mean velocity, the area under the accelerative component is usually slightly greater than under the decelerative component. The difference noted, while varied, is in the order of 10 newtons-seconds. This aspect has not been studied thoroughly as such small values may contain uncertainties.*

(2) <u>Loading Rates</u>. Both the vertical and longitudinal shear force components of the ground reaction forces are characterized

---

*Parenthetical statement: The dynamic range of the longitudinal shear force measurement is ± 250 newtons. System accuracy is approximately plus or minus two percent yielding an uncertainty of about $\pm$ five newtons.

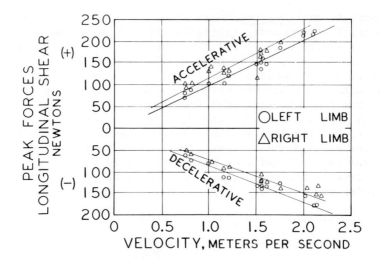

Figure 4. Peak values of the longitudinal shear ground reaction forces, both the accelerative and decelerative aspects, vary directly as the average velocity of walking.

by a rapid increase in force during the weight acceptance period for the limb and subsequently, by a rapid decrease as the weight is removed from the limb (see Figure 7). Figure 5 shows the temporal pattern of both the vertical force and the longitudinal shear forces. The vertical force increases sharply during the double support period and less rapidly as the contralateral limb leaves the ground. The rates at which these forces change are dependent upon several factors, e.g., the mass and height of the subject, the velocity of walking, and the stride frequency. Nominally, the rate of decrease for each variable is approximately the same as the rate of increase.

When relations among variables such as mass, velocity of walking, and stride frequency are analyzed, an empirically determined expression is derived indicating that the loading rates vary directly with the product of mass, velocity, and stride frequency:

$$\frac{\Delta F_y}{\Delta t} = 95 \text{ mVf}$$

$$\frac{\Delta F_x}{\Delta t} = 21 \text{ mVf}$$

where $\frac{\Delta F_y}{\Delta t}$ = rate of change of vertical force, newtons per second

$\frac{\Delta F_x}{\Delta t}$ = rate of change of longitudinal shear force, newtons per second

m = mass of subject, kilograms

V = average velocity, meters per second

f = frequency, strides per second

B. Temporal Phenomena

The stereotyped program for steady-state locomotion can, in part, be described by analyzing: (1) the duration of the stride cycle (e.g., toe-off to the following toe-off of a single limb); and (2) the duration of the components of the stride cycle, namely, swing (toe-off to heel strike), stance (heel strike to toe-off), and double support (both lower limbs in contact with the ground).

The stance period constitutes two functional time intervals, the double support and single support periods. The single support

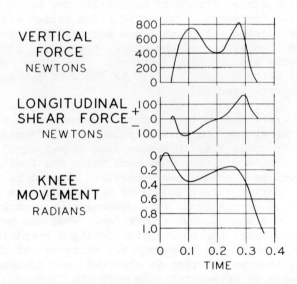

Figure 5. Time relation of the vertical and longitudinal shear ground reaction forces and the simultaneously occurring flexion-extension movement of the knee.

period of the stance limb is equal to the swing period of the contralateral limb (assuming symmetry in the step performance between the two lower limbs). Hence, the time interval between stance and swing (i.e., interval between the sample linear regression lines in Figure 6) can be utilized conveniently to depict the double support period.

There is little need to classify further the stride cycle into smaller units of measurement by employing either (1) the "Philippson step cycle" model, which is used to describe characteristics of steady-state locomotion in the hind limb of the dog and cat and of initiation of gait in man, (Wetzel et al., 1974; Herman et al., 1974; Grillner, 1975), or (2) the floor contact or reaction model, a schema conventionally used to assess single and interlimb movement in man, (Eberhart and Inman, 1947; Drillis, 1959). Neither model can effectively be utilized to describe the synergism or pattern of the complex act of locomotion as a multidimensional process or as a series of distributed interactive events which require precise analysis.

The duration of both the swing and stance periods is assessed with respect to the stride cycle time (Figure 6) by digital processing of foot switch data. In all subjects studied, the relationship between stance and swing periods with cycle time indicates that a large linear change of values for the stance period and a considerably smaller linear change of values for the swing period occurs as cycle time increases. The correlation values signify a strong co-variance between both stance and swing with cycle time. Among all subjects, the slopes of the sample regression lines for both stance and swing are strikingly uniform. The slope ratio between stance and swing ranges from 4.2 to 7.0 with higher values for the female subjects studied.

The double support period (the interval between data points for swing and stance at any one stride cycle period) increases exponentially as the cycle time increases (not illustrated). This relation, however, becomes linear when each value for double support time is treated by a square root function. As can be observed by the assessment of time values in Figure 6, the decrease in the double support period as the cycle time decreases is associated with the dramatic decrease of stance duration. When the percent of time the double support period occupies during a complete stride cycle is related to the frequency (or velocity) of the movement, a strong inverse relation is observed (not illustrated). This relation appears to extend reliably over the frequency range of 0.2 to 1.2 cycles per second. In all subjects studied, the

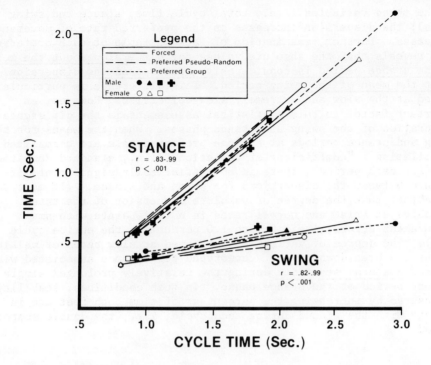

Figure 6. Sample regression lines for the relationship between stance and swing periods and cycle time period. Note the time interval between values of stance and swing at any one cycle time is indicative of the double support period.

ratio of double support period to stride period is approximately 0.25 at 0.9 cycles per second and approximately 0.50 at 0.4 cycles per second. Stated differently, double support comprises 50% of the total cycle period at a frequency of 0.4 cycles per second, and 25% of the total cycle period at a frequency of 0.9 cycles per second. These findings suggest that as frequency or velocity of walking decreases, the subject must rely more on bipedal contact for stability and control.

The swing-to-stance ratio progressively increases as cycle time decreases; when the ratio reaches 1.0, that is, when there is no longer a double support period, the transition from walking to running occurs. At this point, the mean stride frequency (intersection of the sample regression lines in Figure 6) is 1.7 cycles per second ±0.1 (1 standard deviation).

In all subjects there is a direct influence of the preferred rate on dispersion (standard deviation) about the mean of each of the three variables (i.e., total cycle time, stance and swing times); the dispersion increases as the preferred rate of movement decreases. Further, examination of each variable at each preferred rate reveals that the absolute values for <u>dispersion about the mean of the stance period is considerably greater than the dispersion about the mean of the swing period</u>. This difference is particularly marked at the slow and extremely slow velocities. However, an important factor in this statistical assessment is the difference in duration of the swing and stance phases. When the means for the swing and stance periods at any one preferred rate are normalized by utilizing a "coefficient of variation" (i.e., standard deviation/mean for each period), there is <u>no</u> statistically significant difference between the dispersions for swing and stance. (It must be emphasized that the degree of absolute dispersion of the three variables at brisk and normal rates is not disparate with small measurement errors, i.e., 0.5 to 1.0 percent of the entire cycle time). The degree of variability observed at slow rates of walking suggests a breakdown in the stereotyped performance associated with altered <u>postural stability</u> during the relatively prolonged single support period of the stance phase. In such conditions, stability is ensured by increasing the percentage of time both feet are in contact with the ground during each cycle, i.e., the double support period.

III. JOINT MOTION

In all subjects studied, measures of angular displacement of the hip, knee and ankle reveal a precise temporal and spatial ordering, i.e., angular motion is sequentially determined in both the swing and and stance periods. As previously stated (Herman et al., 1974), this behavior is not disturbed by differential block of the tibial nerve, by suppression of excitation-coupling of muscle or by constraint of ankle joint motion.

A. <u>General Comments (Figures 7a and b)</u>:

(1) <u>The Swing Phase</u>. Toe-off of a single limb usually occurs immediately (e.g., 10 - 20 milliseconds) prior to the termination of ankle plantar flexion. At this moment, the hip is in an early state of flexion while the knee is virtually completing its flexion motion. The <u>early</u> swing phase (sometimes referred to as the F phase in the dog and cat, Grillner, 1975) is featured by completion of knee flexion. During this interval, the ankle is rapidly dorsiflexing and the hip is in its mid-range of flexion. The subsequent period of the swing phase is characterized by the respective completion of hip flexion and of knee extension prior to heel strike; ankle dorsiflexion terminates at or immediately following heel strike. This systematic arrangement of joint motion

is not disturbed by varying rates (or frequencies) of movement (e.g., from 0.6 meters per second to 2.0 meters per second).

(2) <u>The Stance Phase</u>. Following heel strike, the <u>early</u> stance phase is characterized by rapid plantar flexion of the ankle, by the yield of the knee and by acceleration of extension of the hip. During the ensuing period, the hip continues to extend, the knee initiates an extension movement, and the ankle

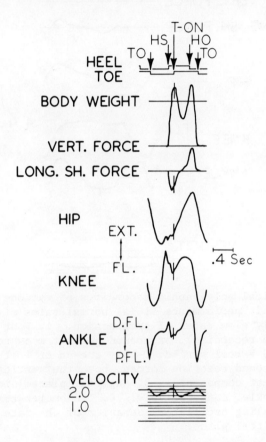

Figure 7a. Typical polygraphic recordings of various single-limb parameters of gait performance at (a) normal rates of progression, and (b) extremely slow rates of progression. 1. Foot switch occurrences (top recording) for ground contact measures where: TO = toe-off; HS = heel strike; T-ON = toe-on or foot flat; HO = heel off. 2. Ground reaction forces, both the vertical and the longitudinal shear components. 3. Sagittal plane joint motion of hip, knee and ankle, consecutively. 4. Tachometer traces (lower recording) revealing oscillatory behavior of the rate of progression in the sagittal plane.

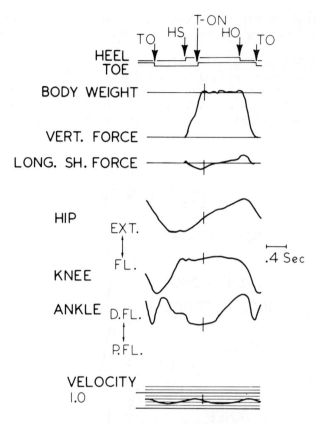

Figure 7b. Typical polygraphic recordings of various single-limb parameters of gait performance at (a) normal rates of progression, and (b) extremely slow rates of progression. 1. Foot switch occurrences (top recording) for ground contact measures where: TO = toe-off; HS = heel strike; T-ON = toe-on or foot flat; HO = heel off. 2. Ground reaction forces, both the vertical and the longitudinal shear components. 3. Sagittal plane joint motion of hip, knee and ankle, consecutively. 4. Tachometer traces (lower recording) revealing oscillatory behavior of the rate of progression in the sagittal plane.

yields in dorsiflexion. Peak displacements of each of the three joints occur in the following sequence: knee extension, ankle dorsiflexion, and hip extension - across a wide range of velocities (Figure 7a). While all three peaks maintain a constant temporal relation with each other at brisk and normal rates of walking, the knee extension peak departs from its temporal relation with both peak ankle dorsiflexion and hip extension at slow and very slow rates (Figure 7b). Although knee flexion commences prior to the initiation of ankle plantar flexion and hip flexion, the pre-swing period is distinguished by a co-varying temporal relation of the ankle-hip complex with the subsequent toe-off at all frequencies and velocities (see below).

B. Kinematic Phenomena

Both computer average processing and cycle-by-cycle analysis (digital processing) of data, derived from the electrogoniometers placed about the hip, knee, and ankle joints, are used to assess the time intervals between prominent flexor and extensor peaks of any one joint and between the peaks of adjacent joints. The following phenomena are observed in all subjects studied and are not influenced statistically by the various protocol conditions (i.e., forced and preferred) utilized in this study.

(1) Time Intervals Within Each Joint. Time intervals between peak displacements within each joint (e.g., knee flexion and knee extension in swing, knee extension in swing and knee extension in stance, hip flexion and hip extension, ankle dorsiflexion at heel strike and ankle dorsiflexion in late stance) co-vary linearly with cycle time. The relation between the two variables is extremely strong as evidenced by correlation coefficient values (greater than 0.9) and by the significance of correlation values ($p < .01$) in all subjects.

Assessment of the relation between the two knee extension peaks deserves special comment. As indicated above, the precise and orderly pattern arrangement of the knee extension peak during stance with the preceding knee flexion and with the subsequent ankle dorsiflexion and hip extension peaks is disturbed during slow rates of movement. Consequently, in many subjects, the correlation between the time intervals of knee extension in swing and knee extension in stance with cycle time is relatively poor (correlation values range from 0.3 to 0.5 at slow rates, while values greater than 0.9 are observed at normal and brisk rates).

(2) Time Intervals Among Joints. Similarly, time intervals between peak displacement of adjacent joints (e.g., knee extension in swing and hip extension in stance, knee extension in swing and ankle dorsiflexion in stance, knee flexion in swing and ankle dorsiflexion at heel strike) co-vary directly with the duration of the stride cycle. The calculated correlation coefficient values are greater than 0.9 and are statistically significant at a 0.01 level in all subjects studied.

In contrast, time intervals between ankle plantar flexion and knee flexion in early swing, knee extension and ankle dorsiflexion in late swing, and ankle dorsiflexion and hip extension in late stance <u>appear</u> constant throughout an entire range of cycle times. As these time intervals are considerably small (range 20-80 milliseconds), the influence of the measurement error introduces uncertainty in interpreting this data on a statistical basis. (The temporal relation between ankle plantar flexion, or toe-off, and peak knee flexion during swing has also been described in lower vertebrates as constant, Grillner, 1975).

(3) <u>Phase Relations</u>. The phase or percentage of the gait cycle that each peak-to-peak interval occupies within each <u>knee</u> and <u>hip</u> joint and between the two adjacent joints are relatively constant across a wide range of frequencies or velocities. Exceptionally low slope and correlation values indicate that the two variables, namely, the phase between peak deflections and frequency, are unrelated or invariant.

In contrast, observations of peak-to-peak intervals of <u>ankle</u> displacement with cycle time indicate that the phase alters with changes in frequency or velocity. The gradual shift in phase (e.g., increasing peak-to-peak deflections during swing and decreasing peak-to-peak deflections during stance as frequency increases) is similar in magnitude and direction to the phase shift of the ground contact measures of swing and stance. For example, the phase for the interval between ankle dorsiflexion and heel strike and ankle dorsiflexion in late stance and for the stance period interval decreases from 70% to 60% as the frequency range increases from 0.55 to 0.95 cycles per second.

(4) <u>Amplitude of Joint Motion</u>. The analysis of peak-to-peak displacement of knee flexion and knee extension in swing, knee extension in stance and knee flexion in swing, and hip flexion and hip extension, discloses a strong inverse linear relationship between the amplitude of displacement and stride cycle time (compare with increasing time intervals between similar peak events as cycle time increases). Ankle displacement is often difficult to interpret as the amplitudes are less predictable.

(5) <u>Systematic Biological Variance</u>. At any one of the four preferred rates of progression, the systematic biological error (variance) observed among populations of time intervals between peak displacements <u>within any one joint</u> and <u>among adjacent joints</u> is statistically insignificant, suggesting a unifying principle of joint coupling (compare with "functional synergy" concept discussed in the Introduction). The variance calculated from the time intervals between peak displacements during either swing or stance is statistically insignificant when compared to the variance observed from the swing interval as measured from the foot switch

data. On the other hand, the variance of the stance period interval is statistically significant when compared to the variance exhibited by analysis of joint displacement in swing or stance, particularly at slower rates of locomotion. At these velocities, the stance period error may be a product of the distributed errors among many joints, including the lumbosacral and intervertebral joints, a sign of increasing instability in the synergic construction of joint motion.

The effect of the various protocol conditions (i.e., preferred, grouped and ungrouped, and forced) on the systematic biological variance observed among a population of phase relations of peak displacements of any one joint or among adjacent joints is statistically insignificant, indicating, again, a rather rigid rule to the organization of locomotor movements.

C. <u>Kinetic Phenomena: Knee Modulus of Flexional Rigidity</u>

One of the numerous functions of the knee is to serve as a shock absorber for the body. The effect may be noted during flexion and extension of the knee between heel strike and mid-stance. Figure 8 illustrates knee movement in its relation to the ground reaction forces. Knee flexion and extension is associated with a concomitant increase and decrease in negative moment acting about the knee joint due to the externally imposed (ground reaction) forces. The relation of the knee movement and knee moment is inversely correlated (Figure 8). As may be observed, the array of scatter points yields an approximately straight line which is used

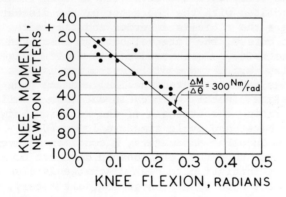

Figure 8. Between heel strike and mid-stance, the moment due to ground reaction forces acting about the knee varies directly as the magnitude of knee flexion-extension. Data points represent 0.025 second intervals. The line depicting the slope was determined by visual estimation.

as an estimate of the stiffness resisting the movement. The stiffness is represented by the ratio of the change in moment (ΔM) to the change in knee angular position ($\Delta\theta$) and may be expressed in terms of newton-meters per radian.

While the magnitudes of both knee movement and knee moment vary as a function of the velocity of walking, the relation between the two tends to be constant (except at very slow speeds where the relation becomes indiscernible). Further, if the performance for the left and right limbs are averaged and then divided by the mass of the subject, the result is consistent among both male and female subjects. This value, termed the modulus of flexional rigidity, is approximately five (5) newton-meters per radian per kilogram body mass.

IV. EMG ANALYSIS

In the subjects studied, the spatial and temporal arrangements of EMG discharges specify a predictable and orderly motor score during both the swing and stance periods. Certain features of the EMG patterns deserve particular emphasis:

A. Features of EMG Bursts

The principal EMG bursts of both flexors and extensors of the hip, knee and ankle during initiation (Herman et al., 1974) and steady-state condition commence during the swing phase. Generally, the flexor discharges in this phase occur earlier than the extensor discharges. The total duration of firing of both flexors and extensors is associated with the character of the discharge, i.e., in muscles (e.g., the tibialis anterior and the rectus femoris) eliciting short bursts, the interval between onset and termination of electrical activity is unchanged across a wide range of cycle times; in muscles (e.g., the medial gastrocnemius, the soleus, the medial hamstrings and the gluteus medius) evoking long bursts, the intervals correlate linearly with cycle duration. While correlation values depicting the strength of the relation between the duration of the burst and cycle time are lower (0.72 - 0.94) than those observed with other parameters, a confidence level of greater than 0.99 indicates that the relationship does not occur by chance. The duration from initiation to peak discharge is also arranged in accordance with the type of discharge and, ultimately, with functional demands. For example:

(1) The peak discharge of the tibialis anterior (Figure 9) appears at heel strike to guide the lengthening contraction of the tibialis anterior during rapid plantar flexion in early stance and to assist tibial rotation on the talus during the

early part of yield (dorsiflexion) of the ankle. A second peak is observed at toe-off to initiate phasic dorsiflexion of the ankle during early swing (cf. Figures 9a and b).

(2) The peak activity of the medial gastrocnemius occurs <u>60 - 120 milliseconds prior to the termination of yield (i.e., maximum dorsiflexion) of the ankle, and 100 - 140 milliseconds prior to the termination of hip extension.</u>

These two phenomena are observed across a wide range of velocities.

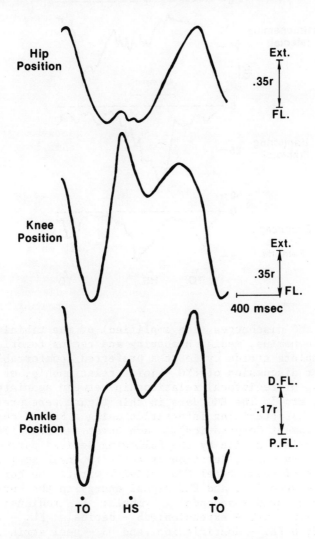

Figure 9a. Angular displacement of the hip, knee and ankle joints. (Figure and legend continued on next page.)

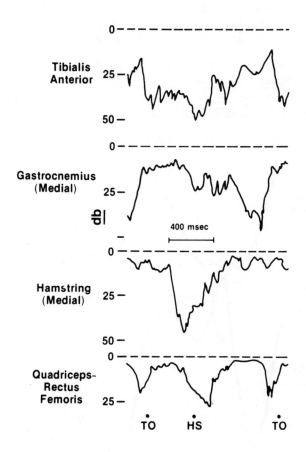

Figure 9b. EMG discharges (log amplified) of the tibialis anterior, medial gastrocnemius, medial hamstring and rectus femoris muscles during a complete stride cycle at a preferred "comfortable" rate. [See text for discussion of EMG - joint relationship, of functional implications, of reciprocal relationship between agonists and antagonists, etc.] The EMG data in this figure were presented to yield burst characterizing information using a PDP-12 computer (Digital Equipment Corporation). Each burst was described relative to the appropriate stride by the following derived parameters; time of activity onset, duration of activity, peak amplitude, time of occurrence of peak, total EMG activity within the burst interval (micro-volt - seconds), and EMG signal energy in the burst interval (micro-volts squared - seconds). Symbols: r - radians; P.FL. - plantar flexion; Ext. - extension; db - decibels; FL. - flexion; TO - toe-off; D.FL. - dosriflexion, and HS - heel strike.

(3) The medial hamstring (Figure 9) reaches its peak activity prior to heel strike at normal and brisk rates of progression in order to decelerate knee motion; at slower rates, the peak occurs after heel strike in association (co-contraction) with the quadriceps peak in order to stabilize the knee and, perhaps, to assist hip extension in early- and mid-stance. Hence, a "switch" in function occurs as the rate of progression changes, a feature common to a number of upper and lower limb muscles (Craik and Herman, this volume).

(4) The peak of the rectus femoris (Figure 9) firing occurs after heel strike to control the yield of the knee in early stance; a second peak is observed in late stance during the yield of the knee in the pre-swing phase of the gait cycle. At rates of knee motion below three radians per second, this second burst is no longer noted. Grieve and Cavanagh (1974) report a similar phenomenon in other muscles.

Quantitatively, whether measuring the magnitude of the discharge in terms of peak voltage, of integrated area within the envelope of the discharge, or of electrical "energy," there is an increase of activity in both flexors and extensors with increasing frequency or rates of movement.

B. Reciprocal Activity

Alternate, reciprocal behavior of agonist and antagonist groups, such as the tibialis anterior with the gastrocnemius, is a common feature of the EMG "program."

During each cycle, there are periods in which the agonist gastrocnemius and antagonist tibialis anterior muscle groups demonstrate an alternating EMG pattern. A particularly striking relationship occurs between the peak and the termination of gastrocnemius activity (at or just prior to the termination of yield of the ankle) when the tibialis anterior demonstrates a rapidly rising discharge (see Figure 9b). The abrupt termination of the gastrocnemius EMG with the accompanying reciprocal action of the tibialis anterior and the following initiation of ankle plantar flexion and hip flexion are defined, in this presentation, as the pre-swing complex. All four events in this pre-swing complex are phase-locked to the toe-off event as well as to the subsequent peak ankle plantar flexion and knee flexion during the swing phase across a wide range of velocities.

C. Gastrocnemius and Ankle Motion

After the gastrocnemius begins firing in the late swing period, its level of EMG activity increases gradually until ankle yield takes place; the firing level increases abruptly, usually within 60 to 80 milliseconds of ankle motion, reaching a peak prior to maximum dorsiflexion (see Figure 9). As indicated above, peak voltage increases with increasing velocities of movement and, hence, with increasing rates of muscle stretch. When ankle motion is constrained by an "ankle-foot" orthosis, the EMG pattern, i.e., the temporal sequencing of the muscle discharge with respect to the time of initiation, peak activity and termination of the gastrocnemius remains unaltered; further, its reciprocal relationship with the tibialis anterior in the pre-swing complex is similar to the behavior observed under natural conditions. This suggests that the response of the muscles governing ankle motion is not determined by angular displacement. Further, it does not appear that the motor discharge is influenced by the duration and magnitude of applied force. Dorsiflexion of the ankle during stance is associated with an increase in plantar flexing torque, a function of the lengthening contraction of the extensor muscles of the calf. The force of the lengthening contraction is derived from both the inherent passive (rheologic), and active (i.e., torque-time, torque-position torque-velocity) properties of muscle and the frequently observed increase in EMG activity in the gastrocnemius and soleus muscles during the yield stage. More specifically, the force associated with the EMG response is related to the dynamics of the movement, i.e., force is considerably higher, at any one angular position, during the lengthening contraction than during an isometric or shortening contraction (Joyce et al., 1969; Freedman, 1971; Grillner, 1972).

The EMG discharge in the gastrocnemius muscle is maximum immediately prior to peak dorsiflexion and, as a result of the lag in electro-mechanical coupling of muscle, should effect the development of torque during the subsequent plantar flexion; however, plantar flexing torque decreases precipitously at the termination of ankle dorsiflexion and during the brief period of rapid plantar flexion prior to toe-off; this is due to both a rapid shortening of the extensor muscles following ankle yield associated with the reciprocal firing patterns between the tibialis anterior and gastrocnemius and the rapidly falling vertical force. On the other hand, when the ankle is constrained by an orthosis, the magnitude and duration of active torque is markedly altered; given a similar peak value and time of initiation of the phasic EMG discharge in the gastrocnemius, muscle force developed during mid-stance is less than the force developed during yield under natural unrestrained conditions. However, prevention of rapid displacement of the ankle permits torque (isometric) values to be sustained for considerably longer periods, due to differences in the torque-position and torque-velocity relationships. Thus, it appears that neither ankle motion nor the magnitude and duration of force is related to the pre-swing complex (i.e., the phase relationship between the EMG events, ankle plantar flexion and hip flexion with toe-off, or the duration of the stance and swing periods).

During any one of the preferred rates of walking, the variability of both the magnitude and duration of the discharge of each of the four muscles described is substantially greater (e.g., four-six fold) than the variability of amplitude and duration of peak-to-peak displacement of joint motion (see preceding section). While the errors (variance) observed with respect to time intervals between peak displacements range from one to three percent of the stride cycle from brisk to extremely slow rates, the variance relating to the duration of the discharge range from four percent (at brisk rates) to eighteen percent (at extremely slow rates) of the stride cycle.

D. Statistical Assessment

The strong tendency for the magnitude of the EMG discharge to increase with increasing frequencies (and velocities) of movement is always accompanied by considerable dispersion about the mean. Frequently, the degree of dispersion encroaches upon the distribution about the mean at a higher or at a lower preferred range of velocity. Consequently, while mean values are statistically different at various rates of walking, a statistical assessment of variability indicates an insignificant relationship among sample populations at each of the preferred rates.

V. DISCUSSION AND CONCLUSIONS

Single limb analysis during steady-state locomotion reveals a precise, predictable, and orderly repertoire of events measured from the input neural stage to the final output stage of cyclical behavior. Given both the "forced and preferred" conditions of this protocol, the locomotion synergy at each of the four operational levels (i.e., cyclical behavior, ground reaction measures, joint motion and EMG analysis) exhibits little variability and adaptability (also, Herman et al., 1974). The ultimate resolution with respect to the strategy that the nervous system adopts to solve a specific motor problem appears to support the hypothesis developed by the Russian workers (see Introduction) and the proposed hypothesis set forth in the design of this study. Further, the constancy of motor output patterns (e.g., joint motion) from cycle-to-cycle seems to confirm that averaging techniques can be utilized to describe numerous parameters of locomotion behavior (see Hildebrand, this volume).

A. Neural Considerations

It is conceivable that, under normal environmental conditions, both supraspinal (or intraspinal) programs and the inherent passive and active properties of muscle are adequate to control the dynamic performance of stepping with sufficient stability without further compensation by reflex activity.

(1) EMG discharges commencing during the swing phase most likely represent a pre-programmed process, presumably preparing the limb for the succeeding stance phase. (Peak discharges of certain muscles (see above) occurring prior to, at, or immediately following heel strike are also indicative of a pre-programmed motor act.) In contrast to observations of locomotion in the cat and dog (Grillner, 1975), it appears that a considerable number of flexors _and_ extensors demonstrate this behavior over a wide spectrum of velocities (Herman et al., 1974). Analogous responses in man are noted during hopping and stepping activities, e.g., the extensors of the calf fire prior to floor contact (Melville-Jones and Watt, 1971a, b; Freedman and Herman, Unpublished observations). However, this latter action, at least in a naive situation, is _suppressed_ by modifying feedback from visual, auditory and/or somatosensory channels suggesting that appropriate sensory cues are necessary for this precise expression of intended performance (Freedman and Herman, Unpublished observations).

(2) In each protocol, the maintenance of the precise and orderly temporal and spatial arrangement of EMG discharges and angular displacements of the hip and knee when the ankle is immobile also implies that the neural program for steady-state locomotion is essentially centrally determined. Adjustments in the program due, for example, to changes in environmental conditions, are readily available from reafferent sources when new tactics and strategies must be employed. In such situations, alterations of reafferent information (e.g., muscle stiffness) lead to readjustment in motor discharge intensity, but not to modification of the basic pattern of EMG, joint, and ground reaction measures (Herman et al., 1974). Hence, the contractile state of muscle is an extremely important feature of the control system in that changes in its inherent properties (e.g., when muscles' ability to develop force during lengthening-shortening is reduced) can bring about a compensatory "rearrangement" with respect to increased motoneuron output. Such compensatory behavior of the nervous system leads to a stable performance as judged by the observed spatio-temporal behavior (rhythm and periodicity) of motor discharges and the torque-time, torque-position and angular motion relationships (Herman et al., 1974).

(3) It remains unclear whether, in the presence of such stereotyped performance, the central nervous system treats information derived from peripheral sources as intermittent signals, as trigger signals or as redundant information (Vaughan et al., 1970; Evarts et al., 1971). The latter concept is unlikely despite certain conclusions, namely developed from investigations of deafferented primates (Taub, this volume). Certainly, the increased errors (variability) in performance (angular displacement, stance periods, forces) noted at slow, unaccustomed rates of progression suggest a continual adjustment (i.e., to match the actual movement to the intended movement) of the central program by ongoing modulation of supraspinal centers (Phillips, 1969; Evarts, 1973). On the other hand, the astonishingly limited biological variance noted in all parameters under each condition (i.e., preferred and forced rates of walking) and in each parameter under all conditions during normal and brisk walking implies that the locomotor synergy is operating with "least interaction" among sensorimotor systems (e.g., Gelfand et al., 1971). Subtle adaptations can most likely be provided by the inherent properties of muscle which can serve as an autogenic regulator to obviate the undue time delays accompanying short and long loop reflexes and electro-mechanical coupling of muscle. However, if required, such reflex events demonstrate a time course (e.g., 30 - 40 milliseconds for myotatic responses and 90 - 110 milliseconds for cortical responses of muscles controlling the ankle) which can provide further stabilizing (e.g., load compensatory) effects during the same stance period (range: 0.71 - 0.84, and 0.49 - 0.70 seconds at normal and fast rates, respectively). This is in contrast to observations in the cat where reflex activity may not provide effective load compensation during the same stance phase at fast velocities (Grillner, 1975).

It is of special interest to note that while temporal and force data of joint movement show little variation from cycle to cycle during constant frequency of walking, both the amplitude and duration of motor discharges display a considerable variability ("neural noise"). The control of torque/angular displacement values at any one frequency or of the rate of change of stiffness associated with a range of frequencies between 0.7 and 0.9 Hz. is somewhat insensitive to this "neural noise" in the motor discharge; this characteristic is most likely due to the filtering and damping properties of muscle which tend to average or smooth the performance, thereby stabilizing the mechanical response (Davis, 1969; Rosenthal et al., 1970; Herman et al., 1973). As Davis (1969) points out, fewer restrictions need be placed upon the operation of the central nervous system control mechanism. In his estimate, a central oscillator need only provide for a general inverse relationship between the amplitude and the period of the oscillation and for the activation of motor discharges in the same sequence during each movement cycle.

B. Biomechanical Considerations: What is Controlled?

During the process of walking, the human body is a complex system of moving segments, each linked to another by articulating joints. At any given time, each moving segment has a certain amount of mechanical energy, part of which may be potential and part of which may be kinetic. Additionally, the kinetic energy may exist in two forms, namely, translational and rotational. During normal walking, the cyclic movements of the segments give rise to continual exchanges of energies across the joints. These include changes in potential and kinetic energies, and within the realm of kinetic energy, exchanges between translational and rotational components. While the energies in this process may not be conserved totally because of frictional losses, they, nonetheless, constitute a ready reserve of energy for adjoining segments to draw upon on demand (e.g., such as introducing rapid corrections or adjustments in movement patterns). By drawing from this reserve of energy, either by lowering a segment (decreasing potential energy), or by reducing its velocity in either the translational and/or rotational mode (decreasing kinetic energy), the sensorimotor system coordinates movements to produce a precise and orderly functional synergy.

The principal mechanism for energy transfer among body segments appears to be one of conservation of momentum. Furthermore, since body segments are physically linked and cannot travel independently for different distances, it appears that the critical issue is the periodicity required to effect the necessary transfer of energies. To achieve smooth, coordinated bodily movements, the requirement is for accurately timed phasic activity among appropriate muscle groups to develop the forces to the necessary extent. Such a strategy appears to be responsive to both the _mass_ and the _velocity_ of the adjoining body segments. Accuracy in timing is further reinforced by the pendulous characteristics of both the upper and lower limbs. A pendulum is, by nature, a time regulatory mechanism. To alter the swing frequency of a pendulum from its natural resonant frequency requires additional forces and consequently, greater energy expenditure rates. Therefore, it seems that the body must adjust its synergistic activities to the _frequency_ and _amplitude_ of segmental movements best suited to its energy transfer requirements. Accordingly, the narrow range in stride frequencies demonstrated by the subjects walking at their preferred comfortable rate is attributed to the natural resonant frequency of the swinging limbs.

The regulatory process for intersegmental activity during various rates of walking is made evident in the cyclic characteristics of the whole body velocity in the forward direction. The fundamental cyclic variation in the whole body velocity during each step cycle is independent of the mean velocity of walking. The average velocity change (ΔV) is approximately 0.28 meters per second. Taking into consideration the effect of the subject's mass and height, the span of ΔV is quite small, ranging from 0.22 to 0.34 meters per second. The corollary to this phenomenon appears in the temporal pattern of the longitudinal shear ground reaction force where the areas under the accelerative and declerative force-time relations are relatively independent of velocity of walking. Both the constancy of ΔV and the force-time relations should be consistent since they are directly related to impulse (product of force and time) and to change in momentum (product of mass and change in velocity), i.e., $\Delta ft = m\Delta V$. The significance of this observation is that the biomechanical process of locomotion operates on a constant incremental change in momentum during each step cycle.

The rates at which the ground reaction forces change during a transitory event reflect the precision by which muscle groups must be recruited to stabilize and to mobilize various joints in the body. The loading and unloading of the limbs represent a perturbation which must be reacted to by the body. As the coefficients for the loading rate tend to be consistent among the subjects, so are the moduli of flexional rigidity (defined as newton-meters/radian/kilogram) of the joints (e.g., knee). Both represent the body's accommodation to a change in externally applied forces.

A direct relation is shown to exist between certain ground reaction force characteristics and the average momentum of the body. It seems clear that these forces, registered at the walking surface, must reflect the manner in which muscles serve intersegmentally. If conditions of pain, loss of joint mobility, or sensorimotor loss exist, the patient may not develop normal muscular forces. The result is a diminution of movements and a functional reduction in walking speed, hence, a reduction in the momenta of body segments. When walking speed is reduced to the point where stability is threatened, both normal subjects and patients systematically increase their ratio of double support period-to-stride period and consequently, rely more on bipedal contact for control. It is evident that less postural reaction is required to walk fast. In attempting to walk slowly, normal subjects experience a disturbance or alteration (i.e., increased variability) in their stereotyped movement patterns, presumably because of unaccustomed diminution of momenta. The resultant movements are suggestive of the altered control seen in patients with certain pathologies who have functionally reduced velocities of walking.

In considering the effect of reduced walking speed upon the quality of performance, it is evident that energy and momentum play an important role. Conceivably, energy is the agent by which efficient control of movement is achieved.

Acknowledgment

The authors wish to acknowledge with deep gratitude the time and effort contributed by many staff members of the Krusen Research Center. This investigation was supported, in part, by Grants No. 23P-55518 and 16P-56804 from the Rehabilitation Services Administration, Department of Health, Education and Welfare, Washington, D.C., U.S.A.

REFERENCES

Barnes, W.J.P., (1975) Leg co-ordination during walking in the crab, Uca pugnax. J. Comp. Physiol. 96, 237-256.

Davis, W.J., (1969) The neural control of swimmeret beating in the lobster. J. Exp. Biol. 50, 99-118.

Drillis, R.J., (1959) The use of gliding cyclograms in the biomechanical analysis of movements. Human Factors. 1, 1-11.

Eberhart, H. and Inman, V., (eds.), (1947) Fundamental Studies of Human Locomotion and Other Information Relative to Design of Artificial Limbs. Report to the National Research Council, Vol. 1, College of Engineering, University of California at Berkeley.

Evarts, E.V., Bizzi, E., Burke, R.E., Delong, M. and Thach, W.T., Jr., (1971) Central control of movement. Neurosci. Res. Progr. Bull. 9, 1-170.

Evarts, E.V., (1973) Motor cortex reflexes associated with learned movement. Science. 179, 501-503.

Finley, F.R., (1969) Pattern recognition of myoelectric signals in the control of an arm prosthesis. J. Can. Phys. Therap. Assoc. 21, 19-24.

Finley, F.R. and Cody, K.A., (1970) Locomotive characteristics of urban pedestrians. Arch. Phys. Med. Rehab. 51, 423-426.

Freedman, W., (1971) Systems analysis of the myotatic reflex in normal, hemiplegic and paraplegic man. Ph.D. Thesis, Drexel University, Philadelphia.

Gelfand, I.M., Gurfinkel, V.S., Tsetlin, M.L. and Shik, M.L., (1971) "Problems in the analysis of movements," In Models of the Structural-Functional Organization of Certain Biological Systems. (Gurfinkel, V.S., Fomin, S.V. and Tsetlin, M.L., eds.), M.I.T. Press, London, (330-345).

Grieve, D.W. and Cavanagh, P.R., (1974) "How electromyographic patterns and limb movements are related to the speed of walking," In Human Locomotor Engineering. The Institution of Mechanical Engineers, London. (9-15).

Grillner, S., (1972) The role of muscle stiffness in meeting the changing postural and locomotor requirements for force development by the ankle extensors. Acta Physiol. Scand. 86, 92-108.

Grillner, S., (1975) Locomotion in vertebrates: Central mechanisms and reflex interaction. Physiological Reviews. 55, 247-304.

Gurfinkel, V.S., Fomin, S.V. and Tsetlin, M.L.,(eds.), (1971a) Models of the Structural-Functional Organization of Certain Biological Systems. M.I.T. Press, London.

Gurfinkel, V.S., Kots, Ya.M., Paltsev, E.I. and Feldman, A.G., (1971b) "The compensation of respiratory disturbances of the erect posture of man as an example of the organization of interarticular interaction," In Models of the Structural-Functional Organization of Certain Biological Systems. (Gurfinkel, V.S. et al., eds.), (382-395).

Hellebrandt, F.A., (1964) Living anatomy. The Wisc. Med. J. 63, 525-535.

Herman, R., Freedman, W., Monster, A.W. and Tamai, Y., (1973) "A systematic analysis of myotatic reflex activity in human spastic muscle," In New Developments in Electromyography and Clinical Neurophysiology. S. Karger, Basel, (556-578).

Herman, R., Cook, T., Cozzens, B. and Freedman, W., (1974) "Control of postural reactions in man: The initiation of gait," In The Control of Posture and Locomotion. (Stein, R. et al., eds.), Plenum, New York, (363-388).

Joyce, C.G., Rack, P.M.H. and Westbury, D.R., (1969) The mechanical properties of cat soleus muscle during controlled lengthening and shortening movement. J. Physiol., Lond., 204, 461-474.

Kots, Ya.M., Krinskiy, V.I., Naydin, V.L. and Shik, M.L., (1971) "The control of movements of the joints and kinesthetic afferentation," In Models of the Structural-Functional Organization of Certain Biological Systems. (Gurfinkel, V.S. et al., eds.), (371-381).

Melville-Jones, G. and Watt, D.G.D., (1971a) Observations in the control of stepping and hopping movements in man. J. Physiol. 219, 709-727.

Melville-Jones, G. and Watt, D.G.D., (1971b) Muscular control of landing from unexpected falls in man. J. Physiol. 219, 729-737.

Molen, N.H., (1973) Problems on the Evaluation of Gait. Ph.D. Thesis, Vrije Universiteit, Amsterdam, Holland.

Morton, J., (1952) Human Locomotion and Body Form. Williams and Wilkins Co., Baltimore.

Murray, M.P., (1967) Gait as a total pattern of movement. Am. J. Phys. Med. 46, 290-333.

Phillips, C.G., (1969) Motor apparatus of the baboon's hand. Proc. R. Soc. B. 173, 141-174.

Rosenthal, N.P., McKean, T.A., Roberts, W.S. and Terzuolo, C.A., (1970) Frequency analysis of stretch reflex and its main subsystems in triceps surae muscles of the cat. J. Neurophysiol. 33, 713-749.

Steindler, A., (1953) A historical review of the studies and investigations made in relation to human gait, J. Bone Joint Surg. 35a, 540-543 and 728.

Szentágothai, J. and Arbib, M.A., (1974) Conceptual models of neural organization. Neurosciences Research Program Bulletin. 12, 313-479.

Vaughan, H.G., Gross, E.G. and Bossom, J., (1970) Cortical motor potential in monkeys before and after upper limb deafferentation. Exp. Neurol. 26, 253-262.

Wetzel, M.C., Atwater, A.E., Wait, J.V. and Stuart, D.G., (1975) Neural implications of different profiles between treadmill and over-ground locomotion timing in cats. J. Neurophys. 38, 492-501.

Wirta, R.W. and Taylor, D.R., Jr., (1970) "Development of a multiple-axis myoelectrically controlled prosthetic arm," In Advances in External Control of Human Extremities, Belgrade. Proceedings of the Third International Symposium on External Control of Human Extremities, Dubrovnik, Yugoslavia, (245-253).

HUMAN SOLUTIONS FOR LOCOMOTION

II. INTERLIMB COORDINATION

        Rebecca Craik, Richard Herman and F. Ray Finley

        Department of Rehabilitation Medicine, Temple University

        Health Sciences Center, and *

Interlimb coordination is presented as a temporal and phase relationship between peak <u>hip extension</u> (HE) in one limb and (1) peak hip extension in the contralateral limb (LHE to RHE), and (2) peak shoulder flexion (SF) and extension (SE) in both the homolateral (e.g., LHE to LSF) and the diagonal (e.g., LHE to RSE) limbs. In the normal subjects studied, a cycle-by-cycle analysis of a series of strides at various rates of free walking reveals an abrupt shift of the frequency relationship between the motion of upper and lower limbs from 1:1 to 2:1 at approximately 0.75 Hz. This relation persists through the complete range of slow frequencies investigated (i.e., 0.5 - 0.75 Hz). The commonly observed characteristics of in-phase coupling between the homolateral HE-SF and the diagonal HE-SE limbs and of alternate phase coupling between both upper limbs are altered at frequencies below 0.75 Hz. At these rhythms, HE couples in-phase with homolateral SE and remains coupled in-phase with the diagonal SE. Thus, below 0.75 Hz, each hip extensor peak during a stride cycle is "locked" to SE bilaterally. The analysis of shoulder motion and associated EMG discharges (e.g., M. posterior deltoid, M. anterior deltoid) indicates that each muscle may change its functional role when the rhythm of upper

---

*Krusen Center for Research and Engineering, Moss Rehabilitation Hospital, Philadelphia, Pa.

limb motion is altered. Additional investigation of muscle function suggests that one or more muscles with common segmental innervation have differing functional activities during the gait cycle, activities that vary in accordance with the tactics and strategies employed.

Introduction

Formulation of a neural control model for the locomotion of man necessitates establishing a detailed description of the sequencing of interlimb events in man prior to investigating similarities in animals. The literature previously has not focused on the area of interlimb coordination in man. Since man can progress at various speeds of walking in the absence of arm swing, some reports state this phenomenon is a passive pendular motion. However, in 1939, Elftman concluded that arm swing was useful, if not necessary, to counteract pelvic rotation during stance and that the force required for arm swing was generated by muscular action with gravity playing only a minor role (Elftman, 1939). Electromyographic activity for both shoulder and arm muscles during locomotion was found related to arm swing (Fernandez-Ballesteros, 1965; Hogue, 1969). Therefore, assuming that arm swing was an active component of the integrated activity of locomotion, Murray described the sequencing of upper to lower limb events during free-speed and faster walking (Murray, 1967). In order to determine if the same pattern of movement is maintained across a wide range of velocities or if there are a number of interlimb patterns as described in the animal literature, data expanding Murray's study must be gathered.

The investigation reported here is preliminary in design and describes the relationships between shoulder and hip motion and related muscle activity for three females and two males who ranged in age from 19 to 28 years.

Subjects were instructed to select a walking speed in response to each of the following verbal commands: very fast, brisk, comfortable, slow and super slow. Instrumentation included bilateral shoulder and hip electrogoniometers and surface electromyographic (EMG) assemblies (Research Engineering Center Progress Report #4, 1975).

The data were examined for time of occurrence of peak flexion and peak extension of bilateral shoulder and hip motions. Additionally, the times of onset of the bursts of myoelectric activity were examined. In order to compare among the mechanical variables, the arbitrary reference time of left peak hip extension was selected. For comparison among the electrical events, the onset of left hamstring firing was selected as the reference for electrical events. Finally, a comparison between mechanical and electrical events was made using left peak hip extension as the reference.

Time of displacement and phase relation were chosen to describe these interlimb relationships. Time of displacement was defined as the time interval between the reference and a mechanical or electrical event; the phase relation was defined as this time interval divided by the cycle time*. A standard error of the mean performance for time of displacement was computed for each subject's performance in each velocity category. The standard error was within the visual reduction (systematic) error or ten milliseconds.

A mean and standard deviation for the time and phase relationship was computed for each subject's performance in each of the five velocity categories. Only those mechanical events which demonstrated an <u>in-phase</u> or <u>alternate</u> phase relationship with peak hip extension were selected for this presentation. The criterion for an in-phase relationship was 0 ± 0.1 and the limits for an alternate phase relationship was 0.5 ± 0.1.

Categories of walking speeds (e.g., brisk, slow) rather than an average velocity or frequency will be described because the subjects individually demonstrated a wide range of preference for both velocity and frequency in response to verbal commands. In fact, the same absolute velocity was found in different categories for different subjects; for example, the velocity selected by one subject to represent slow walking was the same velocity selected by another subject to represent comfortable walking. These differences did not correlate with anthropometric differences, e.g., sex, mass, height, leg length. More importantly, the performance of each subject within a category was consistent and each subject's performance within any given category differed significantly from any other when examined by an analysis of variance.

CYCLICAL BEHAVIOR OF UPPER AND LOWER LIMBS

A change in the shoulder to hip relations is evident when examining performance across the five velocity categories. At the

---

*<u>Definition of terms</u>: Cycle time is defined as the time between two consecutive peak extensions of the left hip, i.e., one stride. Frequency is the reciprocal of this relationship and is expressed in Hertz (stride cycles per second). Homologous refers to paired like joints such as the hips or the shoulders; diagonal refers to the relation of the left hip and the right shoulder or the right hip and the left shoulder and, homolateral refers to the relation of hip and shoulder on the same side of the body.

speeds of comfortable, brisk, and very fast there is a 1:1 relationship between the time intervals of two successive shoulder events; peak extension to the succeeding peak extension, for example, compared to two consecutive hip events (i.e., peak extension to the following peak extension). However, analysis of these events for the slow and super slow speeds demonstrates two complete shoulder cycles for each complete hip cycle (Figure 1).

A composite plot of the shoulder extension frequencies against hip extension frequencies, depicted in Figure 2, shows that as the frequency decreases, the linear 1:1 relationship between the frequencies for shoulder and hip changes to 2:1 at approximately .75 Hertz (cycles per second). This corresponds to slow walking for the subjects studied. A linear regression for both the 1:1 and 2:1 relationship shows intercepts of 0.0 with a correlation coefficient of 0.99, significant at the 0.001 level.

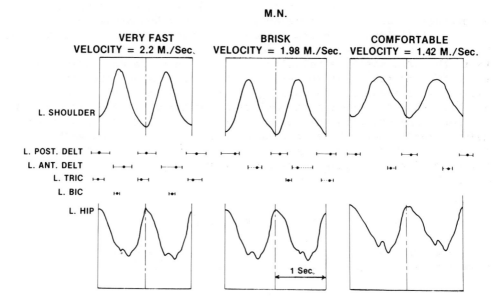

Figure 1A. Characteristic excursions of left shoulder and hip are seen in this figure for walking at comfortable, brisk and very fast speeds. The time base goes from right to left and flexion is the downward deflection. The 1:1 alternate phase relationship between shoulder and hip is apparent. The duration of activity for the muscles recorded in the upper limb is shown. The dotted line represents inconsistent firing.

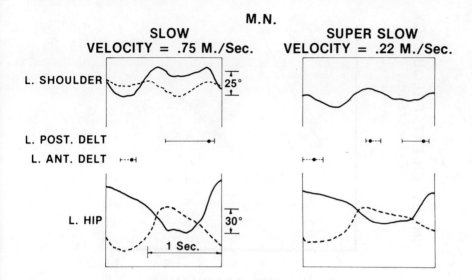

Figure 1B. This figure shows characteristic excursion of shoulder and hip for the speeds of slow and super slow. The dashed lines represent right shoulder and hip motion. The change to a 2:1 in-phase relationship for shoulder to hip events is evident at these walking speeds (see text).

The performance at slow walking for two of the five subjects represents a transition zone in the ratio from 1:1 to 2:1 between upper and lower limbs (Figure 2). The transition was characterized by a prolonged period of shoulder extension rather than two alternating peaks of extension and flexion demonstrated by other subjects at the slow walking speed.

MECHANICAL SEQUENCING

The time interval between peak extensions of the <u>homologous</u> hips increases linearly with an increase in cycle time across the five velocity categories (Figure 3). Although variability increases with an increase in cycle time, the correlation values between the events are above 0.95 which is significant at the 0.01 level for all subjects.

The phase relation of those homologous events ranges from 0.47 to 0.52 for all five subjects for the various walking speeds; these values satisfy the criterion for an alternate phase relation (0.5 ± 0.1) between right and left peak hip extensions.

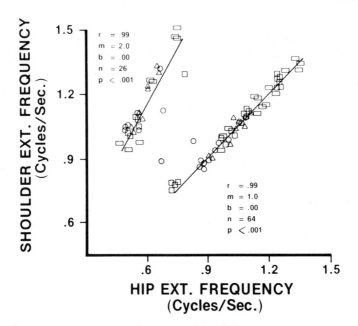

Figure 2. The relation between shoulder to hip frequency for four subjects across the preferred walking speeds. The change from a 1:1 to 2:1 relationship between these two events occurs at approximately .75 Hertz (cycles per sec). Note that two subjects demonstrate neither a 1:1 nor 2:1 ratio at this frequency.

In contrast to homologous lower limb relationships, the time interval between <u>diagonal</u> peak shoulder extension and the reference, contralateral hip extension, remains consistent with an increase in cycle time. While the time of shoulder extension remains within ± 0.12 sec. of the reference for all five subjects across all walking speeds, the shoulder tends to <u>lag</u> behind the contralateral hip (Figure 3). A second shoulder extension peak occurs when the cycle times range from 1.35 sec. to 2.43 sec., i.e., at slow and super slow walking speeds (Figure 4). As was demonstrated for the homologous lower limb events, there is an increase in variability for both time of displacement of the diagonal event and cycle time as the subjects walked slower.

The phase relation of the diagonal events ranges from -0.03 to +0.07 across the five speeds of walking, a finding well within the criterion ($0 \pm .1$) for an in-phase relationship. At slow rates, the phase relations range from 0.50 to 0.66. The higher values, i.e., > 0.6, do not conform to the criterion for alternate phasing due to the transitional state from 1:1 to 2:1 ratio of shoulder to

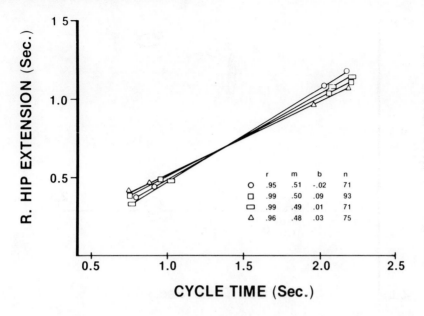

Figure 3. The abscissa is the time between two consecutive left hip extensions. The ordinate represents the interval between right and left peak hip extension. As the four subjects represented in this figure walked slower or increased the cycle time, there is a linear increase in this interval.

hip events in two subjects. However, the phase relationship at super slow walking, ranging from 0.49 - 0.59, is within the limits defined for alternate phasing.

In comparison to the close time relationship across all cycle times for the diagonal pattern, the homolateral relationship of peak shoulder flexion to the reference, ipsilateral peak hip extension, occurs very close in time (+0.09 sec.) only at very fast, brisk, and comfortable speeds of walking (Figure 5). At slow and super slow walking, there is a gradual increase in the interval between the homolateral shoulder flexion peak and the reference. Further, in contrast to the diagonal event, homolateral peak shoulder flexion tends to lead the reference across all velocities. A second shoulder flexion peak occurs within one hip extension period at the cycle times of 1.35 sec. to 2.43 sec., which correspond to the slow and super slow walking speeds. As with the homologous and diagonal events, the variability of time of displacement increases as cycle time increases for the homolateral shoulder event.

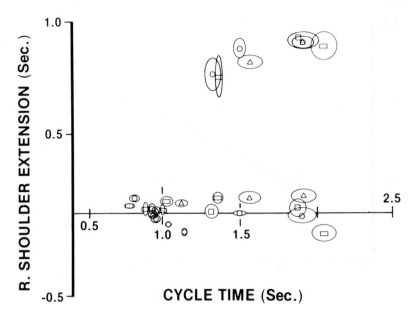

Figure 4. The interval between diagonal peak shoulder extension and the reference of peak hip extension for four subjects is shown. The ellipses represent one standard deviation about the mean of time of displacement as well as one standard deviation about the mean of the cycle time. The occurrence of the second shoulder event is seen at the higher cycle times (see text for description).

The phase of peak homolateral shoulder flexion to peak hip extension is -0.08 to +0.04 which is within the criterion for an in-phase relationship at the comfortable and faster speeds of walking. An abrupt phase shift occurs with the onset of the double flexion during slow and super slow walking. At slow and super slow walking the phase relationship of the two shoulder flexion peaks is -0.15 to -0.28 and +0.23 to +0.38, respectively, which is defined as out-of-phase with peak hip extension. However, these phase relations are similar to those computed for the phase relations of right and left peak hip flexion, respectively, to the reference. Therefore, a change occurs in the phase relationship of homolateral peak shoulder flexion across the five speed categories from "coupling" with peak hip extension at comfortable and faster walking to "coupling" with bilateral peak hip flexion at the speeds of slow and super slow. Hence, at the slower rates of walking both homolateral and diagonal shoulder flexion occur with each hip flexion and each homolateral and diagonal shoulder extension occurs with each hip extension.

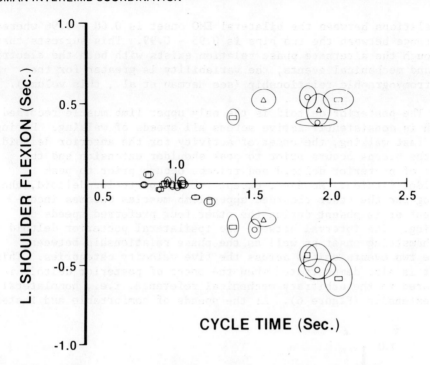

Figure 5. Time of displacement for homolateral peak shoulder flexion is represented. There is an increase in the interval between this event and the reference at the higher cycle times as well as a second shoulder flexion peak.

There is no significant difference when comparing the variance about the mean time of displacement for the homologous, diagonal, and homolateral events. Since the variance about all means is similar, one can then compare the absolute time relationships. Examination of the diagonal and homolateral events, then, suggests that homolateral peak shoulder flexion occurs closer in time to the reference than does diagonal shoulder extension.

EMG ANALYSIS

The interval between onset of right and left hamstring firing was examined as an example of homologous lower limb muscle activity. The reference of left hamstring onset was used to compare electrical events. The interval between right and left hamstring onset increases linearly with cycle time, which is the same relationship observed for the homologous mechanical relationship in the lower limbs. A comparison of the correlation coefficients obtained for these electrical and mechanical events shows that both correlations are significant at the 0.01 level. However, the actual range of correlations between the bilateral EMG onset is 0.68 - 0.96 whereas the range between the two hips is 0.95 - 0.99. This suggests that although the alternate phase relation exists with both the electrical and mechanical events, the variability is greater for the electromyographic relationship (see Herman et al., this volume).

The posterior deltoid is the only upper limb muscle recorded which is consistently active across all speeds of walking. During very fast walking, the onset of activity for the anterior deltoid and the biceps occurs prior to peak shoulder extension and the onset of posterior deltoid and triceps occurs prior to peak shoulder flexion. However, except for the posterior deltoid, the firing for the other recorded upper limb muscles becomes inconsistent or is absent during the other four preferred speeds of walking. The interval between the ipsilateral posterior deltoid and hamstring onset as well as the phase relationship between these two events shifts across the five velocity categories. This shift is also demonstrated when the onset of posterior deltoid is compared to the arbitrary mechanical reference, i.e., homolateral hip extension (Figure 6). At the speeds of comfortable and faster,

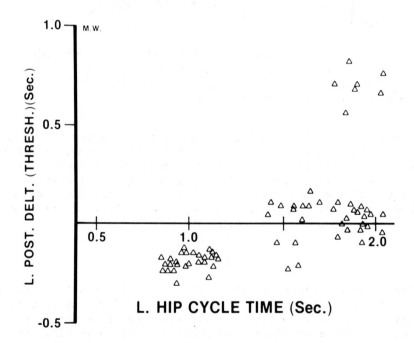

Figure 6. Represented is the time interval between initiation of the posterior deltoid and the mechanical reference of peak hip extension against hip extension cycle time for one subject across the five walking categories (see text for description).

posterior deltoid initiation occurs significantly prior to peak hip extension while at the slow speeds this event occurs very close to the time of the reference. The initiation of the posterior deltoid precedes peak shoulder flexion by an average of 0.15 secs for comfortable walking and above. At the slower speeds of walking the initiation of firing follows shoulder flexion by an average of 0.38 secs. In addition, the posterior deltoid demonstrates two patterns of firing among the subjects: at slower walking speeds the duration of activity is very long and is in-phase with the reference of left hip extension; or, there are two short bursts, one in-phase and the other in alternate phase with left hip extension.

Discussion

The shoulder and hip sequencing measured for walking speeds within the comfortable, brisk, and very fast categories confirms the findings of Murray who studied 30 healthy men (Murray, 1967). While the findings compare favorably, the analyses are based on different event references. Left heel strike was used by Murray, while peak left hip extension was used in this study. Joint motion rather than a ground reference was selected for this analysis to rule out terrain accommodation by the foot and to obtain absolute joint sequencing intervals. The hip extension peak was selected since it provided the most clearly identifiable and reliable reference. The diagonal event, i.e., peak shoulder extension, is more consistently in phase lag of peak hip extension with a tendency for time and phase-locking across all walking speeds. On the other hand, the homolateral event, i.e., shoulder flexion, tends to be in phase lead of peak hip extension with time and phase-locking only at comfortable and faster walking speeds. However, it is recognized that neither the number of subjects nor the number of trials for each subject was large enough to determine unequivocally the consistency of these findings and the limits of the time intervals for these relationships.

These data indicate that the abrupt shift in shoulder to hip sequencing pattern is due to a shift in the shoulder events. The homologous lower limb relationship remains consistently in alternate phase across the five velocities as demonstrated by both mechanical and electrical events. However, at the slower walking speeds, the arm swing frequency doubles with respect to the hip swing frequency and the homolateral shoulder flexion shifts its phase relationship. At these speeds, shoulder extension is locked homolaterally and diagonally to the reference of left hip extension. Since this upper limb change in movement characteristics is accompanied by a shift in EMG, the generation of two separate locomotor sequences is suggested.

The lack of activity in the upper limb muscles is consistent with the studies of Fernandez-Ballesteros (1965) and Hogue (1969). Using wire electrodes, Fernandez-Ballesteros recorded from 12 muscle sites in both shoulder and upper limb while subjects walked at a comfortable speed. Only the upper one-third of the latissimus dorsi, the terres major and the posterior deltoid consistently fired in the 23 subjects examined. In addition, Ballesteros found that the initiation of posterior deltoid preceded peak shoulder flexion by 0.16 sec. which is in agreement with this study's finding of 0.15 secs. for comfortable and faster speeds of walking.

A comparison between the findings in this study and similar studies conducted with healthy cats (Stuart, 1973; Grillner, 1975; Miller, 1975) suggests that this method of investigation may be useful in the development of physiologic correlates between man and other species. Although man's arms serve prehensile functions, their purpose in locomotion appears similar to that described in the quadruped: to counteract the turning couple produced by the hindlimb (Grillner, 1975). A qualitative comparison of the interlimb relationships in man and cat (Miller, this volume) suggests similarities between the cat trotting on the treadmill and man's walking at speeds of comfortable and faster. It is qualitative because: 1) the phase criteria are not yet defined for the cat; 2) differences between cats walking on the treadmill as compared to cats walking overground have been observed (Wetzel, 1975); and 3) the protocols of this particular human study and those designed for the cat are dissimilar, e.g., the human subjects selected a walking speed while in both treadmill and overground studies for the cat, the investigator selects a speed or introduces a stimulus to condition speed selection. Therefore, further investigation is necessary in both quadruped and biped locomotion prior to a quantitative comparison between locomotor movement patterns. However, if such a comparison is valid, the development of a neural model in other species could be used to further understand interlimb coordination in man. Such an understanding may, in turn, be brought to bear on the human locomotor dysfunctions which result from various pathological states.

Summary

The results of this study revealed two distinct synergic patterns of locomotion across the five velocity categories. At comfortable and faster speeds of walking, the upper to lower limb sequencing was an alternating pattern characterized as diagonal shoulder extension and homolateral shoulder flexion occurring in-phase with each peak hip extension. The slow and super slow speeds of walking showed an in-phase form of walking with bilateral shoulder extension in-phase with each peak hip extension and bilateral shoulder flexion occurring with each peak hip flexion. This change in mechanical sequencing was noted as a doubling in frequency of shoulder events at slower walking speeds. This mechanical shift was accompanied by a change in EMG activity suggesting two distinct locomotor synergies. The similarity of shoulder to hip phase relation in man to that described in the cat needs further investigations to determine whether the neural model developed through the study of animal preparations is applicable to man.

Acknowledgment

This work was supported, in part, by Grants No. 16-P-56804 and RD23-P-55118 from the Social Rehabilitation Services, Department of Health, Education, and Welfare, Washington, D.C. The authors wish to express their appreciation to the many people who supported this project with special thanks to Shelley Bampton, Harry Kenosian, Serge Minassian, and Roy Wirta.

REFERENCES

Elftman, H., (1939) The functions of the arms in walking. Hum. Biol. 11, 529-536.

Fernandez-Ballesteros, M.L., Buchtal, F. and Rosenfalck, P., (1965) The pattern of muscular activity during the arm swing of natural walking. Acta Physiol. Scand. 63, 296-310.

Grillner, S., (1975) Locomotion in vertebrates: control mechanisms and reflex interaction. Phys. Rev. 55, 247-304.

Hogue, R.E., (1969) Upper extremity muscular activity at different cadences and inclines during gait. Phys. Ther. 49, 963-972.

Miller, S., Van der Burg, J. and Van der Meché, F.G.A., (1975) Coordination of movements of the hindlimbs and forelimbs in different forms of locomotion in normal and decerebrate cats. Brain Research. 91, 217-257.

Murray, M.P., Sepic, S.B. and Barnard, E.J., (1967) Patterns of sagittal rotation of the upper limbs in walking. Phys. Ther. 47, 272-284.

Stuart, D.G., Withey, T.P., Wetzel, M.C. and Goslow, Jr., G.E., (1973) "Time constraints for interlimb co-ordination in the cat during unrestrained locomotion," In Control of Posture and Locomotion, (Stein, R.B. et al., eds.), Plenum Press, N.Y., (537-560).

The Rehabilitation Engineering Center Progress Report #4., RSA Grant No. 23P-55118 to Temple University, Moss Rehabilitation Hospital, Drexel University, December 1, 1974 - November 30, 1975, pp. 60-65.

Wetzel, M.C., Atwater, A.E., Wait, J.V. and Stuart, D.G., (1975) Neural implications of different profiles between treadmill and over-ground locomotion timing in cats. J. of Neurophy. $\underline{38}$, 492-501.

HUMAN SOLUTIONS FOR LOCOMOTION

III. THE INITIATION OF GAIT

Thomas Cook and Barbara Cozzens

Department of Rehabilitation Medicine, Temple University

Health Sciences Center, and *

The control of human locomotion was examined from the perspective of the transition from quiet upright stance to steady-state locomotion. Ten healthy subjects were studied as they achieved normal walking speeds and speeds faster and slower than their normal rates.

Measured parameters included: velocity of the approximate center of mass; magnitudes and locations of the floor reaction forces under each leg; bilateral hip, knee, and ankle joint changes in the parasagittal plane and myoelectrical activity from five sites in each lower limb. Calculated parameters included torque and stiffness about the ankle joints.

Findings indicated a linear relationship between the steady-state locomotor velocity and the velocity of the center of mass at various events within the first stride. The horizontal impulse (i.e., force x time) generated prior to the first toe-off was directly related to the measured change in velocity of the mass center. Changes in joint angles and moments about the ankle joints showed regular patterns. The data also indicated the manner in which the stance-side ankle stiffness was modified during initiation at different rates.

---

*Krusen Center for Research and Engineering, Moss Rehabilitation Hospital, Philadelphia, Pa.

Gait initiation is defined as the transition from upright standing to steady-state locomotion. In 1966, Carlsoo described this behavior in regard to the "distribution and size of the pressure of the feet against the floor as well as of the activity in certain leg muscles" (Carlsoo, 1966). Herman et al., later investigated gait initiation by examining joint displacements, myoelectrical activity, ankle torques, and floor reaction forces and attempted to characterize the role of central and/or peripheral mechanisms in the control of this behavior (Herman et al., 1973). The object of the present investigation was: 1) to further describe the process of gait initiation at comfortable, normal speeds; and 2) to describe how this process is altered to achieve speeds faster and slower than normal.

The test sample consisted of 10 healthy, normal subjects, six males and four females. The mean age for the sample was 35.8 years and the range was from 23 to 52. During the test procedure, a computer-generated auditory signal was used as a cue for the subject to begin walking. The subject was free to begin walking on whichever foot he or she chose. The first foot to leave the ground was designated the original "swing" limb, the other limb being referred to as the "stance" limb. The three major floor contact events of interest were swing-side toe-off, swing-side heel-strike, and stance-side toe-off. The subject began from a position of quiet standing on the walkway force plates. Twelve trials were conducted on each subject and time between trials was random to minimize anticipation of the auditory cue. For the first three trials the subject was instructed to walk at a comfortable, normal speed. During trials 4 through 8 the subject was instructed to walk progressively slower with each trial and during trials 9 through 12 the instructions were to walk progressively faster with each trial.

Four classes of variables were monitored during each trial. These included: 1) velocity of the approximate center of mass (using a tachometer); 2) magnitudes and point of application of the floor reaction force (three orthogonal components) under each limb; 3) bilateral hip, knee, and ankle joint changes in the parasagittal plane (using electrogoniometers); and 4) myoelectrical activity (using surface electrode assemblies) from the tibialis anterior, gastrocnemius, rectus femoris, medial hamstrings, and gluteus medius muscles in each lower limb. Each of these parameters was sampled by a digital computer at a rate of 200 times per second for a total of 3-1/2 seconds (1/2 second prior to and 3 seconds after the auditory cue). By referencing the starting position of the ankle axis to the force plate coordinate system, the computer calculated moment (or torque) about the ankle joints in the parasagittal plane.

HUMAN GAIT INITIATION

Although the process of gait initiation is, obviously, a behavior involving factors in three dimensions, the present discussion will be limited to a consideration of examples of movements and force changes in the (forward) plane of progression. For reasons to be mentioned shortly, the discussion will be limited further to a consideration of some factors operative about the stance-side ankle.

Figure 1 is a randomly-selected example of the findings regarding floor contact events. The velocity values represent the mean steady-state velocity (and stride frequency) that the subject later achieved once steady-state conditions were attained.

Figure 2 shows examples of the typical changes in velocity of the approximate center of mass following the auditory signal for one of the subjects. It is evident that the pattern depends on the final locomotor velocity. The triangles indicate the time of swing-side toe-off, the circles indicate swing-side heel-strike, and the squares indicate stance-side toe-off. Examining the amplitude of the velocity trace at these three major floor contact events during initiation, typical relationships are found as depicted in Figure 3. The relationships between instantaneous velocity at these floor contact events and the eventual steady-state velocity appear quite linear and predictive although the

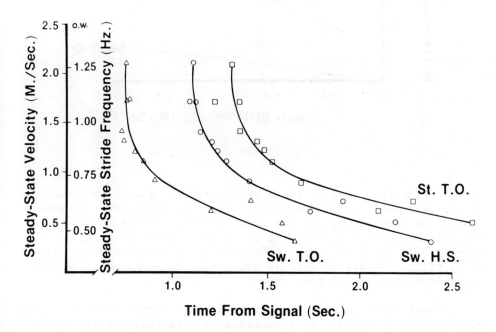

Figure 1

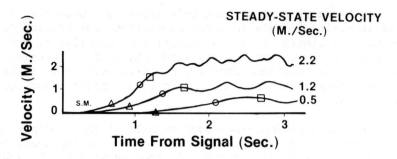

Figure 2

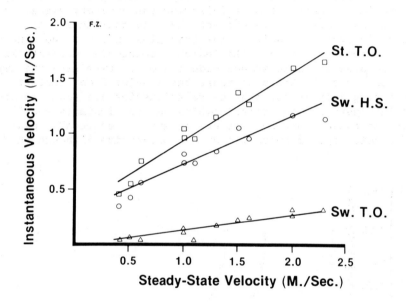

Figure 3

actual velocity at any of these events may be as small as one-tenth of the mean steady-state velocity. These relationships are essentially the same across all ten test subjects for stance-side toe-off, for swing-side heel-strike and even as early as swing-side toe-off. If the magnitude of the velocity of the center of mass is so tightly regulated even at swing-side toe-off, what are the possible mechanisms of regulation?

A basic principle of mechanics is that force equals mass times acceleration (F = ma). If both sides of this equation are multiplied by time (and since a = $\Delta V/t$) it is found that the change in

HUMAN GAIT INITIATION

momentum (mΔV), or, in this case, the change in velocity of the mass center, must be the result of the impulse (Ft) applied to the floor in an equal and opposite direction (Ft = mΔV).

Figure 4 shows examples of changes in the longitudinal shear component of the ground reaction force following the signal for various steady-state velocities. The longitudinal component is the component opposite to the intended direction of progression. The triangles, circles, and squares are representative of the same events as mentioned previously. The cross-hatched areas indicate the longitudinal impulse (or force-time product) generated prior to swing-side toe-off. An example of the relationship of impulse applied to measured change in momentum for a typical normal subject is illustrated in Figure 5.

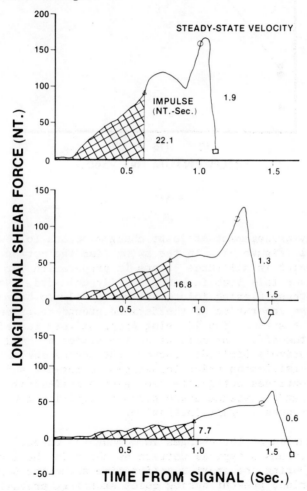

Figure 4

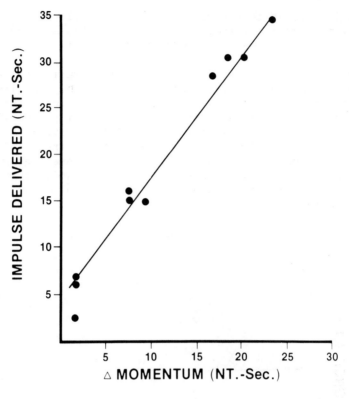

Figure 5

An orderly arrangement of joint changes occurs following the auditory signal (Figure 6). In the swing limb the motion is predominantly flexion in all three joints in preparation for toe-off and swing. Since this limb is being rapidly unloaded, the moments acting across these flexing joints are not likely to be major contributors to the generation of the desired ground reaction forces. In contrast, the stance limb is being progressively loaded prior to swing-side toe-off. The motions of the stance hip and knee joints are relatively limited; however, the stance ankle shows considerable dorsiflexion prior to swing-side toe-off and this dorsiflexion continues until after swing-side heel-strike. It seems likely that the stance ankle plays a significant role in controlling the process of gait initiation.

An examination of the changes in moment (or torque) about the stance ankle reveals a typical pattern as shown in the lower portion of Figure 7. During the short reaction time after the signal, the moment equilibrium, present during quiet standing, prevails (the balance stage, see Herman et al., 1973). Changes in myoelectrical

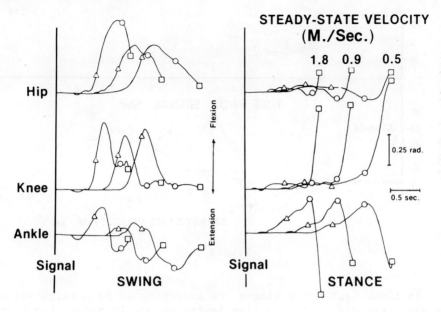

Figure 6

activity (see below) are accompanied by a posterior shift of the center of pressure and the associated decrease in plantar-flexing moment (the imbalance stage). At about the time of swing-side toe-off, the center of pressure is shifted forward again, increasing the torque (the reaction stage), until swing heel-strike. The pre-swing stage occurs as the limb is unloaded and prepared for toe-off. A point of note is that the imbalance stage is absent, or, at least, markedly diminished when the intended steady-state velocity is low.

In examining the relationship of ankle joint motion to moment about the ankle joint, the ankle joint (or any joint) may be described in terms of stiffness, defined as the change in moment per change in angle. Figure 8 is a plot of stance-side ankle moment against ankle angle for a typical subject at normal speed. Increasing plantar-flexing torque is up and decreasing plantar-flexing torque is down; dorsiflexion is to the left and plantar flexion is to the right. The samples are at 50 millisecond intervals following the signal. The clustering at the intersection of the axes corresponds to the balance stage. The imbalance stage (showing dorsiflexion and decreasing plantar-flexing torque) becomes, at swing-side toe-off, the reaction stage (showing continuing dorsiflexion but rapidly increasing torque). This gives way, at swing-side heel-strike, to the brief pre-swing stage (with plantar flexion and torque decreasing).

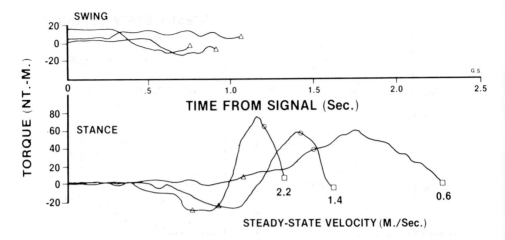

Figure 7

If these last three stages are approximated by straight-line segments, the slope of each line is the change in moment per unit of change in angle, or ankle stiffness, as defined above. The units are newton-meters-per-radian.

When the subject is asked to walk at a faster-than-free speed, the torque-angle plots are as shown in Figure 9 (2.2 m/sec). The basic pattern is the same but somewhat expanded. The four stages are still clearly discernible. The line-segment approximations show differences in stiffness in the middle two stages but the pre-swing stiffness seems to be the same. At slower rates (0.6 m/sec in Figure 9), the basic pattern seems to undergo substantial alterations. It will be recalled that there was only a negligible reduction in plantar-flexion torque at these very slow rates. In terms of the stages referred to here, the imbalance stage is, effectively, absent. (The segment approximations are not as accurate as at normal and faster rates but the pre-swing stiffness is very similar.) These stiffness adjustments must be regulated by the only actuators the nervous system has at its disposal, namely, muscles.

Figure 10 shows typical examples of the myoelectrical activity in the stance-side tibialis anterior, and medial gastrocnemius muscles during gait initiation. (The figure shows the envelopes of the EMG signal after full-wave rectification.) At the normal speeds (middle of the figure) it can be observed that the gastrocnemius activity, usually present during quiet standing, persists for two to three-hundred milliseconds after the signal (the balance stage). The gastrocnemius then shows a marked decrease in activity

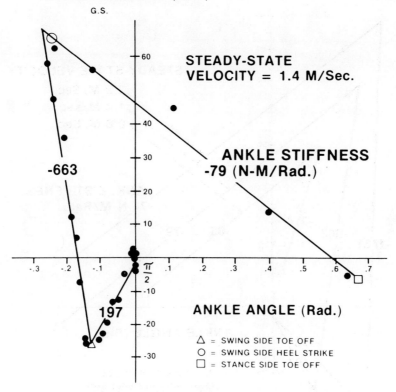

Figure 8

as the tibialis anterior is activated to shift the center of pressure posteriorly. This creates the imbalance needed to cause the body to begin dorsiflexing about the foot. At approximately the time of swing-side toe-off, the gastrocnemius again becomes dominant and controls the rate of dorsiflexion until swing-side heel-strike. The tibialis anterior then assumes its usual pre-swing function of lifting the forefoot. At faster than normal speeds (top of Figure 10), the tibialis anterior shows a relative increase in activity. The function here seems to be to create the largest possible imbalance, thus accelerating the rotation of the body about the ankle. The gastrocnemius activity is correspondingly reduced. At slower than normal speeds (bottom of Figure 10), the tibialis anterior activity (and the resulting imbalance stage) is all-but-absent and the gastrocnemius activity is expanded. As evidenced from the stiffness plots, the major task at these slow speeds is to restrain the rotation of the body about the ankle (i.e., dorsiflexion) during the reaction stage.

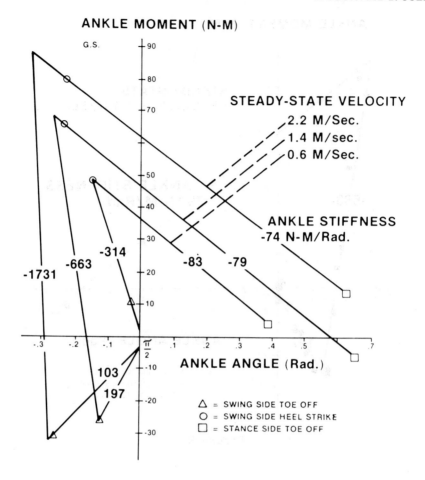

Figure 9

The implications from these findings (see Figure 11) are that verbal instructions regarding locomotor velocity cause the human subject to form a mental image of the desired locomotor system output. Adjustments in the central programming of this behavior are evidenced by changes in muscular activity which combine with gravitational forces to effect adjustments in the torque-angle relationships, or stiffnesses, about the supporting joints. These stiffness adjustments in turn result in the generation of a horizontal impulse which is related to the change in velocity of the approximate center of mass even prior to the first foot leaving the ground.

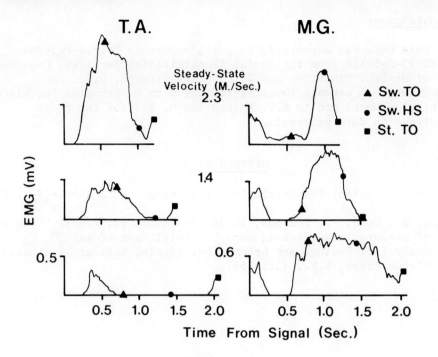

Figure 10

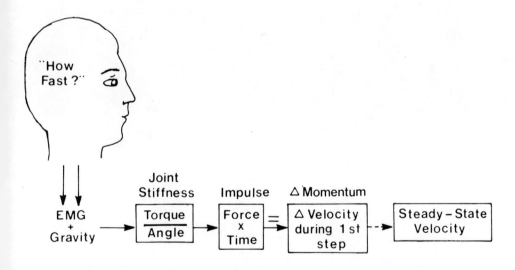

Figure 11

Acknowledgment

This work was supported, in part, by Grants No. 16-P-56804 and RD 23-P-55118 from the Social Rehabilitation Services, Department of Health, Education and Welfare, Washington, D.C. The authors wish to express their appreciation to H. Kenosian for his technical support and to A.R. Bolton and M. Wolfson for their assistance in data processing.

REFERENCES

Carlsoo, S., (1966) The initiation of walking. Acta Anat. <u>65</u>, 1-9.

Herman, R., Cook, T., Cozzens, B. and Freedman, W., (1973) "Control of postural reactions in man: The initiation of gait," <u>In</u> Control of Posture and Locomotion. (Stein, R.B. et al., eds.), Plenum Press, N.Y., (363-388).

ENERGETICS OF HUMAN WALKING

H.J. Ralston

Biomechanics Laboratory, University of California

San Francisco, California

Equations predicting the energy expenditure during human walking are discussed. It is shown that a hyperbolic equation adequately predicts energy expenditure over the entire range of normal walking speeds, while an equation of quadratic form is adequate for speeds up to about 100 m/min, and also predicts energy expenditure at competition speeds.

The energy expenditure per unit distance traversed is shown to be the fundamental parameter, and is a minimum at the natural walking speed.

It is shown that the gross, rather than a "net" energy expenditure, must be used in the formulation of the appropriate energy equations.

The average mechanical power output of an average adult female subject, walking at a natural speed, is about 50 W, while the burst of power output occurring in each step is about 150 W. There is a large margin of tolerance in the energy requirement of natural walking.

The meaning of the term "efficiency" is discussed. As here defined, the efficiency of human walking is in the range of 20%-25%.

Use of the terms "internal work" and "external work" is discussed. It is recommended that the term "external work" refer only to work done by the body on the environment.

It is sometimes important, particularly when dealing with the effects of various types of disability, to compare the energy demand of walking with that of an immobile (or relatively immobile) state, such as occurs in quiet, "at ease" standing.

Molen and Rozendal (1967) studied 10 males and 10 females under conditions of quiet standing in the laboratory. Their results, expressed as gram-calories per minute per kilogram (cal/min/kg) were virtually identical with those obtained by the present author on 7 males and 4 females (Unpublished data). The 17 males and 14 females may therefore be lumped together for the following considerations.

The average value for the males is 22.1 cal/min/kg, with a S.D. of 1.68, and 19.9 cal/min/kg for the females, with a S.D. of 2.60, the ratio of means being 1.11.

The odds against this difference being due to chance are greater than 100:1. Molen and Rozendal noted the lower cost of standing in females, and pointed out that this entered into their finding that the cost of walking was slightly less in females than in males. However, the difference is so small that Molen and Rozendal lumped men and women together in formulating a regression equation relating energy expenditure to speed of walking.

It is of interest that the difference in the cost of standing of males compared with females is eliminated if lean body weight instead of gross body weight is used. The results in this case, using the values given by Durnin and Passmore (1967) for body fat content, become 24.5 cal/min/kg for males and 24.8 cal/min/kg for females.

If males and females are lumped together and values expressed as cal/min/kg, the ratio of standing to "basal" rate (determined from published norms based upon either body surface area or body weight) is 21.0/16.8, or 1.25. The low cost of standing reflects the low activity of the postural muscles during standing, as revealed by electromyography.

RELATION BETWEEN ENERGY EXPENDITURE AND SPEED IN LEVEL WALKING

As soon as muscles are called upon to move the body during walking, a great increase in energy expenditure occurs, compared with standing, reflecting the metabolic cost of moving the body against gravity, and of accelerating and decelerating the various body segments.

Ralston (1958) showed that a quadratic equation of the form

$$E_w = b + mv^2$$

where E_w is the energy expenditure in cal/min/kg, v is speed in m/min, and b and m are constants, predicted adequately the energy cost of walking at speeds up to about 100 m/min (Figure 1, lower--dashed--curve).

In deriving this equation, data from various investigators were used. Since that time, a number of investigators have found that an equation of the same form provides an acceptable basis for predicting energy-speed relations (Cotes and Meade, 1960; Bobbert, 1960; Molen and Rozendal, 1967; Corcoran and Brengelmann, 1970). Combining their equations, using weighted averages of the constants according to number of subjects (86 total, 57 males, 29 females) yields a grand average equation

$$E_w = 32 + 0.0050\ v^2 \quad\quad\quad\quad \text{Equation (1)}$$

The above studies included walking on treadmill, floor, firm path, and grass. There was no indication of systematic differences for the various walking surfaces. In such studies, however, close attention must be paid to the nature of the footwear. A heavy boot or slippery sole may alter the energy demand of walking.

Since the quadratic equation (1) above implies that the energy cost is proportional to body weight, Wyndham et al., (1971) examined this relation in detail, and found that any error involved in such relation could not be proved to be significant in the case of walking.

Durnin and Passmore (1967) provide a table for relating weight of subject, speed, and energy expenditure, over a range of approximately 55-110 m/min. Table 1 shows the predicted values from equation (1) and average values from Durnin and Passmore.

It is clear that over the range 55-90 m/min, the predicted results in the last two columns are virtually identical. At 110 m/min the Durnin and Passmore figures underestimate the average experimental values (E_w = 95) by 11%, while equation (1) is still in good agreement with experiment. At still higher speeds, up to about 145 m/min, which is about the top speed for natural walking, equation (1) also fails.

Zarrugh et al., (1974) deduced a more general, hyperbolic equation for predicting energy expenditure during walking, which takes into account step length and step rate:

$$E_w = \frac{E_0}{(1 - \frac{s^2}{s_u^2})(1 - \frac{n^2}{n_u^2})}$$ Equation (2)

where s is step length in meters, n is step rate in steps/min, E_0 is value of E_w when s = n = 0, s_u is upper limit of s as E_w approaches infinity, and n_u is upper limit of n as E_w approaches infinity.

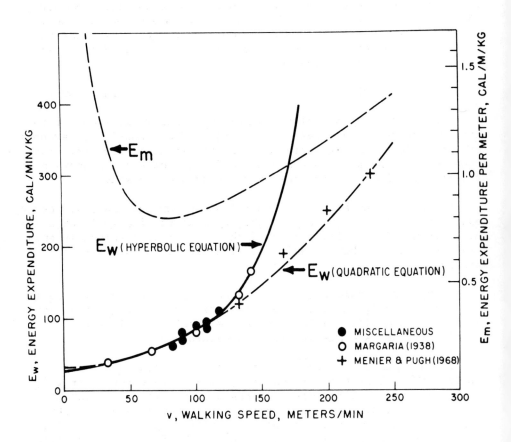

Figure 1. Energy expenditure during walking: Top curve, E_m, cal/m/kg, calculated from equation (4); middle curve, E_w, cal/min/kg, calculated from equation (3); bottom curve, E_w, cal/min/kg, calculated from equation (1). (After Zarrugh et al., 1974.)

Table 1. Predicted values for energy expenditure (cal/min/kg) from equation (1) and average values from Durnin and Passmore.

m/min	$E_w = 32 + 0.0050 \, v^2$	D & P
25	35	--
40	40	--
55	47	46
70	56	55
80	64	64
90	73	74
110	93	85

In natural walking, where the subject adopts his own natural cadence for a particular speed, it can be shown that equation (2) reduces to

$$E_w = \frac{E_0}{(1 - v/v_u)^2} \qquad \text{Equation (3)}$$

where v is speed and v_u = upper limit of v, equal to $n_u s_u$ (Figure 1, middle--solid--curve).

E_0 has an average value of approximately 28 cal/min/kg, and v_u an average value of about 240 m/min.

Thus, for a top natural speed of 145 m/min,

$$E_w = \frac{28}{(1 - 145/240)^2} = 179 \text{ cal/min/kg}$$

which is in good agreement with average experimental values.

Up to speeds of about 100 m/min, equations (1) and (3) predict virtually identical values. Therefore equation (1) may be used for most cases of natural walking.

Rather unexpectedly, equation (1) predicts energy cost within about 10% for competition walking, which is quite unlike natural walking. Menier and Pugh (1968) provide data on 4 male Olympic walkers studied during treadmill walking at an altitude of 1800 m. They state that their findings agreed with studies made at sea level by other investigators.

At a speed of 14 km/hr (233 m/min) the walkers clustered closely at about 60 ml oxygen/min/kg, corresponding to 300 cal/min/kg. Equation (1) predicts

$$E_w = 32 + 0.0050 \, (233)^2 = 303$$

which is within 1% of the observed value.

At a speed of about 135 m/min, the walkers clustered about a value of about 25 ml oxygen/min/kg, corresponding to 125 cal/min/kg. Equation (1) predicts

$$E_w = 32 + 0.005 \, (135)^2 = 123$$

which is within about 2% of observed value.

None of the studies mentioned in this section suggest any great or systematic effect of age upon the energy expenditure of natural walking, at least within an age range of about 20-59 years, when expressed as cal/min/kg.

Booyens and Keatinge (1957) found significantly lower values of E_w for women than for men at speeds of 91 and 107 m/min, and the equation of Cotes et al., (1957) predicts lower values for women than for men, as a result of the smaller step length in women. Ralston (1958) did not find a significant sex difference, and suggested a possible source of error in the study of Booyens and Keatinge.

Durnin and Passmore (1967) state that sex is not a factor. Molen and Rozendal (1967) found a small difference in favor of the female when gross energy expenditure is measured, but not when net energy expenditure (i.e., gross cost minus cost of quiet standing) is used. Since it has been shown in an earlier section that the cost of quiet standing is slightly (but significantly) lower in the female than in the male, it might be expected that at lower speeds this would account for a slight difference between males and females. As Molen and Rozendal (1967) state, however, the difference is not important enough to justify use of different equations for males and females.

It is of interest that Corcoran and Brengelmann (1970) found a somewhat higher value for females than for males during floor walking, which agrees with the present writer's experience (Ralston, 1958).

Significance of Constant = 28 in Equation (3)

At speed v = 0, equation (3) yields E_w = 28 cal/min/kg. This is much higher than the value for resting or quiet standing. Ralston (1958) found that the value for extremely slow walking, in five subjects, averaged 28.6. He concluded that the constant represented the cost of maintaining the body in motion at a barely perceptible speed.

It is clear, therefore, that in calculating the "net" cost of walking, as some investigators have done, subtracting the resting value, or the quietly standing value, from the gross value is neither theoretically nor practically justified. This point will be further amplified in the next section.

It is of added interest that use of "net" values for energy expenditure during walking does not permit formulation of equation (2) and its derivative equation (3), which appear, thus far, to be the only equations successfully predicting energy expenditure throughout the entire range of natural walking speeds.

ENERGY EXPENDITURE PER UNIT DISTANCE WALKED

A number of authors have discussed the energy cost of walking, per unit distance, per unit body weight. The curve of such energy cost, plotted as ordinate against speed as abscissa, is concave upward, and while fairly flat over quite a wide range of speeds (65-100 m/min), still exhibits a minimal value (Figure 1, top curve).

Ralston (1958) showed that the mathematical form of such a curve could be deduced from a quadratic equation of type (1) as follows:

$$E_m = \frac{E_m}{v} = \frac{32}{v} + 0.0050\, v \qquad \text{Equation (4)}$$

where E_m is expressed as calories per meter per kilogram (cal/m/kg).

Differentiating E_m with respect to v, and equating to zero, yields a minimal value of E_m equal to 0.80 cal/m/kg, corresponding to an optimal speed equal to 80 m/min.

Similarly, Zarrugh et al., (1974) showed that by dividing equation (3) by v:

$$E_m = \frac{E_0}{v(1 - v/v_u)^2} \qquad \text{Equation (5)}$$

Differentiating E_m with respect to v and equating to zero yields an optimal speed:

$$v_{opt} = v_u/3 = 240/3 = 80 \text{ m/min} \qquad \text{Equation (6)}$$

as in the case of equation (4).

As shown by Ralston (1958) a person in a natural walk tends to adopt a speed close to this optimal speed. This finding was confirmed by Corcoran and Brengelmann (1970) in a study of 32 normal human subjects during floor walking. These authors found a natural average speed of 83.4 m/min, which differs from the above optimal speed of 80 m/min by only 4%.

Since speed = n · s, where n = steps/min and s = step length (here expressed in meters), it follows that an optimal speed must be based upon a choice of step rate and step length such that minimal energy expenditure per unit distance is achieved. This is an example of a fundamental feature of human motor behavior, which applies to many activities in addition to walking. It may perhaps be expressed in a form such as:

"In freely chosen rate of activity, a rate is chosen that represents minimal energy expenditure per unit task."

In the case of natural walk, the unit task is the traversing of 1 m of ground. A speed is adopted such that each meter is covered as cheaply, from the energy standpoint, as possible.

The principle enunciated above is a sort of biological "Conservation of Energy" law. The celebrated French physiologist E.J. Marey, who was one of the great pioneers in the study of animal locomotion, anticipated the statement of this principle a century ago.

Only the use of *gross* energy values leads to equations (5) and (6), which establish a condition of optimality for free walking pattern and predict an experimentally verifiable optimal speed. The top curve of Figure 2 shows the gross energy expenditure per meter as a function of walking speed v for a typical subject when walking naturally at different speeds. The lower curves show that as larger amounts of energy are subtracted from the gross amount, the optimal speed corresponding to minimal E_m becomes smaller and smaller. It becomes essentially zero when an amount equal to E_0

is subtracted. It is not very enlightening to find that the best way to avoid energy cost during walking is not to walk!

It is suggested that in other types of exercise the gross rather than the net energy expenditure may be the fundamental energy parameter.

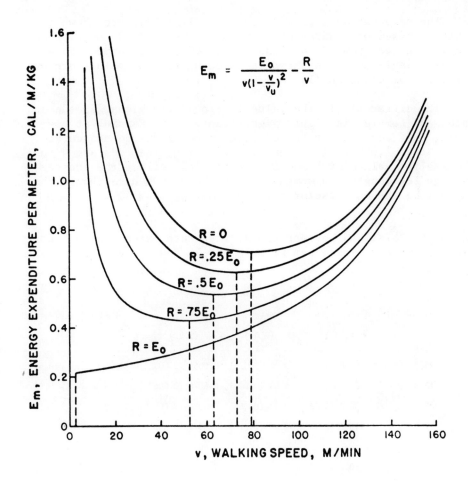

Figure 2. Gross energy expenditure per meter as a function of walking speed v for a typical subject when walking naturally at different speeds. The lower curves show that as larger amounts of energy are subtracted from the gross amount, the optimal speed corresponding to minimal E_m becomes smaller and smaller.

ENERGY EXPENDITURE PER STEP

In the preceding section it was shown that minimal energy expenditure, per meter traversed per kilogram of body weight (E_m), occurred at a speed of 80 m/min. The relations will now be considered between speed, step rate, step length, energy expenditure per minute per kilogram (E_w), and energy expenditure per step per kilogram (E_n).

The relevant data from Zarrugh et al., (1974) are shown in Table 2, based on studies of 10 normal males and 10 normal females, ranging in age from 20 to 55 and 20 to 49 years, respectively. Standard deviations are shown in parentheses. E_w values are calculated from equation (1).

The difference in the value of s/n at the higher speeds in males and females is highly significant. For example, at v = 73.2 m/min, the odds against the difference being due to chance are better than 1,000 to 1. Even at 48.8 m/min, the odds are 4 to 1. The smaller values of s and s/n in females reflect the shorter average leg lengths in women as compared with men, although this may not be the only factor involved.

Table 2. Relations between speed, step rate, step length, energy expenditure per minute per kilogram (E_w), and energy expenditure per step per kilogram (E_n).

	v meters/min	n steps/min (S.D.)	s meters (S.D.)	s/n meters/step/min (S.D.)	E_w cal/min/kg	E_n cal/step/kg = E_w/n
Males	24.4	59.5 (4.33)	0.41 (0.025)	0.0069 (0.00065)	35.0	0.59
	48.8	84.4 (6.48)	0.59 (0.041)	0.0070 (0.00072)	43.9	0.52
	73.2	102.2 (5.43)	0.72 (0.056)	0.0070 (0.00066)	58.8	0.58
	97.6	116.3 (3.13)	0.84 (0.020)	0.0072 (0.00026)	79.6	0.68
Mean				0.0070 (0.0012)		
Females	24.4	60.0 (3.95)	0.41 (0.029)	0.0068 (0.00065)	35.0	0.58
	48.8	86.7 (7.26)	0.57 (0.037)	0.0066 (0.00070)	43.9	0.51
	73.2	109.0 (5.49)	0.67 (0.037)	0.0061 (0.00045)	58.8	0.54
	97.6	126.8 (8.98)	0.77 (0.031)	0.0061 (0.00050)	79.6	0.63
Mean				0.0064 (0.0012)		

ENERGETICS OF HUMAN WALKING

The values shown in Table 2 do not strictly correspond to a natural walk, since the treadmill imposed the speeds. However, the speed 73.2 m/min is fairly close to the natural speed (80 m/min), and therefore should yield values of s/n close to those of a natural walk.

Molen and Rozendal (1972) studied 309 males and 224 females in natural walk along pavement and path. At an average speed of 83.4 m/min, the average value of s/n for males was 0.0072, in good agreement with the values 0.0070 and 0.0072 at speeds of 73.2 and 97.6 m/min in Table 2. At an average speed of 76.1 m/min, the average value of s/n for females was 0.0060, in excellent agreement with the value 0.0061 at 73.2 m/min in Table 2.

The near-constancy in the value of s/n in walking has deep physiological significance. Not only does the value of s/n characterize the walk of the male and female, but it also plays a fundamental role in determining the optimal speed of walking. At any given speed, imposition of an unnatural step rate (or step length) for that speed results in a higher than normal value of the energy expenditure per unit distance traversed (Atzler and Herbst, 1927; Molen and Rozendal, 1972; Molen et al., 1972; Zarrugh et al., 1974).

In the following mathematical treatment, the average values of s/n = 0.0070 and s/n = 0.0064 (Table 2) will be used for males and females, respectively.

In a manner similar to that followed in the preceding section, equation (1) can be used to determine the step rate corresponding to minimal energy expenditure _per step_:

Since v = s · n, and s = 0.0070 in males,

$$v = 0.007 \, n \cdot n = 0.007 \, n^2$$

Substituting in equation (1),

$$E_w = 32 + 0.0050 \, v^2 = 32 + 0.0050 \, (0.007 \, n^2)^2$$

$$E_n = \frac{E_w}{n} = \frac{32}{n} + 0.005 \, (0.007)^2 \, n^3$$

$$= \frac{32}{n} + 2.45 \times 10^{-7} n^3$$

Differentiating E_n with respect to n and equating to zero, E_n (minimum) occurs at n = 81.2 steps/min. Interpolating from Table 2, this would correspond to a speed of about 46 m/min, which is not the speed (80 m/min) corresponding to minimal energy expenditure per meter.

Clearly, the optimal step rate for minimal energy per step does not represent a desirable method of walking, at least in normal subjects.

The great significance of this conclusion will be made clear in a later section dealing with walking of certain disabled human subjects. Here, because of inability to walk at higher speeds, energy expenditure per step greatly exceeds normal values. The saving feature, in such cases, is that energy expenditure per minute is kept low because of low step rate.

Using the value s/n = 0.0064 for females, E_n (optimum) occurs at n = 85 steps/min. Interpolating from Table 2, this would correspond to a speed of about 47 m/min, again representing an undesirable method of walking, unless one is just interested in looking at the flowers!

SLOPE WALKING

A number of investigators have formulated equations relating energy expenditure to speed and grade during slope walking, but such equations tend to be complex and rather cumbersome to use. It is preferable to use representative data in tabular form, such as those provided by McDonald (1961), based on experiments by Margaria (1938) on three normal adult males (Table 3). The calculated values of E_w and E_m have been added, based on a body weight of 72 kg.

As has been noted earlier, a person during level walking tends to adopt a speed such that energy expenditure per unit distance is a minimum. In Table 3 it is evident that minimal values of E_m occur at slopes of +10% to -20%, at speeds of 80-100 m/min. It may therefore be expected that persons will adopt such speeds at those grades. Above +10%, and below -20%, E_m minima do not occur within the range of speeds shown.

STAIR CLIMBING

Unlike the situation in level walking (and, to a limited degree, in slope walking) no general rules relating energy expenditure to speed can be formulated for stair climbing. This is due to a number of factors, including wide variations in height of risers and width of treads, and the great changes in body posture that occur during stair climbing. However, it is instructive to

Table 3. Estimates of E_w (cal/min/kg) and E_m (cal/meter/kg) at various speeds and grades (modified from McDonald, 1961, after Margaria, 1938).

Grade		Speed, m/min								
		40		60		80		100		120	
%	Degrees	E_w	E_m	E_w	E_m	E_w	E_m	E_w	E_m	E_w	E_m
-40	-21.8	54	1.35	70	1.17	85	1.06	98	0.98	106	0.88
-35	-19.3	47	1.18	60	1.00	72	0.90	84	0.84	93	0.78
-30	-16.7	46	1.15	56	0.93	67	0.84	79	0.79	90	0.75
-25	-14.0	44	1.10	53	0.88	63	0.79	74	0.74	87	0.73
-20	-11.3	41	1.03	46	0.77	53	0.66	63	0.63	77	0.64
-15	-8.5	33	0.83	37	0.62	44	0.55	54	0.54	69	0.58
-10	-5.7	30	0.75	35	0.58	43	0.54	54	0.54	70	0.58
-5	-2.9	32	0.80	38	0.63	48	0.60	61	0.61	82	0.68
0	0	41	1.03	49	0.82	62	0.78	81	0.81	110	0.92
5	2.9	59	1.48	74	1.23	95	1.19	125	1.25	167	1.39
10	5.7	69	1.73	93	1.55	121	1.51	155	1.55	193	1.61
15	8.5	82	2.05	114	1.90	150	1.88	187	1.87		
20	11.3	98	2.45	140	2.33	185	2.31				
25	14.0	114	2.85	170	2.83						
30	16.7	133	3.33	204	3.40						
35	19.3	153	3.83								
40	21.8	174	4.35								

compare normal subjects with persons having motor disability, during normal climbing of steps with low (10 cm) and high (19 cm) risers. For this purpose, normal are compared with hemiplegic subjects, both using alternating gait (Table 4), based on data from Hirschberg and Ralston (1965).

As might be expected, the energy expenditure per step is much greater, of the order of 40%, in the hemiplegic than in the normal

Table 4. Average values of energy expenditure, per minute and per step, of normal and hemiplegic subjects during normal stair climbing, using alternating gait.

Experimental Conditions	Steps/min		Energy Expenditure			
			cal/min/kg		cal/step/kg	
	Normal	Hemiplegic	Normal	Hemiplegic	Normal	Hemiplegic
Rest (sitting)			18.9	15.6		
Low stairs	72	47	53.3	49.1	0.74	1.04
High stairs	70	44	70.3	61.5	1.00	1.40

subjects. Without doubt, if the hemiplegic patients were to attempt to maintain the same speed as the normal subjects, hazardous metabolic and cardiovascular demands might well result.

It is seen, however, that the energy expenditure <u>per minute</u> is not very different in the two types of subject, because of the fact that the hemiplegic patients adopt a low step rate. It is important to note that such physiological variables as heart rate, blood pressure, and respiratory rate are linked to the metabolic rate <u>per minute</u>, rather than to the metabolic rate <u>per meter</u> or <u>per step</u>. These physiological variables, while slightly higher in the hemiplegic than in the normal subjects, were still within acceptable limits.

Also, as might be expected, the high stairs made greater physiological demand, <u>per minute</u> and <u>per step</u>, than the low stairs, in both normal and hemiplegic subjects.

Not shown in Table 4, but worthy of note, is the fact that <u>alternating</u> gait required 20%-30% less metabolic demand <u>per step</u> than did unilateral-leading gait, in both normal and hemiplegic subjects.

WORK, POWER, AND EFFICIENCY

An adequate analysis of human walking requires knowledge of the mechanical energy levels (kinetic and potential) of the various segments of the body during successive moments of the walking cycle, along with measurements of energy expenditure. In this section is briefly described a direct method of measuring such energy levels; it is suitable for the study of certain motions of the body at moderate walking speeds, up to about 100 m/min, corresponding to cadences up to about 130 steps/min. A more complete account will be found in Ralston and Lukin (1969).

Cords are attached to the principal segments of the body: HAT (head + arms + trunk), thigh, shank, and foot. The attachments to thigh and shank are at the center of mass of each segment, placed according to the specifications of Dempster (1955). The attachment to the foot is at the heel, and the attachment to the HAT at approximately the level of the second sacral vertebra. The cords attached to the thigh, shank, and foot run horizontally backward, and drive potentiometric transducers. Two cords are attached to the trunk, one horizontal and one vertical, which drive transducers recording motions of the HAT in both horizontal and vertical directions. The electrical signals drive a suitable recorder, such as the Honeywell Visicorder No. 1508.

The method ignores certain motions that are of relatively small magnitude, such as horizontal rotational motions of the HAT and limb segments, and arm swing.

Masses of the body segments are determined from volumes measured by water displacement and from values of the specific gravities provided in the literature (Dempster, 1955).

From the motions and masses, potential (gravitational) and kinetic (inertial) energy levels of the body segments are calculated for intervals of 0.02 sec during the walking cycle.

Figure 3 shows the changes in mechanical energy levels of the various body segments, as labeled, of a young female adult of 58.6 kg, 169 cm, walking at a speed of 73.2 m/min.

The striking features of this record are (1) the approximate mirror-imaging of the HAT potential and HAT horizontal kinetic curves; (2) the flatness of the HAT total curve during about two-thirds of each step; (3) the disturbance of the HAT total curve during transition from stance to swing phase, which, on the basis of electromyographic evidence, coincides with the major muscle activity during the walking cycle; (4) the large change in energy level of the swinging leg, due almost entirely to velocity (kinetic energy) changes in the limb segments; and (5) the large positive work peak in the body total curve after heel contact.

The positive work per step, measured by the increase in the "body total" level, averaged 7.13 cal. The subject walked at an average cadence of 100 steps/min, so that positive work per minute

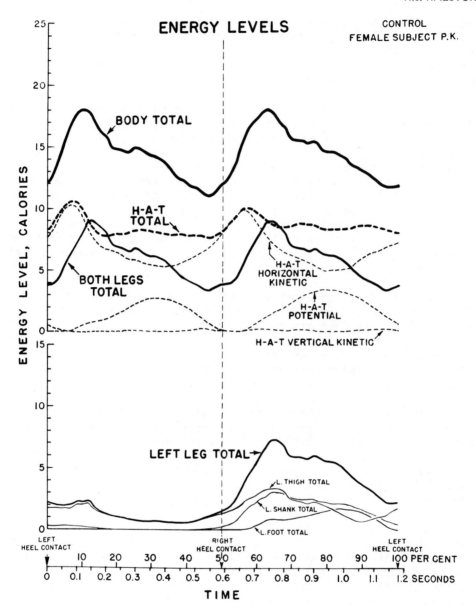

Figure 3. Female, 19, 169 cm, 58.6 kg. Walking speed 73.2 m/min. Energy levels of body segments and of whole body, gm-cal, as labeled. (After Ralston and Lukin, 1969.)

ENERGETICS OF HUMAN WALKING

equals 100 x 7.13 or 713 cal/min, about 0.07 horsepower (HP), or 52 W.

The burst of positive work in each step occurred during an average interval of 0.19 sec. Hence 7.13/0.19 = 37.5 cal/sec, or about 0.21 HP (157 W), was the maximal positive power output during each step.

Wilkie (1960) deduced that in single movements, of duration less than 1 sec, the usable "external power output" of the body is limited to a value somewhat less than 6 HP (4476 W). In brief bouts of exercise of about 6 sec, he gives the value 2 HP (1492 W) and for long-term work lasting all day, 0.2 HP (149 W).

Wilkie's figures are based upon exercise by champion athletes, and he states that ordinary healthy individuals can produce less than 70%-80% as much power. Even so, it is evident that the power demand in the female subject walking at a "natural" rate of 73.2 m/min is much below her maximal capability.

It may be concluded that there is a large margin of tolerance in the power expenditure during normal levels of walking speed.

Statements regarding the "efficiency" of biological processes are notoriously confused and confusing. Spanner (1964) says: "The notion of efficiency with which a process involving energy transformation is carried out is one which comes up fairly frequently. However, the word is not always employed in precisely the same sense, and this is apt to cause some confusion." Spanner is referring to relatively straightforward chemical processes. The situation in such a complex process as human walking is vastly more difficult.

Here a definition of efficiency will be used that is unambiguous, and that has been found by the present writer to be of considerable practical usefulness, especially in the comparison of work done by normal and disabled human subjects during walking. Gross efficiency is defined as the total positive work per minute, cal/min, determined from the "body total" curve of Figure 3, divided by the total metabolic expenditure, cal/min. For the female subject of Figure 3, the positive work was 713 cal/min, and the metabolic expenditure 3010 cal/min. The gross efficiency therefore is 713/3010 or 0.24. This value is consistent with values of efficiency in the literature for various kinds of work, such as bicycling on a bicycle ergometer. The data of Silverman et al., (1951) for exercise on the bicycle ergometer yield values of efficiencies very close to those found by Ralston and Lukin (1969) in normal walking speeds, which commonly range from 0.20 to 0.25.

At first sight, a troublesome feature of using the total metabolic energy expenditure in the calculation of gross efficiency is that at very low levels of walking speed the metabolic energy cost is so much greater compared with total positive work that values of efficiency seem too low. For example, in the female subject of Figure 3, walking at 48.8 m/min, the total positive work was 350 cal/min, and the total metabolic expenditure 2544 cal/min, yielding an efficiency of 350/2544 or 0.14. But there is no real difficulty here, so long as the "gross efficiency" is not confused with "muscle efficiency." We are dealing in this case with a slow walk which is of an ungainly and rather clumsy type, so we should expect a lower gross efficiency.

It is instructive to compare normal subjects with patients with motor disability. Two below-knee amputees, wearing conventional prostheses, were studied by the means described above. As is usual in cases of such disability, the subjects could walk comfortably only at lower speeds, up to about 50 m/min. As would be expected, the gross efficiencies were relatively low, ranging from about 0.08 to 0.14. However, the last figure is about the same as that for the normal female subject, described above, when walking at 48.8 m/min. It may be concluded that one factor involved in the low value of the gross efficiency is simply the low speed, quite apart from the lack of normal muscle coordination associated with the disability.

As is usual in cases of motor disability, the energy expenditure _per step_ or _per meter_ was high compared with normal subjects, while the energy expenditure _per minute_ was moderate, due to the low speeds used. It is the moderate energy cost _per minute_ that is linked to moderate changes in heart rate, blood pressure, and pulmonary ventilation.

Many authors have used a "net" energy figure in calculating efficiency. Such a "net" figure is obtained by subtracting a "resting" energy value from the energy expenditure, or by subtracting an extrapolated value from the energy expenditure. Such a procedure can lead, and frequently has led, to very misleading conclusions. It has already been noted that use of a "net" figure for mathematical formulation of the relation between walking speed and energy expenditure leads to nothing.

Wilkie (1974), in speaking of human muscular exercise, says: "The usual procedure of subtracting the resting oxygen consumption from the total does not correspond to any clear hypothesis about what is being estimated. In order to determine the efficiency of the working muscles themselves one should also subtract the extra oxygen used by heart, respiratory muscles, etc. (Hill, 1965, p. 153)." With this remark the present writer heartily agrees.

It seems desirable to end this section, dealing with the work of walking, with some comments on the use of the terms "external work" and "internal work."

Fenn (1930) refers to the elevation of the center of gravity as "part of the external work" of running.

Müller (1950) refers to work against external resistance as external work. Snellen (1960) states that the external work in level walking is negligible, and describes as external work the work involved in climbing a hill. Wilkie (1960) refers to work in overcoming air resistance as external work, while work associated with raising and lowering the center of gravity, and with changes in kinetic energy of the limbs, he refers to as work "dissipated internally." He evidently regards the work done in climbing a hill as external work.

Cavagna et al., (1963) refer to "external" work during locomotion as that associated with displacement of the center of gravity of the body and to "internal" work as work not leading to a displacement of the center of gravity. Ralston and Lukin (1969) refer to "total external positive work" as work measured by increases in the sum total of the energy levels of the body segments.

During walking the reaction forces at the ground can do no work because the points of application of the forces are fixed (it is assumed that no slippage occurs). The ground reaction forces can cause only <u>acceleration</u> of the center of gravity of the body. Conversely, the forces produced by the muscles of the body can cause no acceleration of the center of gravity, but <u>are</u> responsible for the changes in energy level of the body, both kinetic and potential.

Consequently, the "internal work" during walking is that done by muscles, while there is no "external work" unless one means by this work <u>done by the body on the environment</u> (as in pushing molecules of air out of the way) or, alternatively, <u>by the environment upon the body</u>.

It is suggested that the term "external work" be used only to refer to work done by the body upon the environment.

For a lucid discussion of what is meant by "internal work" and therefore by implication what may legitimately be called "external work," the reader is referred to Sears and Zemansky (1963).

In conclusion, it may be remarked that the student of human locomotion enjoys a great advantage over one who studies locomotion in lower animals, since the experimental conditions in the case of human locomotion may be so easily standardized. As a consequence, as discussed in the preceding pages, it has proved to be relatively easy to formulate quantitative relationships between energy expenditure and such variables as speed and cadence, and to measure such important quantities as work, power, and efficiency. In addition, and very importantly, it is only in human locomotion that one can make direct quantitative comparisons between normal locomotion and a wide variety of unusual or abnormal types of locomotion.

REFERENCES

Atzler, E. and Herbst, R., (1927) Arbeitsphysiologische Studien. Part 3. Pflügers Arch. ges. Physiol. 215, 291-328.

Bobbert, A.C., (1960) Energy expenditure in level and grade walking. J. Appl. Physiol. 15, 1015-1021.

Booyens, J. and Keatinge, W.R., (1957) The expenditure of energy by men and women walking. J. Physiol. (London) 138, 165-171.

Cavagna, G.A., Saibene, F.P. and Margaria, R., (1963) External work in walking. J. Appl. Physiol. 18, 1-9.

Corcoran, P.J. and Brengelmann, G.L., (1970) Oxygen uptake in normal and handicapped subjects, in relation to speed of walking beside velocity-controlled cart. Arch. Phys. Med. Rehab. 51, 78-87.

Cotes, J.E. and Meade, F., (1960) The energy expenditure and mechanical energy demand in walking. Ergonomics. 3, 97-119.

Cotes, J.E., Meade, F. and Wise, M.E., (1957) Standardization of test exercise. Fed. Proc. 16, 25.

Dempster, W.T., (1955) Space requirements of the seated operator. Geometrical, kinematic, and mechanical aspects of the body with special reference to the limbs. U.S. Wright Air Development Center. Technical Report 55-159. (Wright-Patterson Air Force Base, Ohio).

Durnin, J.V.G.A. and Passmore, R., (1967) Energy, Work and Leisure. Heinemann Educational Books Ltd., London.

Fenn, W.O., (1930) Work against gravity and work due to velocity changes in running. Movements of the center of gravity within the body and foot pressure on the ground. Am. J. Physiol. 93, 433-462.

Hirschberg, G.G. and Ralston, H.J., (1965) Energy cost of stair-climbing in normal and hemiplegic subjects. Am. J. Phys. Med. 44, 165-168.

McDonald, I., (1961) Statistical studies of recorded energy expenditure of man. Part II. Expenditure on walking related to weight, sex, age, height, speed and gradient. Nutr. Abstr. Rev. 31, 739-762.

Margaria, R., (1938) Sulla fisiologia e specialmente sul consumo energetico della marcia e della corsa a varie velocità ed inclinazioni del terreno. Vol. 7. Atti Reale Accad. Naz. Lincei. Giovanni Bardi, Rome.

Menier, D.R. and Pugh, L.G.C.E., (1968) The relation of oxygen intake and velocity of walking and running, in competition walkers. J. Physiol. 197, 717-721.

Molen, N.H. and Rozendal, R.H., (1967) Energy expenditure in normal test subjects walking on a motor-driven treadmill. Proc. kon. ned. Akad. Wet. C 75, 192-200.

Molen, N.H. and Rozendal, R.H., (1972) Fundamental characteristics of human gait in relation to sex and location. Proc. kon. ned. Akad. Wet. C 75, 215-223.

Molen, N.H., Rozendal, R.H. and Boon, W., (1972) Graphic representation of the relationship between oxygen-consumption and characteristics of normal gait of the human male. Proc. kon. ned. Akad. Wet. C 75, 305-314.

Müller, E.A., (1950) Der Wirkungsgrad des Gehens. Arbeitsphysiologie. 14, 236-242.

Ralston, H.J., (1958) Energy-speed relation and optimal speed during level walking. Int. Z. angew. Physiol. 17, 277-283.

Ralston, H.J. and Lukin, L., (1969) Energy levels of human body segments during level walking. Ergonomics. 12, 39-46.

Sears, F.W. and Zemansky, M.W., (1963) University Physics. Ed. 3. Addison-Wesley Publishing Co., London, (179-180).

Silverman, L., Lee, G., Plotkin, T., Sawyers, L.A. and Yancey, A.R., (1951) Air flow measurements on human subjects with and without respiratory resistance at several work rates. A.M.A. Arch. Ind. Hygiene Occup. Med. 3, 461-478.

Snellen, J.W., (1960) External work in level and grade walking on a motor-driven treadmill. J. Appl. Physiol. 15, 759-763.

Spanner, D.C., (1964) Introduction to Thermodynamics. Academic Press, London and New York.

Wilkie, D.R., (1960) Man as a source of mechanical power. Ergonomics. 3, 1-8.

Wilkie, D.R., (1974) The efficiency of muscular contraction. J. Mechanochem. Cell Motility. 2, 257-267.

Wyndham, C.H., van der Walt, W.H., van Rensburg, A.J., Rogers, G.G. and Strydom, N.B., (1971) The influence of body weight on energy expenditure during walking on a road and on a treadmill. Int. Z. angew. Physiol. 29, 285-292.

Zarrugh, M.Y., Todd, F.N. and Ralston, H.J., (1974) Optimization of energy expenditure during level walking. Europ. J. Appl. Physiol. 33, 293-306.

MOVEMENTS OF THE HINDLIMB DURING LOCOMOTION OF THE CAT

Mary C. Wetzel[x], Anne E. Atwater[x] and Douglas G. Stuart[o]

[x]Departments of Psychology and Physical Education for Women, University of Arizona

Information about vertebrate stepping has proliferated enormously in the past few years, but a comprehensive frame of reference is still lacking to describe control mechanisms for the step cycle of even one limb, let alone of the four-legged interlimb sequence. The first section of the chapter surveys previous models and data for the outputs of locomotion. The concept is justified that central, afferent input, and motor output variables are integrated over a broad domain. Events in individual afferents, muscles, and joints merge to a smooth unitary ensemble, the step cycle. From kinematics it is learned that a variety of subcomponents of the step remain relatively constant as velocity changes. These subcomponents include the duration of the swing (interval while the foot is off the surface), the flexion epoch, and the sequence of joint excursions. Under a number of environmental manipulations, these "constants" may shift their values somewhat, perhaps to smooth the cycle. More flexible are the duration of the stance (interval when the foot is on the surface), amplitude of joint angles for the limbs, and excursions of the spine. Interlimb timings and footfall patterns are even more permutable than single limb events, and the relative flexibility of <u>all</u> measures has made it difficult to generate enough data to determine the limits of their

---

[o]Department of Physiology, College of Medicine, University of Arizona

operating ranges. Electromyographic and kinetic data are sparse, but reinforce the concept that the electrical and mechanical activities of different muscles are summed to build an uninterrupted step cycle.

We suggest that the whole-limb concept be used to subsume empirical information that has been gathered previously and to describe new features of stepping, in the hope of moving theory closer to such mathematical equations for locomotion as are also emerging. In particular, the previously useful Philippson step-cycle model is shown to be deficient because it treats stepping in temporal but not in spatial, environmental referents. In the second part of the chapter, simple measures of hip height above the substrate, the direct distance from hip to toe, and the trajectory of the toe in relation to the pivot point at the hip and the substrate are applied to overground and treadmill locomotion by the cat. The results show that the whole hindlimb integrates the four epochs of the Philippson step to two smooth waves. Hip-height excursions depend upon footfall patterns of both hindlimbs, suggesting that applications of the whole-limb methods to interlimb coordination are possible. During the swing phase the toe describes rather complicated trajectories that emphasize a long preparation for landing, since the toe reaches its peak height soon after liftoff. At slow speeds the toe is maintained at a constant altitude. At faster speeds a variety of trajectories may be seen, that expand at liftoff and touchdown, presumably to soften the foot's transitions between ground and air and vice versa. These whole limb measures, therefore, are shown to be new, comprehensive definitions of variables which the nervous system controls to build a step.

Introduction

It has been natural to seek economical theoretical approaches to locomotion since so many elements (afferent neurons, interneurons, motoneurons, muscles, joints, limbs) potentially are available for interaction. A reasonable conceptual unit of stepping has been a cycle of a single limb. In practice consideration has been restricted usually to the alternate actions of flexion and extension in a hindlimb. To a large extent this choice has been dictated by neurophysiological theory and findings for the dog and, especially the domestic cat. The usual flexion-extension model is two-dimensional, with lateral movements in these animals being presumed, at least in initial work, to be less important than

excursions in the vertical and parasagittal directions. It is
conceded that this two-dimensional approach may be too simplistic
for even these animals, and it is certainly invalid for other
vertebrates such as fish or lizards.

The flexion-extension model for stepping by the hindlimb has
enjoyed considerable success in terms of central neurophysiology.
It has been possible to extend and verify in part Sherrington's
(1910) view that the flexion and crossed extension reflexes are
similar to spinal stepping and stepping-like movements, and Graham
Brown's (1911, 1912) suggestion that paired half-centers in each
side of the lumbar cord serve flexors and extensors of each hindlimb. Jankowska et al., (1967a, b) located a set of lumbar interneurons with reciprocal inter-connections, that were suitable to
control rhythmic activation of flexors and extensors. In addition,
hindlimb stepping has been found to be relatively autonomous in a
long series of experiments in which low spinal cats have been made
to walk and even run (Freusberg and Goltz, 1874; Shurrager and
Dykman, 1951; Grillner, 1973b).

While a stepping model seems justified that emphasizes the
single hindlimb's movements as an ensemble repertoire, the flexion-
extension dichotomy is simplistic. Each hindlimb has approximately
29 muscles, each with presumably specialized function. These muscles normally are activated sequentially in a complex pattern that
is retained largely in the high decerebrate cat, in which locomotion is elicited by brainstem stimulation (see Figure 2). Nor is
the main sequence altered markedly after limb deafferentation in
amphibians (Székely et al., 1969) or cats (Grillner and Zangger,
1975). The intrinsic control machinery operates, then, in a much
more detailed time scale than would be necessary for just oscillation between flexor and extensor motoneuron activations. No sub-
stantive suggestions have emerged from cat neurophysiological data,
however, as to how this central machinery might be organized (for
review see Gurfinkel and Shik, 1973).

If the possibility of programming for the single limb's step
cycle is examined from the point of view of the periphery, then
the normal nervous system receives abundant afferent feedback,
which may be assumed to be particularly strong while the foot is
on the surface and bearing weight (for arguments, see Goslow et
al., 1973c; Stuart et al., 1973). At least five species of segmental afferents may be involved in stepping: skin afferents,
joint afferents, group Ia and group II from muscle spindles, and
group Ib from tendon organs. Their central connections are as yet
largely elusive (for review see Matthews, 1972), but a variety of
studies has suggested that at least for extensors there is not only
a co-activation of alpha and gamma motoneurons (Severin et al.,
1967; Severin, 1970) but also a co-activation of motoneurons, Ia

and group II spindle afferents, and Ib afferents as well (Stuart et al., 1972; Goslow et al., 1973a, b). Furthermore, activities of flexor and extensor muscles and their afferents must overlap extensively (Severin et al., 1967), and the behavior of joint afferents is influenced by muscle tensions as well as by changes in angles between bones (Millar, 1973). In addition, the same stimulus (a tap to the dorsum of the foot) can enhance extensor EMG activity if applied during the stance and flexor EMG activity if applied during the swing phase (Forssberg et al., 1975). These various findings imply that afferent information is integrated over a far broader domain than within only individual muscles or joints. As with the central machinery, the status of a whole limb (or perhaps all four limbs) is probably the relevant cluster or ensemble of peripheral information (Wetzel and Stuart, 1975).

Central and afferent neural components have been mentioned above in a context that emphasized their aggregate or collective nature in preparation for a discussion of the stepping movements themselves. It will be shown that the movements are also organized in terms of whole limb and interlimb ensembles. The first section of the chapter will outline the kinematic, electromyographic, and kinetic data that have been used previously to characterize the outputs of stepping. While each experiment has given valuable insights, many details are missing about operating ranges and mutability of different events. Nor has there been to date any comprehensive treatment of the unique goals of a whole limb, let alone how these might interact with goals of other limbs and the body as a moving vehicle. It is suggested here that each data base should be enriched, but not by random proliferation of isolated findings.

A logical starting point is with kinematics of the normal animal, since the electrical and mechanical activity of muscles ultimately is translated into coordinated movements of bones. One way to describe the interplay of intralimb and whole-limb goals is by means of limb trajectory plots that are referred to the external substrate. The second portion of the chapter reports the results of some trajectory analyses, as a preliminary attempt to provide a kinematic context for the single, whole hindlimb into which actions by individual muscles at different joints may be fitted and eventually interpreted. Interlimb trajectories are deferred but could follow logically by extrapolation of the methods.

KINEMATICS

An early model of the single hindlimb's step cycle by Philippson (1905) indicated that while one flexion or F epoch of the limb could be discerned during the early part of the swing phase, there were three definable extension epochs for the knee and ankle and one for the hip. Figure 1 displays hip, knee, and ankle joint angles through one walking step cycle (upper portion of the figure) as aligned with the four Philippson epochs (lowermost portion of the figure). It was implied, of course, that flexion and extension epochs were closely associated with activity of the flexor and extensor muscles. In the knee and ankle joints first extension (the E^1 epoch) begins about halfway through the swing (closed circle after the upward arrow at the bottom of Figure 1) but is interrupted in these joints by the onset of E^2 at touchdown (downward arrow in Figure 1). The onset of E^1 in the hip is less clear (note flat-topped profile in Figure 1) but is usually found to follow shortly after the onset of E^1 in knee and ankle (e.g., Engberg and Lundberg, 1969). During E^2 the knee and ankle yield, or flex, before they extend together with the hip in a final epoch, E^3, during the remainder of the stance.

A variety of experiments has shown that in natural, unperturbed locomotion some stepping variables are quite invariant across velocities. The F epoch's duration, in fact the entire swing phase, does not change much as speed increases (for reviews, see Grillner, 1975; Wetzel and Stuart, 1975). Also relatively fixed across velocities is the order of joint angle changes at hip, knee, ankle, and digits (or the corresponding movements of the forelimb), as has been documented in a number of studies (cf. Arshavsky et al., 1965; Gambaryan et al., 1971; Goslow et al., 1973a). Finally, rather surprisingly, when muscle lengths are expressed as percentages of lengths associated with quiet standing or the in situ maximum and minimum lengths, the amplitudes of most excursions are not highly speed-dependent. Only the lengthening contractions of knee and ankle extensors increase markedly with forward speed (Goslow et al., 1973a, b).

In contrast, the total stance phase and, especially, E^3, decrease markedly with speed (Arshavsky et al., 1965; Goslow et al., 1973a); implying a fundamental difference between the goals to be met in the swing for transferring and positioning the limb, and the goals to be met in the stance for supporting and thrusting off the limb. Even casual inspection of a normal animal's step cycle shows that there are no pauses in the limb's circular movement, and data by Orlovsky and Shik (1965) have supported the contention that there are no sharp breaks in angular velocities during changes in joint angles. A major experimental goal, therefore, is to determine how the seemingly discrete and quite different roles of swing-stance and flexion-extension overlap and merge to build a smooth circular movement.

Smoothing may occur partly because control linkages even for so-called constant features of the cycle are flexible to some

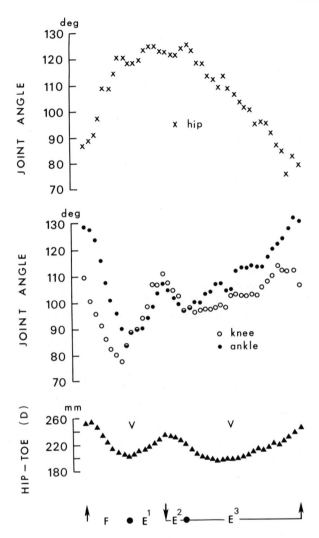

Figure 1. Synthesis of joint angle information (upper 3 curves) for walking with information about direct hip-to-toe distance (D). Joint angles were redrawn from Goslow et al., (1973a), and D was taken from left lowest curve in Figure 4 (with data fitted proportionally into swing and stance, whose order is reversed from that which was measured, Figure 4). Above the D curve are "V's" to indicate the time at which D is vertical (X = zero, see Figure 3B). The conventional epochs of the Philippson (1905) step cycle (F, E^1, E^2, and E^3) are shown at the bottom of the figure, together with swing duration (beginning at upward arrows) and stance duration (beginning at downward arrow). See text for details.

extent. Some supportive evidence has been gained on this point, although the limits of flexibility have not been tested in any systematic way. It is known, however, that the timing of changes in joint angles is not completely fixed. Statistically, the knee and ankle turn to E^1 together in normal overground locomotion by cats, but the knee leads the ankle by some msec on the treadmill (Wetzel et al., 1975). Another measure, F duration, is shorter during overground than treadmill stepping (Wetzel et al., 1975). Ankle angles decrease (the animal crouches) under aversive conditions of shock or a noisy air jet, when compared to angles that are obtained when an animal is working for food alone. Swing duration also can differ subtly as a function of reinforcement contingencies (Lockard et al., 1975). The swing is not a standard element under loaded conditions, when an animal is restrained while trying to run on a treadmill, for then the amplitudes and speeds of movement of the joints change (Orlovsky et al., 1966).

The issue of how wide and mutable are the operating ranges of stepping variables also has been considered for interlimb events. Across limbs, a coupling in time between movements of flexion and extension has been reported in a specific instance; from the onset of extension, E^1, in a hindlimb to the onset of flexion, F, in the ipsilateral forelimb of the cat (Miller and van der Burg, 1973; Miller et al., 1975a, b). A constant coupling time (first reported to be near 40 msec) does not appear to be obligatory, however, in any of three senses. The same value is not necessarily found for an individual cat across all speeds (or in both left and right limbs). The value also differs somewhat from cat to cat and, finally, depends upon whether stepping is on a motor-driven treadmill or over open terrain (Miller et al., 1975a, b; Wetzel et al., 1975).

Indeed, mutability appears to be the dominant feature of interlimb events. Unlike the sequence of joint changes in the single limb step cycle, the four-limb footfall sequences change dramatically with forward velocity. It should be remembered, however, that velocity is not a very sensitive independent variable for assessing interlimb coordination. The periodically repeating movements of one stride (a complete cycle of movement by the four limbs) recur only after approximately 600 msec in slow walking and 300 msec in galloping. Within 300 msec a cat or dog can decelerate from its peak galloping speed to a standstill or, alternatively, can accelerate rapidly. However, in most species the characteristic movements of walking, in which at least two limbs always touch the surface, do give way to trotting (touchdowns of diagonal limbs alternate and there are periods of no support) at somewhat faster speeds. The trotting pattern is replaced in turn at high speeds by galloping, in which movements of the hind- and forelimbs are uncoupled, and there may be long periods of nonsupport (Hildebrand, 1974, for recent review).

In interpreting the shifting interlimb footfall patterns as velocity changes, it is important to remember that movements of the spine may be at least as significant as those of the limbs in some vertebrate species. While ungulates and proboscidians have a quite rigid spine; the successive flexions and extensions of the spine in carnivores provide marked forward accelerations that permit high speeds (Hildebrand, 1961; Gambaryan et al., 1971; Goslow et al., 1973a; Gambaryan, 1974). In addition, movements in the lateral direction may be important for the spine, limbs, or whole body. Lateral undulations of the spine are the rule, of course, in fish, salamanders, newts, and lizards. For these species the usual two-dimensional view from the side is completely inadequate, although a view from the top may suffice. Two-dimensional models may be rejected eventually even in mammals, since lateral movements are visible readily in locomoting skunks, porcupines, cats, and also humans (for discussion of lateral movements in the primitive Australian echidna, a monotreme, see Jenkins, 1970).

Superimposed on speed-dependent patterns of stepping are species adaptations, the variety of which is self-evident in vertebrates. An elephant does not gallop, and some lizards (e.g., agamids, in Sukhanov, 1974) have almost lost the ability for slow movement. An individual member of a species may exhibit a wide range of stepping movements, even at the same velocity. Variability of interlimb footfall patterns has been emphasized for salamanders and lizards (Sukhanov, 1974), and is notable during galloping by the cat (Stuart et al., 1973) and cheetah (Hildebrand, 1961), when no obvious environmental changes are present.

Even more profound alterations in interlimb coordination are attendant upon environmental manipulations, a few of which are now being specified. Moose and caribou, for example, use different footfall patterns on flat as opposed to rough terrain (Dagg and de Vos, 1968). Treadmill locomotion by the cat differs from that over ground. A shorter interval between ipsilateral touchdowns, as well as shifts in the ipsilateral hindlimb-forelimb E^1 - F interval are present with treadmill locomotion. The animal tends to pace, or to move first one pair of ipsilateral hind- and forelimbs and then the opposite pair (Miller and van der Burg, 1973; Miller et al., 1975a, b; Wetzel et al., 1975). Several studies have been directed to interlimb relations. One or more limbs have been subjected to an increased load by restraint or by tilting a treadmill (Orlovsky and Shik, 1965; Orlovsky et al., 1966). In other experiments, limbs have been lifted from a treadmill belt in different combinations while remaining limbs continued to run (Shik and Orlovsky, 1965). In general, these experiments showed that manipulations of one limb have measurable but not large effects on movements of other limbs.

In summary, kinematic data have proliferated recently over a wide spectrum of experiments but in little depth. Therefore statistical boundaries are unavailable for most events. Kinematic data have not as yet provided a basis for strong inferences about the rules of stepping control. To the contrary, most of the single- and interlimb timings studied in any detail appear more or less flexible. The existence of any firm constraints upon them remains unproven.

ELECTROMYOGRAPHY

The electrical activity of muscles is still the best available gauge of overall motoneuron output in the intact limb during stepping. Our knowledge of the profiles and variability of EMG patterns, however, may well be embryonic. Certainly EMG patterns have been studied in fewer situations and in even less depth than have locomotor kinematics. There are few simultaneous recordings from different muscles and few records from substantial numbers of consecutive steps. Step to step variation may be considerable, as was reported in a recent account of EMG activity in the ankle extensor, soleus (Prochazka et al., 1974).

Figure 2 summarizes the generalities of cat and dog EMG records that have been obtained in some of the most complete studies of hindlimb muscles (Engberg, 1964; Engberg and Lundberg, 1969; Gambaryan et al., 1971; Tokuriki, 1973a, b; 1974). But it should be remembered that every horizontal bar (denoting electrical activity) actually represents a waxing and waning intensity profile. Variations in EMG amplitude reflect largely unknown contributions by different motor units operating at different regions of their length-tension curves. Figure 2 shows that extensors of the hip (gluteus medius or semimembranosus anterior), knee (vastus lateralis), and ankle (gastrocnemius) display more uniform activity in all three situations (intact dog on a treadmill, intact cat moving overground, and decerebrate cat under controlled locomotion on a treadmill) than do the flexors at the same joints (iliopsoas, semitendinosus, and tibialis anterior). The active period of semitendinosus anterior varies substantially both with speed and also across the three situations (note the silent period in the decerebrate cat during the F epoch), perhaps because its origin from the ischium allows it to extend the hip in addition to its main action of knee flexion. Iliopsoas also displays quite a different pattern between walking and running in the intact cat and dog.

Even greater departures from simple, alternating flexion and extension are seen in two-joint muscles. Sartorius (a hip flexor and knee extensor) exemplifies the speed dependence of actions by two-joint muscles. Its flexor action is denoted during walking by

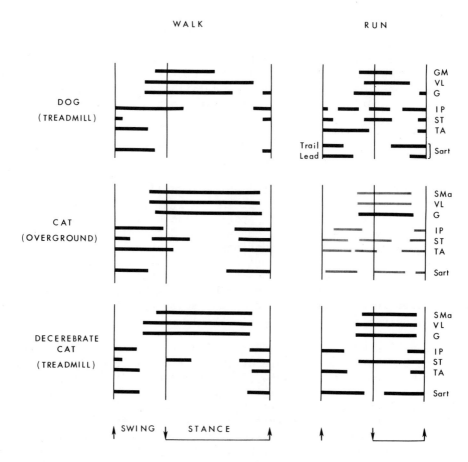

Figure 2. Summaries of EMG activity in hindlimb muscles through a single walking and running (trotting or galloping) step cycle. Normalized displays redrawn or based on data from previous reports for the normal dog on a treadmill (Tokuriki, 1973a, 1974), the normal cat moving across open terrain (Engberg, 1964; Engberg and Lundberg, 1969), and the high decerebrate cat whose treadmill locomotion was elicited by brainstem stimulation (redrawn from Gambaryan et al., 1971). Periods of electrical activity (solid horizontal bars for walk and gallop; hatched bars for trot) or inactivity (open spaces) were fitted proportionally into each of swing (foot off surface) and stance (foot touching surface) phases. To simplify the figure, swing duration is shown as invariant across gaits; with stance duration 2 X swing duration in the walk and equal to swing duration in the run. In each of the three displays, 7 muscles appear from top to bottom: extensors of the hip (SMa, semimembranosus anterior; or GM, gluteus medius), knee (VL, vastus lateralis), and ankle   (Legend continued on next page.)

activity at the end of the stance and during the early swing (F); while during running there is also activity early in the stance, as is true for one-joint extensors.

The EMG patterns of Figure 2 attest that considerable smoothing of the forces of flexors, extensors, and two-joint muscles may derive from overlap in their activity profiles. Even more overlapping and cascading profiles would be expected if all rather than seven limb muscles were represented. Co-activation of flexors and extensors is particularly evident just prior to touchdown, which is an appropriate point to brake the limb and ensure its smooth placement (see Engberg and Lundberg, 1969). Further, the soleus muscle, which presumably extends the ankle exclusively, can be active during a time of flexor activity, the F epoch (Prochazka et al., 1974). Finally, the muscle activity of a single limb is not independent of the behavior of other limbs. As Figure 2 shows for sartorius during galloping by the dog, many muscles of a trailing hindlimb were shown to have stronger activity in the stance phase than did those of the leading limb (Tokuriki, 1974). Moreover, the forelimbs exhibited extensor activity later in the stance phase than did the hindlimbs. This arrangement is appropriate since the hindlimbs strike almost directly under the center of gravity and must act as a weight-bearing brace sooner in their cycles than the forelimbs which strike well in front of the body (Tokuriki, 1973a, b; 1974). As EMG data continue to accumulate, presumably their patterns will be found to be tailored to accompany each variant of stepping movement.

KINETICS

Much less is known about the kinetics of locomotion by animals than about either kinematics or EMG's. A remarkable early study by Manter (1938) remains the best but almost the sole source of information about net limb forces during natural locomotion, and even

---

(G, gastrocnemius); flexors of the hip (IP, iliopsoas), knee (ST, semitendinosus), and ankle (TA, tibialis anterior); and a bifunctional muscle (Sart, sartorius, that flexes the hip and extends the knee). GM was considered to be a more typical hip extensor for the dog than semimembranosus, which showed little activity. Extensor actions were more uniform across preparations and gaits than were those of flexors or the two-joint sartorius (semitendinosus may also have a hip extensor action in addition to knee flexion). Sartorius is representative not only of a two-joint muscle whose action changes with velocity, but also of a muscle whose activity differs depending on whether its limb leads (is frontmost) or trails during galloping.

his analysis was restricted to walking. It was notable that smooth curves of vertical force (see portions of his data that are redrawn in Figure 6), horizontal force, and torques at individual joints were generated by measurement of displacements as the cats walked across spring-supported platforms.

Speculations are scarcely worthwhile at present about how the forces exerted by individual muscles at different joints are summed to produce smooth curves such as Manter (1938) drew (see however an analysis by Grillner, 1972, and Figure 4 in his 1975 review). Until transducers can be miniaturized and made to record forces accurately in small animals, there can be little understanding of the moment-to-moment locomotor mechanics. Promising approaches, only, are available. Yager (1972), for example, was able to measure the force at a single joint by means of a transducer that was implanted on the tendons of insertion of ankle extensors of the cat. This technique would be valuable if combined with recordings from implanted length gauges, that have been developed by Prochazka et al., (1974).

A few general statements about force summation may be warranted, however. First, it is obvious that tensions of muscles with very different strengths are added at a single joint and between joints, from the powerful hip extensors to the small digit muscles. Second, the smooth whole-limb kinetics argue against an overall flexor-versus-extensor dichotomy of control. Third, it may be assumed that the whole-limb force curves are smoothed, in part, by inertial forces, but still the nervous system must be cognizant of these forces in order to provide appropriate muscle tensions throughout the step cycle.

Synthesizing Kinematic and Kinetic Information

The most comprehensive view of the single limb step cycle to date has been provided by matching, over time, records of swing duration, stance duration, EMG, and joint angles according to the four Philippson (1905) epochs. The important discovery was made that extensor activity does not follow touchdown but leads it. This means that afferent feedback is unnecessary for this function in normal stepping (Engberg and Lundberg, 1969). A later study (Goslow et al., 1973a; see also Gambaryan et al., 1971) incorporated muscle length measurements into the schematic and emphasized the magnitude of the yield during E^2, implying that it is a time of strong proprioceptive input. The analysis also tied together conceptually the E^2 and E^3 epochs as a single mechanical event: an <u>active stretch-shorten</u> cycle for knee and ankle extensors and an <u>isometric contraction-shorten</u> cycle for pure extensors of the hip.

It is doubtful that the Philippson (1905) step cycle can be extended to subsume many additional findings about basic characteristics of the single limb movements. It is deficient, in part, because of its emphasis on the angular excursions at knee and ankle joints, and in its neglect of activities of the whole limb or of differences between fore- and hindlimbs (see Miller and van der Meché, 1975, for discussion of forelimb movements). A particularly serious limitation of the Philippson model is that the only connection between the controlled variables of joint angle and contact with the substrate is in the time dimension (Figure 3A).

A quantitative theory of how movements are tied to spatial and temporal coordinates of the environment has been launched fairly recently from the perspective of engineering by McGhee, Frank, and others (see, for example, McGhee and Frank, 1968; and for a recent review, Wetzel and Stuart, 1975). The approach itself is not new, since in the late 1800's Marey plotted locomotor movements by means of stick figures positioned with respect to the ground at successive time intervals (Marey, 1901). No quantitative aspects of limb trajectories were abstracted, however, from Marey's data. Snyder (1952, 1962) later published loop trajectories for lizards that marked displacements from the acetabulum for hip, knee, and ankle joints throughout the step cycle. The acetabulum was assumed to be a stationary point at the hip pivot. The animal was viewed only from above, however, and there was no point of reference on the ground surface. Only a few trajectory data are available for felidae (see Hildebrand, 1961, for the galloping cheetah).

The recent engineering work has not treated the experimental data of vertebrate stepping in any detail but, rather, is helping to provide a mathematical context for the analysis of trajectories. Both the control mechanism and the controlled kinematic variables must be considered in order to realize, for example, a four-legged vehicle (Frank and McGhee, 1969). Energy requirements dictate idealized trajectories for the center of gravity of the body or its parts (i.e., horizontal straight lines). Then computer simulations may determine how these requirements are compromised to promote static and/or dynamic stability of the locomotion under various external conditions and perturbations.

As data from animals accrue, they can be compared directly with such models. One important finding is mathematical proof that the almost universal order of footfalls in slow creeping (left hindlimb, left forelimb, right hindlimb, right forelimb; if the cycle starts with the left hindlimb) is the single one that maximizes static stability (McGhee and Frank, 1968). Faster gaits, in which the stability characteristics cannot be derived a priori have been subjected to computer simulation to yield equations of motion that describe, for example, quadrupedal trotting (Frank and McGhee, 1969).

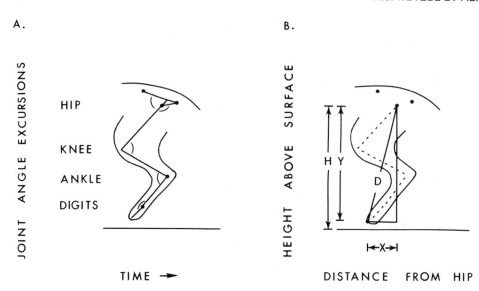

Figure 3. Schematic models for measurement of limb status according to the Philippson procedures (A) and whole-limb events (B). A. Measurements are referred to elapsed time as the cat moves from right to left along a surface (horizontal line). Pivot points at joints are marked with closed circles if they can be measured fairly reliably, without slippage of the skin across the bones during stepping. The pelvic angle does not appear, in order to simplify the figure, but two measurements are depicted for the hip joint. The anterior curve marks a hip angle defined by a line extended along the long axis of the femur, that intersects a line drawn between the ischial tuberosity (posterior) and the crest of the ilium (anterior) on the pelvis. The second, posterior, measure of hip angle lies between a line drawn from the ischial tuberosity to the hip pivot point, that intersects a line along the long axis of the femur. This latter measure was used by Goslow et al., (1973a), since it was considered to be most accurate, but it is unconventional (Engberg and Lundberg, 1969) since it records hip flexion as an increasing angle. All other angles decrease in flexion and increase in extension. B. Measurements that define the status of the entire hindlimb as based on the hip pivot point, tip of the longest toe, and vertical displacement from the surface. H = ordinate of the hip pivot (H = zero at the "surface," which corresponded to a measured displacement from the visible edge of the treadmill belt or solid running surface to the contact line of the foot); Y = vertical distance between hip and toe; X = abscissa of toe (X = zero when toe is directly below hip pivot); D = direct distance between hip and toe, as toe moves forward and backward under the pelvis.

The trajectories described in this chapter summarize the limb movements in space and time so as to emphasize natural collections or ensembles of stepping behavior. The findings are preliminary in their restriction to the hindlimb and to two spatial dimensions, as in previous schematics for mammals. The concepts are so simple, however, that they could be generalized readily to almost any species, and similar formulations could be used for both single- and interlimb events.

Two kinds of measurements were made. The first kind was merely an extension of the Philippson (1905) view of events at different joints (which he had expressed in angular degrees) to a whole-limb measure that was based on the direct distance, D, from the hip pivot point to the tip of the toe, as shown in Figure 3B. The simple length measure describes the limb's successive foldings and unfoldings, excursions that are realized by changes in hip, knee, ankle, and metatarsophalangeal joints. The second kind of measure was based on external coordinates. Two such measurements were plotted. The first was just the height (H) of the hip pivot point from the surface (Figure 3B). The second was a somewhat less conventional view of the step cycle in which the vertical distance (H) between the tip of the toe and the substrate surface was plotted against horizontal distance from the hip (X), as depicted in Figure 7.

METHODS

The present measurements cut across our previous work with cats, that compared single and interlimb values between treadmill and overground stepping (Stuart et al., 1973; Wetzel et al., 1975), different types of reinforcement contingencies during training (Lockard et al., 1975) and different variants of galloping (Wetzel and Norgren, Unpublished data). This wide range of conditions was sampled in order to determine whether or not common trajectory features would emerge from stepping situations that were known to produce differences in several other single- and interlimb measurements. General methods of training animals on a motor-driven treadmill or an open terrain, filming performance with 16 mm movie film at 64-100 frames/sec, recording velocity with a continuously variable tachometer, and recording interframe interval with a free-running timing device (accurate beyond .01 sec) have been described previously (Goslow et al., 1973a; Lockard et al., 1975; Wetzel et al., 1975). Measurements of stride duration, swing and stance durations, joint angle estimates, the times of junctions between the four epochs of the Philippson (1905) step cycle (F, E^1, E^2, and E^3), and representations of gait by the diagram Hildebrand (e.g., 1959) developed were made conventionally (for details, including the difficulty of determining joint angles accurately, see

Goslow et al., 1973a; Lockard et al., 1975; Miller et al., 1975a; Wetzel et al., 1975).

Two cats were trained to work on the treadmill. The contingency for one 2.3 kg cat was avoidance of an air jet that was directed toward the hindquarters if a forward position on the belt was not maintained, in addition to food reward if the position was maintained. Another 2.5 kg cat worked for food alone. The animals learned to move at an almost invariant velocity, at whatever setting was determined for the belt, for filmed sequences of at least 8-10 consecutive strides. A third 2.5 kg cat was trained to walk and trot overground for a food reward and gallop to avoid a large dog, as described previously (Goslow et al., 1973a). A total of 9 strides for each cat were selected for analysis; 3 consecutive strides from within sequences at walking, trotting, and galloping speeds, with the exception that only 2 overground galloping strides were available. The walking velocities were 1.0 m/sec on the treadmill and .95 m/sec overground. Trotting velocities were 2.3 m/sec on the treadmill and 1.5 m/sec overground; the difference reflecting the limited number of velocities available for overground trotting, when forward speed cannot be controlled well. Galloping was obtained at 4.8, 4.1, and 5.0 m/sec for the three cats, respectively. No galloping more rapid than approximately 4 m/sec has been obtained as yet on the treadmill, when the sole incentive is food (M.C. Wetzel, F.G. Postillion and K.S. Norgren, Unpublished data).

Additional data (M.C. Wetzel and K.S. Norgren, Unpublished) for normal versus LH (left hindlimb) deafferented performances by the first cat yielded similar preoperative data at somewhat different velocities than those given here. In general, successive strides by a cat at constant velocity can be very similar (e.g., Miller et al., 1975a, b; Wetzel et al., 1975), and this uniformity proved to be true for the whole-limb measures as well. The data could be read with considerable precision, as was shown by interobserver comparisons and by repeated measurements of a given stride by the same observer.

As viewed by a specially designed projection system, the x, y, coordinates of selected left hindlimb points on the film were digitized and fed directly into a programmable calculator which computed and printed the various X, Y, and H measures that are shown in Figure 3B, as well as the direct distance, $D(D^2 = X^2 + Y^2)$. All 3 cats were of similar height, as defined by the vertical distance, Y, at X = 0 (the point at which the toe passes directly under the hip pivot). These heights were 188 mm, 208 mm, and 205 mm for the 2 treadmill animals (food + air, and food alone) and the overground animal, respectively.

The Philippson Step Cycle and Whole Limb Status

Hip-Toe Distance and Previous Kinematic Timings

Figure 1 aligns in time the four Philippson (1905) epochs, with joint angles at hip, knee, and ankle (data from a walk at .7 m/sec, Goslow et al., 1973a) in conjunction with the whole limb measure, D. The latter data were from the overground cat, whose data at 1 m/sec were fitted proportionally (for both swing and stance) into the display. Note that the net effect of folding and unfolding the limb resulted in two minima in D that fell, as a first approximation, near the $F \cdot E^1$ junction and (less closely) to the $E^2 \cdot E^3$ junction, respectively. An even closer correspondence can be seen between the minima and the time at which the toe passes directly under the hip pivot point (X = 0, and D is vertical, as marked by large V). In this way the four Philippson epochs are translated in the whole limb direct hip-toe measurement to two smooth cycles within each step.

Figure 4 gives details of the direct hip-toe measurement for representative strides, together with LH and RH (left and right hindlimb) footfall data (stance durations marked with horizontal lines). Minimum and maximum values and the ranges between them appear in Table 1 for all strides in the total sample. Locations of the $E^2 \cdot E^3$ junction during the LH stance and $F \cdot E^1$ junction during the LH swing are indicated by an open circle (knee) and arrowhead (ankle) for the overground and one treadmill cat.

In every record the greatest hip-toe distance came after LH liftoff, and the distance at liftoff exceeded that at touchdown. In these ways strong thrustoff force was documented. While the difference between the shortest and the longest lengths of hindlimb muscles does not exceed a range of approximately 20-25 mm during overground walking, trotting, and galloping (Goslow et al., 1973a) the combined actions of muscles at different joints resulted in a length change (D) of from 51-118 mm (range of means in Table 1). Small changes in muscle length at individual joints, then, sum to a large excursion of the whole limb.

The excursion of the whole limb (D) was not only marked, but also could be quite individualized. First, total excursion was not necessarily dependent on velocity changes, since in the uppermost record of Figure 4 the longest distance was slightly greater during galloping than during trotting. Second, the two gallops at approximately 5 m/sec (treadmill, uppermost, and overground,

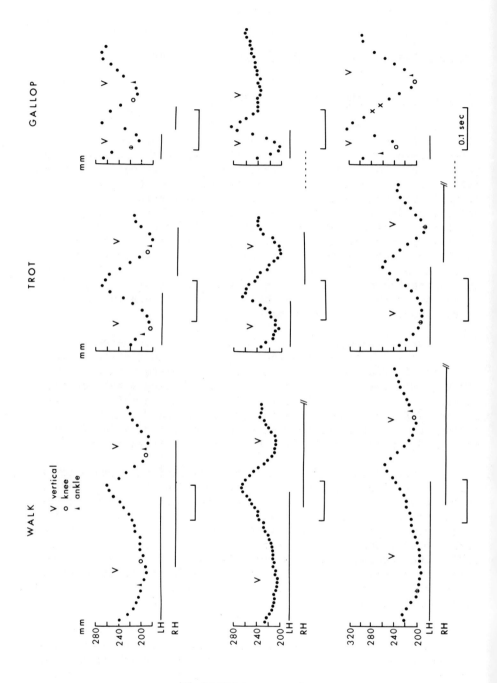

Figure 4. Representative step cycles for walking, trotting, and galloping as measured by direct hip-toe distance (D) for three cats: upper records for cat on a treadmill, food reward + air avoidance; center records for cat on a treadmill, working for food reward alone; lowest records for cat stepping on open terrain (see text for details). Left hindlimb (LH) and right hindlimb (RH) stance durations appear as horizontal lines, with the exception of the dashed lines for RH in the two lower galloping records (to emphasize that RH descended prior to the onset of the LH step cycle). Crosses in the right lowest record indicate estimated values, since the galloping stride was incomplete. For the uppermost and lowermost records, turning points of the knee (open circle) and ankle (arrowhead) to E^3 ($E^2 \cdot E^3$ junction during LH stance) and to E^1 ($F \cdot E^1$ junction during LH swing) are indicated. The same step cycles for each cat were used to illustrate other whole-limb values and appear in the same upper, center, and lower positions in Figures 5, 8, 9, and 10.

---

lowermost records) had quite different excursions. Third, the treadmill gallop near 4 m/sec (center record) had as great an excursion as that near 5 m/sec (top record), but the configuration of D through the step cycle was quite different. These various differences are interpreted to show the relative flexibility with which the neural control program for stepping can utilize forces in different muscles and/or joints to provide for movements of quite different spatial excursions without, presumably, sacrificing efficiency. That excursions at 5 m/sec were smaller for treadmill than overground galloping could have been due to difficulty in balancing on the treadmill (see discussion in Wetzel et al., 1975). Alternatively, escape from a dog may have been more aversive (overground) than avoidance of an air jet (treadmill), and therefore led to larger overground excursions. The flat profile through the latter portion of the swing in the treadmill gallop at 4.1 m/sec probably was an idiosyncratic pattern that characterized the easy galloping by this cat, which worked for food alone.

Hip Height, Forces, and Implications for Hindquarter Progression

The direct measure of hip-toe length is an extension of the Philippson (1905) step cycle model, and as such gives no information about the relation between limb excursions and the ground surface. One measure which does so is the simple one of hip height above the treadmill surface or floor, as described in Figure 5 and Table 2. There were marked differences in this measure as a function of velocity, at least between trotting and galloping. All three curves for the walk were of almost constant amplitude. Only the lowest record suggests two cycles. Trotting showed two

Table 1

Direct Hip-Toe Distance (D in mm)

	Stride	Walk Max.	Walk Min.	Trot Max.	Trot Min.	Gallop Max.	Gallop Min.
1. Treadmill (air, food)	1	260 -	185 = 75	269 -	175 = 94	270 -	203 = 67
	2	266 -	184 = 82	271 -	183 = 88	274 -	189 = 85
	3	259 -	178 = 81	274 -	174 = 100	276 -	189 = 87
	$\bar{X}$		79		94		80
2. Treadmill (food)	1	267 -	206 = 61	270 -	201 = 69	287 -	202 = 85
	2	270 -	202 = 68	267 -	200 = 67	278 -	204 = 74
	3	266 -	202 = 64	269 -	202 = 67	288 -	202 = 86
	$\bar{X}$		64		68		82
3. Overground	1	257 -	193 = 64	251 -	186 = 65	319 -	204 = 115
	2	247 -	206 = 41	261 -	187 = 74	324 -	203 = 121
	3	255 -	208 = 47	254 -	195 = 59		
	$\bar{X}$		51		66		118

pronounced cycles, with the peaks as each foot (left and right) lifted off the supporting surface. During galloping there was only one cycle that was displaced as a function of footfall order. Note that in the topmost record (Figure 5) the right hindlimb leads (is frontmost), while the left hindlimb leads in the other two records. In each case the left hip continues to descend as the trailing limb strikes the surface. The lowest point corresponds to touchdown of the leading foot, with a sharp rise thereafter.

During galloping, then, the height of one hip cannot be predicted without examination of the footfalls of both hindlimbs. In

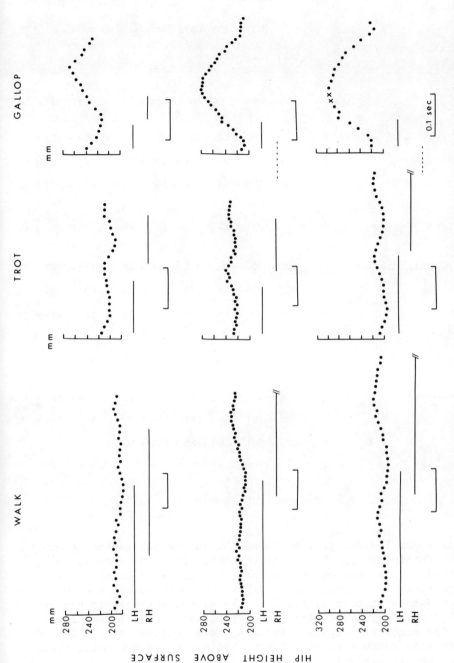

Figure 5. Excursions of hip height above substrate, with conventions for labels the same as in Figure 4.

Table 2

Hip Height Above Surface (H in mm)

		Stride	Walk Max.	Walk Min.	Trot Max.	Trot Min.	Gallop Max.	Gallop Min.
1.	Treadmill (air, food)	1	214	- 196 = 18	217	- 194 = 23	270	- 213 = 57
		2	214	- 202 = 12	214	- 194 = 20	267	- 201 = 66
		3	217	- 188 = 29	214	- 193 = 21	264	- 203 = 61
		$\bar{X}$		20		21		61
2.	Treadmill (food)	1	229	- 206 = 23	238	- 218 = 20	283	- 207 = 76
		2	225	- 206 = 19	239	- 220 = 19	280	- 209 = 71
		3	228	- 204 = 24	240	- 222 = 18	288	- 211 = 77
		$\bar{X}$		22		19		75
3.	Overground	1	218	- 193 = 25	225	- 197 = 28	298	- 195 = 103
		2	231	- 197 = 34	217	- 196 = 21	295	- 214 = 81
		3	223	- 207 = 16	224	- 205 = 19		
		$\bar{X}$		25		23		92

addition, no feature of the hip height curves corresponds closely to the Philippson epochs in the LH limb. In these senses, then, hip height is a measure of the vertical displacement of the entire hindquarters in locomotion, and shows how constraints are set upon the movements of both hindlimbs. The marked increase in hip height as the forward velocity increases probably is associated closely with the successive spinal flexions and extensions that allow the cat to gain speed (cf. Hildebrand, 1961; Gambaryan et al., 1971; Goslow et al., 1973a).

HINDLIMB IN CAT LOCOMOTION

Figure 6 further illustrates, for overground walking, how hindquarter events are represented by different whole-limb measures. The lowermost record of vertical forces exerted by the hindlimbs has been redrawn from Manter (1938). The peak of the LH curve in this record may be taken to represent the peak of the yield at the $E^2 \cdot E^3$ junction (Grillner, 1972, 1973a) and probably would align fairly closely with the same junction from the present data (top record in Figure 6 of hip-toe distance and center record of hip height, from Figures 4 and 5). Note that as force diminishes sharply toward the end of the LH stance, the hip-toe distance continues to increase, presumably because the movement becomes ballistic. Further, at the point where LH force starts to decrease, that exerted by the RH limb after its touchdown has risen

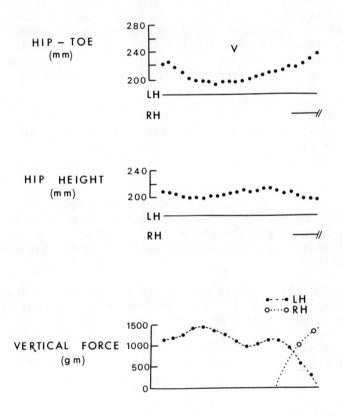

Figure 6. Comparison of whole-limb values (direct hip-toe distance and height of hip from surface) during overground walking (from Figures 4 and 5), with vertical forces developed by the left and right hindlimbs during their stance phases, as redrawn from Manter (1938).

appreciably. In this way the hip (center record) is maintained at a relatively constant height, and excessive upward and downward excursions of the body's center of gravity are avoided.

The Hindlimb Trajectory and The Stepping Surface

Figure 7 is a schematic that includes stick figures of the hindlimb in order to show how the limb trajectory may be plotted in vertical and horizontal dimensions relative to the substrate. Swing and stance phases are subdivided with respect to the point at which the toe passes under the hip (D is vertical, and X = 0). The data are for a walk at 1 m/sec (also shown in the top curve of Figure 8) and emphasize the forward and backward path of the toe in relation to the ground. Lines are drawn from the hip pivot point to the surface at the point of touchdown, at X = 0, and at liftoff, to allow comparison at a glance of swing and stance excursions with hip elevation. Figures 8-10 (in which swing and stance trajectories are superimposed) and Tables 3 and 4 describe the results.

Figure 8 reveals a fundamental characteristic of stepping that has not been seen in previous schematics. For walking at 1 m/sec, irrespective of different animals and stepping conditions, the mission, or primary controlled variable, of the whole limb through the swing is altitude. The adjustment of angles at individual joints is subordinated to the cause of maintaining the toe at a low and relatively constant height above the surface. Attainment of this goal will reduce energy expenditure by minimizing vertical excursions of the center of gravity of the limb, just as holding the center of gravity of the whole body at a constant height will minimize transfer between potential and kinetic energy (e.g., Frank, 1970).

While minimizing vertical excursions of the limb is presumably desirable during trotting and galloping, as well as during walking, Figures 9 and 10 show that some sacrifice is necessary, particularly at the beginning and end of the swing. Trotting records (Figure 9) were very similar for the two treadmill strides, and the lesser excursions of the overground (lowermost) stride probably reflect its low speed of 1.5 m/sec. This record resembles closely those during walking (Figure 8), as well as those for an additional three strides that were measured for the treadmill cat (food, air reinforcement) at 1.5 m/sec but are not shown. For the two records at 2.3 m/sec, the peak vertical excursion appeared just as the foot began to come forward after liftoff. It might have been expected that the peak would coincide with the $F \cdot E^1$ junction (near X = 0, compare Figures 9 and 4), but such was not the case. During most of the swing, in fact, the foot was descending as if already

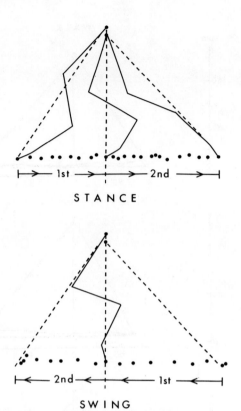

Figure 7. Schematic to show how the trajectory of the toe was measured in relation to spatial coordinates and the hip pivot point (data from Figure 8, uppermost record). Arrows indicate direction of movement of the toe. Upper drawing shows stick figures of the cat's LH limb as it retracts during the stance, while the foot is on the surface of the treadmill. The data points depart from a horizontal straight line during the stance only because of slight errors in measuring the distance from the arbitrary "surface" line. Total excursion (X distance, see Figure 3B) is marked by dashed lines (D) between the hip pivot point and toe at the first frame in which touchdown occurred (left side) and the last frame before liftoff (right side). A vertical dashed line indicates X = zero and separates first and second support periods. Lower drawing (same scale) indicates the direction of movement of the toe as it moves forward during the swing phase, with the first swing period beginning at liftoff and the second period marked as the toe passes under the vertical line drawn from the hip pivot point.

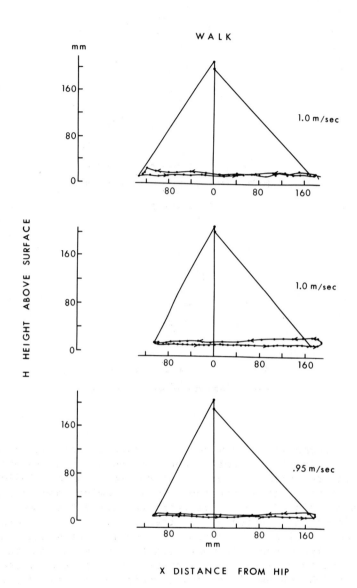

Figure 8. In this and subsequent figures (Figures 9 and 10) of whole-limb trajectories, the swing and stance phases of a single step cycle have been superimposed (same cats and strides as in Figures 4 and 5). Zero on the vertical axis marks the nominal surface (visible edge of treadmill belt or floor). At a walking speed near 1 m/sec the trajectory is flat for cats on a treadmill (upper 2 records) and overground (lowermost record).

HINDLIMB IN CAT LOCOMOTION

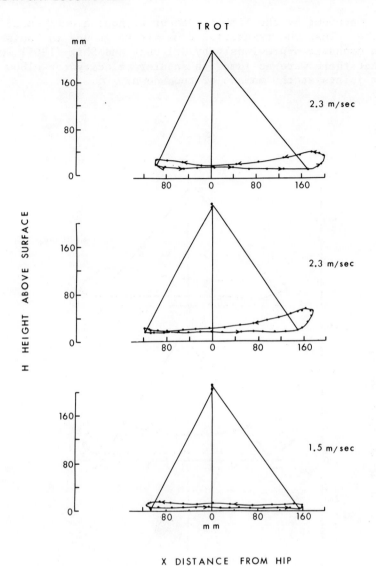

Figure 9. Trotting trajectories of the whole limb for treadmill (upper 2 records) and overground (lowermost record) conditions.

preparing to land. At the end of the swing, note that the foot did not touch down at its furthest forward excursion, but only after the limb started to return backward, an effect that would allow the limb to gather speed before becoming subject to frictional forces between the foot and the ground. In this way a truly smooth cyclical movement by the limb can occur without a noticeable pause at landing. That the transition from air to ground is "noiseless" has been emphasized previously by Orlovsky and Shik (1965), who found that there were no jumps in angular velocity for elbow or shoulder joints at the moment of touchdown.

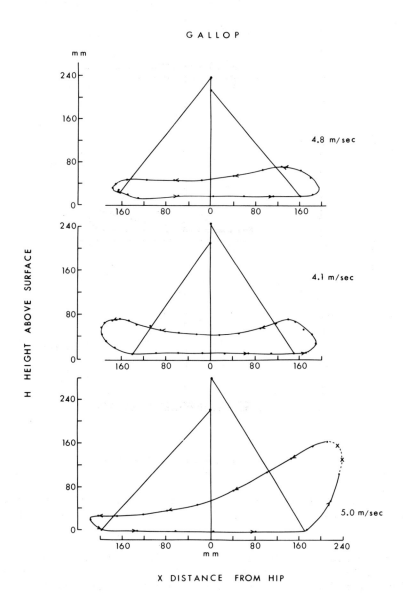

Figure 10. Galloping trajectories of the whole limb for treadmill (upper 2 records) and overground (lowermost record) running.

Table 3

Toe Height Above Surface (H in mm)

		Stride	Walk Max. − Min.	Trot Max. − Min.	Gallop Max. − Min.
1.	Treadmill (air, food)	1	26 − 11 = 15	36 − 11 = 25	72 − 13 = 59
		2	24 − 11 = 13	42 − 11 = 31	72 − 15 = 57
		3	29 − 12 = 17	50 − 12 = 38	78 − 15 = 63
		$\bar{X}$	15	31	60
2.	Treadmill (food)	1	25 − 8 = 17	56 − 16 = 40	75 − 10 = 65
		2	23 − 8 = 15	54 − 16 = 38	73 − 10 = 63
		3	26 − 11 = 15	59 − 14 = 45	80 − 9 = 71
		$\bar{X}$	16	41	66
3.	Overground	1	17 − 9 = 8	14 − 5 = 9	195 − 0 = 195
		2	20 − 9 = 11	16 − 5 = 11	165 − 0 = 165
		3	24 − 9 = 15	20 − 7 = 13	
		$\bar{X}$	11	11	180

During galloping (Figure 10) there were large vertical excursions of the toe, with an exaggeration of the trotting pattern in which the forward movement of the toe was arrested well before touchdown, and a large upward swing appeared shortly after liftoff (documenting the follow-through that was also seen in the hip-toe direct length measure, Figure 4). The trajectory for the toe, further, was different for each cat, with a pronounced upswing before touchdown appearing for the animal that worked for food alone on the treadmill (center record). During this time of hesitation the toe swung forward and then back with no lengthening of the whole limb (see D measure, Figure 4). The pattern may have been an idiosyncrasy particular to that cat, a playful behavior that resulted from performing under non-aversive conditions, or a characteristic feature of slow galloping per se. A similar pattern, however, has been seen in the cheetah (Hildebrand, 1961). It seems surprising that the most variable patterns in every whole-limb measure were seen not at slow speeds, but when the animals were running at 4 m/sec or faster, when demands on every muscle would be greatest.

Table 4

Toe Forward-Backward Excursion (X)

		Walk		Trot		Gallop	
		Pos.*	Neg.	Pos.	Neg.	Pos.	Neg.
1. Treadmill (air, food)	1	133 + 181 = 314		102 + 202 = 304		179 + 195 = 374	
	2	125 + 187 = 312		98 + 199 = 297		185 + 197 = 382	
	3	125 + 191 = 316		105 + 203 = 308		191 + 212 = 403	
	$\bar{X}$	128 + 186 = 314		102 + 201 = 303		185 + 201 = 386	
2. Treadmill (food)	1	91 + 187 = 278		123 + 176 = 299		197 + 190 = 387	
	2	105 + 190 = 295		119 + 173 = 292		198 + 180 = 378	
	3	104 + 188 = 292		114 + 175 = 289		181 + 192 = 373	
	$\bar{X}$	100 + 188 = 288		119 + 175 = 294		192 + 187 = 379	
3. Overground	1	106 + 177 = 283		109 + 147 = 256		219 + 245 = 464	
	2	139 + 144 = 283		112 + 158 = 270		220 + 240 = 460	
	3	139 + 163 = 302		111 + 156 = 267			
	$\bar{X}$	128 + 161 = 289		111 + 154 = 265		220 + 243 = 463	

*Distance (X) was considered positive (Pos.) if the toe was in front of a vertical line dropped from the hip pivot point, and negative (Neg.) if the toe was behind the hip vertical.

Grillner (1975) has reviewed evidence and logic to suggest that since swing duration changes little as a function of velocity, the distance traversed by the limb during the swing increases linearly with speed. The generality may be true for net forward distance, but it fails to take into account the backward movements of the limb while it is in the air. While mean vertical distances of the toe above the substrate increased, in general (Table 3), mean total excursions in the horizontal, X, direction were far from linear with respect to velocity and even fell during trotting for two cats (Table 4). The horizontal excursions are controlled within relatively narrow boundaries during trotting so that movements of a pair of diagonal limbs may be conceptualized as a series of brief stiff-legged falls, which are successively caught as the other pair of diagonal limbs descends.

Perhaps the chief advantage of plotting the toe trajectory in x, y coordinates is the perspective it provides of the positioning tasks of the limb during the swing phase. When duration alone is considered, the swing has long been regarded as a relatively standard element of the step cycle (see reviews by Grillner, 1975; Wetzel and Stuart, 1975). The analysis of trajectories shows that the events at liftoff and, especially, touchdown are at least as significant functionally as the turning points between flexion and extension. Extensor tension will still be marked at liftoff (see discussion in Grillner's 1975 review) as the flexors become active, and co-contraction of flexors and extensors characterizes touchdown (Engberg and Lundberg, 1969). Detailed EMG recordings from a variety of muscles at these two times in the step cycle would separate active from passive braking and accelerating forces and help to reveal exactly how the limb converts its direction of movement from forward to backward, and vice versa.

Conclusions

The various whole limb measurements are simple yet informative. They are less susceptible to errors of measurement than are angular excursions at individual joints, since the skin has minimal slippage over the hip pivot point during running as compared with the skin slippage over the knee (see, e.g., Goslow et al., 1973a). The difficulty in calibrating movement of the knee introduces substantial problems in reading knee, ankle, and hip angles.

Besides being easy to measure, trajectories of the whole limb show how discontinuous events in individual muscles or joints (flexor versus extensor activity, a sharp yield in the knee and ankle, the foot being on or off the surface) are smoothed to produce a "noiseless" step. Measurements integrated over a whole limb can further remind us to interpret the status of one limb in relation to footfalls of both hindlimbs. Preliminary work for the deafferented hindlimb (Wetzel and Norgren, Unpublished data), suggests that the various measures for a whole limb can also readily differentiate pathological from normal limb movements.

In review of the present findings, a number of overall characteristics of stepping can be defined in the following manner:

(1) Excursions at individual joints and the F, E^1, E^2, and E^3 epochs are translated to two uncomplicated and unbroken cycles of shortening and lengthening of the whole limb.

(2) Hip height always increases markedly from walking to galloping, but the hip-toe length of the limb does not necessarily display a similar increase over the same velocity range.

(3) Preparation for landing consumes a major portion of the swing phase, since the toe's highest distance from the surface is attained soon after liftoff.

(4) At slow speeds, toe altitude is maintained at a constant, low level.

(5) With increased speed, trajectory configurations are adjustable within broad limits. The foot can still leave and return to the surface with precision, because the toe can slow in the air to describe a fuller loop (expanded vertical and horizontal excursions both at liftoff and touchdown).

By adjusting individual muscle outputs to fit these various movements, the neural control program builds the step.

Acknowledgment

The work was supported by USPHS Grant NS 11491, USPHS General Research Support Funds to the University of Arizona (FR 07002) and its College of Medicine (FR 05675) and a grant from the University of Arizona Foundation. We should like to thank our colleague Dr. George E. Goslow, Jr., for his generous contribution of the overground data.

REFERENCES

Arshavsky, Yu.I., Kots, Y.M., Orlovsky, G.N., Rodionov, I.M. and Shik, M.L., (1965) Investigation of the biomechanics of running by the dog. Biophysics. $\underline{10}$, 737-746. (Translated from the Russian journal, Biofizika).

Dagg, A.I. and de Vos, A., (1968) The walking gaits of some species of Pecora. J. Zool. (Lond.). 155, 103-110.

Engberg, I., (1964) Reflexes to foot muscles in the cat. Acta Physiol. Scand. 62, (Suppl. 235), 1-64.

Engberg, I. and Lundberg, A., (1969) An electromyographic analysis of muscular activity in the hindlimb of the cat during unrestrained locomotion. Acta Physiol. Scand. 75, 614-630.

Forssberg, H., Grillner, S. and Rossignol, S., (1975) Phase dependent reflex reversal during walking in chronic spinal cats. Brain Research. 85, 103-107.

Frank, A.A. and McGhee, R.B., (1969) Some consideration relating to the design of autopilots for legged vehicles. J. Terramech. 6, 23-35.

Frank, A.A., (1970) An approach to the dynamic analysis and synthesis of biped locomotion machines. Med. Biol. Engng. 8, 465-476.

Freusberg, A. and Goltz, Fr., (1874) Ueber den Einfluss des Nervensystems auf die Vorgänge während der Schwangerschaft und des Gebärakts. Pflüg. Arch. ges. Physiol. 9, 552-565.

Gambaryan, P.P., Orlovsky, G.N., Protopopova, T.G., Severin, F.V. and Shik, M.L., (1971) Muscle work during different forms of locomotion in the cat and adaptive changes of the locomotory organs in the family Felidae. Proc. Inst. Zool. Acad. Sci. U.S.S.R. 48, 220-239. (In Russian).

Gambaryan, P.P., (1974) How Mammals Run: Anatomical Adaptations. Halsted (Wiley), New York. (Translated from the Russian edition, Nauka, Leningrad, 1972).

Goslow, G.E., Jr., Reinking, R.M. and Stuart, D.G., (1973a) The cat step cycle: Hind limb joint angles and muscle lengths during unrestrained locomotion. J. Morph. 141, 1-41.

Goslow, G.E., Jr., Reinking, R.M. and Stuart, D.G., (1973b) Physiological extent, range and rate of muscle stretch for soleus, medial gastrocnemius and tibialis anterior in the cat. Pflüg. Arch. ges. Physiol. 341, 77-86.

Goslow, G.E., Jr., Stauffer, E.K., Nemeth, W.C. and Stuart, D.G., (1973c) The cat step cycle: Responses of muscle spindles and tendon organs to passive stretch within the locomotor range. Brain Research. 60, 35-54.

Graham Brown, T., (1911) The intrinsic factors in the act of progression in the mammal. Proc. R. Soc. 84b, 308-319.

Graham Brown, T., (1912) Studies in the physiology of the nervous system. XI. Immediate reflex phenomena in the simple reflex. Quart. J. exp. Physiol. 5, 237-307.

Grillner, S., (1972) The role of muscle stiffness in meeting the changing postural and locomotor requirements for force development by the ankle extensors. Acta Physiol. Scand. 86, 92-108.

Grillner, S., (1973a) "Muscle stiffness and motor control - forces in the ankle during locomotion and standing," In Motor Control. (Gydikov, A.A., Tankov, N.T. and Kosarov, D.S., eds.), Plenum, New York, (195-215).

Grillner, S., (1973b) "Locomotion in the spinal cat," In Control of Posture and Locomotion. (Stein, R.B., Pearson, K.G., Smith, R.S. and Redford, J.B., eds.), Plenum, New York, (515-535).

Grillner, S., (1975) Locomotion in vertebrates: central mechanisms and reflex interaction. Physiol. Rev. 55, 247-304.

Grillner, S. and Zangger, P., (1975) How detailed is the central pattern generation for locomotion? Brain Research. 88, 367-371.

Gurfinkel, V.S. and Shik, M.L., (1973) "The control of posture and locomotion," In Motor Control. (Gydikov, A.A., Tankov, N.T. and Kosarov, D.S., eds.), Plenum, New York, (217-234).

Hildebrand, M., (1959) Motions of the running cheetah and horse. J. Mammal. 40, 481-495.

Hildebrand, M., (1961) Further studies on locomotion of the cheetah. J. Mammal. 42, 84-91.

Hildebrand, M., (1974) Analysis of Vertebrate Structure. Wiley, New York, (487-515).

Jankowska, E., Jukes, M.G.M., Lund, S. and Lundberg, A., (1967a) The effect of DOPA on the spinal cord. 5. Reciprocal organization of pathways transmitting excitatory action to alpha motoneurons of flexors and extensors. Acta Physiol. Scand. 70, 369-388.

Jankowska, E., Jukes, M.G.M., Lund, S. and Lundberg, A., (1967b) The effect of DOPA on the spinal cord. 6. Half-centre organization of interneurones transmitting effects from flexor reflex afferents. Acta Physiol. Scand. 70, 389-402.

Jenkins, F.A., Jr., (1970) Limb movements in a monotreme (Tachyglossus aculeatus): A cineradiographic analysis. Science. 168, 1473-1475.

Lockard, D.E., Traher, L.M. and Wetzel, M.C., (1975) Reinforcement influences upon topography of treadmill locomotion by cats. Physiol. Behav. (In press).

Manter, J.T., (1938) The dynamics of quadrupedal walking. J. Exp. Biol. 15, 522-540.

Marey, E.J., (1901) La locomotion animale. Traité de physiologie biologique. 1, 227-287.

Matthews, P.B.C., (1972) Mammalian Muscle Receptors and Their Central Actions. Arnold, London.

McGhee, R.B. and Frank, A.A., (1968) On the stability properties of quadruped creeping gaits. Math. Biosci. 3, 331-351.

Millar, J., (1973) Joint afferent fibres responding to muscle stretch, vibration and contraction. Brain Research. 63, 380-383.

Miller, S. and van der Burg, J., (1973) "The function of long propriospinal pathways in the co-ordination of quadrupedal stepping in the cat," In Control of Posture and Locomotion. (Stein, R.B., Pearson, K.G., Smith, R.S. and Redford, J.B., eds.), Plenum, New York, (561-577).

Miller, S. and van der Meché, F.G.A., (1975) Movements of the forelimbs of the cat during stepping on a treadmill. Brain Research. 91, 255-269.

Miller, S., van der Burg, J. and van der Meché, F.G.A., (1975a) Co-ordination of movements of the hindlimbs and forelimbs in different forms of locomotion in normal and decerebrate cats. Brain Research. 91, 217-237.

Miller, S., van der Burg, J. and van der Meché, F.G.A., (1975b) Locomotion in the cat: Basic programmes of movement. Brain Research. 91, 239-253.

Orlovsky, G.N. and Shik, M.L., (1965) Standard elements of cyclic movement. Biophysics. 10, 935-944.

Orlovsky, G.N., Severin, F.V. and Shik, M.L., (1966) Effect of speed and load on coordination of movements during running of the dog. Biophysics. 11, 414-417.

Philippson, M., (1905) L'autonomie et la centralisation dans le système nerveux dex animaux. Trav. Lab. Physiol. Inst. Solvay (Bruxelles). 7, 1-208.

Prochazka, V.J., Tate, K., Westerman, R.A. and Ziccone, S.P., (1974) Remote monitoring of muscle length and EMG in unrestrained cats. Electroenceph. clin. Neurophysiol. 37, 649-653.

Severin, F.V., Orlovsky, G.N. and Shik, M.L., (1967) Work of the muscle receptors during controlled locomotion. Biophysics. 12, 575-586.

Severin, F.V., (1970) The role of the gamma motor system in the activation of the extensor alpha motor neurons during controlled locomotion. Biophysics. 15, 1138-1145.

Sherrington, C.S., (1910) Flexion-reflex of the limb, crossed-extension reflex, and reflex stepping and standing. J. Physiol. 40, 28-121.

Shik, M.L. and Orlovsky, G.N., (1965) Co-ordination of the limbs during running of the dog. Biophysics. 10, 1148-1159.

Shurrager, P.S. and Dykman, R.A., (1951) Walking spinal carnivores. J. comp. physiol. psychol. 44, 252-262.

Snyder, R.C., (1952) Quadrupedal and bipedal locomotion of the lizard. Copeia. 1, 64-70.

Snyder, R.C., (1962) Adaptations for bipedal locomotion of lizards. Amer. Zoologist. 2, 191-203.

Stuart, D.G., Mosher, C.G. and Gerlach, R.L., (1972) "Properties and central connections of Golgi tendon organs with special reference to locomotion," In Research in Muscle Development and the Muscle Spindle. (Banker, B.Q., Pryzbylsky, R.J., van der Meulen, J.P. and Victor, M., eds.), Excerpta Medica, Amsterdam, (437-464).

Stuart, D.G., Withey, T.P., Wetzel, M.C. and Goslow, G.E., Jr., (1973) "Time constraints for inter-limb co-ordination in the cat during unrestrained locomotion," In Control of Posture and Locomotion. (Stein, R.B., Pearson, K.G., Smith, R.S., and Redford, J.B., eds.), Plenum, New York, (537-560).

Sukhanov, V.B., (1974) General System of Symmetrical Locomotion of Terrestrial Vertebrates and Some Features of Movement of Lower Tetrapods. Amerind, New Delhi. (Translated from the Russian edition, Nauka, Leningrad, 1968.)

Székely, G., Czéh, G. and Vöros, G., (1969) The activity pattern of limb muscles in freely moving normal and deafferented newts. Exp. Brain Research. 9, 53-62.

Tokuriki, M., (1973a) Electromyographic and joint-mechanical studies in quadrupedal locomotion. I. Walk. Jap. J. vet. Sci. 35, 433-446.

Tokuriki, M., (1973b) Electromyographic and joint mechanical studies in quadrupedal locomotion. II. Trot. Jap. J. vet. Sci. 35, 525-533.

Tokuriki, M., (1974) Electromyographic and joint-mechanical studies in quadrupedal locomotion. III. Gallop. Jap. J. vet. Sci. 36, 121-132.

Wetzel, M.C. and Stuart, D.G., (1975) Ensemble characteristics of cat locomotion and its neural control. Progr. Neurobiol. (In press).

Wetzel, M.C., Atwater, A.E., Wait, J.V. and Stuart, D.G., (1975) Neural implications of different profiles between treadmill and overground locomotion timings in cats. J. Neurophysiol. 38, 492-501.

Yager, J.G., (1972) The electromyogram as a predicator of muscle mechanical response in locomotion. Unpublished Ph.D. Dissertation. University of Tennessee.

ARTHROPOD WALKING

Graham Hoyle

Department of Biology

University of Oregon, Eugene, Oregon

The neurophysiological mechanisms underlying the generation and control of walking in representative crustacean and insect species are currently being researched in several laboratories, at the level of identified neurons and their interactions. Presuming a common ancestral form, this research holds out the prospect of understanding how a neural machinery has evolved and adaptively radiated. The ancestral form was multi-legged, but now insects use three, or only two, pairs of legs and crustaceans from about a dozen pairs, but commonly four pairs down to a single pair. Living representatives of ancestral forms use different gaits at different speeds or under different loads, showing that there are either several different central programs for step control and inter-limb coordination, or continuous computing from sensory input but limited motor expressions. The fundamental basis is a metachronal wave progressing anteriorwards. Inhibitory mechanisms ensure that successive legs on one side do not interfere with each other. The simplest method of control is that the next anterior leg does not start its step until the foot behind it has been set in place.

All possible combinations of step sequences of all legs have been found in the few crustaceans and insects that have been intensively studied, but some have much greater probabilities than others, especially at prescribed speeds or loading. Two forms of response to loss of limb occur, an instantaneous one that suggests in-built adaptiveness, and a slow one that suggests learning to cope with changed proprioception.

The conclusion is that each limb can step independently of the others, but is also subject to several different, even opposed, patterns of inter-limb coordination control, as the occasion demands. From one to a few overall control mechanisms are superimposed by independent pattern generators. In adaptively radiating, rather than modifying a single ancestral basic central pattern, different animals have evolved new patterns that form facultative alternates to the old ones.

Introduction: why arthropods?

Arthropods demand our attention if only because their numbers of species greatly exceed those of all other animals combined. Their rich diversity of form and behavior have provided a special stimulus to studies by evolutionists. On the physiological side, they have long been the objects of successful investigations of muscular and neuromuscular physiology. Now it is the turn of neuroethology, since quite recently electrophysiologists have begun to master technical difficulties that earlier impeded progress in studying arthropod central nervous physiology. Selected, diverse arthropods, are now amenable to analysis at the level of identified neuron activity, and in a few instances it has become possible to record intracellularly during walking and some other behavioral activities. The prospect before those who will enter this field is therefore enticing indeed, for some of the mysteries of neural integration, including the means by which nerve networks program complex outputs, and perhaps the basis for plasticity of neural function, are about to fall within the purview of the skillful investigator.

The time is ripe, then, to take stock of the existing knowledge of locomotion among arthropods. But is there a common framework on which to hang the results of physiological studies? What do evolutionists say about arthropods? To what extent can we hope to extract general principles by studies on representative species of say a crab, a cockroach and a spider?

Modern authorities are by no means agreed about arthropod evolution, or even that the arthropods are descended from a single common annelidan stock, as was the passionate belief of Ray Lankester (1904) and more recently the great insect anatomist Snodgrass (1950), though that possibility remains valid. Persons who differ from this view usually subscribe to only a diphyletic origin (Tiegs and Manton, 1958; Sharov, 1966).

Recent interest in evolutionary questions of arthropods has been promoted largely by Russians. Their views turn out to be especially favorable for our quest. Sharov (1966) and Beklemishev (1969), both subscribe to the Cuvierian classification that links annelids and arthropods in a single phylum, the Articulata (Figure 1). Sharov (1966) states that the phylum Articulata is unanimously accepted as a genuine animal phylum. Acceptance of this view leads to the type of evolutionary tree proposed by Beklemishev (1969) shown in Figure 2. On this scheme it is quite likely that there are homologous neurons in insects, crustaceans and other major lines. Whichever line of arthropods we choose to examine, within single orders there are forms that are richly diverse morphologically and in behavioral repertoire that everyone can agree have indeed been generated from a common stock. In ther reverse consideration: the insight physiological findings can give to resolving evolutionary questions, it is high time the comparative physiologist made his contribution, by offering critical comparisons of physiological events in diverse groups. Within a Class, it is certainly justifiable to homologise some of the more conspicuous muscles and the neurons that innervate them. Examples are coxal levators and depressors, and tibial extensors and flexors of insects. For some species the motor neuron somata supplying these muscles have been located, a few antecedent interneurons have been located, and more are being sought. Functionally comparable neurons of crustaceans are also known.

We shall eventually have available a treasure-house of identified neurons of many species from different families, orders and even classes of arthropods that could be homologous right back to a common ancestral form. Even if it turns out to be invalid to make insect/crustacean comparisons at the level of homology there remains the intrinsic interest in understanding how control of functionally equivalent parts such as a crab meropodite and a locust femur has been achieved in parallel development.

The central question a majority of recent investigators has addressed, has been the extent to which the control is generated endogenously, i.e., entirely within the central nervous system, without detailed reference to ongoing sensory inputs. The work of the Sherrington and Pavlovian schools had led, by the late '30's to the dogmatic view that all behavior is reflex in origin and control. By this time, the work of the ethologists was getting to the point where it suggested that many instinctive behaviors are inherited as discrete units. Whilst they could not directly contribute to the physiological questions as to how such units are programmed, they hinted at a central nervous determinism for complex sequences of motor act. Support for these views came from the developmental studies of Paul Weiss (1950), and physiological/behavioral studies such as those of Wells (1950) on the lugworm.

Phylum. Articulata

    Subphylum I. Annelida
        Classes: Polychaeta; Echiurida; Oligochaeta; Hirudinea; Sipunculida

    Subphylum II. Stelechopoda
        Classes: Tardigrada, Pentastomida

    Subphylum III. Malacopoda
        Class: Onychophora

    Subphylum IV. Arthropoda

        Superclass 1. Proboscifera, supercl. nov.
            Classes: Dicephalosomita, cl. nov. Pycnogonida

        Superclass 2. Trilobitomorpha
            Classes: Tetracephalosomita, cl. nov.; Trilobita; Trilobitida

        Superclass 3. Chelicerata
            Classes: Merostomata; Eurypterida; Scorpionida; Pedipalpida; Arachnida; Solifugida; Acarina

        Superclass 4. Crustacea
            Classes: Gnathostraca; Ostracoda; Maxillopoda; Malacostraca

        Superclass 5. Atelocerata
            Classes: Chilopoda; Diplopoda; Pauropoda; Symphyla; Protura; Collembola; Diplura; Insecta

Figure 1. Modern classification that groups Arthropoda and Annelida in the same phylum, termed (after Cuvier) the Articulata. (From A.G. Sharov, Basic Arthropodan Stock, Pergamon, 1966.)

Sensory input normally would serve to trigger complex behaviors, but examples were uncovered in which whole chunks of important behavior occurred as "vacuum activities" in the absence of any specific stimulus, or that normally occur spontaneously (Tinbergen, 1951).

ARTHROPODS AND NEUROETHOLOGY

Arthropods, especially insects, had been extensively utilized by some ethologists, notably Tinbergen (1951), Thorpe (1963), Hinde (1970) and Manning (1972), in making the important generalizations of ethology. Because of the ease of handling arthropods at the neurophysiological level, it occurred to several investigators that a direct analysis of the neuronal basis of behavior -neuroethology- was possible for the large species. By intensively studying a given act of behavior, then interfering with the

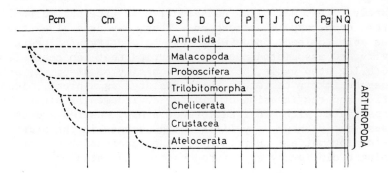

Figure 2. Phylogenetic interrelations and geological distribution of annelids and arthropods. Atelocerata includes insects and myriapods (from Sharov, 1966). Pcm, Pre-Cambrian; Cm, Cambrian; O, Ordovician; S, Silurian; D, Devonian; C, Carboniferous; P, Permian; T, Triassic; J, Jurassic; Cr, Cretaceous; Pg, Paleogene; N, Neogene; Q, Quaternary period.

controlling neuronal circuits, for example, by ablating proprioceptive sense organs, it was anticipated that an evaluation of the latter's role in the generation of behavioral acts would be possible. The most recent developments are on the nearly intact animal that is still capable of behaving nearly normally though tethered or restrained, whilst parts of the nervous system are exposed. Neurons are penetrated with microelectrodes and their roles in the programming of behavior directly examined.

This is not as far-fetched as it may sound. There are only some 10^5 neurons in the entire arthropod central nervous system, and as many as 10^4 of these are probably concerned with reducing data from sense organs. Only two or three hundred are the motor neurons that determine all of the behavior and they are driven by quite a small number of interneurons.

As already stated, one of the easiest behaviors to obtain repeatedly and therefore the most intensively studied, is locomotion. Already there is abundant evidence from diverse arthropods (see Wilson, 1966; Kandel and Kupferman, 1970; Evoy and Cohen, 1971; Hoyle, 1975a, b, for reviews) that locomotion is not controlled simply by sequences of reflex actions as was earlier throught. Their central nervous systems can generate appropriately-programmed outputs, on the basis of the connectivity and physiological properties of central cells, without patterned sensory inputs. Cell properties and connectivity are largely genetically-determined, and the requisite features may well have been evolved but rarely. They may be as conservative as external morphological features, or perhaps even more so. Thus, having evolved the genetic systems able to lead, in development, to a pattern of neural connectivity and associated neural properties that can control, say, a smooth leg step cycle, or a metachronicity of stepping, an organism may be likely to retain these features as a block. Adaptive radiation and the invasion of diverse environmental niches has been associated with modifications of limbs, muscles and peripheral synapses, but perhaps very little with changes in the neural programs for locomotion. Herein, lies the fascinating potential for combining evolutionary and neurophysiological studies.

Eventually, the genetic control of the developing circuitry can be expected to become an object of study, resulting in dual developments of intense biological interest affecting our understanding of both neurobiology and evolution. It is just very unfortunate that the slow pace of neurophysiological and neuroanatomical progress, dictated by the technical skills required, is such that we cannot expect these studies to make much headway in the Twentieth century. But eventually mankind will enjoy the intellectual benefits of this type of knowledge.

Evolutionists are generally agreed that one of the common ancestors was of similar form to existing polychaete worms, even those that do not believe the ancestor was an actual annelid. Compare the forms of existing annelids and arthropods: some of them are very similar (Figure 3). A "living-fossil" terrestrial form of arthropod, the Onychophoran Peripatus, that has tracheal respiration, resembles in gross morphology extinct marine forms (Figure 4) that also look annelid-like. This ancestor was able not only to swim, but also to burrow in mud and soft sand and to crawl along a hard surface. The primitive form was a metamerically segmented tube with lateral protuberances on each segment. Locomotion was achieved by sinusoidal movements, or undulations, of the body wall, the protuberances providing thrust without themselves participating by moving independently of the body (Figure 5). Such undulations have generally been considered to start posteriorly and move headwards. However, this question has not been given sufficient critical attention. The wave may be seen to start at one end, but visual observation alone cannot resolve the way the thrust, or power-wave, is made. It could be either when the undulation is moving forwards, or when it is moving backwards. But the most probable pattern was thrusting backwards as the wave travelled forwards.

Within the protuberances discrete pairs of antagonistic muscles eventually developed, attached distally to a cuticular thickening, or bristle. They enabled forwards and backwards movements of individual protuberances to occur independently of body-wall

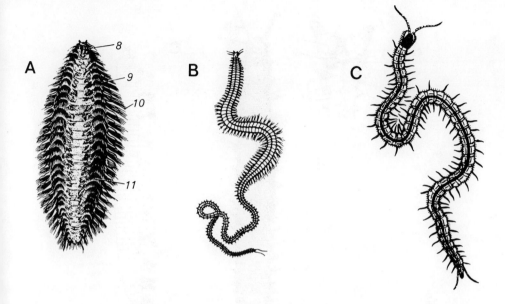

Figure 3. Two modern polychaetes (A,B) compared with a Centipede (C) to show the resemblance in overall form to polychaete worms. A) Chloeia (Amphinomidae); B) Japanese Palolo worm, to show the simple lateral projections from which legs evolved; C) Pachymerium (Chilopoda). (From Beklemishev, 1969.)

contractions. Probably later, an additional pair was added, permitting dorsal/ventral oscillation. The protuberances thereby evolved to become legs; the initial two pairs of antagonistic muscles operating between the base of each protuberance remain the principal ones providing for leg movement. They cycle forwards (protraction) to prepare for a thrust, being lifted (levation) off the ground to avoid backwards thrust, then lowered (depression) to press against the ground and form a strut through which propulsive force (thrust) is provided whilst the leg is moved backwards (retraction). This complete cycle is repeated unchanged in a rhythmic procession.

The appendages probably were initially simply lateral, in all forms; then in the arthropod direction they came first to lie under the body as simple props (Onychophora), until an increase in leg length occurred, after which they became bent out laterally, finally reaching the stage in which the body is suspended from the legs (Figure 6). Undulations of the body wall gradually were phased out as producers of movement, and replaced by co-ordinated movements of individual legs (Figure 5). The great modern classes,

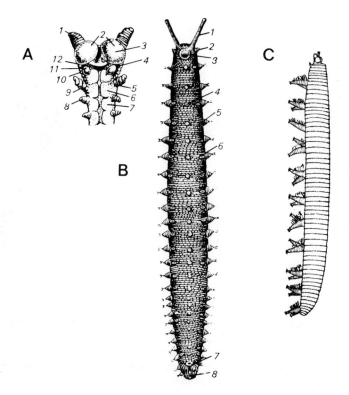

Figure 4. Modern Onychophora (A,B) compared with marine fossil creature from the mid-Cambrian (C). A, Peripatus edwardsii - embryo, B, Eoperipatus weldoni, both ventral, C, Reconstruction of Aysheaia pedunculata. (From Beklemishev, 1969.)

Arachnida, Crustacea and Insecta have all evolved after reducing the number of pairs of legs in the adult. How that was achieved, in detail, remains a matter for endless speculation. Some notion as to the insect route seems to be provided by the Pauropods and the notion of neotony. Although the adults have multiple pairs of legs, the larva, like adult insects, has only three pairs (Figure 7).

It is our task to analyze the fundamentals of stepping and co-ordination between legs, and also the relationship of the legs to other body movements that contribute to locomotion. We must try to do this for a variety of arthropods within each major class. Our choice must be guided by the ultimate feasibility of the chosen organism for neurophysiological examination, especially at the level of identified neurons that are involved in locomotory behavior.

ARTHROPOD WALKING   145

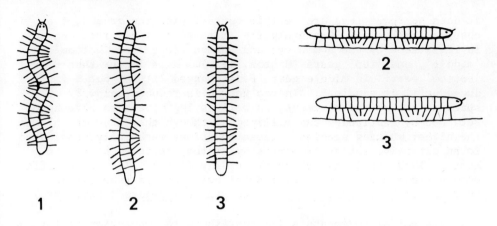

Figure 5. Evolution of arthropod locomotion. Stage 1, locomotion was first produced by undulation of the body by waves initiated posteriorly and progressive anteriorly, the thrust being generated by lateral projections located on each segment as they are moved backwards against the substrate. Stage 2, undulations are reduced in amplitude and movements of the leg are superimposed. See also side view. Stage 3, the body is held stiffly and the legs themselves move, but with the same pattern as previously produced by undulation. See also side view. Crustaceans and insects advance beyond this point by reducing the number of legs and by increasing the number of neural patterns controlling the legs.

Figure 6. Evolution of legs and posture of terrestrial arthropods. The legs are at first simple struts: eventually the body is suspended from them.

   To describe the cycle of a single leg we need to know: the protraction time (P), the retraction time (R), the ratio between them P/R, the cycle duration or period (T) and its reciprocal, the frequency of stepping (f) and the step or span (s). In the most primitive arthropods, body wall movements are still used to locomote in addition to leg movements. The soft-bodied Onychophoran Peripatus, like polychaetes, elongates as it walks faster and as

it does so, fewer legs are left in contact with the ground, i.e., it
changes its "gait." Certainly the impression of discrete changes
is registered by the human eye and Manton recognized 'bottom',
'middle', and 'top' gears (Manton, 1950). R is longer than P in
'bottom' gear. In middle gear, P is changed little, but R is reduced until it equals P. In top gear R is reduced below P. Conspicuous sinusoidal movement, as well as leg movements have been
retained by centipedes and millipedes, though they show up only at
the higher walking species. These questions were fully considered
in an extensive series of papers by Manton, reviewed by her in
Manton, 1953 and in two chapters in Gray (1968). The phase difference between a given leg and its following one is greatest in
"middle gear", in Peripatus, but not in Scolopendra (Gray, 1968).

Sinusoidal movements still contribute to locomotion of larval
forms of holometabolous insects and to a limited extent in Isopod
crustaceans. They are apparent in centipedes at fast running
speeds (Figure 8). There must be interesting transitions in either
the wiring or the network properties of adult moths and butterflies
compared with the first three pairs of legs of caterpillars. When
legs alone are used, the gaits of most forms with a large number of
legs involve the passage of a wave of movement from posterior to

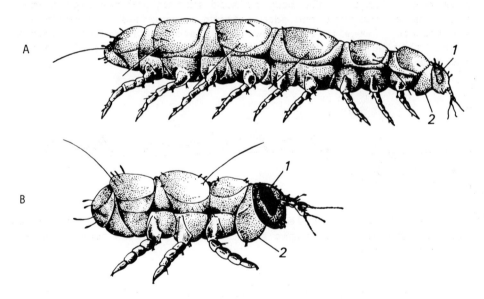

Figure 7. Adult (A) and larval (B) Pauropus. Thought to be related to myriapods. Modern forms live in soil, and have lost their
compound eyes, but their ancestors lived in the open and had eyes.
The things that look like eyes on the larval form are not. But
note the remarkably insect-like hexapod form of the larva. (From
Beklemishev, 1969.)

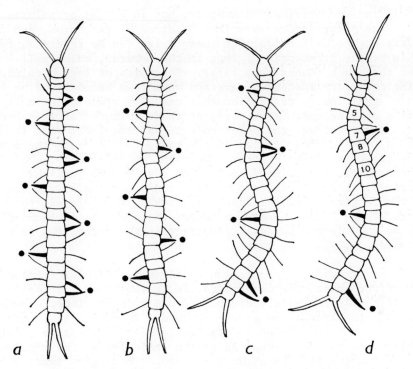

Figure 8. <u>Scolopendra</u>. Gait of the centipede <u>Scolopendra</u> as it runs progressively faster (a - d). Note the appearance of undulations, in addition to leg movements, at the faster speeds. (From Manton (1968) in Gray (1968).)

anterior. In only a few, such as <u>Cryptops</u>, does the wave start anteriorly. The phase difference between successive legs is small. Many thrust-providing, retracting legs are on the ground at the same time. As many as five may be participating in a given cluster.

All the evolutionarily more advanced forms of arthropods are ones that have few pairs of walking legs; there are only three in insects and four in the higher crustaceans, spiders and scorpions. The fields of movement of the legs are greater and the legs themselves are longer, than in the multi-legged forms. Could the basic neural control mechanisms have evolved simply from those used in the multi-legged forms by, as it were, just dropping out connections to the posterior pairs of legs?

ARTHROPOD LEGS AND THE MECHANICS OF WALKING

The legs of arthropods are themselves segmented, and in the forms with few legs the muscles in each segment play major roles not only in leg position and support, but in adding propulsive force. The power stroke is primarily aided by extension, and the recovery stroke by flexion, but these roles can be reversed, for example by the front pair of legs in the lobster, which pull rather than push (Macmillan, 1975). The roles of legs of sideways walking crabs are instantly reversed when they become leading instead of trailing, a transition that can take place in a small fraction of a second. Also, many of these animals can walk backwards. Nevertheless, the major role in walking locomotion is still played by the muscles located in the body, that operate the whole leg as if it is a simple strut. Vertical and horizontal whole leg movements have, in all hard-bodied and some of the soft-bodied arthropods, become separately manipulated structurally, in the form of a double gimbal articulation (Figure 9). That closest to the body is always large and always termed the coxa and it is hinged so as to provide the swings of protraction and retraction in most arthropods.

In crustaceans, articulation to the body skeletal framework of the first joint, the coxopodite, is in two places, at opposite poles of the approximately circular profile of the inner margin of the coxa. In Homarus the axis of movement is almost horizontal and perpendicular to the antero-posterior/dorso-ventral plane. This restricts the movement imparted to the whole leg by the muscles operating on the coxa to about a 55° arc. The leg is swung to and fro under the body by this movement, primarily a swimming motion. In order to protract and retract fully in walking, Homarus has to also flex and extend at the basipodite (B)/ischiopodite (I) joint. Individual legs of Homarus differ in the angle of attachment and hence in the extent to which B/I movement is needed. The more vertical the coxal articulation the greater the P/R component. In spiny lobsters, crayfish, Squilla and all heavy-bodied crabs, the axis of movement of the coxa has become almost vertical, though tilted anteriorly about 25°. Furthermore, in association with this modification, coxal swing is greater, almost 90°. The coxal movement alone can adequately account for almost the whole of protraction.

By a simple expedient, the insect leg has achieved a major advance over the crustacean condition in regard to articulation of the leg with the body. The ventral articulation of the coxopodite has been lost and the dorsal one changed to a peg and socket so that enormous flexibility of coxal position is possible. Almost 360° rotation can be achieved by some insect hind legs. The rest of the joints of insect legs all have two articulations, as in crustaceans. The ultra-flexible, most proximal joint has not affected the mechanics of major stepping movements, so they are essentially identical for insects and most of the large decapod crustaceans. It is possible that the muscles responsible for these

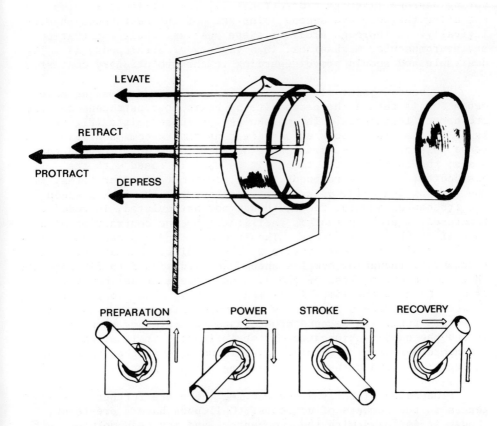

Figure 9. Drawings illustrating the functioning of the arthropod leg to body articulation which is perhaps, the most significant single structural engineering feature of the phylum. It consists of a gimbal, or pair of rings pivoted on axes at right angles to each other. The first ring is the coxa; the second has a variety of names in different arthropod classes. Two pairs of antagonistic muscles (sometimes in synergistic clusters) combine to produce the major movements of stepping: levation, opposed by depression, and protraction, opposed by retraction. In the drawings, you are looking at a left leg from the front. The circular rotation sequence is illustrated in the four smaller drawings below. The principal power stroke of all arthropod locomotion occurs when the depressed leg, having been set-down anteriorly is retracting. A major accompaniment, and the principal power stroke of some hind legs, is extension of the tibia, which occurs at the same time as retraction.

movements in insects and crustaceans are homologous and the innervating neurons likewise. Unfortunately, the intrinsic muscles operating the coxa and second joint are not the best known physiologically. A thorough study of these muscles, their innervation and neuromuscular mechanisms, in a variety of arthropods, is desirable and should prove rewarding in the evolutionary context.

The second joint always serves for levation/depression movements and is termed the basipodite or basis in crustaceans and the trochanter in insects. The morphology of these joints (Figure 10) is so similar in insects and crustaceans that a common origin seems possible. Crustaceans have an additional joint beyond the basis; the ischiopodite, though it is usually fused with the basipodite. This corresponds to the pre-femur of other forms. In the jumping leg only, of saltatory orthoptera, the link between the trochanter and the femur, though not articulated, is readily fractured, a process that is facilitated by the contraction of a special trochanteral muscle. The fracture, and contraction of this muscle result in a neat closure of the opening, the margins of the trochanter, though of complex shape, being designed to fit together snugly, preventing loss of blood. The process of deliberate, drastic loss of the limb, is termed autotomy. The close similarity between this process in various arthropods also suggests a common ancestral form. In decapod crustaceans the autotomy plane is around the basi-ischiopodite whilst in insects it is at the trochantero-femoral articulation: this may aid in the process of analogising - and perhaps homologising - leg segments.

Proceeding distally along the leg there are further divergences in the numbers of segments. Chilopods have a pre-femur, though it is fused with the trochanter, and may be homologous with the crustacean ischiopodite. The next joint is the most ubiquitous, after the coxa, namely the femur, as it is called in a majority of arthropods, or the meropodite, in crustaceans. This contains the muscles that are the best-known physiologically. Their task is to provide a combination of levator tone, by passive extension of the remainder of the leg, and powerful thrusts for forwards or lateral locomotion. In saltatory forms the extensor muscle located in this joint is almost entirely responsible for the jumping thrust. The femur, in most arthropodan forms, is tilted upwards at an angle of $30-45°$ to the horizontal. In crabs that run on land, such as the ghost crab, the similarity in use to the insect femur is very clear.

Beyond the femur, variations are numerous and homologies are difficult to recognize. Insects and centipedes have a long, thin joint, the tibia, on which the femoral extensors and flexors act. But crustaceans have a double-segmented strut before the dactylopodite, which is the most distal, and may be thought of as the

Figure 10. Typical crustacean walking leg (above) compared with corresponding insect leg (below). Directions of movement at the articulations indicated by arrows. The insect articulation with the body is at a single point, permitting extra degrees of freedom.

equivalent of the tarsus. The most proximal of the two joints is the carpopodite, the other the propopodite. Spiders have a pre-tibial segment termed the patella, and a pre-tarsal segment termed the metatarsus. The pointed dactyl of crustaceans is used as a strut when extended or as a skid, when flexed. It does play a postural role and it may participate in rocking movements in spiny lobsters, but it is never used like a true foot, as is the insect tarsus.

The insect tarsus has three jointed segments plus the terminal claw or unguis. A centipede tarsus has as many as thirty segments. There are a great many variations on the basic themes. The first pair of legs is the one most commonly modified, as large claws in several crustaceans or as prey-catchers in the praying mantis. The second pair forms small claws in most Macrura and the astonishing prey-capture weapons in mantis shrimps, leaving the latter with only three pairs of walking legs, like insects. Swimming crabs have the last pair of legs modified as paddles, which they use with extraordinary versatility for swimming control.

One of the most primitive of living arthropods, Limulus, has five pairs of walking legs, and these have unique articulations. Scorpions, mites, pycnogonids, harvestmen and spiders all have four pairs of legs, and widely different patterns of locomotion. If these Arachnids started out with common basic patterns of motor output circuitry and programming, their different behavior patterns suggest that they have all undergone drastic modifications during the course of their radiation.

The best prospects for serious homologization and comparison reside within insects or crustaceans, but having worked extensively with both classes I find that I am constantly comparing them and being impressed by physiological similarities. A serious comparison between them is feasible. If we assume that insects, in simplifying their leg structure, have lost the segment corresponding to the crustacean ischiopodite, then the femur is at least the analogue of the meropodite. Structurally, and functionally, they are indeed closely similar.

RESTRAINTS THAT MAY GOVERN LOCOMOTOR PROGRAMMING

Manton's studies (1953), especially those on myriopods, led her to make some generalizations that define practical requirements of locomotion, as follows:

(1) To provide continuous, well-balanced support for the body.

(2) To avoid excessive overlap between legs.

(3) To keep the number of retracting legs providing thrust approximately constant.

(4) To keep the animal poised for sudden shifts in speed or direction.

(5) To relate power output to functional requirement.

(6) To avoid distorting the body during locomotion.

The multi-legged arthropods clearly all use a metachronous wave mechanism. Out of the basic type, have evolved arthropods that specialize in speed, or in strength, or that have compromised between the two, whatever their leg number. The faster ones have the greater range of speeds and have adopted more than one gait, whilst the slower, force generators, can get by with but a single gait. In association with their habit, the force generators have fewer muscles that are intrinsically slower, shorter, thicker. A program of motor output and neuromuscular transmission well-suited to the latter would not also be able to operate the former, except at the slower speeds.

The first requirement, that successive legs on one side of an animal do not interfere with each other, requires coupling between them to prevent movement of the anterior leg during the initial part of the movement of its posterior neighbor, assuming the primeval pattern to have been initiation of walking at the posterior end, with postero-anteriorwards progression of waves of leg movement. Even when a hexapod in which one pair of legs is specialized, say the praying mantis, walks using only two pairs of legs, it commonly starts walking with a single hind leg. The leg in front does not start to step until the hind leg has completed its cycle and placed the foot down close behind the foot in front. This may seem a self-evident requirement, since otherwise the insect would for a time have no support on one side and keel over. Such a requirement was not present in the multi-legged ancestor. It could, in principle, be determined by proprioceptive reflexes arranged so that levation and protraction motorneurons of the anterior leg are inhibited when the posterior leg is moving. No such connections have ever been found for an arthropod, however; nor have any reciprocal inhibitory neural connections built into the circuitry that links their motorneurons, since each muscle can be fully activated independently of the others.

The inhibition is associated with the motor neural program that is sent to these motorneurons <u>when walking is called for</u> and not at other times (see Barnes et al., 1972).

In this respect arthropod research has been instrumental in leading us away from certain erroneous views of nervous system function that arose out of Sherrington's classical work on cat preparations. These issues are no longer subjects of controversy. Nervous systems develop in considerable measure according to genetically-determined rules that lead to their having not only some relatively fixed "wiring," but also a variety of built-in, i.e., endogenous, neural programs. These programs are able to generate complex movements without reference to detailed sensory input from proprioceptors (Wilson, 1961; Hoyle, 1964; Clarac and Coulmance, 1971; Pearson, 1972; Dorsett, Willows and Hoyle, 1973; Evoy and Fourtner, 1973). Several sets of programs share the same motorneurons. Exogenous reflex patterns are there as well. The difference between modern and classical views of behavior control is the recent realization that the reflex inputs can, at one extreme, be totally centrally inhibited during the play-back of endogenous programs, e.g., Barnes et al., (1972). Every degree of involvement from zero to 100% is possible. The intermediate effects are collectively referred to as "modulation" of the central program.

I have suggested elsewhere (Hoyle, 1976) that the sensory input is probably not only utilized for on-going adjustment of central programs to insure optimal performance in the way of compensations for load changes and sundry environmental shifts. It is probably also used to effect long-term changes in the motor programs that will compensate for changes associated with growth, aging and specific habitat variations. Also, during development, sensory input probably plays a major role in determining the score for some motor programs. However, these are all in the realm of speculations. In no case do we actually have sufficient evidence even to allow us to determine the relative importance of the various factors.

Now let us consider some specific aspects of locomotion control, beginning with "gait." We have already referred to Manton's suggestion of 'low', 'middle' and 'high' gear in regard to locomotion of multi-legged arthropods. Does this mean that there are separate motor programs for each speed range? As locomotory speed increases there is progressively less time for reflexively-generated sensory inputs to influence motor output and at high speeds it is certainly not possible for them to directly influence the next cycle. Perhaps sensory input is averaged over a few cycles and alters average motor output only after a delay of a few cycles? There are unfortunately only a few studies where walking legs have been examined from sufficiently high-speed movies providing sufficient accuracy to permit a thorough analysis. None of them has been on an animal with a large number of legs. They include a stick insect (Wendler, 1964, 1966), a cockroach (Delcomyn, 1971a, b), treadmill walking by a lobster (Macmillan, 1973, 1975), normal walking by a crab (Barnes, 1975) and crayfish (Barnes, In preparation). Delcomyn states emphatically that earlier, much-quoted studies by Hughes (1952, 1957) on the cockroach were made at camera speeds too slow for accurate measurements at the higher speeds of locomotion and the same criticism may apply to the analyses of various multi-legged arthropods made by Manton (1950, 1953, 1968). She claims discrete changes in P/R and also marked changes of phase, between successive legs at different speeds. That would suggest the possibility of different motor 'tapes' or 'scores'. Yet she stresses that there is merging rather than abrupt transition, between them. It is thus possible that specific gaits are more apparent than real, a consequence of body elongation, greatly decreased retraction times, and optical illusion.

Body undulation does not _facilitate_ fast walking: it happens, presumably, because of a historical relict of neural pattern generation. On the contrary, fast-moving multi-legged arthropods such as millipedes must use a variety of skeletal and muscular tricks to increase body rigidity at the same time speed increases. Since undulation also leads to head swaying, loss of visual input stability is associated with it.

The medium in, or on, which locomotion occurs is also of relevance. Insects that swim underwater or on the surface film, like those that jump, tend to use synchronous contralateral leg kicks in a sort of rowing motion. When the same insects are on land they adopt a traditional tripodal gait.

Such transitions are not necessarily only accidental. Water may provide a means of escape from a predator. A marching band of locusts may naturally encounter a stream whilst on the march and swim right across it (Kennedy, 1945). Their technique is to use repeated synchronous kicks with their hindlegs as if hopping or jumping. Of greater interest is the fact that the middle legs, which normally alternate strictly during walking, act synchronously also. Therefore swimming is a characteristic movement that is different from hopping, and which can be executed perfectly even though it has never previously been tried or practiced. A highly-specialized insect, that walks extremely rarely, and is hardly a prime candidate for swimming skills, is the praying mantis. In fact a praying mantis can swim superbly both on the surface and under water (Miller, 1972). The forelegs are held pointed frontwards and so do not take part in swimming. This is in contrast to walking in which they are often used alternately, along with the other two pairs of legs, in tripodal gait. Propulsion is made equally by the 2nd and 3rd pairs of legs each making synchronous, wide-angle, strokes. The 3rd pair leads, the phase difference increasing with increasing swimming speed. Accurate steering is achieved, the prothorax and extended prothorax being rotated as an anterior rudder. It is not the liquid medium as such, but contact with water on the tarsi that initiates this complex action pattern. Whilst this behavior may seem like a striking pre-adaptation of potential, if rarely required, survival value, it is possibly no more than a utilization of commonly made movement. Mantids commonly make a lunge with their whole body when capturing prey. They can also leap. These movements utilize synchronous backwards leg thrusts.

Let us now consider specific examples in greater detail.

INSECT WALKING

The smallest number of legs that any arthropod can use for a stable base is three, in the form of a tripod. It has long been believed that insects use an alternating triangle gait, with the first and third legs of one side extended at the same time as the middle leg of the opposite side. With three legs firmly planted, the corresponding legs of the opposite side are then lifted and moved forwards at the same time. They are progressively lowered after moving anteriorly and placed on the ground in an extreme anterior position. This mode of operation leads to progression because the body is carried forwards at the same time as the legs are held on the ground: these legs are turning posteriorly (retraction) and at the same time they are extending. An additional thrust may be produced by rotation of the whole, or part of the leg in the anticlockwise direction (pronation). (See Figures 9 and 10.)

The only way to find out if a given insect really does use this mode of progression is to film it with a high speed camera and analyze the film frame by frame. Both the shutter action and the frame cycle time must be fast relative to leg cycle time. For really satisfactory results a rotating-prism device such as the Hycam, operating at 500 frames per second, is required to resolve movement at the higher speeds of, say, a cockroach running. The range of speeds for cockroaches is almost a hundred-fold, from 1 to about 80 cm/s. at 29° C. To accomplish this the leg frequency has to increase from 1 Hz. to 23 Hz.

The movement patterns may be conveniently diagrammed by a simple method utilized by Donald Wilson (1966). Protraction is illustrated by a solid bar, and retraction by a broken line, or gap. The three legs of one side being represented above, and of the other side below them. Scanning from left to right shows the succession of steps, whilst glancing from top to bottom shows the location of all legs at any given instant. Changes in gait of the kind proposed by Manton, should they occur, would be apparent as abrupt discontinuities in the pattern of bars. If no such disruptions occur at points which are markedly transitional to the human eye, then we would know that it is a perceptual phenomenon. One minor criticism of Wilson's scheme for representing walking suffers from making the eye-catching, dark bands equal to protraction. Protraction merely serves to prepare a leg for the power stroke of retraction, except in backwards walking, which is quite rare, and normal operation of the front legs of decapods. Retraction times are the significant ones.

I, at least, intuitively grasp walking much better by reversing the emphasis (Figure 11). What is really most needed for understanding locomotion is an indication of when propulsive force (power) is being applied. However, for consideration of the generation of co-ordination, plots of power/recovery could only confuse the issue. Hughes' (1952, 1957) diagrams using linked numbers

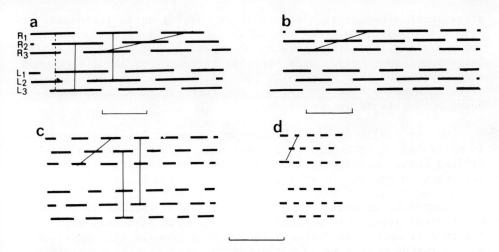

Figure 11. Stepping in the cockroach at different speeds. The traditional scheme, using a black bar to indicate protraction, has been reversed, to emphasize the power stroke and how its duration changes as speed increases. a. Slow speed (1.5 Hz.), bar = 0.5s. The solid vertical bars, linking the middle leg of one side with the fore and hind legs of the other, show that the basic pattern is alternating triangles, although at times all six legs are on the ground at once (broken line), and at other times four or five. b. Medium slow speed (4 Hz.), bar = 0.25s., shows shortening of protraction relative to retraction. There are still periods when all six legs are on the ground. c. Average running speed (9 Hz.), bar = 0.2s. Almost perfect alternating triangles. d. Fast running (22 Hz.), bar = 0.2s. The diagonal bars follow the progression of movements of legs on one side. R1 is prothoracic, R2 is mesothoracic and R3 is metathoracic. (Data from Delcomyn, 1971a.)

to indicate the order of lifting are easy to grasp, though less useful than Wilson's scheme since they give no indication as to P and R durations. Indeed, gaits that have identical number sequences can be totally different from each other. The P/R bar illustration immediately makes this apparent. When only two pairs of legs are used, they are especially confusing. In script I shall use a letter to indicate side and a number to indicate leg; order is in the sequence printed. Where legs are moved synchronously they are linked with a bar above. The tip of the abdomen is used by a few insects routinely as a support, and in others occasionally. A description of these should add something like +A (const) to indicate this aspect. At very low speeds of walking in the cockroach (Delcomyn, 1971a) there is a great deal of variation in P and R for each leg. At times, all six legs are on the ground at the same time (Figure 11). However, scanning reveals the predominant presence of alternating triangles. As walking speed increases from the slowest, to about 4 Hz., both P and R decrease at first and the alternating triangles become clearer, but there are still large variations in P. At a stepping frequency of about 8 Hz. and above, variance in P is greatly reduced. Thereafter, with increasing speed P does not reduce, the increase in walking speed being accomplished by reductions in R.

Manton (1953) found large changes in phase with walking speed in multi-legged arthropods. By contrast, in the cockroach Delcomyn (1971a) found that at all aspects except the slowest, the phase of one leg with respect to another remained constant, at 0.5 for succeeding legs. It follows that there is but a single gait for the cockroach, unless it is conceded that slow walking represents a different gait. The walking control program possesses some of the ancient ancestral features, being metachronal, with postero-anteriorwards moving waves. The dominant ganglion in determining the rhythm is the more posterior one, the metathoracic.

However, for each motorneuron there is a lot of reflex input. When the cockroach is walking slowly this input affects P and R and also phase between legs. As speed increases, this input must still be present but is apparently ignored; we may speculate that it is inhibited centrally by the neural program that is generating the fast walking.

Slow walking is as variable as it is, probably not because of any switch in the central programming, but just because central programming is "interfered with" by reflex actions to a much greater extent at slow walking speeds. There is good evidence that basic stepping in the cockroach is made by a motor "tape" i.e., independently of sensory input (Pearson, 1972).

Comparable analyses of walking in the stick insect Carausius morosus (Wendler, 1966; Graham, 1972) revealed some differences from the cockroach. At its higher walking speeds, the stick insect walks much more slowly than the cockroach (maximum speed being about 1/3 that of the cockroach running at top speed) and the same alternating triangles commonly occur. But at its slower speeds a bilateral asymmetry of the step pattern becomes apparent that represents a true change of gait. The right legs protract soon after the left on the same segment (phase 0.3). If one side is followed it is relatively constant. When the other is compared with it, it is also constant, but shifts smoothly with respect to the first, in a form of 'gliding co-ordination' (Wendler, 1966). This indicates that coupling between the two sides is very weak. The legs protract in diagonal pairs and retract in sets of four, so that at least four legs are always in contact with the ground. The stick insect is long, with an especially long abdomen and a propensity

for swaying from side to side. Thus its stability on a tripodal
base is weak, so the four-legged stance may offer considerable
advantages. Instantaneous velocity fluctuates within each cycle,
the insect progressing by a sequence of lunges.

Graham (1972) analyzed the walking not only in the adult
insect but also in a juvenile, 1st instar. He found two distinct
gaits in the juvenile, one of which, the faster one (Gait II) was
the familiar alternating tripod; it appeared over the whole speed
range, and was the only one at high speeds. At slow speeds a
different gait (Gait I) is always used. A transition between the
two gaits was sometimes present, but the juvenile insect more
often used the slow-speed gait at high speeds also. In juvenile
Gait II and adults, legs L2 and R3, L1 and R2, and L3 and R1, are
protracted together. Coupling between the two sides is weak. In
Gait I only two legs at a time are protracting, L3 and R2, L2 and
R1, or L1 and R3. Protraction times are short, retraction times
are long. During the transition between gaits I and II the principal change is a decrease in retraction time, but protraction times
actually increase. There is a strong tendency for simultaneous
protraction of the diagonal leg pairs, as noted by Bethe (1930), as
well as a phase of 0.35 for right on left, that increases to 0.5 as
speed increases. Small weights added to the insect increase the
step period. An interesting abnormality that occurred was a doubling of protraction, i.e., interposition of a very quick retraction during the protraction phase. The same thing happens whenever
a leg slips. Failure of protraction to occur at all, at the appropriate time, can also occur. An R3 R2 R1 / L3 L2 L1 rhythm occurs
in a number of insects either occasionally at normal temperatures,
or commonly, at low temperatures. Insects with a specialized pair
of legs, such as the first pair of the mantis, for prey capture,
or the hindlegs of grasshoppers for jumping, walking often utilizes
only the other four. The mantis then uses either L3 L2 / R3 R4 or
L3 L2 L2 R3. It also uses the standard tripod R3 L2 R1 / L3 R2 L1
and Roeder (1937) has found R3 L1 L3 R2 L2 R1. The jumping
Orthoptera do not use their specialized hind legs much during walking. When they do, the stop appropriately in the typical insect
tripodal mode. But they do not follow the usual protraction/retraction cycling. The hind legs have only a very limited ability
to protract because the coxa is directed to point backwards.
Walking movements are produced mainly by flexion and extension of
the tibia, and to a lesser extent by raising and lowering the
femur by rotation of the coxa. The latter is equivalent to normal
levation and depression, and extension and flexion of the tibia
have, in most modern arthropod walking legs, been added to the
retraction phase as major power-producing elements. What has
happened to the hind legs is that retraction and protraction have
been lost as elements in walking. We now know the location of the
relevant motorneurons, at least in the locust, and can observe them

electrically, with intracellular microelectrodes, during walking on a turntable. We also know the locations of the corresponding pro- and mesothoracic neurons. A cellular analysis of the differences between neural control of the stepping mechanism of front, middle and hind legs is now a feasible research project.

CRUSTACEAN WALKING

Crayfish. In spite of a hundred years of study there is still no general agreement about the walking patterns of crayfish. The most recent studies utilize high-speed films and computer analysis of phase data. A variety of nomenclatures has been utilized, with both letter symbols or numbers for the legs, and sequences starting with 2, or B, taking note of the probable homology of the claws with walking legs. We shall, however, exclude the chelae, and use 1 for the first pair of walking legs. A probable reason for the absence of a simple story is versatility in patterns related in part to environmental conditions, but difficulties in cinematographic analysis may be, in part, to blame. It was clearly claimed that crayfish use simultaneously linked pairs out of water (Voelkel, 1922). A bilateral symmetry may occur in the stepping order, but the crayfish can easily step out of a symmetrical mode, especially when it is walking in water and buoyant (Parrack, 1964). Parrack found all possible gaits at various times in walking out of water, but since he didn't allow for turning, etc., they may not all have been valid observations. The ability of the gaits to be shifted to permit turning is further evidence of the looseness of neural coupling. There are metachronous waves of motion and a strong tendency for diagonally opposite legs to work together. But whilst the first pair of legs always alternate the posterior pair often work together. He found 4 2 3 1 to be the commonest gait out of water. In water a sideways, crab-like motion often occurs. In forwards walking in water the dominant gait is 1 4 2 3 (Barnes, In preparation). The stepping pattern producing this dominant gait suggests metachronal waves passing down the body from the front. The basic sequence is 1 2 3 4; the gait is 1 4 2 3 because the waves overlap, i.e.

Walking Gait of Crayfish in Water (sequence reads in time from left to right)

ARTHROPOD WALKING

This is the same as that found by Parrack for walking out of water, allowing for overlap, except that Barnes points out that his findings suggest a metachronous wave that starts anteriorly rather than posteriorly.

Lobster. In the lobster Homarus americanus a systematic, modern study with computer analysis has been made by Macmillan (1975). There are six possible gaits depending on stepping order:

GAIT	ORDER (sequence)
1	4 1 2 3
2	4 3 1 2
3	4 1 3 2
4	4 2 1 3
5	4 2 3 1
6	4 3 2 1

The only alternatives are associated with one leg missing a cycle, or making two cycles in rapid succession. A large number of sequences obtained on a treadmill was considered, two legs at a time, and the number of occurrences of any given leg being followed by any other given leg scored. Probabilities of each leg following each other leg were then calculated. All six sequences were found, but the probabilities determined were much lower than actual occurrences. Transitions from one "gait" to another occurred smoothly, and were always of the simplest kind i.e., one leg at a time changed with respect to the preceding one, such as 4 2 3 1 to 4 3 2 1. The dominant gait observed by Macmillan was 5 (4 2 3 1):

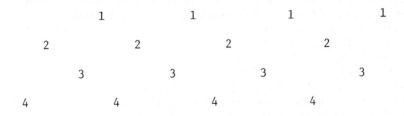

The net result is similar to crayfish walking (see previous page).

Macmillan's analysis has been criticised by Barnes (personal communication, in preparation) on the grounds that he only considered two legs at a time, and assumed that at any particular point in time the determination of a leg stepping depends only on which leg has just stepped. This method minimizes programmed stepping sequences and gives very low probabilities for different gaits because the whole past history of the stepping has not been taken into account.

Macmillan found that the dominant gait is equally dominant on the two sides, i.e., the walking is never asymmetrical, or crab-like, as occasionally happens in crayfish.

Crab. The fiddler crab Uca pugnax walking on land moves in the familiar sideways manner, with different dominant sequences on the two sides. On the leading side it is 1 3 2 4 and on the failing side 1 4 2 3 (Barnes, 1975). A metachronal coordination sometimes occurred in trailing legs at slow speeds of walking. Power stroke times decrease as speed increases, and return stroke also, though to a lesser extent but there is no change in sequence. At various times Barnes observed all possible available gaits, though rarely a departure from an alternating tetrapod, whatever the gait.

The basic mechanism of sideways walking in the crab on land is closely similar to that of a crayfish walking forwards in water, suggesting strongly that they are based on similar neural pattern generation inherited from a common ancestor. One major difference, however, is that the crab holds the trochantero-coxal joint steady, and uses the meropodite-carpopodite for propulsion, whereas the crayfish holds the latter steady and moves the former. Should we take this as evidence that the joints may not be assumed homologous?; or that a fixed neural program has been shifted from operating one set of joints to operating another?; or that similar neural programs have evolved independently, to operate different joints in different animals?

The ghost crab (Ocypode ceratophthalma) runs so fast at top speed (4 m/s - 10 mph) in relation to its ability to contract its muscles, that it must be leaving the ground between each propulsion thrust (Burrows and Hoyle, 1973a). Direct tests showed that this is indeed the case. At the start of a run the crabs use all eight walking legs. Then, as the sideways run accelerates, the crab lifts up the hind legs and runs with an alternating-tripod "gait", like a forwards-moving insect. Next, the first and third walking legs on the leading side become tonically raised and the second leg of the side is lowered, with its dactyl pointing under the body. The dactyl in this position serves as a skid. At the same time the body is tilted, with the leading edge upwards. Thus the crab is leaping, rather than running, with the trailing legs only

ARTHROPOD WALKING

providing the power, 1 and 3 alternating with 2, the most powerful. At highest speeds, 2 and 3 alone alternate - the crab has become a bipedal runner! If it overbalances the recurved dactyl of leading leg 2 skids along the ground briefly.

This dramatic running behavior requires no special neural program except tonic ones that result in tilting of the body and raising of the leading legs. The muscles of the leading legs are all still receiving alternating burst patterns of excitation, but in the absence of any great load are in a state of fused tetanus. If the crab tips over lightly the skid action is seen, but if it tips strongly the added force cuts down the flexion, the leading legs instantly become the driving side, a reverse tilt occurs, and the crab runs off at equal speed in the opposite direction.

Thus no change is needed in the neural programs controlling stepping, or in ipsilateral and contralateral alternation.

Spider. A modern analysis is available for a large tarantula spider (Wilson, 1967). He found all possible stepping orders, with 4 2 3 1 predominating, 4 1 3 2 and 4 3 2 1 common. The legs are often stepping at the same frequency, leading to phase drifts between pairs of legs during steady locomotion. Allowing for there being four pairs of legs instead of three, Wilson was impressed by how basically similar it is to insect walking. Legs on opposite sides tend to alternate, and also adjacent legs, resulting in the simple pattern, with a tendency for the legs linked by lines to act together.

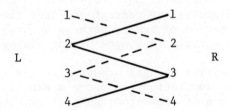

Compression of an overlapping series of waves starting at the posterior and moving forwards gives rise to the 4 2 3 1 as follows:

```
1st wave      4    3    2    1
2nd wave           4    3    2    1
3rd wave                     4    3    2    1
              ─────────────────────────────────
              4 2  3 1  4 2  3 1
```

AMPUTATION

Loss of one or more legs by arthropods is a common natural accident from a variety of causes: molting problems, getting stuck, fighting, especially among crustaceans and during escape from predators. It has been used by many investigators, starting with Bethe (1930), attempting to understand the neural control of locomotion. All have been astonished by one aspect of their results: the rapidity with which adjustments occur and an alternative gait established that permits reasonably effective walking. Even a quick amputation in mid-stride hardly seems to daunt a walking cockroach. It is as if the nervous system had adaptive alternative control programs built into its neural organization that can take care of any possible emergency. But this is obviously a nonsense possibility. What these experiences suggest is that the altered sensory input is the direct cause of the changes and therefore, that sensory input is a major factor in the neural program that generates normal stepping. But that argument falls down because inter-limb co-ordination is not at all affected by altered sensory input in other instances. In some experiments, hours or days may pass before a suitable new walking pattern is developed. The latter show an element of learning to adjust to the new input. Perhaps we are looking at a range of learning speeds, from 1-trial "instant" learning, to multiple-trial, slow learning. The learning hypothesis proposes that the altered sensory input causes a stable change in the functioning of the central pattern generator. Since we know that 1-trial learning to hold a leg in an abnormal posture can occur (Tosney and Hoyle, In preparation) this possibility needs to be given serious consideration. Perhaps this was what Bethe had in mind when he introduced the concept of Plastizitat. This concept has obviously not found favor, perhaps because it was not explained in terms acceptable to a modern neurobiologist.

Interesting adaptive changes occurred in the <u>fiddler crab</u> <u>Uca pugnax</u> following amputations in Barne's experiments (Barnes, 1975). Removal of one of the second pair of walking legs was followed by a switch from synchrony to alternation, of walking legs 1 and 3 on contralateral side. Even more interesting was the frequent use of the claws in walking after the amputation. The claws are not uncommonly used in whole crayfish and lobsters, but in ghost and fiddler crabs they are never used in intact animals, so no experience by the animal in using them can have been gained. How can we resolve these paradoxes? It is abundantly clear that not a single one of the numerous attempts made to answer such questions solved the problem even for a particular animal. Nor did any of them do what all were supposed to: further our understanding of the general questions of neural integration and the generation of motor output programs. We hope that some direct answers will be obtained now that we have entered a new era, the analysis

ARTHROPOD WALKING

of underlying neuronal events by direct analysis of identified neuron activity during walking.

Wilson (1967) removed a first leg and a contralateral hind leg from the Tarantula spider. If it followed its basic rule for all 8 legs present only two legs would support weight part of the time and so the pattern, though nice for co-ordination of stepping would not be stable. Instead, they frequently use a diagonal rhythm of alternating tripods, that should be compared with the normal one,

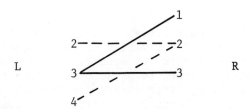

which is quite stable. It resembles insect walking, but it must be more difficult to produce at the neural programming level. It requires the second and third pair of legs to step in synchrony. Thus in walking the spider with these two legs missing flips between alternation and synchrony for two pairs of legs so intrinsic couplings and extrinsic proprioceptive reflex controls must both be weak, perhaps dominated by different neural pattern generators. Here the presumed ancestral metachronism seems completely absent.

FUNDAMENTAL TRANSITIONS IN THE USE OF LOCOMOTORY MUSCLES

We are fortunate in having a wealth of data concerning the switching of functional roles in muscles that serve at least one function in addition to locomotion. In insects there is the conspicuous switch from utilization in walking to serving as flight muscles, by the anterior and posterior tergo-coxal muscles and by the subalar and 2nd basalar. The anterior and posterior tergo-coxals serve synergistically as elevators in flight, but are antagonists serving protraction and retraction, respectively, for leg movements. The 2nd basalar is a protractor and the subalar a retractor, but both depress the wing. They exert additional effects independently on the wing though, the former pronating and the latter supinating.

The last pair of legs of some swimming crabs are used entirely independently in gentle and medium-speed swimming, usually alternating (Hoyle and Burrows, 1973c). But in escape swimming they are used synchronously. At the same time as they are applying power the paddles are also twisted to alter the direction of swimming. In adult males of Portunus sanguinolentus the paddles are held aloft, for up to a few days, without moving, as a sexual signal (Ryan, 1966). The hind legs of many crab species are held indefinitely in the raised position, sometimes clasping a piece of sponge for camouflage, or a sea anemone for protection, above the back.

Grasshoppers and locusts can switch abruptly from alternation of extension of the hind legs, for walking, to synchrony for jumping.

Moths raise the temperatures of their abdomens prior to flight by an asynchronous form of activation of the muscles that differs from the orderly co-excitation of synergists, reciprocally with that of antagonists, resorted to as soon as flight commences. A katydid in order to sing, first raises its thoracic temperature. This it does by co-contracting antagonists. The same muscles abruptly switch to a strict alternation, for sound production (Josephson et al., 1975).

Sound production in insects, except where executed by vibrating a tympanum as in cicadas, or clicking a specially-sclerotized joint as in some beetles, is achieved by rubbing legs against the body, each other, or wings, or by rubbing wings together. In all cases the muscles used are also ones used in walking though the rhythms are much faster than those used in walking. Rates may exceed 200 Hz in singing, compared with about a 10 Hz maximum in walking.

Extraordinary transitions occur between larval and adult stages of many arthropods. Consider, for example, that which occurs in barnacles, from a free-swimming larva to a sessile form in which what might have become legs walking with a complex gait actually start to serve as rhythmically-acting synchronous food gatherers. What do such different patterns from the primeval locomotory one require in the way of neural system changes? Fortunately, the cellular neural mechanisms underlying barnacle feeding are amenable to modern, direct, analysis (Gwilliam, 1973), so we may eventually be able to give a satisfactory answer to this question.

Interesting questions are raised by the ability of many arthropods to walk quite well backwards as well as forwards, and sideways also in some. Some muscles that are synergistic for forwards walking are antagonistic for backwards walking and similarly, muscles that are synergistic for lateral walking in one direction are antagonistic for lateral walking in the opposite direction (Ayers, In preparation). Bowerman and Larimer (1974) have recently shown that there are interneurons in the crayfish anterior connectives that drive co-ordinated backwards walking. To me this strongly suggests that each of the four directions of movement has a different neural program generator.

The mouthparts, antennae, antennules, etc., are probably modified elements of the original chain of locomotory appendages, so it will be interesting, some day, to compare their neural control with that shown by the waling legs. In Limulus, the walking legs participate directly in feeding. At the base of each coxa is a spine, directed inwards, that shreds food and also pushes it into the mouth, which is situated at the base of the legs. To do this, opposite legs work synchronously, but out of phase with the adjacent legs (Manton, 1964; Wyse and Dwyer, 1973). Two muscles act radially to the arc of adduction-abduction and displace the pivot of that arc dorsolaterally. When these muscles lag behind adduction ingestive chewing results, but when they are in advance egestive chewing occurs. Here, then, a major behavioral alteration has been shown to result from simple phase-shifting of the action of two muscles (Wyse and Dwyer, 1973).

The basic program of chewing is a novel one that has been added to the locomotion repertoire.

The Muscle Fiber Contents of Arthropod Muscles in Relation to the Evolution of Neural Control Mechanisms

Until the late 50's it had been taken for granted that a muscle was a homogeneous population of similar fibers and that a few random samples of neuromuscular physiology would serve to delineate mechanisms applicable across class lines. Anomalies experienced in a comparative survey of crustacean neuromuscular transmission by Hoyle and Wiersma (1958) sounded the alarm for this naive approach and led to studies in which selected muscles were intensively examined, one muscle fiber at a time (see Hoyle, 1967; Atwood, 1972). Light microscopical, electron microscopical and enzyme studies were added. Only very recently have comparable approaches been used on insect muscles, with their smaller, tightly-bound fibers (Hoyle, In preparation).

What has transpired is the realization that arthropod muscles are highly complex mixtures of fibers having very different individual physiological properties, associated with equally diverse ultrastructural and biochemical features. No rules have yet been established to guide us in interpreting the relation between the particular mix of fiber features and the functional roles of specialised muscles and indeed, there may not be any. A common difference between fibers is in the lengths of their sarcomeres, which is usually short in fast muscle fibers and long in slow ones.

But for every rule we find an exception. We know, for example, that the power-stroke muscle of the swimming leg of Portunus sanguinolentus shows as wide a range of speeds and tonic properties as any known arthropod muscle (Hoyle and Burrows, 1973a). They comprise distinct slow, fast and intermediate muscle fibers. Yet the sarcomere lengths of the fibers are substantially similar (Hoyle, 1973). The common structural route used by most known crustacean muscles to achieve speed differentiation was not used by this muscle. An evolutionary line may at one time have required a given muscle to be very fast and to be operated phasically, but at another time to serve a slow, tonic function. At one time a large, but imprecise length change may have been called for, at another a short, but highly accurate, one. A species must use whatever mutated variants are available. Since the same goal can be achieved by different routes, or combinations of routes, we can expect a lot of diversity. The list of basic elements on which variations in muscle fiber type occur includes: sarcomere length, T(E)-system, nature of sarcoplasmic reticulum (SR), extent of SR, nature of T(E) system/SR contacts, I/A filament ratio, A filament thickness, M-band, Z-band, mitochondrial type, mitochondrial number, variety and proportions of enzymes, etc. (see Hoyle, 1967; Atwood, 1972; Huddart, 1975). The number of possible permutations and combinations of these parameters is so large that we may ask if any two muscles need ever be truly alike. "No", looks like an appropriate answer - as our sense of taste confirms when we eat meat from different animals or parts of the same animal.

Yet as an investigator becomes familiar with a given muscle, he comes to know precisely where to look in it for a particular type of fiber. Thus in any locust or grasshopper extensor tibiae, to find a fast muscle fiber he goes to the outside margin 1/3 from the proximal articulation. To find a slow one he looks in the most distal or the most proximal bundles. And to find an intermediate he goes to the inner margin about 2/3 from the proximal articulation. There is therefore genetic control over the locations that crosses family lines. The same is true for fibers in the power-stroke swimming muscle of the Portunid crabs.

Whilst no two muscles of one animal can be considered to be alike, each muscle, however complex in composition, is like the same muscle taken either from different specimens, or ones that are within the same family.

Crustacean muscle fibers have a sarcomere length range from below 1μ to about 20μ (Hoyle, 1967; Hoyle, McNeill and Selverston, 1973). Their T(E) system ranges from a dense feltwork to nil. Thin to thick filament ratios are from fewer than 2:1 to more than 7:1. Some fiber types, such as the giant barnacle retractors have no M-bands, others have a single M-band, but double M-bands are common.

Insect muscle fibers have a narrower range than crustacean, with sarcomere lengths from about 2μ - 10μ. They have single M-bands, thin: thick filament ratios from 3:1 to 5:1 and a wide range of T(E) systems. Both insect and crustacean muscle fibers have a wide range of mitochondrial contents, from almost none to more than 60% of fiber volume. Crustaceans have not evolved the giant mitochondria seen in some insect flight muscles.

All this diversity is staggering. Whilst much of it is clearly adaptive, there are many instances in which a clear relation to function is not apparent. What is clear is that diversity at this level is much greater than in innervation patterns and neuromuscular transmission or in central patterns. The implications will be considered in the discussion.

Innervation Patterns of Arthropod Muscles

Another feature that we may expect to have been conservative is the pattern of innervation. Arthropods utilize very few neurons to innervate a muscle; functionally important muscles receiving a single excitor, like the insect adductor coxae are common. Equally common are muscles receiving three axons: a single "slow" excitor, a single "fast" excitor and an inhibitor. This is the pattern of all known insect extensor tibiae. Curiously, it is also the pattern for the functionally equivalent crustacean extensors - of the carpopodite. What is even more interesting is that all known insect flexors of the tibia have more than two excitors, probably seven in all. The corresponding crustacean flexors also have more than two, commonly four.

The pattern of innervation of muscle fibers of a cockroach extensor tibia is of triple innervation proximally, and also in the extreme distal portion. Fast only innervated fibers are found in the central region. Slow and inhibitor only innervated fibers are found especially distally with some proximal ones also (Atwood et al., 1969; Hoyle, 1974). Now these are precisely the same locations for similar innervation patterns in the locust extensor (Cochrane et al., 1972; Hoyle, In preparation). Cockroaches and locusts are very different anatomically and behaviorally, and they have evolved independently at least since the Jurassic, but the detailed pattern of innervation of some of their muscles appears to have been set in some common orthopteran stock just as have the common anatomical features that resulted in their being classified in the same Order. It is now becoming apparent that there are also close similarities in one important neural control parameter - the motor output patterns - to functionally analagous muscles. In extensive recordings from several insects (<u>Locusta</u>, <u>Romalea</u>, <u>Periplaneta</u>, <u>Schistocerca</u>) and crustaceans (<u>Cancer</u>, <u>Ocypode</u>, <u>Cambarus</u>,

Portunus) I have found (Hoyle, Unpublished) that the detailed patterns of motor impulses in slow and fast extensor axons during walking show close similarities. In particular, the modes of utilization of frequency variation in the slow axon and the way the fast axon comes in to supplement the slow, are remarkably similar in detail. The axons use the same transmitter substances, 1. glutamic acid for excitation, and γ-amino butyric acid for inhibition.

Conclusions

It will be evident that we have a paradox. On the one hand we have evidence for stereotyped, inflexible programs, and on the other hand are contrasted remarkable, instantaneous plasticity. The two sometimes apply equally to the same limb of the same animal. Examples of strong central pattern-generating mechanisms programming behavior independently of sensory input, conflict with experiments that demonstrate instantaneous modification following altered sensory input. Antagonists at times co-contract strongly, but at others are obligatorily reciprocal.

The resolution of this conflict lies in the basic truth that all are true – at certain times. It might be that these properties would all reside in different integrating and programming circuits and that motorneurons are simply followers of instructions. Yet detailed motorneuron studies are now available for the locust, and they show that a great deal of integration occurs at the level of the motorneuron integrating segments themselves (Hoyle and Burrows, 1973a, b). The behavioral work, discussed in the foregoing account, showed that the connections between neurons cannot be physiologically inflexible. For many interneurons and motorneurons, stimulation of many sensory modalities in all parts of the animal leads to added synaptic input and modulated discharge in a preparation. If the animal were to take stock of every input it would need a sizeable computing system for each motorneuron. The motorneurons supplying each leg can, when properly excited, generate oscillations at periods within the walking rate range that do not require specific, timed proprioceptive feedback. Independently, proprioceptive reflexes are able to generate comparable rhythmicity. The dependence of any given leg upon another can vary from zero coupling to any other leg, to strong coupling to one other leg, or coupling to more than one, even all, other legs. There are times when proprioceptive reflex actions may simply augment the motor program (Delcomyn, 1971b). At other times they are opposed.

The ability of the legs to operate under such conflicting controls is explained by the omnipresence of inhibition and, therefore, of gating. It is not surprising that in the only bit of nervous system where a thorough study of the nature of connections between all component neurons has been made, via the stomatogastric ganglion of the lobster (Mulloney and Selverston, 1974; Selverston and Mulloney, 1974), by far the largest proportion are inhibitory. The moment a central program is "turned on", active inhibitions occur at the same time that completely suppress many inputs and partially block others. A particularly strong reflex input still makes its presence felt, and modulates the output. It may do this directly at the motorneuron level or indirectly via the central pattern generator, though this will require to be done over a few cycles when the actions are very fast. Several possible mechanisms can be invoked to explain synergism flipping to antagonism. For example, a large phase shift may occur in related, paired, oscillatory mechanisms antecedent to the outputs. Or, a delay stage may be introduced for one channel. A simple way to model the phenomenon is to have a pair of neuromimes, with fatigue properties, that are excited by a common input. They produce synergistic bursts of firing. Now turn on reciprocal inhibitory links between the two synergists and they alternate (Dagan et al., 1975). This is precisely the kind of switching that a nervous system seems designed to accomplish easily.

Unfortunately, little is yet known about how the neurons generating the step cycle of a single leg are positively linked to others. From the variety of co-ordinations observed, it is evident that excitatory links exist between any given leg and all others. These links appear strong at times, especially between legs that tend to step together. Yet direct tests of such links always give negative results. This could be due to the relative dominance of inhibition. But it could also be because these links occur only during general excitatory driving (="command neuron" activation).

But it is unfortunately true that we know very little about the mechanisms underlying the generation of a step beyond the fact that steps can be entirely centrally-generated but by mechanisms that are greatly modulated by reflexive sensory inputs. Pearson and Fourtner (1975) have recently brought us closer to understanding how a cockroach leg step is brought about. They have recorded intracellularly from a small group of interneurons showing oscillatory membrane potential changes during the making of steps, that may themselves be driving the coxal levator motorneurons. Such neurons would be part of the pattern-generating machinery of the individual leg, that has been rather loosely referred to as the leg "oscillator".

We may ask the important question: "Is it possible, that neural programs generating motor patterns have remained relatively fixed and that it is peripheral (ultimately tension) responses to the fixed outputs that have varied to meet adaptive needs?" An affirmative answer to the question is extremely likely, in my opinion. Why, otherwise, would there be such diversity in the muscles? Most functional variations in muscular contractions that are needed in adaptive radiation could, in principle, be taken care of by having a single broad-spectrum muscle type (say 1/3 fast fibers, 1/3 slow fibers, 1/3 intermediate fibers) with a triple innervation (slow axon, fast axon, inhibitory axon), and then using different motor output programs. However, we are already sure that during the course of evolution this has not been done. The major variations are in muscle fiber types, of which already a large number are known from the few muscles that have been seriously studied on a fiber-by-fiber basis.

During the course of their evolution the arthropods need not have done more than add some new central pattern generators, whilst keeping the old ones. From the start of neural evolution, neural program-generating mechanisms (NPG's) may have been kept separate from reflex ones. There are probably inputs to the NPG's from second or third order integrating interneurons on which sensory inputs converge. Their principal role may not be a direct one, but rather a modification of the NPG over many cycles, i.e., by a form of learning.

All that these complications require is a complex of multiple, independent, excitatory inputs to effector neurons and an associated set of inhibitory connections that can turn each of them off. The latter could be presynaptic, or postsynaptic and close to the excitatory terminals. The excitatory inputs must be sufficiently strong to excite the output element at least when the background level of excitation is high, i.e., they are to be thought of as switches, with a master switch (level of general excitation) that determines whether any of them will be functionally effective or not. The arthropod motorneuron is singularly well-suited anatomically to such a role. Firstly, the cell body is not a site for synaptic excitation and inhibition. These sites are all located on neuropilar branches from the integrating segment. There is a large number of such branches. About 20 major branches, approximately equal anatomically, are the general rule (Selverston and Remler, 1972; Burrows, 1973; Burrows and Hoyle, 1973b). The branches may well be able to conduct impulses as far as the integrating segment. An impulse initiation zone is located close to the point where the single motor axon leaves the neuropile. Thus the branches might be equivalent to each other and each might serve as input channel and switch for some particular behavior. The possibility to test this notion experimentally is almost within our grasp.

All workers on the neurobiology of arthropod locomotion are agreed that the simplest fundamental intrinsic neural program is one that does not require specific sensory input, i.e., that which causes P/R cycles in a single leg. Some refer to this as a leg "oscillator". This was probably historically the second to evolve, the first being the one for causing side-to-side bending which starts posteriorly and moves anteriorwards. As we have mentioned, hints that the latter is still around are to be seen during walking in all arthropods. Anterio-posterior wave generation has not been shown definitively to have an endogenous base though there is evidence against it being a chain-reflex behavior and, allowing for variable central delays, a central program is possible. We are on the threshold of understanding how these ancient neural programs operate at the cellular level, and of having some clues as to how they have evolved.

Leg co-ordination in walking can only be understood on the basis of there being a combination of reflex actions with one of several different central programs. The appropriate program must be selected on the basis of a multitude of factors, involving many environmental aspects, from visual input to pheromone concentration. The multitude of variations in gait and sequence represent the extent to which an endogenous program and reflex actions are interacting. Intermember co-ordination, as all investigators have found, is also based upon an interaction between endogenous programs and reflexes. The neural mechanisms for interaction between legs are weak, and readily modulated, by an endogenous program under many circumstances, or by current sensory input.

What chances do we have for understanding these subtle interactions in a precise manner? It has to be admitted that the vast amount of attention previously paid to the experimental analysis of arthropod walking has not led to an understanding of how behavior is programmed. This can be explained quite simply. The questions at issue are neurophysiological ones and they cannot be resolved by deductive reasoning. A direct attack is essential, so the future lies squarely with the neuroethologist and it looks very promising indeed. Pearson and Fourtner's (1975) remarkable success in finding pre-motor interneurons associated with walking in the cockroach brings us into the realm where a direct neurophysiological analysis of both an NPG and of a switching mechanism, have become possible. But such studies will be technically very difficult, take both a long time and superhuman patience, and rely to a large extent upon luck, as well as investigator skill. Nevertheless, that is the area where arthropod locomotion studies must now head. A necessary preliminary is a thorough knowledge of the basic locomotion of the test object, but neurophysiological feasibility should hereafter determine the choice of subject matter. In the long run we shall then learn not only about locomotion, but neurobiology and evolution as well.

We should at all times take into account the plasticity of arthropod neurons. There is a strong tendency for investigators to ignore this aspect and search for fixed neural properties and connections. Yet we know that the mean firing frequency of tonically-firing insect motorneurons can be stably reprogrammed in either the up or down directions (Tosney and Hoyle, 1973). The learning process can take place, minimally, after a single trial. Furthermore, the learning can be quickly reversed. Some of the changes in gait seen with change in speed might be caused by temporary, but learned, changes in neuron properties occurring in response to specific sensory inputs at critical speeds, that are automatically reversed, also by learning, when slowing occurs.

Acknowledgment

I would like to thank Drs. W.J.P. Barnes, Glasgow; F. Delcomyn, Urbana; and D.L. Macmillan, Melbourne, for their critical comments on the manuscript.

REFERENCES

Atwood, H.L., Smythe, T., Jr. and Johnston, H.S., (1969) Neuromuscular synapses in the cockroach extensor tibiae muscle. J. Insect Physiol. 15, 529-535.

Atwood, H.L., (1972) "Crustacean Muscle," In The Structure and Function of Muscle. Vol. 1, 2nd ed., (Bourne, G.A., ed.), Academic Press, N.Y., (422-489).

Barnes, W.J.P., Spirito, C.P., Evoy, N.H., (1972) Nervous control of walking in the crab, Cardisoma guanhumi. II. Role of resistance refexes in walking. Z. vergl. Physiol. 76, 16-31.

Barnes, W.J.P., (1975) Leg coordination during walking in the crab, Uca pugnax. J. Comp. Physiol. 96, 237-256.

Beklemischev, W.N., (1969) Principles of Comparative Anatomy of Invertebrates. Volume 1, Promorphology, University of Chicago Press, Chicago.

Bethe, A., (1930) Studien über die Plastizität des Nervensystems. I. Mitteilung. Arachnoideen und Crustaceen. Pflüg. Arch. ges. Physiol. 224, 793-820.

Bowerman, R.F. and Larimer, J.L., (1974) Command fibres in the circumoesophageal connectives of crayfish. II. Plastic fibres. J. Exp. Biol. 60, 119-134.

Burrows, M., (1973) Physiological and morphological properties of the metathoracic common inhibitory neuron of the locust. J. Comp. Physiol. 82, 59-78.

Burrows, M. and Hoyle, G., (1973a) The mechanism of rapid running in the ghost crab, Ocypode ceretophthalma. J. Exp. Biol. 58, 327-349.

Burrows, M. and Hoyle, G., (1973b) Neural mechanisms underlying behavior in the locust Schistocerca gregaria. III. Topography of the limb motorneurons in the metathoracic ganglion. J. Neurobiol. 4, 167-186.

Clarac, F. and Coulmance, M., (1971) La march laterale du crabe (Carcinus) coordination des mouvements articulaires et regulation proprioceptive. Z. vergl. Physiol. 73, 408-438.

Cochrane, D.G., Elder, H.Y. and Usherwood, R.N.P., (1972) Physiology and ultrastructure of phasic and tonic skeletal muscle fibres in the locust Schistocerca gregaria. J. Cell Sc. 10, 419-441.

Dagan, D., Vernon, L. and Hoyle, G., (1975) Neuromimes: Self-exciting alternate firing pattern models. Science. 188, 1035.

Delcomyn, F., (1971a) The locomotion of the cockroach Periplaneta americana. J. Exp. Biol. 54, 443-452.

Delcomyn, F., (1971b) The effect of limb amputation on locomotion in the cockroach, Periplaneta americana. J. Exp. Biol. 54, 453-469.

Dorsett, D.A., Willows, A.O.D. and Hoyle, G., (1973) The neuronal basis of behavior in Tritonia. IV. The central origin of a fixed action pattern demonstrated in the isolated brain. J. Neurobiol. 4, 287-300.

Evoy, W.H. and Cohen, M.J., (1971) Central and peripheral control of arthropod movements. Adv. Comp. Physiol. and Biochem. 4, 225-266.

Evoy, W.H. and Fourtner, C.R., (1973) Nervous control of walking in the crab, Cardiosoma guanhumi. IV. Effects of myochordotonal organ ablation. J. comp. Physiol. 83, 319-329.

Graham, D., (1972) A behavioural analysis of the temporal organization of walking movements with the first instar and adult stick insect Carausius morosus. J. comp. Physiol. 81, 23-52.

Gray, J., (1968) Animal Locomotion. W.W. Norton and Co., New York.

Gwilliam, F.W., (1973) Reciprocal bursting patterns in barnacle central neurons. American Zoologist. 13, 1303.

Hinde, R.A., (1970) Animal Behavior. A synthesis of ethology and comparative physiology. 2nd ed., McGraw Hill, New York.

Hoyle, G. and Wiersma, C.A.G., (1958) Coupling of membrane potential to contraction in crustacean muscles. J. Physiol. 143, 441-453.

Hoyle, G., (1964) "Exploration of Neuronal Mechanisms Underlying Insect Behavior," In Neural Theory and Modeling, (Reiss, R.F., ed.), Stanford University Press, Stanford, (346-376).

Hoyle, G., (1967) Diversity of striated muscle. Am. Zool. 7, 435-449.

Hoyle, G., (1973) Correlated physiological and ultrastructural studies in specialized muscles. 3B. Fine structure of the power-stroke muscle of the swimming leg of Portunus sanguinolentus. J. Exp. Zool. 185, 97-110.

Hoyle, G. and Burrows, M., (1973a) Neural mechanisms underlying behavior in the locust Schistocerca gregaria. I. Physiology of identified motorneurons in the metathoracic ganglion. J. Neurobiol. 4, 3-41.

Hoyle, G. and Burrows, M., (1973b) Neural mechanisms underlying behavior in the locust Schistocerca gregaria. II. Integrative activity in metathoracic neurons. J. Neurobiol. 4, 43-67.

Hoyle, G. and Burrows, M., (1973c) Correlated physiological and ultrastructural studies on specialized muscles. IIIa. Neuromuscular physiology of the swimming leg of Portunus sanguinolentus. J. Exp. Zool. 185, 83-96.

Hoyle, G., McNeill, P.A. and Selverston, A.I., (1973) Ultrastructure of barnacle giant muscle fibers. J. Cell. Biol. 56, 74-91.

Hoyle, G., (1974) "Neural Control of Skeletal Muscle," In The Physiology of Insecta, Vol. IV, 2nd ed., (Rockstein, M., ed.), Academic Press, New York, (175-236).

Hoyle, G., (1975a) "Neural mechanisms underlying behavior of invertebrates," In Handbook of Psychobiology, (Blakemore, R. and Gazzaniga, M.S., eds.), Academic Press, Inc. N.Y., (3-48).

Hoyle, G., (1975b) Identified neurons and the future of neuroethology. J. Exp. Zool. 194, 51-73.

Hoyle, G., (1976) "Approaches to understanding the neurophysiological bases of behavior," In Simpler Networks: An approach to patterned behavior and its foundations. (In press).

Huddart, H., (1975) The Comparative Structure and Function of Muscle. Pergamon Press, Oxford.

Hughes, G.M., (1952) The co-ordination of insect movements. I. The walking movements of insects. J. Exp. Biol. 29, 267-284.

Hughes, G.M., (1957) The co-ordination of insect movements. II. The effect of limb amputation and the cutting of commissures in the cockroach (Blatta onentalis). J. Exp. Biol. 34, 306-333.

Josephson, R.K., Stokes, D.R. and Chen, V., (1975) The neural control of contraction in a fast insect muscle. J. Exp. Zool. 193, 281-300.

Kandel, E.R. and Kupferman, I., (1970) The functional organization of invertebrate ganglia. Ann. Rev. Physiol. 32, 193-258.

Kennedy, J.S., (1945) Observations on the mass migration of desert locust hoppers. Trans. R. ent. Soc. (London) 95, 247-262.

Lakester, E.R., (1904) The structure and classification of the Arthropods. Quart. J. Micr. Sc. 47, 523.

Macmillan, D.L., (1973) A physiological analysis of walking in the American Lobster (Homarus americanus). (Ph.D. Thesis), University of Oregon.

Macmillan, D.L., (1975) A physiological analysis of walking in the American Lobster (Homarus americanus). Phil. Trans. R. Soc. of London. 270, 1-59.

Manning, A., (1972) An Introduction to Animal Behavior. 2nd ed., Edward Arnold Ltd., London.

Manton, S.M., (1950) The evolution of arthropodan locomotory mechanics. Part I. The locomotion of Peripatus. J. Linn. Soc. (Zool). 41, 529-570.

Manton, S.M., (1953) Locomotory habits and the evolution of the larger arthropodan groups. Symp. Soc. Exp. Biol. 7, 339-376.

Manton, S.M., (1964) Mandibular mechanisms and evolution of arthropods. Phil. Trans. (B) 247, 1-183.

Manton, S.M., (1968) "Terrestrial Arthropods," In Animal Locomotion. (Gray, J., ed.), Norton, New York, (303-376).

Miller, P.L., (1972) Swimming in mantids. J. Entomol. 46, 91-97.

Mulloney, B. and Selverston, A.I., (1974) Organization of the stomatogastric ganglion of the spiny lobster. I. Neurons driving the lateral teeth and III. Coordination of the two subsets of the gastric system. J. Comp. Physiol. 91, 1-32 and 53-78 respectively.

Selverston, A.I. and Mulloney, B., (1974) Organization of the stomatogastric ganglion of the spiny lobster. II. Neurons driving the medial tooth. J. Comp. Physiol. 91, 33-51.

Selverston, A.I. and Remler, M.P., (1972) Neural geometry and activation of crayfish fast flexor motorneurons. J. Neurophysiol. 35, 797-814.

Sharov, A.G., (1966) Basic Arthropodan Stock. Pergamon Press, Oxford, England.

Snodgrass, R.E., (1950) Comparative studies on the jaws of mandibulate Arthropods. Smithson. Misc. Coll. 110, 1-93.

Thorpe, W.H., (1963) Learning and Instinct in Animals, 2nd ed., Methuen, London.

Tiegs, O.W. and Manton, S.M., (1958). The evolution of the Arthropoda. Biol. Rev. 33, 255-337.

Tinbergern, N., (1951) The Study of Instinct. Oxford University Press, London.

Tosney, T. and Hoyle, G., (1973) Automatic entrainment for a cellular learning study. Society for Neuroscience 1973 Annual Meeting. San Diego, California.

Voelkel, H., (1922) Die Fortbewegungsarten des Flusskrebses. Arch. ges. Physiol. 194, 224-229.

Weiss, P., (1950) Experimental analysis of co-ordination by the disarrangement of central-peripheral arrangements. Symp. Soc. Exp. Biol. 4, 92-111.

Wells, G.P., (1950) The anatomy of the body wall and appendages in Arenicola marina L., Arenicola claparedii Levinsen and Arenicola ecaudata Johnston. J. Mar. Biol. Assn. Plymouth. 29, 1-44.

Wendler, G., (1964) Laufen und Stehen der Stabheuschrecke
   Caransius morosus: Sinnesborstenfilder in den Beingelenken
   als Glieder von Regelkreisen. Z. vergl. Physiol. 48, 198-250.

Wendler, G., (1966) The co-ordination of walking movements in
   Arthropods. Symp. Soc. Exp. Biol. 20, 229-250.

Wilson, D.M., (1961) The central nervous control of flight in a
   locust. J. Exp. Biol. 38, 471-490.

Wilson, D.M., (1966) Insect walking. Ann. Rev. Entomol. 11,
   103-122.

Wilson, D.M., (1967) Stepping patterns in tarantula spiders. J.
   Exp. Biol. 47, 133-151.

Wyse, G.A. and Dwyer, N.K., (1973) The neuromuscular basis of
   coxal feeding and locomotory movements in Limulus Biol. Bull.
   144, 567-579.

ON THE GENERATION AND PERFORMANCE OF SWIMMING IN FISH

S. Grillner[x] and S. Kashin[xx] [o]

For propulsion most fish mainly use undulatory body movements. These movements are due to alternating muscle contractions on the two sides of the body; the rostral segments contract earlier than the caudal ones. The amplitude of these movements vary along the body and differ markedly between species, but are tightly related to the properties of the body with respect to its form and its rigid and elastic properties.

To understand which nervous commands are given to the muscles during swimming, EMG activity from several points along the body has been related to the actual movements recorded on 16 mm film, while the fish has been swimming at different constant velocities in a "swim-mill" (0-3.3 m/s). Such data obtained from different types of fish will be considered in relation to previous data on the spinal generation of the swimming movements as well as to the geometrical organization of the muscle fibers of the body.

---

[x] From the Department of Physiology, University of Göteborg, Göteborg, and, GIH, Stockholm, Sweden
[xx] From the Institute of Oceanology, Academy of Sciences, Moscow, U.S.S.R.
[o] Names published in alphabetical order (Presenter: S. Grillner)

Introduction

The swimming movements of fish can be analysed from many different aspects: (1) What movements are actually carried out during swimming at different speeds; which factors change with speed and which remain constant; how do the movements vary with body form, etc.. (2) How are these movements interacting with the surrounding medium to produce the forward propulsion, i.e., the hydromechanical aspects. (3) How does the central nervous system actually generate the swimming movements.

In this chapter some pertinent data will be presented concerning (1) and (3) and in particular the body movements at different speeds of swimming in relation to the electromyographical activity in different segments along the body and the different neural mechanisms involved. Although we will try to cover this field to some extent we will mainly summarize data that we have obtained in Moscow or Göteborg independently or together with other colleagues. We neglect, however, the hydromechanical aspects (2) for which the reader is referred to the interesting work of Gray (1968) and Lighthill (1969).

I. PARAMETERS CHANGING WITH THE SPEED OF FORWARD LOCOMOTION

Fish swim with the help of (1) undulatory body movements, and/or (2) rhythmic movements of the fins, in particular, the pectoral fins. Some fish, such as the carp, use the pectoral fins for propulsion at low speeds without apparent body movements, but at higher speeds the body movements take over and finally the fins are not used at all. Other species use the fins primarily for changing direction during swimming and as stabilizors (e.g., Gray, 1968; Lighthill, 1969). The body movements are associated with a wave of lateral displacement travelling down the body with increasing amplitude towards the tail and with a constant phase lag between rostral and caudal segments (see below). Three factors vary with the speed of movement, the frequency of alternation between the two sides (i.e., of tail beat), the amplitude of the lateral displacement, and the speed of the travelling wave (Bainbridge, 1958; Gray, 1968; Lighthill, 1969; Webb, 1971).

A. The Relation Between Forward Speed and Frequency of Alternation

That speed increases with frequency (Figure 1A), has been demonstrated in many different species such as eel, trout, dace and goldfish (Bainbridge, 1958, 1961; Gray, 1968; Webb, 1971; Grillner et al., 1975a, b). The actual frequency of tail beat for a trout can go beyond 20 Hz (for methods see legend of Figure 1).

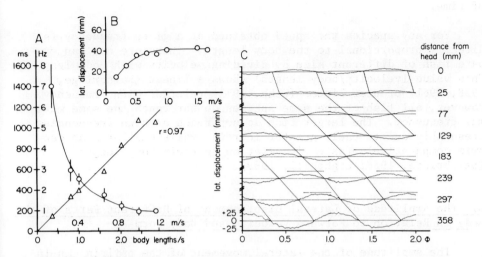

Figure 1. Forward velocity, frequency and amplitude of lateral displacement. The left graph (A) shows the frequency (ordinate, Hz) of alternation (tail beat) of an eel swimming at different forward velocities (abscissa) and also the corresponding cycle duration (ms). The fish were swimming in a closed chamber, 0.15 x 0.15 x 0.70 m, in which the water flow could be varied up to velocities of 3.3 m/s by a motor driven propeller in a closed system. The wires of EMG electrodes were passed through an aperture in the swimming chamber and the fish, on which spots of light reflecting material were sewn, were filmed (80 fr/s) through the transparent chamber with synchronization between EMG and film. The EMG was recorded either on an inkwriter (Mingograph, straight frequency response up to 1200 Hz) or on FM-tape recorder (7500 Hz). The films were enlarged on a digitizer table: and the data (x y coordinates of reflecting spots) were fed into a small computer together with the EMG data. The data in Figure 1, 3-7 were obtained in this experimental set up. The graph in (B) shows the lateral displacement of the tip of the tail fin of a 280 mm long trout swimming at different speeds of forward velocity. (C) shows the amplitude of lateral displacement at different points of an eel (distance from the tip of the head is shown to the right). The amplitude of lateral displacement is increasing from head to tail. The scale (ordinate in mm) given for the lowermost curve applies for all points. The bars below the record for point nos. 2, 4, 5, and 6 indicate the period of EMG activity on one side at the same level as the point where lateral displacement was measured. The crossing points at the different levels are connected by thin lines to emphasize the propagated nature of the wave (Grillner, Kashin, Rossignol and Wallén) (unpublished data).

But such frequencies can be maintained only for very short periods of time.

For any species the speed obtained at a given frequency is directly proportional to the body length. Therefore one can compare fish of different size by dividing velocity with body length. This value (velocity/body length) shows a linear (Bainbridge, 1958, 1961) or near linear (Webb, 1971) relation with frequency. However, all fish of the same species do not reach the same maximal frequency. The larger fish do not reach as high frequencies presumably due to their greater inertia, while smaller fish can swim comparatively faster in relation to their body length (Bainbridge, 1958; Lighthill, 1969).

B. The Amplitude of Lateral Displacement of Different Parts of the Body in Relation to Forward Velocity and Body Form

The amplitude of the lateral movement of the tailfin can increase as the speed progresses from very slow to moderate (Figure 1B). After this point the amplitude remains approximately constant (Bainbridge, 1958; Webb, 1971). The frequency (of tail beat) at which the amplitude becomes stable is presumably dependent on fish size and species but was estimated by Webb (1971) to be as high as 5 Hz (trout). The maximal amplitude of the lateral movement does not generally exceed 20-25% of the body length (Bainbridge, 1958). Under these conditions a certain frequency is paired with a certain amplitude of lateral displacement at any given point of the body. It is, therefore, interesting to note that if a fish has to drag an extra load during swimming it will with the same frequency of alternation increase the amplitude of the movements (Webb, 1971) and thereby the force produced (cf. Gray, 1968).

Not only the tail beat but each part of the body of the fish shows lateral displacement during locomotion (Figure 1C) and different species show different kinds of lateral displacement along the body (Breder, 1926; Gray, 1968). To understand how the amplitude of the lateral displacement varies along the body and how it is related to body form, the movements of several different species with widely different shapes of the body were filmed (Kashin, 1971). On the basis of the lateral displacement it was possible to divide the fish into three different groups; (1) with monotonously increasing amplitude from head to tail and with substantial movements also of the head (e.g., eel), (2) with a point of minimal amplitude about 1/4 - 1/3 from the front end and with increasing amplitude both rostrally and caudally, and (3) with monotonously increasing amplitude backward but with no movements in the rostral one third. These subdivisions correspond largely to Breder's (1926) classifications in anguilliform (eel), carangiform (cod, perch) and ostraciform (triggerfish) locomotion.

Since the form of the body must be expected to influence the lateral displacement, the size of the side projections for the different parts of the body in each fish was measured (indirectly related to the resistance of the water to the lateral displacement) in Figure 2B and also the mass and the cross sections at the different levels (Figure 2C, D). These parameters change together along the body and it is apparent that in the eel they decrease uniformly, whereas in the carangiform (cod) group there is a maximum at one third of the body length and for the ostraciform group this maximum is located more rostral. When comparing Figure 2A with B, C, and D it is apparent that there is an "inverse" relationship between the amplitudes of the movements along the body for the different groups and the distribution of the side projections (Figure 2B). The dorsal and ventral fins, which indeed influence the lateral surface, have only been included in the estimations for eel. For the other groups they have not been considered, since they often are pressed against the body during normal swimming (Bainbridge, 1963).

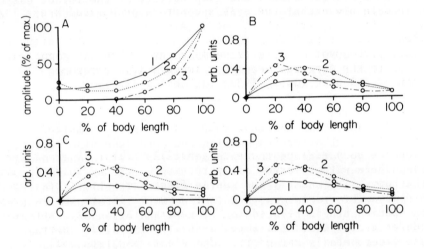

Figure 2. The relation between body form and the amplitude of lateral displacement along the body. (A), Distribution of the amplitude of the relative lateral displacement along the body. The amplitude of the tip of the tail equals 100%. (B), Distribution of the relative area of the side projection along the body from head to tail. The largest side projection equals 1. (C), Distribution of the relative cross-section areas along the body. (D), Distribution of the relative masses along the body. The mass of the whole body equals 1. In each diagram eel is indicated by (1), carp by (2) and Myoxocephalus sp. by (3). (Modified from Kashin, 1971).

The distribution of the amplitudes of any undulating rigid body must be correlated to its elastic and inertial properties (Machin, 1958). The morphological differences in these groups will result in characteristic differences in these properties as well as in resistance to lateral displacements.

The shift of the minimum amplitude to a more caudal point observed in the carangiform group results in a decrease of the relative length of the caudal part of the body, acting as a short undulating lever and allowing a higher frequency of alternation (cf. Bainbridge, 1958). The maximal frequency of a cottus was two times higher than that of an eel of equal length (Kashin, 1971).

C. The Relation Between Forward Velocity and the Speed of the Travelling Wave

During swimming the undulating wave is propagated along the body. The speed of this wave varies with the frequency of alternation in such a way that there is a constant mechanical phase lag between different parts of the body (see Section III). This applies to different types of fish such as eel, trout, cottus and dogfish. The speed of this travelling wave, (V), has a certain relation to the forward speed of swimming, (U), and generally U is between 40-60% of V (Gray, 1933; Lighthill, 1969).

II. THE ACTIVITY RELATED TO THE INDIVIDUAL SEGMENT

On the body surface there is generally a thin layer of red muscle fibers with a larger strip running along the lateral part of the body. The rest of the muscle fibers are white (see legend, Figure 4). The relative amount of red muscle fibers varies greatly among species and is related to the general habits of a species (Boddeke et al., 1959; cf. Bone, 1966). The activity among the muscle fibers easily can be recorded electromyographically (e.g., Bone, 1966; Grillner, 1973; 1974; Hudson, 1973). With the electrodes in the red lateral muscles a period of burst activity can be recorded from each side in each swimming cycle at all speeds. If the duration of the swimming cycle is plotted against burst duration (Figure 3) a linear relation occurs both in trout, dace, eel and spinal dogfish (Grillner, 1974; Grillner et al., 1975a, b, d). Hence the burst always occupies a certain part of the swimming cycle which is independent of the speed. This value can vary somewhat along the body but tends to become shorter caudally (from 50 to 25%, see Figures 5 and 8).

In the red lateral muscles bursts of activity occur at all speeds of swimming. No activity is recorded from electrodes

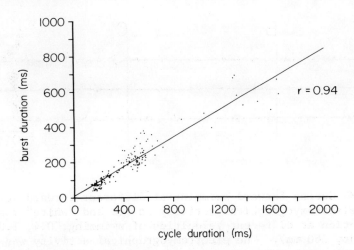

Figure 3. EEL. The relation between the duration of the electromyographical bursts and the cycle length. The EMG was recorded as in Figure 4. The duration of the EMG activity in one electrode located in the red lateral muscle (77 mm from the tip of a 410 mm long eel) was measured together with the duration of the cycle. The records were measured when the eel was swimming at different constant velocities between 0.1-1.3 m/s (Grillner, Kashin, Rossignol and Wallén, Unpublished data).

located among the white muscle fibers at slow speeds (dogfish, dace and trout) (Bone, 1966; Rayner and Keenan, 1967; Hudson, 1973; Grillner, 1973, 1974; Grillner et al., 1975b), but at higher speeds suddenly large burst activity can be recorded (Figure 4). The duration of the burst of activity in the white muscles is then further shortened with increasing speed (trout and dogfish, Grillner, 1974; Grillner et al., 1975b). Thus the white muscle fibers are only recruited at higher speeds and it appears that they are only activated in cycles below a certain duration. For the eel, however, activity can be recorded from "white muscle electrodes" even at the slowest speeds (Grillner et al., 1975a). It remains unclear if this is due to modified properties of white muscle fibers (see Boddeke et al., 1959) or to a participation of "real" white fibers even at low speeds.

It is thus apparent that with increasing speed of swimming new muscle fibers are recruited. This is clearly revealed also by the markedly increased amplitude in the bursts recorded at each electrode as speed increases (Grillner, 1974; Grillner et al., 1975a,

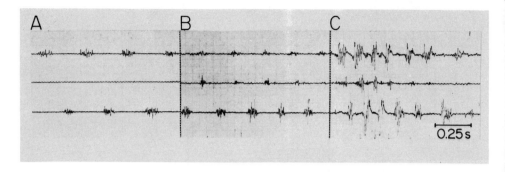

Figure 4. EMG during forward swimming at different forward velocities. The electromyographical activity in red and "white" muscles during locomotion at different velocities of swimming; 0.4, 1.0, 1.3 m/s (trout, 260 mm). The electromyographical activity was recorded bipolarly with thin copper wires insulated except for the tip. The recordings were made with one electrode in the red lateral muscles in the middle part of the body (upper records) and one in the middle of the "white" epaxial muscles at the same level (middle records) and one in the red muscles at the contralateral side. It should be noted that there is no activity in the "white" electrode at the slowest speed but that it appears in (B) at a frequency of about 4 Hz. This applies to trout which has been claimed to have red fibers intermingled among the white fibers (Boddeke et al., 1959) in contrast to dogfish and dace in which the white fibers nevertheless become active (Grillner, Kashin, Rossignol and Wallén, Unpublished data).

b). This amplitude increase is not only due to recruitment, but also very likely to an increased frequency in each active motor unit. This is indicated by the studies on both red and white muscle fibers of the dogfish dorsal fin, performed by Roberts (1969c).

The activity recorded in one muscle segment is related to the activation of the spinal nerves both rostral and caudal to the segment (cf. Bone, 1966). It is interesting to understand how the activation of a segment during swimming relates to the local movements of the body. The angle between the adjacent parts of the body and a particular segment (see insert at Figure 6) can be regarded as the local expression of the swimming movement. During each swimming cycle the onset (midpoint or termination) of the EMG starts at approximately the same part of the mechanical cycle except at the highest speeds when it can be slightly phase advanced (Grillner et al., 1975a, b, c).

III. INTERSEGMENTAL COORDINATION

As a fish swims an undulating wave travels down the body. This is due to a lag between the activation of rostral and more caudal segments. This can be demonstrated by simultaneous recording of the electromyographical activity in several places along the body. Figure 5 shows the period of EMG activity during 10-52 consecutive cycles recorded simultaneously at four different points along the body and at four different velocities of swimming (Figure 5A-D). The bars show at each recording point the period of EMG activity and the dots the duration of each cycle. The left graph in each pair is in ms whereas the right is normalized to cycle duration. At all speeds there is a lag between the activation of the different parts of the body. It is noteworthy that the cycle duration decreases from approximately 1500 ms in A to 200 ms in D and that the burst duration and the lags decrease approximately in proportion (evident particularly from the normalized graphs in A to D). Thus the time lag between the segments varies dramatically. The graph in Figure 6A shows the lag in ms between the rostral and the more caudal recording points, for several different velocities of swimming (eel). The time lag increases monotonously with distance from head to tail and the value of this lag becomes shorter with increasing speed. Figure 6B shows the relation between period length and time lag between two recording points, obtained during swimming at ten different velocities. There is a linear relation between these parameters passing through the origin, which means that the lag between the two points is always a certain proportion of the period length (i.e., there is an electromyographical phase lag along the body). Such a constant phase lag first was recognized in the spinal dogfish (Grillner, 1973, 1974), but now is established for different intact freely swimming fish such as eel, trout, and dace (Grillner et al., 1975a, b). Between each pair of segments there is a phase lag. The phase lag between two points increases approximately in proportion to the number of segments.

Initially it might be expected that the mechanical lag (Gray, 1933; Kashin, 1971) between different parts of the body would follow exactly the electromyographical lag. However, this cannot be assumed, since the propagation of the mechanical wave from the rostral part can be influenced by several, both active and passive factors. Figure 7A shows how the angles change at several points along the body (see insert). The mechanical lag increases monotonously with distance from head at any given speed (Figure 7A, B), as does a constant phase lag. If one compares the EMG lag between the different points with the mechanical lag (Figure 7A, B) it turns out that the former is significantly shorter at any speed. This applies to all investigated fish, i.e., eel, trout, and spinal dogfish (Grillner et al., 1975a, b, c). This emphasizes the great difficulty in predicting the resulting movements from an electromyographical signal without knowing the passive properties of the

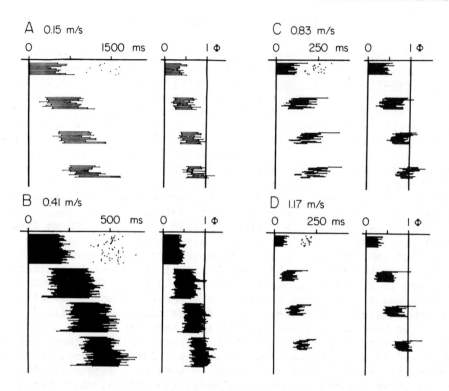

Figure 5. The electromyographical activity at four different locations along the body and at four different velocities of swimming. The left graph in each pair (A-D) shows the period of EMG activity (bars) at four different locations for 11 or more consecutive cycles at four different velocities 0.15-1.17 m/s. The dot to the right shows the termination of each cycle. Note the different time scales in A and B-D. The right graph shows the period of EMG activity in relation to each cycle normalized to 1. The recordings are from an eel (411 mm) and the electrodes are located 77, 183, 239, 297 mm from the head (Grillner, Kashin, Rossignol and Wallén, Unpublished data).

body, the actual distribution of forces in the body and the interaction with the surrounding medium (see below section IV on the complex arrangement of muscle fibers).

This constant phase coupling has one presumably very important function: it assures that the speed of the travelling wave on the body (V) is always proportional to the forward speed of swimming (U) (or vice versa). This becomes clear if one considers Bainbridge's demonstration (1958) of a linear relation between frequency (f) (of tail beat) and forward speed of swimming (U) and our own finding of

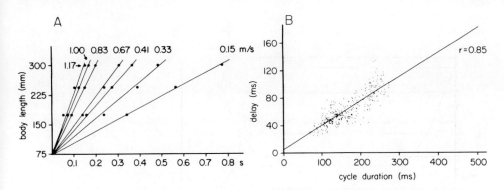

Figure 6. The electromyographical intersegmental lag at different velocities of swimming. The left graph (A) shows the time lag (abscissa) between a rostral recording point (for electromyogram) and several more caudal points on the body (ordinate) for several different velocities of forward swimming (as indicated in the graph in m/s). The records are from a 411 mm eel. The right graph (B) shows the relation between the time lag of the onset of EMG in two recording points (120 mm apart in a 280 mm trout, i.e., approximately 30 segments) and the duration of the cycle. The data were obtained when the fish was swimming at several different constant velocities. Similar graphs have been constructed also for intact eel and dace and spinal dogfish (Grillner, Kashin, Rossignol and Wallén, Unpublished data).

a constant phase lag (ρ). At any given instance and at any speed the distance (L) and the phase lag (ρ) between two points on the body will be constant, and the period length is $\frac{1}{f}$.

$$V = L : (\frac{1}{f} \cdot \rho) = \frac{L}{\rho} \cdot f$$

V is then directly proportional to f as is true for U (Bainbridge, 1958 and Figure 1) which results in a strict proportionality between U and V.

IV. ORGANIZATION OF THE BODY MUSCULATURE

In the preceding section (III) it was shown that the mechanical undulating wave travels more slowly along the body than the electromyographical wave. It thus seems important to consider the organization of muscle fibers in the fish body. The body musculature of fish has a segmental organization. The number of segments is equal to the number of vertebrae. The propagating locomotor waves are

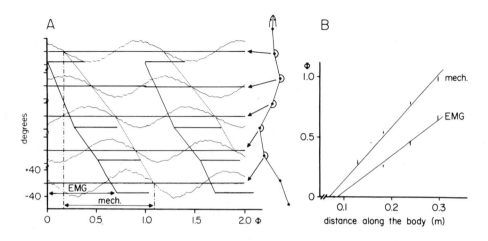

Figure 7. The relation between electrical and mechanical phase lag during swimming. The graph to the left (A) shows the angles at five different points along the body of an eel (77, 129, 183, 239 and 297 mm) during two swimming cycles at 0.41 m/s as indicated by the inserted drawing of an eel. Each horizontal line indicates 180° for the given angle and the excursions are indicated as ±40° from this angle. It can be seen that the amplitude of the angle excursions increase rather little caudally. The mechanical lag between the different parts is indicated by the interrupted line. The smaller bars below each line indicate the period of EMG activity at the same level as the tip of the angle. The solid line connects the onsets of EMG activity. Below the graph the EMG lag is indicated and the mechanical lag between the same two points. The cycle duration is normalized to 1 but would correspond to 504 ms. The graph to the right (B) shows a similar mechanical lag between five different points of the body (10 measurements, the bars indicate ± SE) and from the same cycles the electromyographical lag. The regression lines are calculated with least mean square method (Grillner, Kashin, Rossignol and Wallén, Unpublished data).

due to excitation of successive muscular segments. The muscle fibers are short and do not have tendons, but insert directly on the myosepta separating each segment. Thus there is usually no direct connection between the fibers and the bony parts of the skeleton. If each muscular segment would move only two of the neighbouring vertebrae, the lever arms would be very short and furthermore lateral and medial muscle fibers would have quite different mechanical conditions. However, each muscular segment has a very complex form (Nursall, 1956; Willemse, 1959; Kashin and

Smoljaninov, 1969) with rostral and caudal cones extending over as much as 4 vertebrae in each direction (Trachurus trachurus).

For an understanding of the mechanical advantage of such a non-tendon system, the orientation of the muscle fibers relative to the long axis of the body was investigated (Alexander, 1969; Kashin and Smoljaninov, 1969). It was found that the red muscle fibers, which comprise the superficial aspect of the body musculature, are always directed in parallel with the back bone. The deeper white muscle fibers are organized in a more complicated way. The muscle fibers in different parts of one segment have different directions in relation to the long axis of the body. On the other hand, fibers in the same part of two adjacent segments have approximately the same orientation and can be regarded as continuations of each other with the myoseptum interposed. In fact muscle fibers of successive segments are linked together in a long chain from segment to segment. The course of this chain (e.g., of 15 segments) starts from the plane of the medial (midline) septum near the medial cone of a given segment, goes in the dorsocaudal direction toward the surface of the musculature and then again deep into the dorsal muscles to insert in an acute angle with the dorsal parts of the medial septum.

What is the role of such long chains of muscle fibers? Obviously, if all fibers in a chain contract, it will act to bend the body more effectively than if the fibers had been passing from vertebra to vertebra (Alexander, 1969; Kashin and Smoljaninov, 1969). If only part of the muscle fibers in such a chain contract the result cannot be predicted, since the viscoelastic properties of the adjacent inactive muscle fibers which transmit the force to the "backbone" are not known. If the mechanical "transmission" is effective the contraction of one segment can give a substantial turning couple on the body; if not, a contraction of one segment in isolation will give a very small mechanical effect. It is noteworthy that the activation of the muscle fibers during locomotion proceeds from segment to segment (Grillner, 1973, 1974) with several segments cocontracting at each instance. Finally the arrangement of the segments in cones allows the orientation of muscle fibers in chains, but it might have the additional virtue of allowing more complex body movements (cf. Gray, 1968) than those of swimming. However the role of the cones in this context is largely a field for speculation.

V. THE NEURAL GENERATION OF LOCOMOTION

If the spinal cord of a dogfish or an eel is transected, the lower part of the body can perform continuous swimming movements

(Steiner, 1886; Bethe, 1899; Bickel, 1897; LeMare, 1936; Gray and Sand, 1936; Lissmann, 1946a). These coordinated movements are thus generated by the spinal cord without influence from higher centers. The different findings discussed above for the intact fish were shown earlier for the spinal dogfish (Grillner, 1973, 1974; Grillner et al., 1975c; for spinal eel, Wallen, 1975). These results included linear relation between burst activity and period length; recruitment of white fibers only at higher rates of activity (Bone, 1966); and a phase lag between consecutive segments both electromyographically (Figure 8A) and mechanically. The frequency range displayed can be varied markedly by adding exteroceptive stimuli for the dogfish (between 0.4 to 2.8 Hz, Grillner, 1974) whereas the eel appears to have a more limited frequency range. Even with spinal cord transections at several places, parts of at least 8 segments or more can still generate rhythmic alternating activity with maintained intersegmental coordination. The conclusion is that all basic features of the stereotyped swimming movements can be coordinated within the entire spinal cord including the intersegmental phase coupling.

In most species a transection of the spinal cord leads to inactivity in the caudal part of the body, but usually a pinch of the tail fin can lead to a few undulatory waves. The difference between fish with and without spontaneous spinal swimming is presumably due not to any fundamental difference in organization of the nervous system but more to a difference in degree of excitability in the interneuronal network responsible for generating the movements (Grillner, 1975). Thus in eel or dogfish this network operates continuously in the spinal state whereas in other species the excitability level would need to be increased for movements to be generated.

The intersegmental coordination in fish, and in the spinal dogfish in particular, has been suggested (Lissmann, 1946a, b; Gray, 1968; Roberts, 1969a, b, c) to be dependent on reflex arcs from subcutaneous stretch sensitive receptors. This old hypothesis (e.g., Friedlander, 1888) was tested for eels by von Holst (1935) who made a bilateral transection of several dorsal and ventral roots in the middle part of the body. Under these conditions there was a maintained coordination between the rostral and the caudal parts of the body even when transmission of movements from the rostral to the caudal part was prevented. In the spinal dogfish the intersegmental coordination remains both after a bilateral dorsal root transection over 20-30 segments and also after preventing all movements by curarization (Grillner et al., 1975d). Under the latter conditions rhythmic efferent activity was recorded in the ventral roots. The activity in different roots was coordinated and the frequency was at 0.5 Hz or higher (as in the normal spinal dogfish, Grillner et al., 1975d). Thus Roberts' (1969a) previous

SWIMMING IN FISH

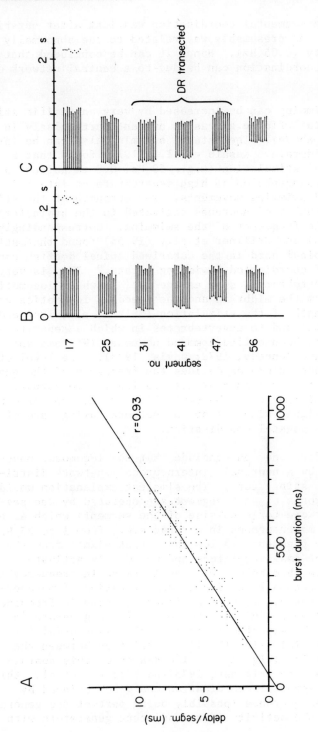

Figure 8. Phase lag and the effect of dorsal root transection in the swimming spinal dogfish. The left graph (A) shows the time lag between onset of EMG between two electrodes being 25 segments apart in the red lateral muscle (normalized to delay/segment) versus time duration of the burst (equivalent to period length) in a swimming spinal dogfish (modified from Grillner, 1974). (B) and (C) show the same type of representation of EMG activity and cycle length as in Figure 4 at six different levels (segment no. indicated to the left). In C the dorsal roots of segments no. 30 to 50 has been transected bilaterally. Spinal dogfish (modified from Grillner et al., 1975d).

conclusion that intersegmental coordination was lost after curarization is erroneous and presumably was related to the abnormally low level of activity (0.05 Hz). Hence it can be concluded that the intersegmental coordination can be due to a central network of neurones.

The rate of swimming can be increased by very unspecific stimuli applied to the tail fin, e.g., as is evident particularly in spinal animals (see above). The rate of swimming also can be influenced from the midbrain. Kashih et al., (1974) found that a continuous electrical stimulation (e.g., 50 Hz, 0.8 ms pulses) of a limited region in mesencephalon of high decerebrate or intact fish elicited coordinated swimming movements. The strength of the stimulation (i.e., the number of neurones activated in the brainstem) was correlated to the frequency of the swimming. Correspondingly, cf. Lissmann, (1946b) and Grillner et al., (1975d) found that stimulation of the cut spinal cord in the curarized spinal dogfish can increase the rate of coordinated "swimming bursts." This is very similar to higher vertebrates, such as the cat, in which locomotion can be initiated from the midbrain and the speed of locomotion controlled by the strength of the stimulation (Shik et al., 1967 and Grillner, this volume) and to invertebrates in which a separate class of neurones has been called command neurones (Wiersma and Ikeda, 1964; Davis and Kennedy, 1972; Davis, 1973). The level of tonic activity in these neurones decides the frequency of the generators and thereby the frequency of alternation in the output. Thus the speed of swimming could be controlled by one type of continuous descending signal with the spinal network taking care of segmental and intersegmental coordination.

From these studies one can conclude that the locomotor movements are generated by a neuronal (interneuronal) network distributed throughout the spinal cord. The simplest explanation would be that the motoneurones of each segment are operated by one segmental generator (or one for each side of the segment) which alternately activate the motoneurones in a range e.g., from 1 to 25 Hz (cf. trout). If the frequency is low the output signal from the generator would be of small amplitude and would only activate motoneurones with low threshold. With an increase in frequency the output amplitude increases resulting in an activation of motoneurones with higher threshold as well as in an increase in frequency of already recruited motor units. The shape of the generator's output signal cannot be expected to be of a pure sinusoidal form since this is incompatible with the linear relation between the burst and cycle duration (Figure 3). The design of this postulated generator is unknown (see Grillner, 1975 and Edgerton et al., this volume). The intersegmental coordination could be achieved by separate coordinating neurones (possibly being part of the generator itself) linking the activity of two adjacent generators with a given phase shift in a similar way to that has been discussed for the crustacean swimmeret system (Stein, 1974). In this way the activity in a whole chain of generators can be coordinated and be made to act together in a coordinated fashion. Also there would be a fixed phase lag along the chain as long as the excitability in the different generators is approximately similar. It should be noted that although the lag is usually in the rostrocaudal direction, fish can swim backwards and then have a reversed phase coupling EMG-wise (Grillner, 1974; Grillner et al., 1975a). Hence the phase coupling between two adjacent segments can be reversed. However, one segment located somewhere in the body (rostral, rostral one third or caudal part) is always the leading one and from this segment the EMG-wave is propagated with the appropriate lag in a caudal, rostral or bilateral direction (Grillner, 1974). The level of activity in the generators can be influenced by stimulation of the midbrain and by unspecific stimuli applied to the body surface (see above). These inputs can act separately or via a common group of neurones (cf. Davis, 1973; Grillner, 1975) to bring all the generators to the same level of activity (i.e., to be active e.g., from 1 Hz to 25 Hz). In this way the "command system" sets the level of activity in the generators and with increasing frequency more motoneurones automatically will be recruited, thereby producing more force. The intersegmental coordination is achieved by interneurones, assuring a given phase lag between the different parts of the body at all velocities. This basic construction (Figure 9) would be very easy to operate. We must recall, however, that we do not know how the generator is constructed or even if there is one in each segment and not a more complex situation with e.g., somewhat "overlapping" generators in nearby segments (Smoljaninov, Personal communication).

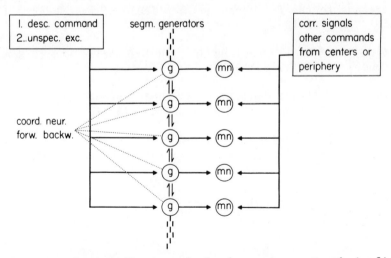

Figure 9. Speculative diagram of the locomotor network in fish.

Of course the motoneurones also are accessible to many other signals, like descending correcting signals, peripheral signals or signals carrying commands to change direction and also those related to the phase dependent reflex reversal (Grillner et al., 1975e; see Forssberg et al., this volume).

In conclusion the swimming movements appear to be controlled by a spinal network, the activity of which could be controlled in a rather simple way to produce the stereotyped swimming movements in their frequency range. The more complex task of making the locomotion purposeful for the individual by adapting it to the environment is left for the higher structures.

Acknowledgment

This work was supported by the Swedish Medical Research Council (proj. no. 3026) and S.K. for five months by the Swedish Medical Research Council Visiting Scientist Fellowship. We are grateful for the permission to discuss unpublished data obtained together with Dr. S. Rossignol and P. Wallén and for their large contribution to these studies. The help of Mrs. M. Svanberg is gratefully acknowledged.

REFERENCES

Alexander, R.M.N., (1969) Orientation of muscle fibres in the myomeres of fish. J. mar. biol. Ass. U.K. 49, 263-290.

Bainbridge, R., (1958) The speed of swimming of fish as related to size and to the frequency and amplitude of tail beat. J. Exp. Biol. 35, 109-133.

Bainbridge, R., (1961) Problems of fish locomotion. Zool. Soc. (Lond.) Symposia. 5, 13-32.

Bainbridge, R., (1963) Caudal fin and body movement in the propulsion of some fish. J. Exp. Biol. 40, 23-56.

Bethe, A., (1899) Die Locomotion des Haifisches (Scyllium) und ihre Beziehungen zu den einzelnen Gehirntheilen und zum Labyrinth. Arch. ges. Physiol. 76, 470-493.

Bickel, A., (1897) Beitrage zur Ruckenmarksphysiologie des Aales. Pflügers Arch. ges. Physiol. 68, 110-119.

Boddeke, R., Slijper, E.J., and Van Der Stelt, A., (1959) Histological characteristics of the body musculature of fish in connection with their mode of life. Proc. K. Ned. Akad. Wet. C 62, 576-588.

Bone, Q., (1966) On the function of the two types of myotomal muscle fibre in elasmobranch fish. J. mar. biol. Ass. U.K. 46, 321-349.

Breder, C.M. Jr., (1926) The locomotion of fishes. Zoologica. 4, 159-297.

Davis, W.J., and Kennedy, D., (1972) Command interneurones controlling swimmeret movements in the lobster. I. Types of effects on motoneurones. J. Neurophysiol. 35, 1-12.

Davis, W.J., (1973) "Neuronal organization and ontogeny in the lobster swimmeret system," In Control of Posture and Locomotion. (Stein, R.B. et al., eds.), Plenum Press, New York, (437-455).

Friedländer, B., (1888) Ueber das Kriechen der Regenwürmer. Biologisches Centralblatt 8, 363-366.

Gray, J., (1933) Studies in animal locomotion. I. The movement of fish with special reference to the eel. J. Exp. Biol. 10, 88-103.

Gray, J. and Sand, A., (1936) The locomotory rhythm of the dogfish (Scyllium canicula). J. Exp. Biol. 13, 200-209.

Gray, J. (1968) Animal Locomotion. Weidenfeld and Nicholson, London W 1.

Grillner, S., (1973) Locomotion in the spinal dogfish. Acta Physiol. Scand. 87, 31-32A.

Grillner, S., (1974) On the generation of locomotion in the spinal dogfish. Exp. Brain Res. 20, 459-470.

Grillner, S., (1975) Locomotion in vertebrates - central mechanisms and reflex interaction. Physiol. Rev. 55, 247-304.

Grillner, S., Kashin, S., Rossignol, S. and Wallén, P., (1975a) Kinematical and electromyographical studies of fish locomotion. I. Eel. (In preparation).

Grillner, S., Kashin, S., Rossignol, S. and Wallén, P., (1975b) Kinematical and electromyographical studies of fish locomotion. II. Trout. (In preparation).

Grillner, S., Kashin, S., Rossignol, S. and Wallén, P., (1975c) Kinematical and electromyographical studies of fish locomotion. III. Spinal dogfish. (In preparation).

Grillner, S., Perret, C. and Zangger, P., (1975d) Central generation of locomotion in the spinal dogfish. Brain Res. (In press).

Grillner, S., Rossignol, S. and Wallén, P., (1975e) Phase dependent reflex reversal during swimming in the spinal dogfish. Acta physiol. scand. (In press).

von Holst, E., (1935) Über den Process der zentralnervosen Koordination. Pflügers Arch. ges. Physiol. 236, 149-158.

Hudson, R.C.L., (1973) On the function of the white muscles in teleosts at intermediate swimming speeds. J. Exp. Biol. 58, 509-522.

Kashin, S.M., and Smoljaninov, V.V., (1969) On the question of the geometrical arrangement of body muscles in fish. Voproci iktiologii. 9, 1139-1142. (In Russian).

Kashin, S.M., (1971) An investigation of the kinematics of swimming in fishes and the structural organization of their motor apparatus. (Available in the Lenin Library, Moscow.)

Kashin, S.M., Feldman, A.G. and Orlovsky, G.N., (1974) Locomotion of fish evoked by electrical stimulation of the brain. Brain Res. 82, 41-47.

Lemare, D.W., (1936) Reflex and rhythmical movements in the dogfish. J. Exp. Biol. 13, 429-442.

Lighthill, M.J., (1969) Hydrodynamics of aquatic animal propulsion. Ann. Rev. Fluid Mech. 1, 413.

Lissmann, H.W., (1946a) The neurological basis of the locomotory rhythm in the spinal dogfish (Scyllium canicula, Acanthias vulgaris). I. Reflex behaviour. J. Exp. Biol. 23, 143-161.

Lissmann, H.W., (1946b) The neurological basis of the locomotory rhythm in the spinal dogfish (Scyllium canicula, Acanthias vulgaris). II. The effect of de-afferentation. J. Exp. Biol. 23, 162-176.

Machin, K.E., (1958) Wave propagation along flagella. J. Exp. Biol. 35, 796-806.

Nursall, J.R., (1956) The lateral musculature and the swimming of fish. Proc. Zool. Soc. Lond. 126, 127-143.

Rayner, M.D. and Keenan, M.J., (1967) Role of red and white muscles in the swimming of the skipjack tuna. Nature. 214, 392-393.

Roberts, B.L., (1969a) Spontaneous rhythms in the motoneurons of spinal dogfish (Scyliorhinus canicula). J. mar. biol. Ass. U.K. 49, 33-49.

Roberts, B.L., (1969b) The response of a proprioceptor to the undulatory movements of dogfish. J. Exp. Biol. 51, 775-785.

Roberts, B.L., (1969c) The co-ordination of the rhythmical fin movements of dogfish. J. mar. biol. Ass. U.K. 49, 357-379.

Shik, M.L., Severin, F.V. and Orlovsky, G.N., (1967) Structures of the brain stem responsible for evoked locomotion. Fiziol. Zh. USSR. 12, 660-668.

Stein, P.S.G., (1974) Neural control of interappendage phase during locomotion. Am. Zool. 14, 1003-1016.

Steiner, I., (1886) Uber das Centralnervensystem der grunen Eidechse, nebst weiteren Untersuchungen uber das des Haifisches. Sitzungsberichten der K. Preuss. Akad. Wissensch. 32, 539-543.

Wallén, P., (1975) Locomotion in the spinal eel. (In preparation).

Webb, P.W., (1971) The swimming energetics of trout. I. Thrust and power output at cruising speeds. J. Exp. Biol. 55, 489-520.

Wiersma, C.A.G. and Ikeda, K., (1964) Interneurons commanding swimmeret movements in the crayfish, Procambarus Clarkii (Girard). Comp. Biochem. Physiol. 12, 509-525.

Willemse, J.J., (1959) The way in which flexures of the body are caused by muscular contractions. Proc. K. Ned. Akad. Wet. C 62, 589-593.

ANALYSIS OF TETRAPOD GAITS: GENERAL CONSIDERATIONS AND SYMMETRICAL GAITS

Milton Hildebrand

Professor of Zoology, University of California

Davis, California

The relative timing of the cyclic contacts that the feet of tetrapods make with the ground in terrestrial locomotion determines the gaits of the animals. This chapter reports on a comprehensive and integrative study that establishes a system for analyzing gaits. The model facilitates description, identifies all possible gaits, permits the simultaneous study of hundreds of locomotor performances, and helps to interpret the selection of gaits by the various animals.

The manner of collecting data and of making footfall formulas and gait diagrams will be reviewed and evaluated. Symmetrical gaits (walks, pace, trot) with fore and hind contact intervals equal have only two variables and can be plotted on a graph. Overlays permit instant determination of the footfall formula and the duration of the support phases for any plotted performance. Overlays to a similar graph give corresponding information for symmetrical gaits with fore and hind contact intervals unequal.

Asymmetrical gaits (gallops, bounds) have more variables, but when selected data are plotted on an appropriate graph, a wide range of these gaits can similarly be analyzed with the use of representative overlays. Such gait characteristics as transverse versus rotary sequence of footfalls, and alternate versus simultaneous placement of the feet of a pair, are differentially distributed on the graph.

All gaits form a continuum. Correlations will be made between gait selection and systematics, body size, body proportions, speed, endurance, habit, and habitat.

Introduction

Many investigators are studying the terrestrial locomotion of tetrapods. Their techniques include cinematography, cineradiography, dissection, force plate analysis, electromyography, and neurophysiology. The results of such studies usually relate to the gaits of the test animals. A gait is an accustomed manner of moving in terrestrial locomotion. The events of a gait are cyclic. They involve the timing of the footfalls, the durations of the contacts that the feet make with the ground, and the sequences of the different patterns of support made by the feet in combination. This discussion will exclude erratic or nonrecurring actions; neurotic or pathogenic behaviors; and rapid changes from one way of moving to another (although these may be of interest, and can be studied by the same methods). Also excluded are motions of the head and spine and the trajectories of the feet in the air.

Analysis: Objectives and Characteristics

This study has four characteristics:

(1) It is comprehensive. When extended to include asymmetrical gaits, all gaits (within reasonable limits), will be identified whether actually observed, probable, or only theoretically possible.

(2) It is integrative. All of the principal variables of gaits are correlated.

(3) It standardizes locomotor performance. The discrepancies among successive corresponding actions are rectified. Whether slight or marked, these irregularities always occur. Unless the peculiarities of each cycle are damped out, generalizations cannot be made.

(4) Finally, it contributes to the interpretation of gait selection by the various tetrapods.

The first three characteristics of the investigation, comprehensiveness, integration, and standardization, make it a system of analysis or model. The model emphasizes graphic, rather than numerical representations. Practice allows one to visualize gaits shown as loci on the graphs. Advantages of the model include

facilitation of data recovery, predictive value, and provision of
a conceptual basis for the study of gaits. The model answers or
at least helps to answer, various questions.

(1) How many gaits are there?

(2) How are gaits interrelated?

(3) Why are some theoretically possible gaits not observed?

(4) How do gaits relate to systematics, body size and proportions, speed, endurance, habit, and habitat?

The general method described is somewhat flexible. If greater
precision and less averaging or standardization are desirable, the
procedures for recording and interpreting the data can be adapted.
Modeling should supplement but not substitute for exact study of
individual cycles if exactness is more appropriate to particular
objectives.

There are numerous options for recording and analyzing gaits.
The variables may be expressed in absolute or relative terms. An
appropriate entry into cyclic events must be determined. Some
variables may better be related as ratios. Terminology must be
selected. The choices may be immaterial, but one solution is often
superior to others. It is desirable for students of vertebrate
locomotion to seek consensus. Adopting uniform usages (where that
would not interfere with the expression of differing insights and
opinions) would eliminate ambiguities and facilitate the rapid
interpretation of an individual work.

Before gaits can be compared and a system of analysis devised,
it is necessary to observe, record, and describe individual performances, such as the particular walk of a particular horse.
Historically, such performances have been captured for study by the
unaided eye, by pressure tubes attached to the feet (Marey, 1874),
and by a series of still cameras triggered in sequence (Muybridge,
1899). Trackways have also served as a basis for study (Peabody,
1948; Gambaryan, 1974). For an historical survey see Sukhanov
(1974).

Since the development of cinematography, motion pictures have
become the standard raw material for the student of locomotion.
Slow motion photography (64 to 150 frames per second) is usually
needed; an accurate record requires about 30 frames per cycle.
Hildebrand (1964) has discussed camera speed, camera angle, range,
parallax, substrate, background, and grids. The film is studied
frame by frame.

SUPPORT FORMULA, AND ENTRY INTO THE CYCLE

Returning now to an individual performance, such as film showing one pass of one animal, what information should one record? First, the body is supported in sequence by different numbers of feet in combination. Thus, a walking horse commonly has three feet on the ground for an instant, followed by two, and then again three, then two, and so on. For a walking horse there are usually eight combinations of supporting feet before the first is repeated. The numbers of supporting feet have been written, for many years, in sequence as the support formula, which in this case is 3-2-3-2-3-2-3-2. The formula has utility. The alligator seemingly selects the gait with the formula 4-3-2-3-4-3-2-3 in preference to another (which also might be used at the same speed) with the formula 3-2-3-2-3-2-3-2 because the former is clearly more stable. Similarly, the running elephant appears to select the gait with the formula 2-1-2-1-2-1-2-1 in preference to another (which also might be used at the same speed) with the formula 1-2-1-0-1-2-1-0 because the former gives more continuity of support for the animal's great bulk.

The support formula has been stressed by several authors, but it has limitations. It does not indicate the relative durations of the various phases of a gait, nor which feet are on the ground in each phase. Consequently, as mentioned below, several different gaits may have the same formula.

Before going on to more useful ways of recording gaits, note that here one comes to a discretionary choice. Since gaits are cyclic activities, it is necessary to isolate one unit of each for study; i.e., all events in sequence until repetition starts. Such a unit is called a stride. The starting point arbitrarily could be the strike of the left forefoot, or the lift off of the right hind foot, or any other identifiable event. Accordingly, the last support formula printed above might have started 2-1-0, or 1-0-1, or 0-1-2 instead of 1-2-1. Surprisingly, various leading authors (Howell, 1944; Gambaryan, 1974; Sukhanov, 1974) have not used the same event consistently as a reference point in their own papers.

This author considers that there are strong reasons to adopt the strike of a hind foot as the point to enter the cycle. The hind feet nearly always have a greater role in propulsion than do the forefeet. If the duration of contacts with the ground made by fore and hind feet are different, then the hind contacts tend to be the longer (exceptions noted below). The duration of contacts by forefeet never markedly exceeds those of hind feet, whereas the converse is not true. Some fast-moving tetrapods (various lizards) barely may touch their forefeet to the ground. Bipeds use the hind legs, never the forelegs. (It is true, however, that the forelegs often play the greater role in support and shock absorption.)

TETRAPOD GAITS

FOOTFALL FORMULA

Muybridge (1899) was probably the first to use the footfall formula which provides all the information of the gait formula and in addition indicates which feet are on and off the ground in each phase of a gait. Figure 1 shows that a footfall formula is a series of stylized sketches that represent an animal as seen from above as it moves from left to right. Light circles represent feet that are lifted and dark circles, feet on the ground. Eight sketches, each representing a different combination of supporting feet, commonly are required before the cycle repeats.

In Figure 1 the second sketch resembles the notation of Muybridge (1899); the subsequent sketches follow the notation of Howell (1944), which has the advantage of greater simplicity. The author (1965 and since) has used the notation shown in Figure 2 on the left, (also in Figure 6). The equivalent notation shown on the right in Figure 2 (also in Figure 13), is introduced here. All three of these representations are satisfactory, but the latter two

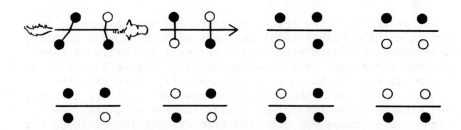

Figure 1. Derivation and notation of a footfall formula.

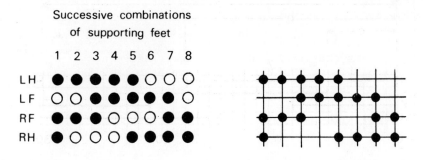

Figure 2. Alternative notations for the same footfall formula. The initials, L, R, F, and H stand for the left, right, fore, and hind feet. Dark circles indicate feet supporting weight; open circles, unweighted feet.

(Figure 2) have advantages over the former (Figure 1). They are more compact and can be scanned more easily by the eyes; they have the character of a stylized gait diagram.

In recording footfall formulas, authors have been inconsistent about where to break into the cycle. It is suggested the strike of a hind foot, preferably the left, be accepted as the initiation point.

GAIT DIAGRAM

The gait diagram was introduced by Goiffon and Vincent in 1779 (cited in Sukhanov, 1974), and has been used, with modifications, by most students of tetrapod locomotion since then (though curiously not by Muybridge or Howell). It provides the information of the footfall formula and, in addition, the relative durations of the phases of a gait. Gait diagrams are prepared on a graph paper as shown in Figure 3, but may be reproduced without a grid (Figure 9). Also, a time scale may be substituted for the count of motion picture frames, or both omitted.

The ordering of the traces representing the respective feet according to Hildebrand (1959) has been adopted by most authors, although over the years others have arranged traces in at least five other sequences. The ordering shown in Figure 3 most closely approximates the trackways of animals and is, therefore, the easiest to visualize. The traces for the left feet are to the left as the diagram (like the animal represented) progresses from left to right. The traces for the hind feet are outside of those for the forefeet, as are the hind tracks of many tetrapods (some

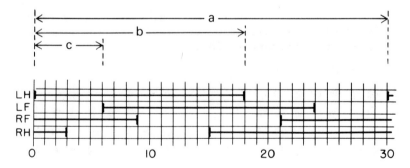

Figure 3. Gait diagram. Horizontal rows of squares are assigned to the respective feet; initials same as in Figure 2. Vertical lines represent successive motion picture frames. Squares are lined in if the respective foot is on the ground during the time interval represented.

lizards, most rodents, rabbits, and galloping carnivores). This reasoning applies equally to the ordering of the symbols for the respective feet in the footfall formula, as illustrated in Figure 2. Again, it is desirable to enter the cycle with the strike of a hind foot.

THE VARIABLES

A gait diagram accurately records the cyclic actions at the ground as an animal moves along. It is the best way to store the information needed for analysis, but does not accomplish analysis directly. In order to do that, one must compare many records and seek generalizations. This requires identification and correlation of the variables.

Since each foot is on the ground once and off the ground once in each cycle, it follows that gaits can have a maximum of nine variables. These can be expressed as the time that each foot strikes the ground, the duration of the contact that each foot makes with the ground, and the time that a reference foot strikes the ground a second time (to establish the duration of a full cycle). In short, there are four footfalls, four contact intervals, and a cycle length.

By using ratios one combines variables. Thus, by designating one footfall as a reference (by entering the cycle at that instant) and by identifying the remaining three footfalls as following the first by specified percentages of the full cycle, the number of variables is reduced to eight. Similarly, by expressing the four contact intervals as percentages of the duration of the cycle, the number of variables becomes seven.

The durations of the contacts with the ground made by the right and left feet of a pair (fore and hind) average the same for the gaits of nearly all tetrapods. They commonly tend to be different for some primates, even those that "walk" normally. It is possible, though not proven, that the contacts of the leading and trailing feet of a galloping animal might average a slight difference, particularly while the animal is turning. Such differences in the use of the feet of a pair may be of functional interest, and may be analyzed by the methods presented here. However, the differences are slight and infrequent and will be disregarded in this report. As a result, there are five remaining variables.

All gaits fall into one of two general classes, asymmetrical or symmetrical. Asymmetrical gaits usually retain five variables; symmetrical gaits have only two or three. These remaining variables are identified further in subsequent sections.

SYMMETRICAL GAITS

The pace, various walks and running walks, and the trot are symmetrical gaits. Such gaits have the footfalls of each pair of feet evenly spaced in time. The interval of time between the left footfall and right footfall of a pair is the same as the interval between the right footfall and left footfall. This is a necessary and sufficient definition of this large class of gaits. Nevertheless, there has been confusion about the definition of symmetrical gaits. In addition to evenly alternating footfalls for each pair of feet (and to the consequent reduction in the number of gait variables) symmetrical gaits have the following characteristics.

1) The spine flexes (if at all) laterally, as in salamanders, rather than vertically, as in galloping carnivores.

2) The support formula has an even number of figures, and those of the second half of the formula repeat those of the first half.

3) The entire gait diagram can be calculated from the record of half a cycle.

4) Footfalls of fore and hind feet alternate (unless they are simultaneous).

Sukhanov (1974, p. 80), (followed by Gambaryan, 1974), considered item (4) above to be primary. Consequently he concluded that "Canters, usually considered as a different type of gallop, should be considered as non-uniform and asymmetrical forms of symmetrical locomotion." When symmetrical and asymmetrical gaits are graphed each as separate classes of locomotion, certain asymmetrical gaits (gaits for which the footfalls of a pair of feet are not spaced evenly in time) also are shown to have the characteristic that the footfalls of fore and hind feet alternate. Canters, therefore, should be classified as asymmetrical gaits, as has been usually accepted.

SYMMETRICAL GAITS WITH ALL CONTACTS EQUAL

The Gait Formula and Graph

For some symmetrical gaits of some tetrapods, the duration of the contacts made with the ground by forefeet average shorter than those by hind feet; in other instances the fore contacts are the longer. Such gaits will be outlined below; but first the more simple and more usual circumstance, fore and hind contacts averaging the same, will be discussed.

It has long been recognized that symmetrical gaits with fore and hind contacts the same have only two variables. Hildebrand (1962 and since) demonstrated that these can be quantified and correlated on a simple graph. His designation lists the first variable as the percentage of the stride interval that each foot (or any foot) is on the ground (100b/a in Figure 3), and the second variable as the percentage of the stride that the footfall of a forefoot follows the footfall of the hind foot on the same side of the body (100c/a in Figure 3). The first variable measures how much of the time the feet are on the ground and varies with speed; the second relates the actions of the forefeet as a pair to those of the hind feet as a pair.

The two percentage figures comprise a gait formula. Formulas are most conveniently calculated from gait diagrams. In order to smooth out the irregularities of individual strides, it is best to average the equivalent events of several consecutive cycles. One or more gait formulas can be prepared for each pass of an animal in front of the camera.

Gait formulas are plotted on the basic gait graph for this class of gaits (Figure 4). The right border of the graph is an absolute limit; there, the feet cease to touch the ground. Similarly, the left border is absolute; there, the feet never leave the ground. The top and bottom borders, in contrast, are identical and are determined by the point of entry into the cycle. Thus, as the gait cycle repeats, it is the same thing for the footfall of a forefoot to follow that of the hind foot on the same side by 0% or 100% of the stride interval, and the position 3 on the ordinate is only 6 percentage points removed from the position 97. One might think of the graph as being rolled into a cylinder such that top and bottom borders coincide.

Sukhanov has independently developed a corresponding gait graph for this class of gaits. His scheme was first published (this author believes) in 1967 and is in his book, which was published in Russian in 1968. He omits from his graph the area corresponding to the lower half of the graph in Figure 4, stating that this portion represents a mirror reflection of the upper portion. Unfortunately this is not so; most gait formulas for primates, and some for other mammals, fall in the lower portion (Hildebrand, 1967). Further, he expresses the variables in different terms and orients the graph differently than the one presented here.

Footfall Formulas and Durations of Phases

Hildebrand (1965, 1966) showed that since gait formulas representing similar gaits fall at similar places on the gait graph,

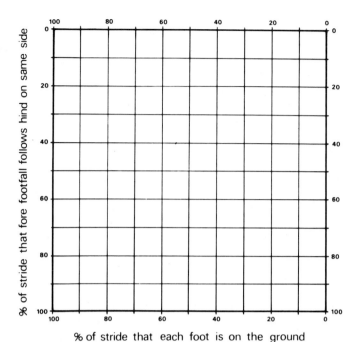

Figure 4. Gait graph for symmetrical gaits having fore and hind contact intervals the same.

their respective footfall formulas are distributed over the graph in a regular pattern (Figure 5). If drawn to scale on tracing paper, the pattern can be superimposed on the graph for immediate determination of the footfall formula for any plotted gait formula. A different footfall formula is represented by each of the 16 triangular areas, each of the 22 lines, and each of the 6 intersects; 44 formulas in all. (Recall that the bottom border repeats the top border, and that right and left borders do not represent possible gaits.) The formulas are shown in Figures 6 and 13. All formulas keyed to areas have eight phases, all keyed to diagonal lines have six phases, all to other lines have four phases, and all to intersects have two or four. (It now can be verified that gaits with different footfall formulas, such as numbers 3 and 7, can have identical support formulas.)

Other overlays, (Figure 7), enable one to determine quickly the percentages of the cycle that an animal is supported by each combination of feet (as shown by the phases of the respective footfall formula). Thus, for the single gait formula plotted in Figure 7, the footfall formula (from Figure 5) is number 4 and the animal is supported by an ipsilateral pair of feet for 20% of the time (10%

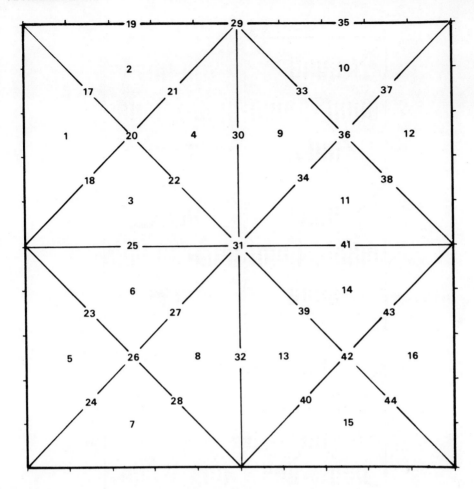

Figure 5. Overlay to the gait graph of Figure 4 showing the relationships of all possible footfall formulas. The numbers are keyed to the respective formulas in Figure 13.

in each combination); by a contralateral pair of feet for 40% of the time (20% in each combination); by three feet for 40% of the time (10% in each combination); and never has only one foot on the ground or all feet either on or off the ground at one time.

Prost (1965); Zug (1972b); and Dagg and de Vos (1968a, b) did not use gait formulas or gait graphs. Nevertheless, they did employ the percentages of the time that certain animals are supported by each combination of feet. Data were measured and tabulated directly from gait diagrams rather than recovered by the

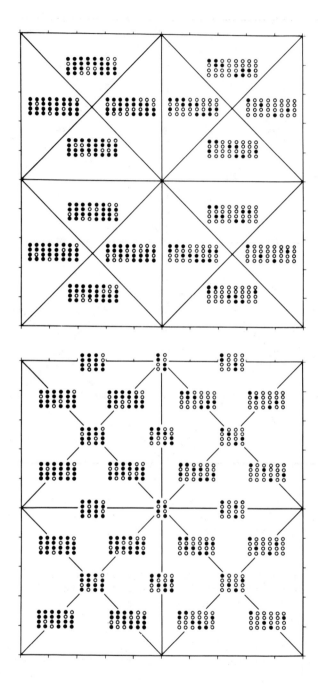

Figure 6. Overlays to the gait graph of Figure 4 showing the footfall formulas and their distribution.

TETRAPOD GAITS 215

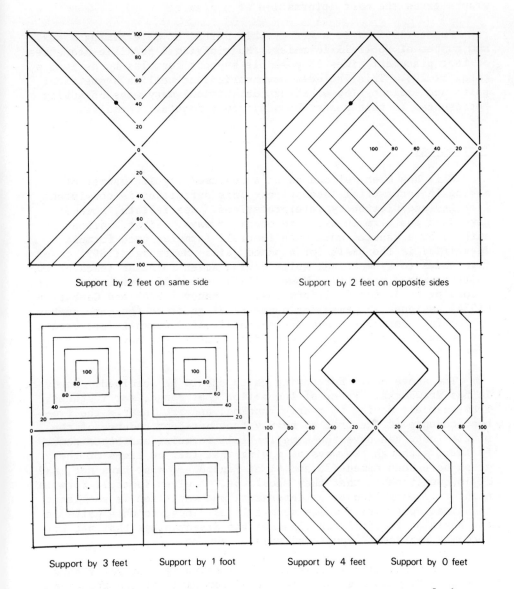

Figure 7. Overlays to Figure 4 showing the percentages of the stride interval that the body is supported by different combinations of feet.

overlay method. As Zug (1972a) has noted, each approach is useful. The author believes that the plotting of gait formulas on gait graphs gives the most information at a glance.

Prost (1965) used a non graphic method of determining a maximum number of footfall formulas. Multiplying 6 possible sequences of foot placement times 18 possible support formulas he derived a total of 108. This includes asymmetrical as well as symmetrical gaits and excludes (by arbitrary definition) many ways of moving represented in this report by additional footfall formulas.

Terminology

Gait has been defined as an accustomed, cyclic manner of moving in terrestrial locomotion. This definition is consistent with lay and dictionary interpretations. Specialists, however, have used the term in two more specific ways. First, various authors have equated gait with footfall formula; each formula identified is the basis for a named gait. Muybridge (1899) gave five names to symmetrical gaits having seven formulas (each of two gaits having a variant). Howell (1944) named 12 symmetrical gaits according to 12 footfall formulas. Sukhanov (1974) and Gambaryan (1974) assigned 13 names for symmetrical gaits to the same number of footfall formulas. About eight additional formulas are used by one or another of the animals studied, but as yet have not been named.

The above basis for naming gaits should be retained. However, it has drawbacks. Types of locomotion that range, for instance, from scarcely moving to nearly running, and from nearly the two-beat cadence of the trot to nearly the four-beat cadence of the single-foot, can have the same footfall formula, and hence name. Further, although two ways of moving that fall on lines of the graph have been named (25 and 41 of Figure 4), most have not, and it is questionable that they should be. A given footfall formula might fall on a line or an intersect, but variation is such that no animal customarily walks such a line. Thus, to name gaits on the basis of footfall formulas has the disadvantage that not all formulas represent accustomed ways of moving.

Hildebrand (1965) presented the terminology for gaits shown in Figure 8. Names are not indicated for parts of the graph on which no gait formula has fallen; however, the appropriate names for such parts could be supplied easily by extension of the naming system. The dimensions of the areas on the graph assigned to each named gait (10 by 12-1/2 percentage points) are appropriate. A trained observer can distinguish by eye between ways of moving that differ on either scale of the graph by 10 to 15 percentage points.

TETRAPOD GAITS

	WALK				RUN					
	Very Slow	Slow	Moderate	Fast	Slow	Moderate	Fast			PACE
		slow walking pace	mod. walking pace	fast walking pace	slow running pace	mod. running pace	fast running pace			
	very slow lat. seq. lat. cpts. walk	slow lat. seq. lat. cpts. walk	moderate lat. seq. lat. cpts. walk	fast lat. seq. lat. cpts. walk	slow lat. seq. lat. cpts. run	moderate lat. seq. lat. cpts. run	fast lat. seq. lat. cpts. run		LATERAL SEQUENCE	Lateral Couplets
	very slow walk. lat. seq. single foot	slow walk. lat. seq. single foot	moderate walk. lat. seq. single foot	fast walk. lat. seq. single foot	slow run. lat. seq. single foot	moderate run. lat. seq. single foot	fast run. lat. seq. single foot			Single-foot
	very slow lat. seq. diag. cpts. walk	slow lat. seq. diag. cpts. walk	moderate lat. seq. diag. cpts. walk	fast lat. seq. diag. cpts. walk	slow lat. seq. diag. cpts. run	moderate lat. seq. diag. cpts. run	fast lat. seq. diag. cpts. run			Diagonal Couplets
	very slow walking trot	slow walking trot	moderate walking trot	fast walking trot	slow running trot	moderate running trot	fast running trot			TROT
		slow diag. seq. diag. cpts. walk	moderate diag. seq. diag. cpts. walk	fast diag. seq. diag. cpts. walk	slow diag. seq. diag. cpts. run	moderate diag. seq. diag. cpts. run	fast diag. seq. diag. cpts. run		DIAGONAL SEQUENCE	Diagonal Couplets
		moderate walk. diag. seq. single foot	fast walk. diag. seq. single foot	slow run. diag. seq. single foot	moderate run. diag. seq. single foot	fast run. diag. seq. single foot				Single-foot
										Lateral Couplets
		slow walking pace	mod. walking pace	fast walking pace	slow running pace	mod. running pace	fast running pace			PACE

Figure 8. Overlay to Figure 4 showing a scheme of names for symmetrical gaits.

Also, gait formulas for several animals of a kind moving together commonly differ by about that amount. The system provides as much differentiation as is useful yet not more than is reasonable.

    Hildebrand designates most of the gaits on the upper half of the graph as lateral sequence gaits, and those on the lower half as diagonal sequence gaits. Howell (1944), Sukhanov (1974) and Gambaryan (1974), refer to the same gaits, respectively, as diagonal and lateral. The discrepancy is unfortunate, but not inadvertent. Neither class is evidently lateral or diagonal unless some event arbitrarily is keyed to another. Howell (1944) adopts a forefoot as the reference foot. For reasons explained above, Hildebrand adopts a hind foot as the reference. In his scheme, a gait has lateral sequence if the forefoot that strikes after the reference foot is on the same side of the body; diagonal sequence, if the forefoot to strike next is on the opposite side of the body. A gait has lateral couplets if footfalls of ipsilateral feet are related in time as a pair; diagonal couplets, if footfalls of contralateral feet are related in time as a pair. The scheme is internally consistent and now is used commonly.

SYMMETRICAL GAITS WITH FORE AND HIND CONTACTS UNEQUAL

Variables and Gait Formula

Symmetrical gaits have three variables when the durations of contacts made with the ground by fore and hind feet are unequal. Fore contacts are the longer ones for some heavy-shouldered artiodactyls (and for most tetrapods during rapid deceleration). The hind contacts are the longer for most salamanders and lizards, and for some rodents (and for most tetrapods during rapid acceleration).

The first variable is the percentage of the stride interval that each hind foot contacts the ground (100b/a in Figure 9). It is desired that the second variable relate the actions of the forefeet as a pair to the actions of the hind feet as a pair. For the previous class of symmetrical gaits (having all contacts equal) it was convenient to express this variable as the percentage of the stride interval that the footfall of a forefoot follows that of the hind foot on the same side of the body. However, here, the scheme for this class of gaits will have more functional significance if one does not measure the interval between hind and fore footfalls, but instead measures the interval between instants of maximum support by the respective legs. To approximate this, the concept of the midtime of a foot contact is introduced, or the instant that is halfway through the duration of the contact. The second variable now becomes the percentage of the stride interval that the fore midtime follows the hind midtime on the same side of the body, (100c/a, Figure 9).

It is desired that the third variable relate the duration of fore and hind contacts. This is best expressed as the percentage that the fore contact is of the hind contact (100d/b, Figure 9).

The gait formula for this class of gaits represents these three percentage figures written in sequence.

The Gait Graph and Footfall Formulas

How is it best to relate these variables? A three dimensional grid could be used. Another method is to plot the first two

TETRAPOD GAITS

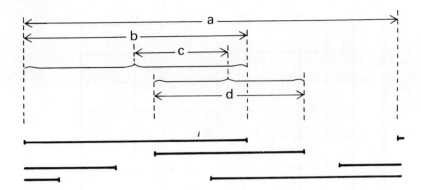

Figure 9. Calculation of the three variables of symmetrical gaits having fore and hind contact intervals unequal.

variables on a two dimensional grid (Figure 10), and to prepare a separate overlay containing the resultant pattern of footfall formulas for each desired value of the third variable. There are several reasons for this preference. It is the simplest procedure. Values of the third variable seldom exceed 110% for any tetrapod; a single overlay adequately describes most gaits having the fore contacts the longer. This variable rarely is significantly less than 100% for the natural gaits of large animals. Gait diagrams for small animals usually show very irregular actions. Much smoothing of the record is necessary before any generalizations can be made. Accordingly, several overlays for representative values of the third variable adequately represent all situations. And, finally, correlation of the third variable with the other two can be made sufficiently clear by this procedure.

Consider first, gaits having the fore contacts shorter than the hind. Overlays showing the distribution of footfall formulas on the gait graph (Figure 10) for representative ratios of fore to hind contacts are shown in Figure 11. The patterns are complex, yet in the unlikely event that overlays are needed for ratios not illustrated here, they are easily prepared. The left border, a, represents the place each hind foot (only) is always on the ground. Vertical line b is constant because it marks the gait transitions where each hind foot is on the ground half the time. Points c, d, and e are also fixed. Line f marks gait transitions where each forefoot is on the ground half the time. Its value on the abscissa is 50/n%, or 5000/n, where n is the percentage that the fore contact is of the hind contact. Thus, f = a when n = 50, and for lower values it is off the graph (forefeet are then never on the ground as much as half the time). Point g = (100-n)/2 on the ordinate, and establishes the slope of line cg. The symmetry of the overlays is such that all other lines can be drawn from this information.

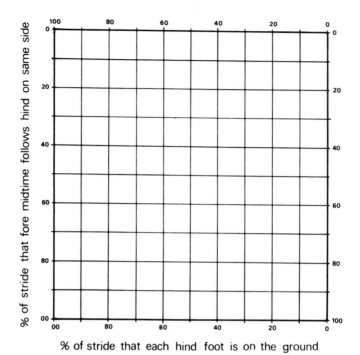

Figure 10. Gait graph for symmetrical gaits having fore and hind contact intervals unequal.

Numbers on unshaded areas of the overlays (Figure 11) are keyed to footfall formulas previously identified for gaits having fore and hind contacts equal. Shaded areas are keyed to 16 additional footfall formulas. (All these formulas are illustrated in Figure 13.) In addition, there are 44 other "new" formulas represented by lines and intersects within and bordering the shaded areas. These are not illustrated, but could easily be calculated from gait diagrams drawn to specification.

Turning to gaits having fore contacts longer than hind, overlays to the gait graph (Figure 10) showing the footfall formulas are shown in Figure 12. Line b, and points c, d, and e are again fixed. Line f (where forefeet are on the ground half the time) is now to the right of b instead of to the left, but is calculated as before. Line h shows the duration of hind contacts when forefeet are on the ground all the time; its value on the abscissa is $10,000/n$. No gait can fall to the left of this line. Point g now equals $(n-100)/2$. Shaded areas represent 14 additional footfall formulas, which are illustrated in Figure 13. An additional 38 formulas are represented by the lines and intersects within and bordering the shaded areas. These are not illustrated. This

TETRAPOD GAITS

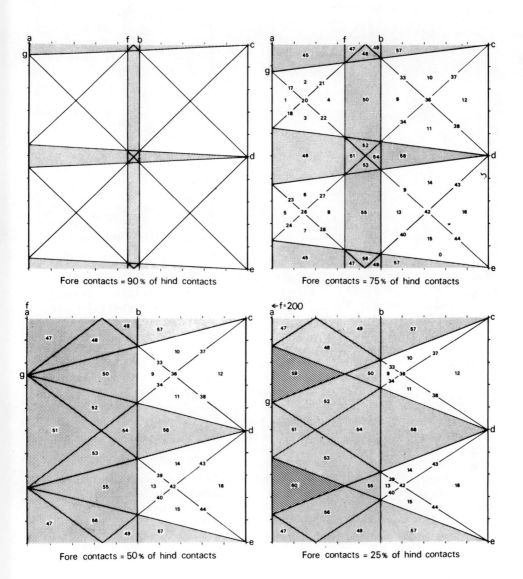

Figure 11. Overlays to the gait graph of Figure 10 showing the relationships of all possible footfall formulas when fore contacts are shorter than hind contacts. The numbers are keyed to the respective formulas in Figure 13.

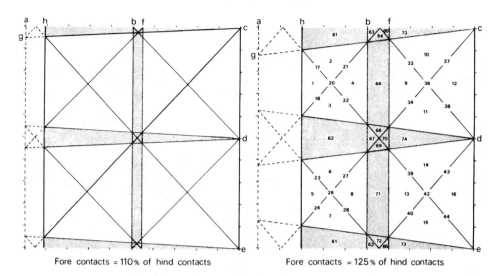

Figure 12. Overlays to the gait graph of Figure 10 showing the relationships of footfall formulas when fore contacts are longer than hind contacts. The numbers are keyed to the respective formulas in Figure 13.

brings to 156 the total number of theoretically possible footfall formulas for symmetrical gaits. (Should h be < b, which is virtually impossible, 8 other formulas would appear.)

Although not included here, overlays could again be prepared to recover from the gait graph the relative durations of the various combinations of supporting feet.

A separate terminology for gaits having fore and hind contacts unequal seems unreasonable. The terms shown in Figure 8 again can be used, supplemented by statements such as "with fore contacts longer than hind," or "with fore contacts about 75 percent of hind."

SYMMETRICAL GAITS OF VARIOUS TETRAPODS

Some symmetrical gaits of various tetrapods are shown by Figures 14, 15 and 16, which summarize 524 plotted gait formulas for 128 genera. The data are original except for 60 plots for turtles, viverrids, and a marsupial as credited in the legends. These summaries do not include the data already published on the horse (Hildebrand, 1965), primates (Hildebrand, 1967), and the domestic dog (Hildebrand, 1968).

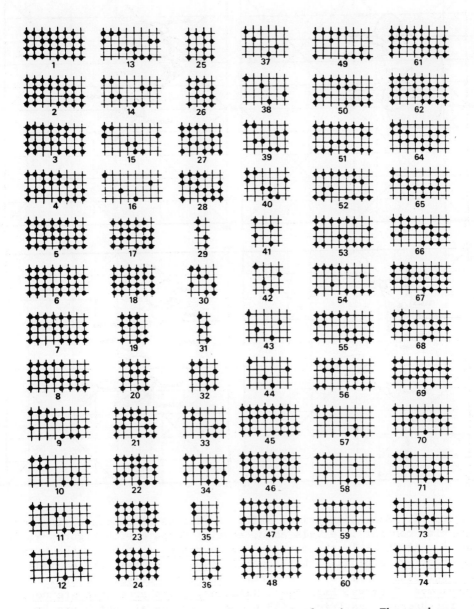

Figure 13. Footfall formulas of symmetrical gaits. The numbers are keyed to Figures 5, 11, and 12 (except that numbers 63 and 72 are omitted). The notation is explained by Figure 2.

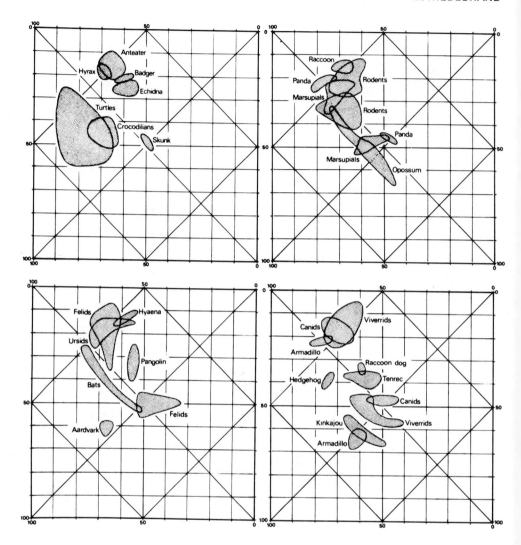

Figure 14. Some symmetrical gaits of various tetrapods having fore and hind contact intervals about equal. Compare with Figures 4, 5, and 8. ABOVE LEFT: the anteater Myrmecophaga (6 plots); the hyrax Heterohyrax (2 plots); the badger Taxidea (2 plots); the echidna Tachyglossus (5 plots); the turtles Testudo, Chelydra, Cuora, Chrysemys, Clemmys, Emydoidea, Malaclemys, Gopherus, Kinosternon, Sternothaerus, and Trionix (59 plots, of which all but 3 are from Zug, 1971); the crocodilians Alligator and Caiman (7 plots); the skunk Spilogale (2 plots). ABOVE RIGHT: the raccoon Procyon (5 plots); the panda Ailurus (5 plots); the rodents (above) Cavia, Dolichotis, Rattus, Cynomys, and Hydrochoerus (16 plots); the rodents (below) Marmota,   (Legend continued on next page.)

TETRAPOD GAITS

Figures 14 and 15 show gaits having fore and hind contact intervals equal, or from the limited data available, not known to have fore and hind contacts consistently more than slightly unequal. The data are figured against the background pattern that indicates the distribution of footfall formulas for such gaits (compare with Figure 5). Figure 16 shows gaits having fore and hind contact intervals unequal, and the data are likewise figured against the background patterns that indicate the respective distribution of footfall formulas (compare with Figures 11 and 12).

The range of individual variation is such that the designated ratios of fore to hind contacts should not be considered precise. Irregularities are particularly marked for the gaits of turtles, yet Zug (1972b) found that fore contacts average 92% of hind contacts. Gambaryan (1974) states that the forefeet may predominate for camels and goats, although unpublished records of Hildebrand do not show this. Also, some domestic cattle walk with forefeet predominating and some not. The gaits of lizards are strikingly irregular, as is clearly shown by gait diagrams in Sukhanov (1974); yet the fore contacts may be considered to range between the limits shown in Figure 16 (33% to 85% of hind contacts) and beyond.

On Figures 14, 15 and 16, species are gathered by genus, and some genera by Family, since available data do not justify further distinction. It is not implied that all species of a genus have the same gaits (and some exceptions are noted below), but the species of a genus usually move in similar ways. Most of the respective shaded areas would be increased, some considerably, by further data. They show ways that the given animals have been observed to move, not _the_ ways that those kinds of animals move.

Castor, Dipodomys, and Dasyprocta (10 plots); the marsupials (above) Dendrolagus, Sarcophilus, Phascolomis, and Petaurus (5 plots, of which one is from Windsor and Dagg, 1971); the marsupials (below) Marmosa, Lasiorhinus, and Petaurus (4 plots); the opossum Didelphis (6 plots). BELOW LEFT: the hyaena Hyaena (2 plots); the felids Felis, Panthera, and Acinonyx (29 plots); the ursids Thalarctos, Melursus, Helarctos, Tremarctos, Selenarctos, and Ursus (18 plots); the pangolin Manis (5 plots); the bats Pteropus and Tadarida (4 plots); the aardvark Orycteropus (4 plots). BELOW RIGHT: the viverrids Arctictis, Paguma, Crossarchus, Suricata, Atilax, Ichneumia, Nandinia; and Herpestes (16 plots, of which 3 are from Taylor, 1970); the raccoon dog Nyctereutes (3 plots); the other canids Lycaon and Urocyon (6 plots); the armadillo (above) Dasypus (2 plots); the armadillo (below) Priodontes (6 plots); the hedgehog Hemiechinus (3 plots); the tenrec Echinops (5 plots); the kinkajou Potos (9 plots).

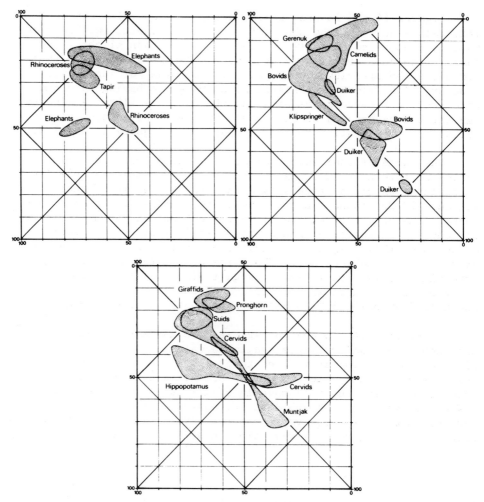

Figure 15. Some symmetrical gaits of various tetrapods having fore and hind contact intervals about equal. Compare with Figures 4, 5, and 8. ABOVE LEFT: the elephants Elephas and Loxodonta (34 plots); the rhinoceroses Ceratotherium, Diceros, and Rhinoceros (14 plots); the tapir Tapirus (8 plots). ABOVE RIGHT: the camelids Camelus, Vicugna, and Lama (26 plots); the gerenuk Litocranius (10 plots); the klipspringer Oreotragus (2 plots); the duiker Cephalophus (16 plots); the other bovids Taurotragus, Boselaphus, Bos, Silvicapra, Kobus, Orynx, Alcelaphus, Connochaetes, Antilope, Aepyceros, Gazella, Antidorcas, Hemitragus, Ovis (67 plots). BELOW: the giraffids Giraffa and Okapia (10 plots); the pronghorn Antilocapra (3 plots); the suids Tayassu, Babirussa, and Sus (13 plots); the hippopotamus Hippopotamus (15 plots); the muntjak Muntiacus (8 plots); the other cervids Alces, Elaphurus, Rangifer, Cervus, and Odocoileus (26 plots).

The Interpretation of Symmetrical Gaits

Figure 17 summarizes data on symmetrical gaits, and serves as a basis for discussion and interpretation. The figure combines material presented in this chapter and that from prior publications of Hildebrand (1965, 1967, 1968). In all, it brings together 1178 gait formulas of 156 genera. Fourteen orders of mammals are represented. All of the plotted formulas (about six erratic records excepted) fall within the marginal outline shown. The plots, or gaits, form a continuum within the entire area; nevertheless, the discussion is assisted by identifying, with shading and lettering, various overlapping and somewhat arbitrary subareas. Arrows show (as qualified below) probable relationships of the gaits in terms of phylogeny, rate of travel, or both.

Area A encloses most of the gaits recorded for amphibians and turtles, and doubtless also represents the gaits of the ancestral tetrapods. (The hippopotamus secondarily has returned to this part of the graph when moving slowly.) This area is relatively large because most of these animals, having short and sometimes weak legs, and usually a wide stance, are unsteady and irregular in their locomotion. As most of them are small, the substrate is uneven for them. They can only move slowly (hippopotamus excepted), and must select gaits having high stability. At slowest rates of travel they use footfall formula 1 (Figure 5), which alternates support by three and four feet, the maximum in stability. When speed is increased from the very slow walk to the slow walk (Figure 8), the gaits of these animals tend to drop down and to the right on the graph toward footfall formula 3. In doing so, simultaneous support by all four feet is not reduced (Figure 7), but support by tripods is partly replaced by contralateral bipods. Since the line of support provided by such bipods runs diagonally under the animal, the tendency to pitch or roll is not great. Gray (1944) showed that for animals that undulate the body from side to side in walking, and that have legs splayed to the sides, lateral sequence gaits (formulas 1 and 3) are superior to diagonal sequence gaits (formulas 5 and 6); the tripods of support provided by the former enclose greater areas and hence the animal can maintain its center of mass more easily over its support. The animals of area A tend to have fore contact intervals shorter than the hind, but only slightly.

As ancestral tetrapods became longer-legged and more agile, they probably shifted to area B. Most lizards and crocodilians fall in this area (and the hippopotamus when walking at moderate speeds). Periods of support by contralateral bipods are now relatively longer, but scarcely longer in actual duration because rate of travel has increased. As the body moves faster, dynamic equilibrium increases and the need for the static equilibrium provided by tripods diminishes. Most of these animals have longer

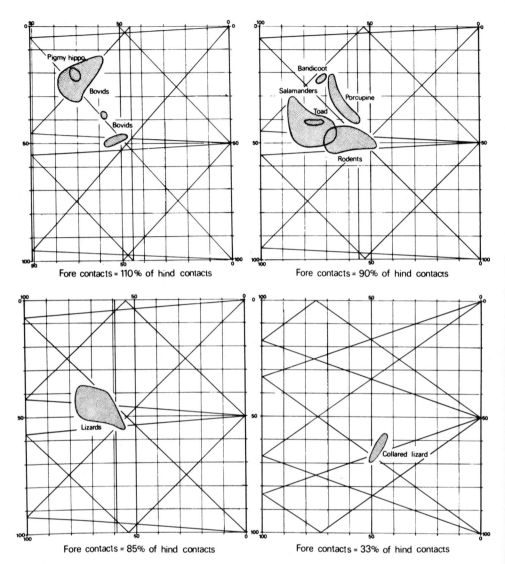

Figure 16. Some symmetrical gaits of various tetrapods having fore and hind contact intervals unequal. Compare with Figures 10, 11, and 12. ABOVE LEFT: the pigmy hippopotamus Choeropsis (4 plots); the bovids Tragelaphus, Bison, Hippotragus, Synceros, Ovibos, and some Bos (21 plots); ABOVE RIGHT: the bandicoot Isoodon (2 plots); the porcupine Hystrix (5 plots); the other rodents Mus, Hypogeomys, and Thomomys (18 plots); the salamanders Ambystoma, Salamandra, Taricha, Plethodon, Dicamptodon, and Rhycotriton (26 plots); the toad Bufo (3 plots). BELOW LEFT: the lizards Varanus, Anolis, and Heloderma (7 plots). BELOW RIGHT: the collared lizard Crotophytus (2 plots).

TETRAPOD GAITS

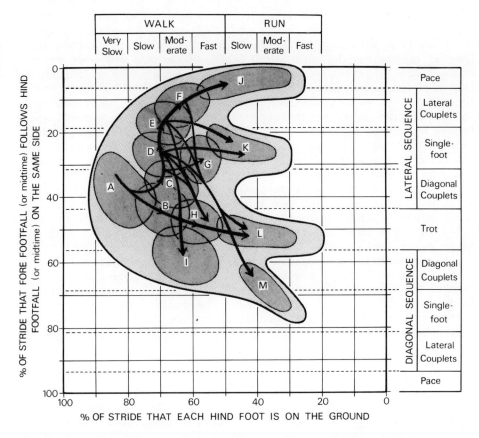

Figure 17. Distribution on the gait graph of 1178 plots of symmetrical gaits for 156 genera of tetrapods. Arrows show probable relationships.

hind legs than forelegs, and the contacts of their hind feet are longer than those of their forefeet. As they move faster, they shift to the right on the graph, and the predominance of the hind feet increases.

It is probable that as the legs of ancestral reptiles came to be positioned more under the body, their gaits shifted to area C. Such gaits are used by the insectivores of this small sample (hedgehog, tenrec), by some rodents (beaver, marmot, agouti, porcupine), and the raccoon dog (which is short-legged). These are all animals of small to moderate size which walk well but usually not fast or far. Most are stocky in build. The contact intervals of their fore and hind feet seem to be about equal.

From area C there evolved, in succession, the gaits of areas D, E, and F, which share the lateral sequence of footfalls, but which shift the timing of the footfalls from the single-foot to lateral couplets. In area D are found my records for some marsupials (tree kangaroo, wombat, phalanger), some monkeys (though this is not their usual gait), some rodents (guinea pig, prairie dog, domestic rat), one armadillo (Dasypus), suids, most cervids, the jaguarondi (which is short-legged for a cat) and the serval, short-legged breeds of domestic dog, some bovids (eland, nilgai, kudu, gnu, cattle, musk ox, water buffalo, sheep, tahr), the tapir, and the horse. These are animals of moderate or large size. Except for some of the ungulates, they tend to be more compact creatures than their relatives in higher positions on the graph. They run well, and may travel far, yet (horse excepted) rarely move at great speed. Many are woodland dwellers (tapir, suids, most cervids) and live in rough terrain. Some support heavy horns or antlers, or large heads. Consequently, stability is important to their locomotion. In comparison with animals with gaits in areas E and F, these animals are supported more by tripods and contralateral bipods, and less by ipsilateral bipods. Like the animals in areas E and F, these animals have the contact intervals of fore and hind feet nearly equal, except that fore contacts may be about 110% of the hind contacts for some of the artiodactyls with heavy forequarters.

In area E are my records for some long-legged rodents (capybara, Patagonian cavi), giant anteater, most felids, most canids, ursids, some viverrids, hyrax, elephants, pronghorn, and some bovids (hartebeest, springbuck, gazelle, impala, kob, bison). (Elephants may also use the slow walking trot. Records presented in Figure 15 for rhinoceroses fall between areas D and E, as do those of the pigmy hippopotamus--which is surprising because of its chunky build). Most of the animals of area E are predators or cursors dwelling in open country. (It was expected the horse would place here instead of in area D, and the hyrax and anteater to be there instead of here.) They are of moderate to huge size. Most are agile and have long legs. The feet are placed near the midline under the body, so the increased support by ipsilateral bipods, and decreased support by contralateral bipods, does not significantly reduce stability. Lateral sequence, lateral couplets gaits have the advantage that long strides can be taken without interference between fore and hind feet. (These records for animals with gaits in areas D, E, and F support the interpretations of Dagg and deVos, 1968a.)

Area F represents my records of the gaits of long-legged breeds of domestic dog, the cheetah, a hyaena, camelids, giraffids, and the gerenuk. For the horse, this gait may be transitional to the pace. These animals are again of moderate to large size, and they continue the trend to long-legged and slender body form. They inhabit open country and are notably cursorial. As for the previous category of gaits, long strides are achieved without interferences between fore and hind feet.

Area G is a somewhat enigmatic category which is probably derived from area C. It includes the records in this study for an echidna and a pangolin. It was expected these short-legged animals would move in area C, and perhaps they do at slower rates of travel. The walk of the echidna is quite unsteady.

Area H falls at the moderate and fast walking trot. This may be the usual slower mode of progression for some rodents (gopher, house mouse). Lizards (and the hippopotamus) shift to area H from area B merely by increasing their rate of travel. Such lizards tend to have markedly longer contacts by hind feet than forefeet; at the right edge of this area some of them are transitional to bipedal locomotion. Area H also may be reached from areas C, D, or E; examples are some marsupials (opossum, murine opossum, gliding phalanger), some viverrids (binturong, banded mongoose), some bats, and probably other rodents. These mammals may shift to area H through transitional gaits, as observed in the records for the opossum and bats in Figure 14, or they may instead change gaits more abruptly.

Area I represents rather highly evolved gaits seemingly derived from area D, possibly by way of areas B and H. These are diagonal sequence, diagonal couplets gaits using footfall formulas 6 and 8. Here are found most of the symmetrical gaits of lemurs, monkeys, and apes. (Their gaits also range, however, into the lower halves of areas B and H, and some of these animals use area D on occasion.) In addition, area I includes gaits of an unexpected assortment, the aardvark, an opossum, kinkajou, and giant armadillo. This area equates roughly to the "meterpetico" gait of Magna de la Croix (1929a, b), to which he ascribes armadillos in general ("tatus," "peludos"), including the nine-banded armadillo ("mulitas") which the few records shown here place in area D.

Why do these few animals use diagonal sequence gaits? Muybridge (1899) considered that the baboon "disregards the law governing walking." Gray's (1944) explanation for the avoidance of diagonal sequence gaits by lower tetrapods (see above) seems less applicable here, since these mammals, placing their feet well under the body, have small triangles of support regardless of sequence of footfalls. Prost (1965, 1972) considers that when primates are climbing, the diagonal sequence is superior because lateral flexion of the body can then increase the reach of the hand; and roll and yaw of the body are better controlled by following the thrust of a hind foot by placement of the hand on the opposite side. This reasoning may also apply to the kinkajou and opossum, since they are also scansorial. It is less clear why these animals (and not other scansors) use these gaits when not climbing. So far, the best answer appears to lie in our inability to say why they should not use diagonal sequence gaits. A principal reason that most tetrapods use lateral sequence gaits is that such gaits better avoid interference between fore and hind feet on the same side, particularly when the legs are long. In the films reported on here, the aardvark and giant armadillo instead avoid interference by taking rather mincing steps. Apes and monkeys avoid interference in a unique way (Hildebrand, 1967). The hind feet are placed beside the forefeet, each hind foot moving to the same side of its respective forefoot. Thus, one hind foot is outside of a forefoot and one inside. The consequence may be a slightly asymmetrical "symmetrical" gait, and somewhat longer contact intervals by one foot of a pair than by its opposite.

The phylogeny postulated above seemingly was recapitulated by the ontogeny of the gaits of a macaque (Hildebrand, 1967). The animal's first steps, taken at 14 days, fell in area A. As it matured, it shifted to area C, whence it moved through area B to area I, but also occasionally up to area D. The ontogeny of gaits is a subject that invites further study. Magna de la Croix (1929a) says in a footnote that an eight-day old kitten was observed to use the gait he calls "prototipico," which, by my analysis of his scheme, would have footfall formula 5. This is very surprising if it really is an accustomed manner of moving. One gait formula for a young puppy has footfall formula 1 (Hildebrand, 1968).

Running gaits have each foot off of the ground more than half the time. The four symmetrical running gaits extend across the gait graph horizontally from left to right like fingers (Figure 17). Area J represents the running and fast walking pace. These are natural gaits of camelids and individual dogs of several large and lanky breeds. Horses can be trained to pace, and perhaps some horses use the gait naturally without training or breeding. Several other mammals have been said to pace, but verification is needed. Harness pacers run slightly faster than trotters at all distances. The gait seems well adapted for sustaining a slow run over open terrain. There is no chance that a hind foot will strike a forefoot. Fore contact intervals of pacing horses average about 95% of hind contacts. The body of a pacing animal sways from side to side. This could be a reason why long legs are needed to perform the gait. A consequence of the swaying is that the pace is less stable than the trot, which seems to correlate with its use in open country. Webb (1972) considers that camels have large, digitigrade feet, with spreading toes, partly to increase lateral stability.

Area K comprises lateral sequence, single-foot gaits. At the fast walk and slow run they are the natural fast gaits of elephants. Show horses are trained to use gaits falling in this area at the slow run (termed by horsemen as the "paso," "slow gait," and "running walk") and moderate run ("rack" and, again, "running walk"). Being unnatural gaits, there is much variation in their execution, even in the same show ring at the same time. Fore and hind contact intervals are of about equal duration for elephants, but the animation the horses are trained to exhibit reduces the fore contacts to 80-90% (Tennessee Walking Horse) or 60 to 80% (American Saddle Horse) of hind contacts. These gaits are tiring for horses. They are selected by elephants because they can speed up (from area E) merely by doing the same thing faster, thus avoiding the jolting and fast changes in oscillations that accompany gait changes represented by vertical shifts on the graph. Also, the single-foot, at these speeds, does not have phases when all feet are off the ground at once, as would gaits falling in the pockets between areas K and J, and K and L (a reason for the "fingers" of the gait distribution).

Area L is the trot. It is commonly used at slower speeds by the more agile lizards, many marsupials, many rodents, most carnivores, the hippopotamus, and most ungulates. The moderate and fast running trot are used by some carnivores, cervids, some bovids, and the horse. The trot probably evolved from areas B and C through area H, and was retained by animals that subsequently evolved to areas D and E. The trot is achieved from area H merely by speeding up, and from areas E or F usually by a more sudden change of gait. From area D it may be achieved either gradually (over several cycles) or suddenly (within one cycle). Camelids and giraffids (area F) apparently do not trot at all. The trot is more stable and maneuverable than the pace. It is preferable, therefore, for animals that are not large and slender; for animals with heavy antlers, horns, or heads; and for animals moving over rough terrain. The trot is less tiring than the single-foot because two feet thrust together instead of one.

Area M is a diagonal sequence gait sloping down on the graph from diagonal couplets to the single-foot. So far, only the muntjak (a cervid) and duiker (a bovid) are known to use this mode of travel. Since these animals are in different families, it is probable that they acquired the gait independently. Both are small, forest-dwelling animals. Possibly other small artiodactyls (Tragulus, Ourebia, Raphicerus, Nesotragus, Neotragus, Madoqua) will be found to use the same gait. The gait has a dainty, tripping, high-stepping appearance. It may enable the animals to step easily over tall grass and sticks. Possibly it is more stable than the single-foot in lateral sequence. The gait appears to be derived from area D; transitional gaits also are used (Figure 15).

No gait formulas fall on the right hand part of the gait graph (values between 0 and 20 on the abscissa) because animals cannot sustain symmetrical gaits having so little support. The left hand part of the graph is empty because animals rarely move so slowly in a regular fashion; an herbivore moving one foot at a time while grazing might qualify. As explained above, footfall formulas 1 and 3 are more stable than corresponding formulas 5 and 6, and formula 3 (using contralateral bipods) is more stable than formula 2 (using ipsilateral bipods). This explains the shape of the left margin of the distribution of gaits (Figure 17), and also the absence of formulas in the lower left part of the graph. Parts of the lower half of the graph corresponding to areas D, E, and F are empty because they are less accessible from primitive area A, and because there would be more interference there between fore and hind feet. The lower central part of the graph is empty (the pace centers below 0 on the ordinate) because it is jolting for a pacing animal to place a forefoot a little sooner than the ipsilateral hind, and probably for other reasons of muscle mechanics (see Hildebrand, 1965). The running trot, likewise, is centered a little below 50 on the ordinate.

Conclusion

It is seen that by using slow motion picture film to make gait diagrams, by calculating gait formulas from these diagrams and plotting them on a gait graph, and by using appropriate overlays to assist in analysing the distribution of plots on the graph, dozens or hundreds of symmetrical locomotor performances can be compared and interpreted. Analysis of asymmetrical gaits (which nearly is complete) indicates that similar graphs and overlays can be used in the identification, comparison, and interpretation of that class of gaits.

REFERENCES

Dagg, A.I. and deVos, A., (1968a) The walking gaits of some species of Pecora. J. Zool. (London). 155, 103-110.

Dagg, A.I. and deVos, A., (1968b) Fast gaits of pecoran species. J. Zool. (London). 155, 499-506.

Gambaryan, P.P., (1974) How Mammals Run: Anatomical Adaptations. John Wiley and Sons, New York, (Translated from the Russian edition, Nauka, Leningrad, 1972).

Gray, J., (1944) Studies in the mechanics of the tetrapod skeleton. J. Exp. Biol. 20, 88-116.

Hildebrand, M., (1959) Motions of the running cheetah and horse. J. Mammology. 40, 481-495.

Hildebrand, M., (1962) Walking, running, and jumping. Amer. Zoologist. 2, 151-155.

Hildebrand, M., (1964) Cinematography for research on vertebrate locomotion. Research Film. 5, 1-4.

Hildebrand, M., (1965) Symmetrical gaits of horses. Science. 150, (3697), 701-708.

Hildebrand, M., (1966) Analysis of the symmetrical gaits of tetrapods. Folia Biotheoretica. 6, 9-22.

Hildebrand, M., (1967) Symmetrical gaits of primates. Amer. J. Physical Anthropology. 26, 119-130.

Hildebrand, M., (1968) Symmetrical gaits of dogs in relation to body build. J. Morphology. 124, 353-359.

Howell, A.B., (1944) Speed in Animals. Univ. of Chicago Press, Chicago.

Magna de la Croix, P., (1929a) Filogenia de las locomocienes cuadrupedal y bipedal en los vertebrados y evolution de la forma consecutiva de la evolution de la locomocion. Anales de la Soc. Cientifica Argentina. 108, 383-406.

Magna de la Croix, P., (1929b) Los andares cuadrupedales y bipedales del hombre y del mono. Semina Medica. 48, 1581-1588.

Marey, E.J., (1874) Animal Mechanism: A Treatise on Terrestrial and Aerial Locomotion. Appleton and Co., New York.

Muybridge, E., (1899) Animals in Motion. Chapman and Hall, Ltd. London. (Republished with minor changes in 1957. Brown, L.S., ed., Dover Publ., New York.)

Peabody, F.E., (1948) Reptile and amphibian trackways from the Lower Triassic Moenkopi formation of Arizona and Utah. University of California Publ., Bull. Dept. Geol. Sci. 27, 295-468.

Prost, J.H., (1965) The methodology of gait analysis and gaits of monkeys. Amer. J. Physical Anthropology. 23, 215-240.

Prost, J.H., (1972) A replication study on monkey gaits. Amer. J. Physical Anthropology. 30, 203-208.

Sukhanov, V.B., (1967) Materialy po lokomotsii nazemnykh pozvonochnykh: I. Obshchaya klassifikatsiya simmetrichnykh pokhodok. Moskovskoe Obshchestvo Ispytatelei Prirody Otdel Biologicheskii Biulleten Novaia Seriia. 72, 118-135.

Sukhanov, V.B., (1974) General System of Symmetrical Locomotion of Terrestrial Vertebrates and Some Features of Movement of Lower Tetrapods. Smithsonian Institution and National Science Foundation Washington, D.C., (Translated from the Russian edition, Nauka, Leningrad, 1968.)

Taylor, M.E., (1970) Locomotion in some East African viverrids. J. Mammology. 51, 42-51.

Webb, S.D., (1972) Locomotor evolution in camels. Forma et Functio. 5, 99-111.

Windsor, D.E. and Dagg, A.I., (1971) The gaits of the Macropodinae (Marsupialia). J. Zool. (London). 163, 165-175.

Zug, G.R., (1971) Buoyancy, locomotion, morphology of pelvic girdle and hindlimb, and systematics of cryptodiran turtles. Museum of Zool., University of Michigan, Misc. Publ. No. 142.

Zug, G.R., (1972a) A critique of the walk pattern analysis of symmetrical quadrupedal gaits. Animal Behavior. 20, 436-438.

Zug, G.R., (1972b) Walk pattern analysis of cryptodiran turtle gaits. Animal Behavior. 20, 439-443.

ROBOT LOCOMOTION

Robert B. McGhee

Department of Electrical Engineering

The Ohio State University, Columbus, Ohio 43210

Recent advances in electronics have permitted the construction of a number of experimental robots which depend upon systems of levers for support and propulsion in place of either the wheels or tracks of conventional vehicles. Although the locomotion processes of such robots resemble those of natural systems, at present their performance is grossly inferior to that of even simple animals. This chapter will summarize the characteristics of all robot vehicles which successfully have exhibited stable legged locomotion and will include certain aspects of their engineering theory. It is hoped that this theory will provide a method for studying animal locomotion which can in turn broaden our understanding of neural control mechanisms.

Introduction

It is generally recognized that cursorial animals, including man, possess off-road mobility characteristics far superior to those of conventional wheeled or tracked vehicles. This advantage can be demonstrated not only by informal observation, but also by careful engineering analysis based upon quantitative knowledge concerning soil mechanics, vehicle dynamics, powerplant characteristics, etc. The principles of legged locomotion (McKenney, 1961; Morrison, 1968; Bekker, 1969) in theory allow for efficient and mobile legged vehicles, but the actual effective realization of such robots is in its infancy.

The failure of machine designers to produce really effective walking machines may be due to a limited theoretical understanding of limb-joint-coordination control, and to a lack of motor units as well adapted to limb joint rotation as natural muscles. Fortunately, however, some progress is being made in both of these areas and a number of functioning, although primitive, legged vehicles have been produced within the past decade. This chapter will summarize the characteristics of these machines and will include certain aspects of their engineering theory.

Although these machines are relatively simple and of limited capability in comparison to living systems, their design theory is complicated since it cannot merely describe, but must permit synthesis. That is, the theory must predict all behavioral aspects of these legged machines in order to permit the construction of systems capable of competing with other forms of transportation. This very demanding requirement places the present engineering theory of legged locomotion within the realms of mathematics and mechanics rather than that of biological science. It is hoped that this mathematical and mechanical nature of the theory may stimulate new investigative approaches to natural locomotion systems.

The following discussion will highlight briefly some aspects of legged vehicle design and construction. The references cited provide additional information on aspects of robot locomotion not covered here.

LEGGED VEHICLES: CHARACTERISTICS AND CAPABILITIES

Present day walking toys achieve joint coordination through mechanical linkages incapable of adapting to terrain variations. Since these constructions bear no resemblance to animal locomotion systems with respect to joint control, they will not be discussed here. Only machines with independently powered joints coordinated by an external agency will be described. Within this general category, two basically different approaches to joint control have been utilized, manual control and computer control. Examples of each are presented in the following discussion.

Manually-Controlled Machines

The largest walking machine constructed to date is also the world's largest off-road vehicle. "Big Muskie," a coal-mining dragline, weighs twenty-seven million pounds and is propelled by four hydraulically powered legs, one at each corner of the machine (Cox, 1970). During normal mining operations, Big Muskie rests on

a cylindrical base 105 feet in diameter. During walking, this
machine utilizes twenty-four electric motors of 600 horsepower
each to provide hydraulic power for raising the base off the
ground while transferring the weight of the machine to four feet,
each having dimensions of sixty-five by twenty feet. Once raised,
a second set of actuators moves the machine in a backward stride
of fourteen feet and then rests it again upon its base. This
walking action is accomplished with the aid of an electronic
sequencer which cycles the legs until the operator commands a halt
via a control panel push button. A legged locomotion system was
chosen for Big Muskie in preference to wheeled or tracked systems
because it provided the greatest degree of flexibility at the
least cost under the expected vehicle operating conditions.

Perhaps the second largest walking machine demonstrated to
date is the General Electric Quadruped Transporter. This vehicle
is approximately the size of an elephant and weighs about 3000
pounds. It possesses four legs, each with three degrees of freedom, and also is controlled manually (Liston and Mosher, 1968).
During vehicle operation, the driver is strapped into a seat and
with his hands and feet controls a system of levers which in turn
direct the motions of the vehicle limbs. Force reflecting servomechanisms provide the operator with an indication of the interaction of the vehicle with the supporting terrain. This machine
first walked in 1968 and later exhibited a significant ability to
climb over obstacles and to traverse difficult terrain. Unfortunately, the task of coordinating twelve independent joints was so
demanding that it required nearly all of the operator's attention
and was so physically exhausting that vehicle operation was limited
to a few minutes at a time. Although the mechanical capabilities
of this machine were impressive, its primary contribution to the
field of walking machines may have been to illustrate the need for
computer control of machines with this many degrees of freedom.

Computer-Controlled Vehicles

The first legged vehicle that walked autonomously under full
computer control was the "Phoney Pony" constructed by Frank and
McGhee at the University of Southern California in 1966 (McGhee,
1966; Frank, 1968). This experimental machine was furnished with
four electrically powered legs, each with a single degree of freedom hip joint and an independent single degree of freedom knee
joint. A passive suspension system also was included for permitting
vertical excursion of each leg relative to the vehicle body. This
machine was approximately one hundred pounds in weight and was
roughly the size of a small pony. It was powered by two twelve-
volt automobile batteries connected to the vehicle by a trailing
cable.

The eight independent joints of the Phoney Pony's legs were coordinated by a small special purpose digital computer. The machine's only purpose was to demonstrate that the joint coordination control problem could indeed be solved by an electronic linkage rather than by a mechanical linkage. Two different gaits were demonstrated, the quadruped trot and the quadruped crawl, and then the machine was retired.

In parallel with the work at the University of Southern California, an affiliated group working under Vukobratovic at Institute Mihailo Pupin in Belgrade, Yugoslavia, developed a powered biped exoskeleton intended for application to the locomotion of paraplegics. This device was powered pneumatically and was coordinated by an analog computer using angle versus time joint commands which were derived from measurements of normal gait. Successful operation of this brace both with and without the inclusion of a patient was reported in 1972 (Vukobratovic et al., 1972).

Both of the above machines were true robots in the sense that they were fully autonomous; human interaction was neither necessary nor permitted. More recently, two experimental systems employing interactive computer control have been reported. One of these, developed at the University of Rome, is a six-legged electrically powered system similar to the Phoney Pony except that a degree of operator interaction eventually is intended (Petternella and Salinari, 1974). Another research program, at Waseda University in Tokyo, Japan, has produced a series of computer controlled biped robots with both stair climbing ability and operator control of direction. One in this series is powered hydraulically, weighs 130 kg., and is able to carry a load of 30 kg. (Kato and Tsuiki, 1972). At their present stage of development, Kato's machines are very slow, and require up to 90 seconds per step. Nevertheless, with respect to computer control, they represent the greatest advancement in terms of functioning systems.

Table 1 provides a summary of the characteristics of the machines described above. Photographs of most of these machines and more engineering details can be found in a research monograph by Vukobratovic (1975).

In addition to these six machines, the author knows of only one other successful legged robot with independently powered joints. This is the hexapod system constructed in the Moscow Physio-Technical Institute (Schneider et al., 1974). This machine makes use of electronic joint control similar to the USC Phoney Pony; however further details are not available at this time.

Table 1. Characteristics and Capabilities of Existing Legged Vehicles

Vehicle	Date of First Test	No. of Legs	Approx. Weight	Approx. Top Speed	Payload	Actuator Type	Coordination Method	Control of Speed and Direction
Phoney Pony	1966	4	100 lbs.	.5 mph	none	electric	computer	none
General Electric Quadruped Transporter	1968	4	3000 lbs.	2 mph	500 lbs.	hydraulic	manual	manual
Big Muskie	1969	4	27,000,000 lbs.	.1 mph	500,000 lbs.	hydraulic	electronic sequencer	manual
Mihailo Pupin Exoskeleton	1972	2	50 lbs.	.5 mph	200 lbs.	pneumatic	computer	none
Waseda Biped	1972	2	290 lbs.	.02 mph	65 lbs.	hydraulic	computer	interactive
Univ. of Rome Hexapod	1972	6	100 lbs.	.5 mph	none	electric	computer	interactive

GAIT SELECTION

The first problem to be confronted in the design of a legged robot is that of determining the number of legs to be used for support and propulsion. This decision relates to the purpose and operating environment of the vehicle and to other factors which will not be treated here. However, once the appropriate number has been decided upon, the next step is the choice of a specific gait suitable for a given operating environment. This selection depends upon several variables such as time constraints governing limb motions, gait stability, adaptation to terrain alterations and accommodation of commands from human operators. These aspects of gait are discussed in the following paragraphs.

Sequential and Temporal Aspects of Gait

The most elementary description of a gait is the <u>footfall formula</u> introduced in 1899 by Muybridge (1957) which describes each successive phase or <u>epoch</u> in terms of feet in contact with the ground (Stuart et al., 1973). An equivalent but more mathematical description of gait is the <u>event sequence</u> defined by McGhee and Jain (1972). In the latter description, the legs of the machine are numbered 1,2,...,k. The event of placing leg i is then denoted event i while lifting of leg i is arbitrarily denoted event i+k. Thus, for example, quadruped gaits can be represented by an ordering of the integers 1 through 8. Such orderings are called event sequences. Examples are provided by Figure 1.

$$G = \begin{bmatrix} 0 & 0 & 0 & 0 \\ 0 & 0 & 0 & 1 \\ 0 & 0 & 0 & 0 \\ 0 & 1 & 0 & 0 \\ 0 & 0 & 0 & 0 \\ 0 & 0 & 1 & 0 \\ 0 & 0 & 0 & 0 \\ 1 & 0 & 0 & 0 \end{bmatrix} \qquad G = \begin{bmatrix} 0 & 0 & 1 & 0 \\ 0 & 0 & 1 & 1 \\ 0 & 0 & 0 & 1 \\ 1 & 0 & 0 & 1 \\ 1 & 1 & 0 & 1 \\ 1 & 1 & 0 & 0 \\ 1 & 0 & 0 & 0 \\ 1 & 0 & 1 & 0 \end{bmatrix}$$

E = 18462735   E = 18356427

a) A compatible matrix   b) An incompatible matrix

Figure 1. Column Compatible and Column Incompatible Quadruped Gait Matrices.

When none of the 2k events associated with a given gait occurs simultaneously, the event sequence is said to be totally ordered. Gaits associated with totally ordered event sequences are called connected gaits while partially ordered sequences correspond to singular gaits (McGhee, 1968). It has been noted (McGhee, 1968) that only totally ordered event sequences or connected gaits can be used in strictly periodic motion by an animal or machine. All other event sequences represent the limit of some totally ordered sequence as the phase relationships between various legs are altered. As a result, most of the following analysis is limited to totally ordered event sequences.

The footfall formula or event sequence descriptions of gait are deficient since the relative duration of successive gait epochs is not considered. To remedy this exclusion, Hildebrand (1966) suggested the use of a gait diagram consisting of horizontal lines plotted against time. Each line in a gait diagram corresponds to a specific leg in contact with the ground while interruption of the line indicates that the corresponding leg is in the air. McGhee (1968) converted this description to an equivalent mathematical definition by defining: 1) the duty factor, β_i, for leg i as the fraction of a locomotion cycle during which leg i is in contact with the ground; and 2) the phase, ϕ_i, as the fraction of a cycle by which the contact of leg i with the ground lags behind the contact of leg 1. Relationships between event sequences and gait duty factor and phase variables were studied first by Hildebrand and later by others (Hildebrand, 1965, 1966; McGhee, 1968; Koozekanani and McGhee, 1972; McGhee and Jain, 1972; Stuart et al., 1973; Sun, 1974) and much is now known about this complex subject. Some of the results most significant to the selection of robot gaits are summarized in the following paragraphs.

McGhee (1968) demonstrated that the number of distinct totally ordered event sequences for a k-legged system is given by

$$N = (2k - 1)!$$ Equation (1)

Thus, for quadrupeds, a total of 7! = 5040 connected gaits are possible. For hexapods, the number of such gaits is 11! = 39,916,800. Obviously, selecting one gait from all of these possibilities is a problem of overwhelming combinatorial complexity. This necessitates finding a simple criterion for rejecting most of them. A possible criterion is furnished by the following observation: if all legs operate with the same duty factor, then no leg can be placed and then lifted while another leg is on the ground. Gaits not containing such an event are said to be column compatible. With few exceptions, animals use only compatible gaits (Wilson, 1966, 1967; McGhee and Jain, 1972).

The notion of gait compatibility is understood more easily by making use of an alternative representation of gaits called the gait matrix. In the analytic model for gait formulated by McGhee (1968), a gait matrix, G, is formally defined as a k-column, binary matrix in which no two successive rows are identical (including the first and last row) and in which each column is composed of a string of zeros and a string of ones with a single change from zero to one and a single change from one to zero (again, the first row is considered to follow the last). Figure 1a provides an example of a connected gait matrix and its associated event sequence. In this figure, the matrix G is derived from the event sequence E by assuming that column i of G represents the state of leg i and that the 0-state corresponds to a leg in contact with the ground while the 1-state represents a leg in the air. Referring to G in this figure, it is obvious that this gait is compatible since none of the zeros of one column overlaps the zeros of any other column. Figure 1b shows a gait which is not compatible since the zeros of column 2 overlap the zeros of column 1.

McGhee and Jain (1972) showed that out of the 5040 theoretically possible connected quadruped gaits, only 492 are column compatible. To simplify their work, they noted that column compatibility is a temporal property of a gait matrix which is not affected by column permutations. They also observed that complementation (exchanging all zeros for ones and conversely) of a gait matrix also does not alter compatibility. By using certain combinations of these two operations they thereby reduced the 492 compatible gait matrices to a set of 45 equivalence classes. More recently, Sun (1974) has used a wider class of transformations to further reduce this set to only 14 equivalence classes. These classes are listed in Table 2 in which each class is represented by one of its members, the minimal row and column canonical form.

For an arbitrary k-column connected gait matrix, G, the minimal row and column canonical form is obtained by the following process.

(1) Permute columns 2,3,...k so that the placing events are in the order of the columns; i.e., in the event sequence, event 2 precedes event 3, 3 precedes 4, etc. This is the row and column canonical form (McGhee and Jain, 1972) for G. Call this event sequence E_1 and the corresponding gait matrix G_1.

(2) Rotate the columns of G_1 to the left by one position; i.e., move column 2 to column 1, 3 to 2, 4 to 3, etc., and move 1 to k. Then rotate the rows so that column 1 begins with a zero and ends with a 1. Call this matrix G_2 and the associated event sequence E_2.

(3) Repeat step 2 a total of k-2 more times (twice more for quadrupeds) yielding $E_3, E_4, \ldots, E_k$.

(4) Complement G_1 and rotate its rows so that column 1 again begins with zero and ends with one.

(5) Repeat steps 1 through 3 to generate E_{k+1} through E_{2k}.

(6) The minimal event sequence, E^*, is given by

$$E^* = \min_i E_i \qquad \qquad \text{Equation (2)}$$

The corresponding matrix, G^*, is the minimal row and column canonical form for G. Figure 2 illustrates this procedure for the gait defined by Figure 1b. Of course the value obtained for E^* in this case is not in Table 2 since, as noted previously, this gait is not column compatible.

Table 2. Equivalence Classes of Column Compatible Connected Quadruped Gait Matrices

Equivalence Class Number	Number of Members	Minimal Event Sequence	Maximum Duty Factor, β_{max}	Minimum Duty Factor, β_{min}
1	24	12345678	1	0
2	48	12354678	1/2	0
3	48	12356478	1/2	0
4	48	12356748	1/2	0
5	48	12385467	1	1/2
6	48	12536478	1/2	0
7	48	12536748	1/3	0
8	24	12563478	1/2	0
9	48	12563748	1/3	0
10	48	12835467	2/3	1/2
11	24	12835647	2/3	1/3
12	12	12853467	---	---
13	12	15263748	1/4	0
14	12	17283546	3/4	1/2

Note: Class 1 and Class 11 are self-complementary classes. Class 12 is column compatible, but not regularly realizable. The duty factors given are for the minimal event sequence for each equivalence class. For classes which are not self-complementary, complementary gaits are realizable with complementary duty factor ranges (McGhee and Jain, 1972).

Step 1: E = 18356427   so E_1 = 17258346

Step 2: E_2 = 18723546

Step 3: E_3 = 12835476

E_4 = 17283654

Step 4: Complement of E_1 = 14782536

Step 5: E_5 = 12863547

E_6 = 17528364

E_7 = 17253468

E_8 = 18235746

Step 6: $E^* = \min_i E_i = E_3$ = 12835476

Figure 2. Computation of Minimal Event Sequence for a Quadruped Gait.

In order that all legs may be constructed identically in robots, it is desirable that the duty factor be the same for all legs. That is, for such regular gaits,

$$\beta_i = \beta_j = \beta, \quad i,j = 1,2,\ldots,k \qquad \text{Equation (3)}$$

Although some exceptions have been noted, regular gaits also are favored by animals (Hildebrand, 1965, 1966; Wilson, 1966, 1967; McGhee and Jain, 1972). In McGhee's analysis (1968), the concept of regular realizability of a gait matrix is introduced. Briefly described, a gait matrix G is regularly realizable if it is possible to assign a time duration to each row of G so that β is the same for all legs. Evidently, regular realizability is a stronger condition than compatibility and it is interesting to determine which of the compatible gaits possess this property. McGhee and Jain (1972) point out that this question can be answered by solving a linear programming problem for each minimal row and column canonical form. Such a calculation shows that 13 of the 14 classes listed in Table 2 are regularly realizable (Sun, 1974). It should be noted that solution of the linear programming problem also determines the allowable duty factor ranges included in this table for each gait (McGhee and Jain, 1972).

The information presented in Table 2 demonstrates the value of mathematical analysis in understanding physical and biological systems. An extension of these results shows that the total number of possible quadruped gaits including the singular gaits, is actually 63,136 (Koozekanani and McGhee, 1973), a number much larger than intuition suggests. Of this total, only the 5040 connected gaits need be considered since every other gait is a singular limit of some connected gait. Out of the 5040 connected gaits, for either mechanical systems or living quadrupeds, all but 492 can be discarded on the basis of gait matrix column incompatibility. Finally, this latter set reduces to just fourteen equivalence classes under column permutations and matrix complementation. Table 2, therefore, represents a complete catalogue of all possible quadruped gaits possessing the column compatibility property. Of the sixteen nonsingular symmetric gaits studied by Hildebrand (1965, 1966), eight are in class 8, four are in class 13, and the remaining four are in class 14. All of the named asymmetrical nonsingular quadruped gaits such as the rotatory gallop, transverse gallop, canter, etc., also are included along with many other theoretically possible gaits never observed in use by any animal. A listing of all named nonsingular quadruped gaits known to the author along with their corresponding event sequences can be found in the work of McGhee and Jain (1972).

The above analysis and classification of quadruped gaits has been extended to hexapods by Sun (1974). While these results are too lengthy and complex for inclusion here, it is noteworthy that his analysis shows within the 39,916,800 theoretically possible nonsingular hexapod gaits, there are only 148 equivalence classes of compatible gait. Of these, exactly 135 classes are found to be regularly realizable, and the duty factor range for each of these classes also is given. All 288 nonsingular regular symmetric hexapod gaits are contained in 7 equivalence classes, a remarkably small number.

Sun's results represent the furthest advance to date in analysis of the combinatorial aspects of gait. Similar results for machines or animals possessing more than six legs are not available as yet.

Gait Stability

While the above discussion has summarized the effect of certain types of time constraints (Stuart et al., 1973) on allowable gait patterns, such considerations alone are not sufficient to define a unique gait for a given speed range for a particular vehicle or animal. Suppose, for example, that a nonsingular quadruped gait suitable for utilization with a duty factor, $\beta = 1/2$ is desired. Such a gait is possible only for equivalence classes for which (McGhee, 1968)

$$\beta_{min} < 1/2 < \beta_{max} . \qquad \text{Equation (4)}$$

Reference to Table 2 shows that this condition is satisfied by the 24 gaits contained in class 1 and 24 gaits contained in class 2. While this represents a considerable narrowing of the original 5040 possibilities, it still does not yield a unique choice for the given duty factor. Some further criterion must be found to permit such a selection. Unfortunately, except for low speed locomotion, specifically for $\beta \geq 3/4$, no such criterion has been found for quadrupeds. The available engineering theory of legged locomotion is neither able to specify the best gait for quadruped robots nor able to explain quantitatively the gait preferences of living quadrupeds in other duty factor ranges.

For low speed locomotion of quadrupeds, and for all hexapod gaits, it has been found that maximization of a criterion called the longitudinal stability margin, $s^*(G)$, accounts for the gait selections made by animals and also leads to a unique choice among the elements of the appropriate equivalence classes of gait for robots. The longitudinal stability margin is computed for a given gait matrix G in the following way (McGhee and Frank, 1968).

(1) Consider all assignments of duty factor β and leg phases $\phi_2, \phi_3, \ldots, \phi_k$ capable of producing G.

(2) For any assumed leg geometry, and for a given assignment of β and ϕ_i, $i = 2,\ldots,k$, determine the closest approach in the direction of travel of the vertical projection of the vehicle center of gravity to the front or rear boundary of the support pattern formed by the feet in contact with the supporting surface. Call this distance s.

(3) The maximum of s over all β and ϕ_i implying G is the quantity $s^*(G)$.

McGhee and Frank (1968) have proved a theorem which shows that if leg mass can be ignored in comparison to body mass, then the unique optimum quadruped gait with respect to maximization of $s^*(G)$ is the quadruped singular crawl. This gait is a regular symmetric gait defined by the single parameter

$$\phi_3 = \beta , \quad \beta \geq .75 \qquad \text{Equation (5)}$$

where ϕ_3 is defined as the fraction of a locomotion cycle by which the placing of the left rear leg follows the placing of the left front. Figure 3 illustrates the successive epochs of this gait.

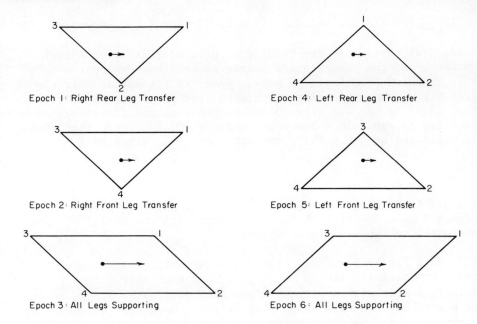

Figure 3. Support patterns for successive epochs of quadruped singular crawl for β = 11/12. Arrows indicate total motion of center of gravity during each epoch.

More recently, Bessonov and Umnov (1973) have shown that for six-legged gait, s*(G) is maximized by a regular symmetric gait in which

$$\phi_3 = \beta, \quad \phi_5 = 2\beta - 1, \quad \beta \geq .5, \qquad \text{Equation (6)}$$

and in which ϕ_3 is the time delay of the left middle leg and ϕ_5 is the delay of the left rear leg. Both are measured as a fraction of a total leg cycle and are relative to the placing of the left front leg as in Equation (5). Both Equation (5) and Equation (6) describe <u>wave gaits</u> in which a wave of placing events runs from the rear to the front along either side of an animal or vehicle with a constant time interval between the action of adjacent legs on the same side. Such gaits are favored by simple animals and have been proposed for use in vehicles (Sindall, 1964; Wilson, 1966, 1967; Bessonov and Umnov, 1973; Okhotsimski and Platonov, 1973).

In his comprehensive study of gait, Sun (1974) considers the general regular symmetric wave gait defined for vehicles with k = 2K legs by

$$\phi_{2n+1} = R(n\phi_o), \quad n=1,2,3,\ldots,k-1, \quad 0 < \phi_o < 1 \qquad \text{Equation (7)}$$

where $R(x)$ represents the fractional part or residue of a real number x and the subscripts n denote successive legs on the left side numbered from front to back. Under an assumption that the 2K legs are arranged in pairs along each side of a central body and that both the spacing between successive legs and the stroke of each leg is equal to a constant distance, d, Sun finds that $s^*(G)$ is maximized by using the phase increment, ϕ_o, given by

$$\phi_o = \beta, \quad \beta \geq \frac{3}{k}. \qquad \text{Equation (8)}$$

The stability margin for such optimal wave gaits is presented on Figure 4 as a function of the duty factor and the number of legs possessed by the vehicle or animal.

Both Figure 4 and Equation (8) show that for a k-legged system, the stability margin is greater than zero only for $\beta \geq \frac{3}{k}$. This is not at all surprising since this constraint simply ensures, at a minimum, a stable supporting tripod. However, the curves of Figure 4 do not merely confirm this obvious fact, but rather quantitatively depict the _degree_ of static stability possible with a given number of legs as a function of leg duty factor.

Further examination of Figure 4 sheds additional light on certain characteristics of animal gait. First, it has been

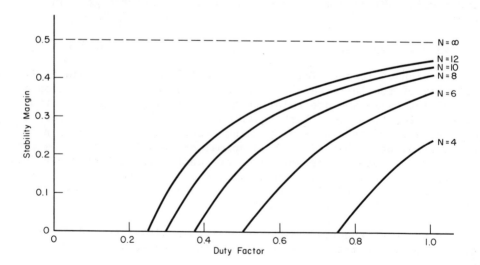

Figure 4. Optimal wave gait stability margin as fraction of body length vs. duty factor for N-legged locomotion systems.

reported that insects never employ gaits with a duty factor less than one-half and moreover they always use wave gaits during steady constant speed locomotion (Wilson, 1966). Spiders appear to utilize the same basic principle, although their greater degree of stability permits less critical control of leg phasing (Wilson, 1967). On the other hand, quadruped animals employ the crawl gait (Figure 3), or wave gait, only for very low speed activities. During high speed locomotion, quadrupeds may use duty factors as low as $\beta = .2$ with resulting gait epochs in which no legs contact the ground (Hildebrand, 1965). Evidently, the stability of such gaits depends upon complex neural sensing and processing networks and cannot be accounted for by simple considerations of static stability. Robots using such principles for gait stability have not been built, although they have been studied by means of computer simulation (Frank, 1968; Frank and McGhee, 1969; McGhee and Pai, 1974).

With bipeds, initially it appears that notions of static stability cannot be applied since the constraint on duty factor expressed by Equation (8) cannot be satisfied with less than four legs. However, this is not an accurate interpretation. The underlying assumptions which produced this curve were that the leg mass was negligible when compared with total system mass and that each foot contacted the ground in a single point. Neither of these conditions is satisfied by most bipeds and certainly not by man. Actually, the large feet and massive legs of human beings facilitate shifting of weight during walking in such a way that the center of pressure or zero moment point of the foot reaction with the ground always lies within the support pattern formed by foot-ground contact and including the space between the feet during double support. It is this action which produces the characteristic side to side sway observed in normal human locomotion (Saunders et al., 1953; Vukobratovic, 1975). Unfortunately, mathematical treatment of this type of gait requires the use of differential equation models. Such analysis is still in an early stage of development, although it is nevertheless already too complex for in depth discussion here. Instead, an overview of the present state of knowledge regarding the dynamic stability of bipeds will be given.

The simplest possible mechanical model for a biped robot is illustrated by Figure 5 (Gubina et al., 1974). In this model, the mass of the legs still is ignored but their movement produces a moment, M, and a force, F, both of which act on the upper body. The balancing principle is dynamic rather than static. For biped robots, Gubina et al., (1974) quantitatively show that overall gait stability can be achieved by proper placement of the feet. This results from feedback control derived from the dynamic state of the body. This stability results from alternate fall and recovery cycles. Again, however, these results were obtained from computer

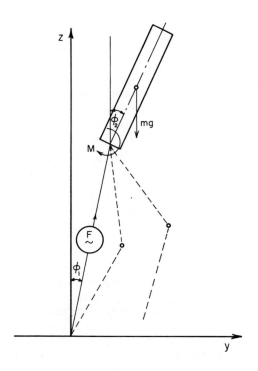

Figure 5. Elementary mechanical model for a biped robot.

simulation rather than from physical experiments. While such simulation studies yield valuable insights into possible robot control schemes, more work and more detailed, accurate, dynamic models are needed before the foot placement principle of gait stabilization can be applied to a physical biped. All successful biped robots demonstrated to date have depended upon zero moment point control for gait stability. Selection of a proper gait in terms of time dependent relationships between joint angles to implement this principle is a complex matter and the reader is referred to the literature for a more complete treatment of this problem (Vukobratovic et al., 1972; Kato and Tsuiki, 1972; Vukobratovic, 1975).

Terrain Variability and Human Interaction

The preceding discussion of gaits relates to constant speed locomotion in a straight line over level hard surfaces. However, a useful robot should acknowledge higher authority commands, such as steering and speed, and also should adapt by itself to non-ideal terrain. Indeed, these are precisely the long range goals of legged robot research. Unfortunately, a great deal remains to be done in this area. While these problems have received some attention via computer simulation (Okhotsimski and Platonov, 1973; McGhee and Orin, 1973), and while at least two six-legged robots intended to incorporate both of these features presently are under construction (Ignatiev et al., 1972; Orin, 1976), as yet no experimental results are available. Much of future robot research undoubtedly will be focused on these problems.

JOINT COORDINATION CONTROL

There are two principal approaches in robot design for achieving joint coordination control in experimental systems. The first of these is called <u>model reference control</u> (McGhee and Pai, 1974) or <u>algorithmic control</u> (Vukobratovic et al., 1972; Vukobratovic, 1975), and depends upon computer solution of the differential equations of motion of a legged system. The joint angle commands generated by the computer are subsequently converted to actual joint positions by electric, hydraulic, or pneumatic servomechanisms. While this approach offers considerable promise for robots, it bears little or no relation to neural control and will not be discussed further.

The second approach to joint coordination, originally proposed by Tomovic and McGhee (1966), is called <u>finite state control</u> and is much more biologically motivated. In finite state control, each limb functions similar to an oscillator with the action of each oscillator being modulated by cross-coupling to other oscillators and possibly also by signals from a higher center. Control of joint rotation is accomplished by simple on-off signals sent to joint motors. These signals are determined not by computer solutions of differential equations or of kinematic relationships, but rather by certain straightforward logical relationships observed in joint positions. In order to explain this approach to control concretely, the functioning of the "Phoney Pony" controller (McGhee, 1966; Frank, 1968) will be detailed.

Finite State Control of the Trot Gait

Since the trot gait involves alternating support by diagonal pairs of legs, it is not statically stable during much of its cycle. To cope with this difficulty without resorting to model reference control, the Phoney Pony was constructed with wide tubular feet. This provided a support zone of considerable size even when only two feet were on the ground. Figure 6 is a photograph of this machine exhibiting this feature. As shown, each leg has a single degree of freedom knee joint and a single degree of freedom hip

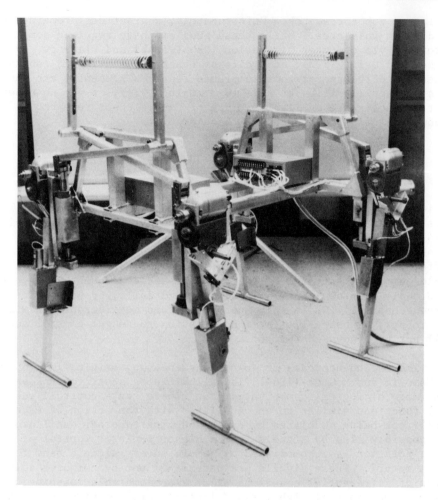

Figure 6. The "Phoney Pony" walking machine.

joint, both powered by electric motors. Figure 7 is a close-up photograph of one leg showing both joints flexed. Each joint of each leg was driven through a worm gear and could be switched electronically to one of three states; forward rotation, rearward rotation, and locked. By appropriate use of limit sensors, each leg was caused to operate as a monostable oscillator. Figure 8 illustrates this action for one leg. Figure 9 presents a state graph for the electronic controller used to produce the desired behavior for each leg. This diagram includes an indication of the conditions governing the change of each state to its successor state. Table 3 gives the joint rotational states associated with each controller state.

Figure 7. Detail of Phoney Pony leg showing both joints flexed.

Figure 9 can be understood with the help of Figure 8 and Table 3 as follows. Starting in controller State 1, the knee is straight and the hip is fully flexed (Figure 8i). As Table 3 indicates, both joints are locked at this time resulting in a stable leg state. The signal S changes from 0 to 1 upon receipt of a start signal from a central oscillator contained in the control computer. At that time, the controller state changes from 1 to 2 and the leg power stroke begins. This stroke continues until the hip reaches its rearward limit of -12 degrees, as shown on Figure 8c. When

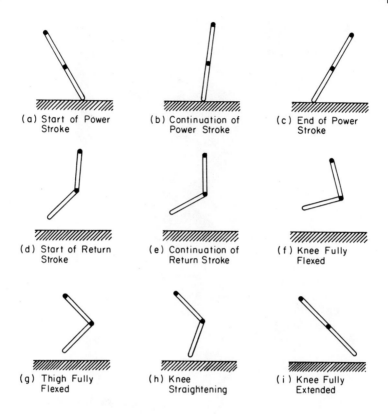

Figure 8. Successive phases of Phoney Pony leg cycle.

this limit is detected by an appropriately placed sensor, the controller enters State 3 and the hip power is turned off until the interlock signal, I, changes from 0 to 1. The purpose of this signal is to insure that the ipsilateral leg of the vehicle is in a supporting position before the given leg is lifted. Normally, this will be the case and State 3 usually will be of very short duration. That is, the desired gait is a slow trot or jog, in which alternately 2 and 4 legs are on the ground, and not the fast trot which includes periods with no feet on the ground (Hildebrand, 1965). When the interlock condition is satisfied, the controller enters State 4 and the knee is flexed to about -70° (Figure 8f) at which time the controller goes to State 5. State 6 is entered when the hip forward limit sensor detects a flexion angle of +30°. At this time the knee is straightened and the controller goes back into State 1, thereby completing one cycle.

It is important to realize that the Phoney Pony control computer contained four electronic circuits corresponding to Figure 9;

ROBOT LOCOMOTION

Table 3. Dependence of Joint Rotational States on Controller States for Quadruped Trot

Control State	Hip Rotation	Knee Rotation
1	locked	locked
2	rearward	locked
3	locked	locked
4	forward	rearward
5	forward	locked
6	locked	forward

i.e., one for each leg. It is also important to note that while four controller state transitions are governed by the interaction of knee and ankle angles on the same leg, each such cycle is also modulated by ipsilateral leg coupling via the interrupt signal while the overall synchronization of leg motion is accomplished by the central "clock" oscillator producing the signal S. Moreover, since in the trot gait the vehicle legs operate in two pairs, there must be two such synchronizing signals, one for the left front and the right rear leg and another for the other two legs. Obviously, these two start signals must be separated from each other in time by exactly one-half of a gait cycle.

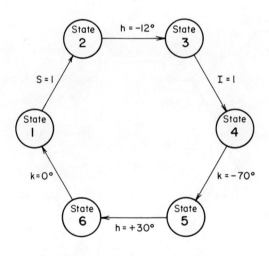

h = hip angle
k = knee angle
S = start signal
I = interlock signal

Figure 9. Control state diagram for trot controller.

Crawl Controller

While the Phoney Pony successfully exhibited computer control of the trot gait in 1966, it subsequently was decided that much could be learned by programming it for some other gait. The optimal stability properties of the singular crawl (Figure 3) were discovered at about this time. Therefore, it was selected for the next gait. The intention was to design the crawl controller as an asynchronous machine; i.e., to operate without the central oscillator used to generate the start signal for the trot controller. After considerable analysis and many experiments, a successful crawl controller was demonstrated in 1968. The leg control algorithm used by this controller is shown in Figure 10. Table 4 presents the leg-joint states associated with each controller state of this figure.

Comparison of Figure 10 with Figure 9 reveals that the trot and crawl controller are similar in design. The obvious differences are an extra state, namely State 2', added to the crawl controller and replacement of the start signal, S, which caused the transition from State 1 to State 2 on the trot controller by the signal h = +12° on the crawl controller. These changes can be

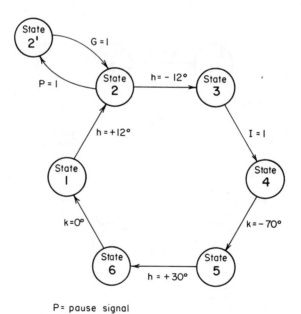

P = pause signal
G = synchronization Signal

Figure 10. Control state diagram for crawl controller.

Table 4. Dependence of Joint Rotational States on
Controller States for Quadruped Crawl

Control State	Hip Rotation	Knee Rotation
1	rearward	locked
2	rearward	locked
2'	locked	locked
3	locked	locked
4	forward	rearward
5	forward	locked
6	locked	forward

explained in the following way. First, since the crawl controller is asynchronous, there is no synchronizing signal available to it. Instead, each leg begins its power stroke immediately upon the completion of knee straightening. This means that the leg state illustrated by Figure 8i is not stable for the crawl. Inspection of Table 4 confirms this, since the hip rotation in State 1 is rearward for the crawl rather than locked as in the trot (also see Table 3).

The addition of State 2' in Figure 10 was for the purpose of maintaining contralateral leg synchronization just as State 4 was included in Figure 9 for ipsilateral leg synchronization. Specifically, the crawl controller was designed to operate with a nominal duty factor $\beta = 3/4$. Thus, the movement of any leg from +12 degrees to -12 degrees during its supporting phase (Figures 8a through 8c) consumed 75 percent of a gait cycle, while the remaining 25 percent of the cycle was allocated to the return phase (Figures 8d to 8a). A consequence of this motion is that whenever a contralateral pair of legs are both in contact with the ground, since their relative phase should be equal to exactly one half of a cycle, the difference between their hip angles should be 16 degrees at all times. In actuality, of course, one leg or the other of such a pair may complete its return stroke too rapidly or too slowly and this condition may not be satisfied. Likewise, even when both legs contact the ground, one or the other may slip on the supporting surface and thereby upset the desired relative angular relationship. To correct for such accidents of performance in the crawl controller, the pause signal, P for each leg, is set to one at h = +12 degrees and again at h = -4 degrees. The signal G then is set to one only when the contralateral leg has reached its proper position. For example, suppose Figure 10 is interpreted as the left front (LF) leg controller. Then the contralateral leg is the right front (RF). Thus, while the LF controller is in State 1 and the LF leg is rotating backwards toward h = +12 degrees and the start of its power stroke, the RF controller should be in State 2 with the RF leg moving backward somewhere between +12 degrees and -4 degrees. Reaching the rearward limit (+12 degrees for LF, -4 degrees for RF) causes the controller of that leg to enter State 2' thereby stopping leg rotation (see Table 4). When the other leg reaches its limit, then its controller will also enter State 2'. At this time, since both legs are stopped, G=1 and both controllers enter State 2, restarting rearward hip rotation.

There is another change in the crawl controller not apparent from a comparison of Figures 9 and 10. This change relates to the functioning of the interlock signal, I. The purpose of this signal is to ensure that at least two legs are in contact with the ground at all times during the trot, but for the crawl gait it must keep at least three legs on the ground. Observing the sequence of events illustrated by Figure 4, note the LF leg must be stopped at -12 degrees unless the LR leg has reached +12 degrees (ipsilateral interlock) but the LR leg must stop at -12 degrees unless the RF has reached +12 degrees (diagonal interlock). Likewise, RF is ipsilaterally interlocked to RR, while RR is diagonally interlocked to LF.

Again, as for the trot, the control computer for the crawl contains four individual leg controllers. Since each of these controllers is realized as an electronic circuit, the switching action, already described, occurs very rapidly and, if the leg phasing errors are small, no perceptible hesitation in motion will be observed. On the other hand, if such errors are too large, thereby allowing jerky motion, the resynchronizing action can be rendered more subtle by setting P=1 at more than just two points in the cycle. This, of course, does not change the state diagram of Figure 10, but merely adds additional logical conditions to the computation of P and G. Still smoother control can be achieved by including three or even all four hip angles in G for each leg. The controller in the Phoney Pony experiments reported by Frank (1968) used some such additional terms, but the controller state graph and its principle of operation were identical to that described and only the four decision points h = +30, +12, -4, and -12 degrees were used.

In summary, two concrete examples of simple finite algorithms for accomplishing quadruped limb coordination control have been proven effective experimentally. The control computer realizing these algorithms was very small, containing only 16 flip flops in each case (Frank, 1968). One algorithm (the trot controller), makes use of an autonomous oscillator in conjunction with an ipsilateral limb coupling signal to control the basic leg cycle. The other (the crawl controller), is self-timing and achieves coordination through ipsilateral, contralateral, and diagonal leg coupling signals. At this time, the author knows of no other finite algorithms for limb coordination which have been similarly demonstrated to be capable of producing stable locomotion.

Conclusions

This chapter has presented an overview concerning design and construction of artificial legged locomotion systems with an emphasis on the gait selection and limb joint coordination control problems. Methods used for classifying and analyzing gaits for robots with four or more legs have been shown to be applicable to studies of animal locomotion. It has been noted that for locomotion systems with six or more legs, the gaits found to be optimally stable for robots correspond to the gait preferences of animals and a quantitative measure of gait stability for such systems has been presented.

The present theory for biped robots, though complicated, is not yet complete enough for realistic treatment of human gait, although stable biped machines have been constructed. On the other hand, for quadruped robots, two simple finite algorithms have been shown to possess sufficient power to accomplish limb coordination. One of these algorithms requires a central clock oscillator to synchronize the individual leg patterns while the other depends solely on intralimb and interlimb coupling signals. Presumably, comparably simple algorithms could be found for six-legged machines. However, the main problems currently under consideration in legged vehicle design are those of adaptation to terrain alterations and the development of limb joint control algorithms capable of responding to human guidance signals. At least two six-legged vehicles specifically intended for such studies presently are under construction.

The theory of robot legged locomotion is quite different from the corresponding theory for animal locomotion because it must be constructive and not merely descriptive. While the man-made vehicles demonstrated to date are very primitive in comparison to living systems, their principles of operation are understood exactly. No hypotheses are involved in descriptions of their function. It is hoped that this simplicity of artificial legged locomotion systems coupled with the exactness of their theoretical basis will provide some new directions for the study of animal locomotion. While only a brief summary of the presently available theory of robot locomotion has been presented, the references cited pursue this subject in considerably greater depth.

Acknowledgment

The support of this research by the National Science Foundation under Grant ENG74-21664 is gratefully acknowledged.

REFERENCES

Bekker, M.G., (1969) <u>Introduction to Terrain-vehicle Systems.</u> University of Michigan Press, Ann Arbor.

Bessenov, A.P. and Umnov, N.V., (1973) The analysis of gaits in six-legged vehicles according to their static stability. Proc. of Symposium on Theory and Practice of Robots and Manipulators, Udine, Italy.

Cox, W., (1970) Big Muskie. News in Engineering, Ohio State University, Columbus, (25-27).

Frank, A.A., (1968) Automatic control systems for legged locomotion. Ph.D. Dissertation, Univ. of Southern California, Los Angeles.

Frank, A.A. and McGhee, R.B., (1969) Some considerations relating to the design of autopilots for legged vehicles. J. Terramechanics. $\underline{6}$, 23-25.

Gubina, F., Hemami, H. and McGhee, R.B., (1974) On the dynamic stability of biped locomotion. IEEE Trans. on Biomed. Engr. $\underline{21}$, 102-108.

Hildebrand, M., (1965) Symmetrical gaits of horses. Science. $\underline{150}$, 701-708.

Hildebrand, M., (1966) Analysis of the symmetrical gaits of tetrapods. Folia Biotheoretica. $\underline{4}$, 9-22.

Ignatiev, M.B., Kulakov, F.M. and Pokrovsky, A.M., (1972) <u>Algorithms for the Control of Robot-Manipulators.</u> Machinostroeniyeh Press, Leningrad. (In Russian).

Kato, I. and Tsuiki, H., (1972) Hydraulically powered biped walking machine with a high carrying capacity. Proc. Fourth International Symposium on External Control of Human Extremities. Dubrovnik, Yugoslavia.

Koozekanani, S.H. and McGhee, R.B., (1973) Occupancy problems with pairwise exclusion constraints--an aspect of gait enumeration. J. Cybernetics. $\underline{2}$, 14-26.

Liston, R.A. and Mosher, R.S., (1968) A versatile walking truck. Proc. 1968 Transportation Engineering Conference, ASME-NYAS. Washington, D.C.

McGhee, R.B., (1966) Finite state control of quadruped locomotion. Proc. of Second International Symposium on External Control of Human Extremities. Dubrovnik, Yugoslavia.

McGhee, R.B., (1968) Some finite state aspects of legged locomotion. Math. Biosciences. $\underline{2}$, 67-84.

McGhee, R.B. and Frank, A.A., (1968) On the stability properties of quadruped creeping gaits. Math. Biosciences. $\underline{3}$, 331-351.

McGhee, R.B. and Jain, A.K., (1972) Some properties of regularly realizable gait matrices. Math. Biosciences. $\underline{13}$, 179-183.

McGhee, R.B. and Orin, D.E., (1973) An interactive computer-control system for a quadruped robot. Proc. of Symposium on Theory and Practice of Robots and Manipulators. Udine, Italy.

McGhee, R.B. and Pai, A.L., (1974) An approach to computer-control for legged vehicles. J. Terramechanics. $\underline{11}$, 9-27.

McKenney, J.D., (1961) Investigation for a walking device for high efficiency lunar locomotion. Paper 2016-61. American Rocket Society. Space Flight Report to the Nation, New York.

Morrison, R.A., (1968) "Iron mule train," Proc. of Off-Road Mobility Research Symposium. International Soc. for Terrain Vehicle Systems. Washington, D.C., (381-400).

Muybridge, E., (1957) <u>Animals in Motion.</u> Dover Publications, New York, (First published in 1899, Chapman and Hall, Ltd., London.).

Okhotsimski, D.E. and Platonov, A.K., (1973) Control algorithm of the walker climbing over obstacles. Proc. of the Third International Joint Conference on Artificial Intelligence. Stanford, California.

Orin, D.E., (1976) Control of a six-legged vehicle with optimization of both stability and energy. (Ph.D. Dissertation), The Ohio State University, Columbus, Ohio.

Petternella, M. and Salinari, S., (1974) Six legged walking vehicles. Report No. 74-31, Istituto di Automatica, University of Rome. Rome, Italy.

Saunders, J.B., Inman, V.T. and Eberhart, H.D., (1953) The major determinants in normal and pathological gait. J. Bone and Joint Surg. 35-A, 543-558.

Schneider, A. Yu., Gurfinkel, E.V., Kanaev, E.M. and Ostapchuk, V.G., (1974) A system for controlling the extremities of an artificial walking apparatus. Report No. 5, General and Molecular Physics Series, Physio-Technical Institute, Moscow, USSR. (In Russian).

Sindall, J.N., (1964) The wave mode of walking locomotion. J. Terramechanics. 1, 54-73.

Stuart, D.G., Withey, T.P., Wetzel, M.C. and Goslow, Jr., G.E., (1973) "Time constraints for inter-limb coordination in the cat during unrestrained locomotion," In Control of Posture and Locomotion. (Stein, R.B. et al., eds.), Plenum Press, New York, (537-560).

Sun, S.S., (1974) A theoretical study of gaits for legged locomotion systems. (Ph.D. Dissertation), The Ohio State University, Columbus, Ohio.

Tomovic, R. and McGhee, R.B., (1966) A finite state approach to the synthesis of bioengineering control systems. IEEE Trans. on Human Factors in Electronics. HFE-2, 65-69.

Vukobratovic, M., Ciric, V., Hristic, D. and Stepanenko, J., (1972) Contribution to the study of anthropomorphic robots. Paper 18.1, Proc. of IFAC V World Congress, Paris.

Vukobratovic, M., (1975) Legged Locomotion Robots and Anthropomorphic Systems. Research monograph. Institute Mihialo Pupin, Belgrade, Yugoslavia.

Wilson, D.M., (1966) Insect walking. Annual Rev. of Entomology. 11, 103-121.

Wilson, D.M., (1967) Stepping patterns in tarantula spiders. J. Exp. Biol. 47, 133-151.

ORGANIZATIONAL CONCEPTS IN THE CENTRAL MOTOR NETWORKS OF

INVERTEBRATES

William J. Davis

The Thimann Laboratories, University of California at

Santa Cruz, Santa Cruz, California

Motor behavior is produced by central networks of neurons that can operate without the aid of sensory feedback. Such networks contain at least four classes of nerve cells; command, oscillator, coordinating and motor neurons. The properties of each of these classes are reviewed briefly.

Current concepts dealing with the organization of these neurons into central motor networks hold that: 1) motor networks are organized hierarchically and unidirectionally (the concept of hierarchy); 2) the functional roles of different neurons are mutually exclusive (compartmentalization of function); 3) network properties such as oscillation are contained completely within single neurons that comprise the network (elemental origin of network properties); and 4) single command neurons elicit complete behavioral acts (unicellular command). It is shown that these concepts are inadequate on two grounds. First, the concepts are unavoidably based upon insufficient data. Second, the concepts conflict with recent data from several motor systems, including the motor system that controls rhythmic feeding behavior in Pleurobranchaea.

On this basis, four revised concepts are proposed. According to these concepts, 1) central motor networks incorporate extensive reciprocal interaction between different classes of neurons (the concept of reciprocity); 2) individual neurons typically play more than one functional role (distributed function); 3) network properties such as oscillation are emergent properties of neuronal interactions (<u>emergent properties</u>); and 4) command neurons normally operate together rather than singly (<u>consensus</u>).

The possible generality of these concepts of motor network organization is discussed, and six theoretical implications are explored. In particular, it is suggested that extensive reciprocal interactions between different classes of central neurons comprising a motor network yield non-hierarchical, cooperative redundancy and consequent variable command loci. Under such circumstances command neurons may exercise their natural function of initiating motor behavior largely by virtue of unique access to the sensory information that normally initiates the behavior.

Introduction

One of the landmark achievements of neurobiologists in the last decade has been the repeated verification of the centralist theory of locomotion. According to this theory, motor output patterns are generated by endogenous central networks, and not by chains of reflexes that are activated by sensory feedback (Wilson, 1966; Kennedy and Davis, 1975). The past few years have witnessed astonishing progress in the cellular analysis of such endogenous central motor networks, such that we can now identify some of the major classes of neurons that comprise them. This chapter provides a brief review of these neurons and their properties, and then discusses how they are organized into networks. It is argued that existing concepts of central network organization are outmoded by recent data, and on this basis revised concepts are formulated and offered as hypotheses. Although these hypotheses are based on data from invertebrate motor systems, they can be applied also to vertebrate locomotion.

CENTRAL COMPONENTS OF MOTOR NETWORKS

At least four general classes of nerve cells have been identified in the central motor networks of both invertebrates and vertebrates; command, oscillator, coordinating and motor neurons. In this section the major properties and identifying criteria of each of these classes of neurons are reviewed briefly.

Command Neurons

Wiersma's discovery of command neurons (Wiersma, 1938, 1952; Wiersma and Novitski, 1942; Wiersma and Ikeda, 1964; Atwood and

Wiersma, 1967) has proved to be one of the most influential developments in the last two decades of invertebrate neurobiology. Kennedy and colleagues extended and refined the concept of the command neuron through experiments on the abdominal flexor system of the crayfish (Kennedy et al., 1966, 1967; Evoy and Kennedy, 1967). In these experiments, single axons were microdissected from the ventral nerve cord and stimulated electrically while motor activity was recorded from peripheral nerves and muscles. A command neuron was defined operationally as any neuron that exerted a widespread and apparently adaptive motor effect, a criterion that necessarily includes intersegmental sensory neurons as well as interneurons. Subsequent research on the swimmeret system of the lobster showed that command neurons exhibit integrative properties expected of interneurons (Davis and Kennedy, 1972a; Davis, 1973). It is now generally assumed that command neurons are interneurons that typically descend from higher centers to activate local, autonomous motor networks.

It is also assumed that command neurons represent the neuronal pathways by which normal behavior is controlled. For example, electrical stimulation of command neurons in the circumoesophageal connectives of semi-intact crayfish elicits repeatable and recognizable postures and behavior, providing persuasive indirect evidence that command neurons actually control behavior (Bowerman and Larimer, 1974a, b). Recently we have obtained more direct evidence for this possibility from experiments on lobsters (Davis and Ayers, Unpublished data). Small-bore capillary suction electrodes were used to make focal extracellular recordings from the lateral circumoesophageal connectives, a region known in crayfish to contain walking command neurons (Bowerman and Larimer, 1974b). When specimens were then made to walk on an optokinetic treadmill (Davis and Ayers, 1972), descending neurons in the lateral connectives discharged immediately before and during induced walking (Figure 1A). Focal electrical stimulation through the same electrode using low current also caused walking movements (Figure 1B), suggesting that the neurons active during the behavior were conventional command neurons.

Command neurons have been studied most extensively in crustaceans, although neurons having analogous properties have been reported in mollusks (Gorman and Mirolli, 1969; Koester et al., 1974). In *Tritonia*, a bilateral population of brain cells is said to play the role of triggering swimming behavior (Willows and Hoyle, 1969). Recent experiments, however, involving intracellular recording and stimulation of five trigger neurons at a time in the isolated brain, suggest that these cells are neither necessary nor sufficient to initiate swimming (Getting, 1975).

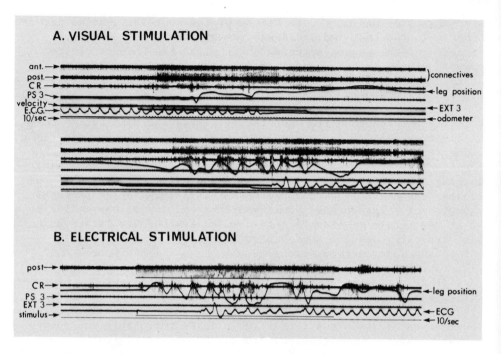

Figure 1. A, activity of walking command neurons in the lobster during optokinetically driven walking. The onset of treadmill is marked by the first step displacement of the velocity trace in the top record (first pulses on odometer trace). In the second record (which is continuous with the first), the treadmill speed was increased to elicit a fresh burst of more vigorous walking. Note that the discharge of a small number of descending central neurons (top two traces) precedes and accompanies such visually-induced walking behavior. B, effects of electrical stimulation at low voltage through one of the two connective electrodes used to record the activity in A. Stimulus is marked by a bar below the top trace. Cyclic walking movements result from the stimulus, signifying that walking command neurons have been activated. Abbreviations and symbols: A; ant. and post., extracellular recordings from anterior and posterior connective electrodes. CR, electromyogram from the coxal retractor muscles of the third right walking leg; PS3, electromyogram from the main power stroke muscle of the third right swimmeret; velocity, treadmill belt velocity; E.C.G., electrocardiogram; 10/sec., time mark; connectives, circumoesophageal connective recordings; leg position, movement about coxobasal joint of third right walking leg, recorded with a capacitative movement transducer; EXT 3, electromyogram from main extensor muscle of third abdominal segment; odometer, monitor of treadmill belt movement (1 pulse per cm.). B, as in A. (From Davis and Ayers, 1975).

Oscillator Neurons

Rhythmic behaviors such as locomotion by definition are controlled by neuronal oscillators. Such oscillators in principle could be single neurons that discharge in rhythmic bursts, such as the parabolic burster (R15) of Aplysia (Arvanataki and Chalazonitis, 1968; Alving, 1968; Strumwasser, 1968; Junge and Stevens, 1973; Parnas and Strumwasser, 1974). Alternatively, such oscillators could be networks containing several neurons, none of which is alone capable of bursting (Selverston, 1974). Mendelson (1971) studied non-spiking neurons whose membrane potential fluctuated rhythmically, alternately exciting and inhibiting crustacean respiratory motoneurons, and similar neurons have been identified and studied more thoroughly in the cockroach walking system (Pearson and Fourtner, 1975). In neither of these examples, however, have the identified oscillator neurons been proven capable of endogenous oscillation. Such neurons may exhibit oscillatory activity largely because they receive cyclic synaptic inputs from the neuronal network of which they are a part.

Coordinating Neurons

Complex behaviors such as locomotion are typically controlled by several widely distributed, autonomous central motor networks. The necessary coupling between these centers is achieved by central nerve cells termed coordinating neurons.

The concept of coordinating neurons may be traced at least as far back as Helmholtz's theory of efference copy (Helmholtz, 1925). Evidence for coordinating neurons in invertebrates stems from the work of Wiersma and colleagues on the crayfish swimmeret system (Hughes and Wiersma, 1960; Wiersma and Ikeda, 1964; Ikeda and Wiersma, 1964), and was amplified and extended significantly by Stein (1971 and this volume). In the swimmeret system, segmental oscillators in each abdominal ganglion are coupled by ascending interneurons, an arrangement that ensures the central coordination of the metachronous swimmeret movements.

Analogous coordinating neurons are found in the feeding system of the mollusk Pleurobranchaea. Cyclic feeding behavior in this carnivorous gastropod is controlled by at least two central nervous structures, the cerebropleural ganglion (brain) and the buccal ganglion. Each of these ganglia alone is able to produce oscillatory motor output in feeding nerves, but their coordinated action depends on coupling between the ganglia. Such coupling is achieved by ascending coordinating interneurons whose axons pass through the cerebrobuccal connectives and whose somata lie in the buccal ganglion (Davis et al., 1973; Davis et al., 1974).

Coordinating neurons promise to be among the most interesting and important members of central motor networks. Data from the cat walking system demonstrate the possibility that local spinal centers are coupled with the cerebellum by means of coordinating neurons (Arshavsky et al., 1972), and such neurons may play equally significant roles in the control of non-cyclic motor behaviors. For example, coordinating neurons may be responsible for the suppression of one motor act during the execution of another, as dictated by behavioral hierarchies (Davis et al., 1974a, b).

Motoneurons

The activity of the above types of neurons converges eventually upon the final common pathway (Sherrington, 1906), the motoneurons. To identify a nerve cell as a motoneuron requires the application of several well developed criteria enumerated elsewhere (e.g., Davis, 1971; Siegler et al., 1974). Such criteria are aimed basically at proving that a given central soma is contiguous with a peripheral axon that in turn directly innervates a muscle.

The identification of motoneuron somata is complicated by the recent discovery in invertebrates that some central somata belong to sensory neurons. The simplest distinguishing test is to determine the direction of impulse conduction under semi-natural conditions (e.g., Siegler et al., 1974); sensory neurons are expected to conduct toward, motoneurons away from the central nervous system.

ORGANIZING CONCEPTS IN CENTRAL MOTOR NETWORKS

The above classes of neurons, i.e., command, oscillator, coordinating and motor, comprise the major known central elements of motor networks in invertebrates and vertebrates alike (Grillner, 1975a, Figure 13). Conventional concepts of how these elements are organized into motor networks are well illustrated in a quotation from a recent textbook of general biology: "...the output of a single command fiber, through one or many synapses, excites a particular collection of motor neurons.... It is speculated that the command neuron excites a particular group of interneurons within ganglia and interneurons leading to other ganglia. These neurons in turn excite motor neurons or other interneurons that eventually excite motor neurons. The hierarchical organization of excitatory and inhibitory synapses could lead to generation of a complex output from the simple input..." (Biology Today, 1972, p. 554-555).

CENTRAL ORGANIZATION OF INVERTEBRATES

Implicit in this quotation are four concepts of network organization: neuronal hierarchy, compartmentalization of function, elemental origin of network properties, and unicellular command. In the remainder of this section each of these concepts will be defined and expanded, using the swimmeret system of the lobster for illustration (Figure 2).

Neuronal Hierarchy

The concept of hierarchical organization is deeply entrenched in Western thought. Tinbergen, for example, described the nervous system as organized into cascading or divergent networks, with "higher centres controlling a number of centres of a next lower level, each of these in their turn controlling a number of lower

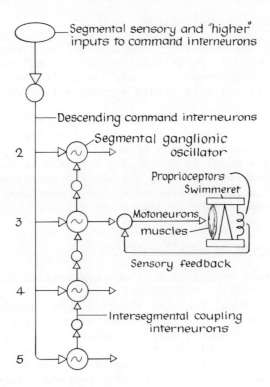

Figure 2. Hypothetical organization of the motor system that controls cyclic locomotor movements of crustacean swimmerets. Numbers on the left signify abdominal segment number. The peripheral components of a swimmeret are schematized for the third segment only. (From Davis, 1973).

centres, etc." (Tinbergen, 1950, p. 307). Indeed, the notions of neuronal convergence and divergence, so fundamental to our view of central nervous organization, imply a cellular hierarchy. A corollary that usually is implicit in the concept of hierarchy is functional polarization; that is, the unidirectionality (or at least strong asymmetry) of information flow through a hierarchically arranged network of neurons.

Earlier interpretations of the lobster swimmeret system (e.g., Davis, 1973; Figure 2) reflect the above concepts. Implicit in Figure 2 is a hierarchy of central elements, in which command neurons excite oscillators, which in turn excite coordinating and motor neurons. Moreover, the corollary of functional polarization also is embodied in Figure 2. That is, there is no provision for internal feedback within the central motor network in a "backward" direction, i.e., from motor to oscillator to command neurons.

Compartmentalization of Function

The concept of neuronal hierarchy implies a separation of function, or neuronal specialization. In the extreme, specialization can imply compartmentalization, i.e., the complete investment of a given functional role within a given structural entity.

Compartmentalization is reflected in Figure 2, in which the swimmeret system is presumed to consist of several discrete structural elements (neurons), each of which has a separate and unique function. Such a model is based upon complete separation of function. Thus a command neuron is not also an oscillator; neither a command nor an oscillator neuron is also a coordinating neuron; and none of these neurons is also a motoneuron.

Elemental Origin of Network Properties

A significant implication of compartmentalization is that properties of a central network are directly reflective of properties of the individual neuronal elements that comprise the network. According to this concept, the cyclic neuronal oscillation of a motor network represents a direct consequence of the capacity of individual neurons within the network to generate oscillatory activity. Figure 2 is not specific on this issue because there is not yet evidence for single oscillator neurons within the swimmeret system.

Unicellular Command

Applying the concept of elemental properties to the special case of initiation of motor behavior suggests the possibility that the function of command is achieved not by groups of neurons, but rather by single neurons, in accord with the original definition of command neurons. In the swimmeret system single command interneurons elicit rhythmic locomotor output, but as elaborated below, the normal output of the swimmeret system requires the simultaneous activity of more than one command interneuron.

A CRITIQUE OF EXISTING CONCEPTS

In the above paragraphs I have briefly sketched four general concepts of motor network organization, and attempted to demonstrate how they are epistimologically interrelated. Neuronal hierarchy implies specialization, i.e., compartmentalized function, which implies the elemental origin of network properties, which in turn implies unicellular command.

These concepts probably represent the best approximations that could be formulated from earlier data, but it is clear that earlier data were incomplete in several crucial respects. In this section I will highlight the limitations of earlier data, in preparation for the proposed conceptual revisions to follow.

Neuronal Hierarchy

Although it is usually assumed that central motor systems are hierarchically arranged, concrete evidence for this assumption is scant. Experiments on command neurons have seldom been executed in a way that could reveal the effects of internal feedback from the elements they excite. Studies on lobster swimmeret command neurons, for example, were performed on detached abdomens (Davis and Kennedy, 1972a, b, c); thus any internal feedback that is routed through higher nervous centers was eliminated by the experimental paradigm. Moreover, in only a few preparations have recordings been made from identified command cells, and in no case have such recordings been made during the execution of the commanded behavior. Similarly, with the notable exception of small motor ganglia in crustacea (cardiac and stomatogastric ganglia), experiments competent to reveal reciprocal interactions between elements of central motor networks have not been performed. Therefore, the data required for a rigorous test of hierarchy within motor systems are simply not available.

Compartmentalization of Function

Because recordings from command, oscillator and coordinating neurons have seldom been made, we do not know whether a single neuron can serve more than one of these functions. We can imagine circumstances in which extreme cellular specialization might have some utility, e.g., in motor systems that perform a single, relatively simple motor act. On the other hand, existing data from at least three motor systems lend support to the alternative view. In the lobster stomatogastric system, for example, the pyloric dilator neurons serve the dual function of oscillator and motor neurons (Maynard, 1972; Selverston, 1974; Mulloney and Selverston, 1974; see also Selverston, this volume). Similarly, the follower cells in the crustacean cardiac ganglion are motoneurons that innervate the heart muscle, and also exhibit endogenous pacemaker properties (Connor, 1969; Tazaki, 1971; Livengood and Kusano, 1972). Finally, most motoneurons innervating the feeding musculature of Pleurobranchaea also exhibit properties expected of neuronal oscillators (Siegler et al., 1974). In other words, few data are available to support the concept of compartmentalization; and a small but significant body of data support the alternative.

Elemental Origin of Network Properties

The hypothesis that the properties of a motor network directly reflect the properties of single neurons within the network has arisen chiefly in the context of oscillation. As mentioned above, single neurons that oscillate have been analyzed intracellularly in motor systems, but such analysis seldom has been performed while the cell was isolated from the motor network(s) of which it was a component. In those few cases in which progressive isolation has been achieved (e.g., by surgical means), the capacity of an oscillator neuron to produce cyclic activity declines as it is progressively separated from other oscillators (e.g., Kovac, 1974). This finding suggests that oscillation is a network property rather than a property of single neurons within the network.

The possibility that oscillation is a network property was mentioned by Sherrington (1906) and developed theoretically by Wilson (Wilson, 1966; see also Dagan et al., 1974), but until recently we have lacked sound evidence for such a possibility. Studies on the lobster stomatogastric ganglion, however, suggest that the gastric rhythm results as a property of synaptic interconnections within a neural network, and not as a property of any single neuron within the network (Mulloney and Selverston, 1974; see Selverston, 1974, for a review). Indeed, when this network is modeled on a computer utilizing empirically-derived physiological parameters, the modeled network exhibits oscillatory output even though no single modeled cell is given the capacity to oscillate endogenously (Perkel and Mulloney, 1974). Therefore, the hypothesis that oscillation is a network property at least exhibits sufficiency.

With regard to the function of command, there is evidence that the trigger network in Tritonia has properties that are not invested in any single neuron within the network (see Willows, this volume, for details). As mentioned earlier, recent data indicate that this network does not play the functional role originally ascribed to it, that of initiating escape swimming (Getting, 1975). Studies on this network have nevertheless uncovered principles that may be expected to apply to motor systems.

In summary, existing data give little reason to believe that the functions of motor networks are accomplished through mechanisms that reside wholly within any single cell comprising the network. On the contrary, a small but growing body of data are in direct conflict with this concept.

Unicellular Command

The notion that single command neurons release complete behavioral acts has an obvious etiology. The first command neurons studied were the giant neurons of crayfish (Wiersma, 1938), and a single giant neuron can indeed cause a normal tail flip (e.g., Larimer et al., 1971; Wine and Krasne, 1972). The tail flip system is controlled by only two pairs of giant neurons, however, and the investment of considerable functional responsibility into single giant neurons may be a direct consequence of this paucity of control elements.

In marked contrast to the tail flip escape system, other crustacean motor systems contain from ten (swimmeret system) to two dozen (walking system; slow abdominal flexor system) command neurons, raising the possibility of a more distributed responsibility for the control of a motor act. In support of this possibility, electrical stimulation of single swimmeret command interneurons elicits a narrow range of the normal behavioral repertoire. Indeed, some swimmeret command neurons drive a distinctly abnormal motor pattern, involving, for example, bursting in powerstroke motoneurons with no antagonistic return stroke discharge (Davis and Kennedy, 1972a). Stimulation of swimmeret command interneurons in pairs, however, elicits a more normal swimmeret rhythm (Davis and Kennedy, 1972b). Similarly, studies on circumoesophageal command fibers in the crayfish have led to the conclusion that a large number of single tonic command fibers elicit postures that are not reminiscent of known behavior patterns (Bowerman and

Larimer, 1974a). Stimulation of single phasic command fibers produced more recognizable behaviors, such as forward and backward walking, but whether command fibres with a basic classification (e.g., forward walking) are used singly or in ensembles is uncertain (Bowerman and Larimer, 1974b).

In summary, it seems clear that the elicitation of complete behavioral acts by electrical stimulation of single command cells is the exception rather than the rule; and there is little or no evidence that command neurons are normally employed singly by freely behaving animals.

THE METACEREBRAL GIANT (MCG) NEURON OF PLEUROBRANCHAEA

The four concepts of motor network organization outlined above are not only unsupported by past data; in addition, they are in direct conflict with recent data obtained from the motor system that controls cyclic feeding behavior in Pleurobranchaea (Gillette and Davis, 1975, 1976). These data, which prompted the present re-evaluation, will be summarized briefly in this section, in preparation for the conceptual revisions they suggest.

Rhythmic feeding behavior in Pleurobranchaea is controlled by a central network whose main neuronal components have been identified and reviewed elsewhere (e.g., Davis et al., 1974). This network includes a group of about thirty putative feeding command neurons on each side of the brain, the largest of which is the metacerebral giant (MCG) neuron (Figure 3). In an isolated, quiescent nervous system (brain plus buccal ganglion), intracellular stimulation of a single MCG neuron can initiate cyclic motor output in feeding nerves of the brain and buccal ganglion, although the effect is usually weak. In an isolated or semi-intact (brain, buccal ganglion and attached buccal mass) preparation, MCG discharge usually increases the frequency of the spontaneous feeding rhythm (Figure 4A, 4B). In addition, the MCG receives potent synaptic inputs from virtually all brain nerves. Finally, MCG discharge usually precedes and accompanies spontaneous feeding behavior in whole-animal preparations.

These data support the view that the MCG neuron is one of several command neurons for feeding behavior in Pleurobranchaea. During feeding output, however, the membrane potential of the MCG neuron fluctuates in phase with the feeding rhythm (Figure 4C). Intracellular analysis of the MCG neuron indicates that this fluctuation has two sources. First, the MCG soma membrane exhibits a depolarization-induced conductance increase, as found also in the feeding motoneurons (Siegler et al., 1974). As detailed elsewhere, this endogenous property can contribute to network oscillation (Siegler et al., 1974). Second, the MCG neuron receives synaptic influences (excitatory and inhibitory, chemical and electrical) from other neurons in the feeding network, including identified motoneurons in the brain and buccal ganglion, as well as a population of coordinating interneurons in the buccal ganglion (the corollary discharge interneurons; Davis et al., 1974). The MCG neuron is therefore arranged to sample the activity of the motor network which it is capable of driving.

Although the MCG can act as a feeding command cell, it is not specialized for this role alone. As mentioned above, the MCG neuron exhibits biophysical properties expected of an oscillator neuron, and its membrane potential in fact oscillates during feeding output. Moreover, the MCG sends numerous axonal branches to the periphery via the mouth nerve and the buccal ganglion nerves, as found also in Aplysia, where the MCG may play the role of priming the feeding musculature (Weiss et al., 1974). The functional role of these peripheral branches in Pleurobranchaea has not been determined, and the possibility that the MCG has a sensory or motor function, in addition to its other demonstrated functions, remains open. Finally, the MCG neuron normally discharges bursts

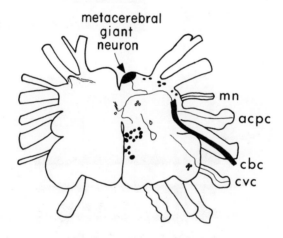

Figure 3. Tracing from a photograph of the dorsal view of the brain of Pleurobranchaea, showing somata of neurons that send descending axons to the buccal ganglion. The somata were filled by back injecting the right cerebrobuccal connective (cbc) with cobalt sulfide (see Siegler et al., 1974). Anterior is toward the top. The brain is about one cm. in width. Abbreviations: mn, mouth nerve; acpc, anterior cerebropedal connective; cbc, cerebrobuccal connective; cvc, cerebrovisceral connective. (Modified from Davis et al., 1974).

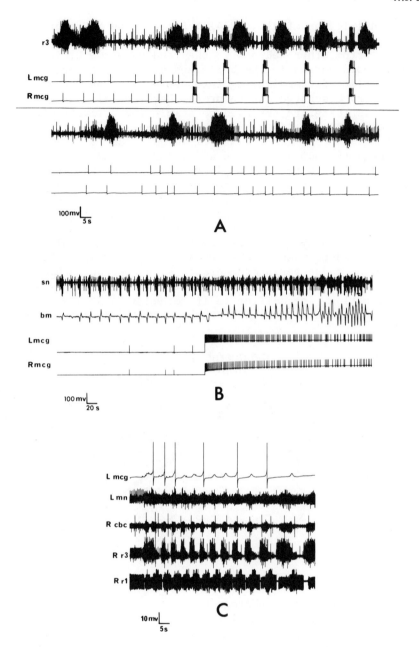

Figure 4. Activity of the metacerebral giant (MCG) neuron of Pleurobranchaea. In A, the left and right MCG were stimulated intracellularly in an isolated brain/buccal ganglion preparation. This stimulus accelerated (Legend continued on next page.)

CENTRAL ORGANIZATION OF INVERTEBRATES

of impulses that descend to the buccal ganglion. Such activity contains timing information that could in principle help to coordinate the feeding rhythms generated independently in the brain and buccal ganglion. In other words, the MCG may also act as a coordinating neuron.

REVISED CONCEPTS OF MOTOR ORGANIZATION

It is possible now to reassess the four concepts of motor organization discussed earlier, with the aim of incorporating the revisions suggested by earlier data and the new data summarized above.

The Concept of Reciprocity

As reviewed briefly above, the metacerebral giant command neuron makes reciprocal connections with oscillator, coordinating and motor neurons. Such findings are not consistent with a model in which the components of a central motor network are arranged in a vertical hierarchy, with the implicit corollary of one-way synaptic transactions from command to oscillator to motor neurons. In other words, the concept of neuronal hierarchy, as elaborated above, is inapplicable. The data suggest an alternative concept, that of reciprocity.

Reciprocal connections _within a given class_ of neurons (e.g., motoneurons) is already well-established in motor systems (e.g.,

---

a spontaneous feeding rhythm recorded from the third buccal root (r3), increased the intensity of feeding bursts, and phase-locked feeding bursts. In the lower three traces stimulation is ceased and the spontaneous but irregular rhythm continues. In B, a similar experiment is performed on a semi-intact nervous system/buccal mass preparation. Feeding activity is monitored by extracellular recordings from the stomatogastric nerve (sn) and from movement of the buccal mass (bm), recorded with a tension transducer attached to the radular sac. Intracellular stimulation (by means of release from imposed hyperpolarization) of both MCG neurons increases the frequency and amplitude of the feeding rhythm. In C, feeding was elicited from an isolated brain/buccal ganglion preparation by stomatogastric nerve stimulation, while intracellular recordings were made from the left MCG (top trace). Note rhythmic depolarizations of the MCG during feeding. Abbreviations, Lmcg, left metacerebral giant neuron; Lmn, left mouth nerve of the brain; Rcbc, right cerebrobuccal connective; Rr3, right third root of the buccal ganglion; Rr1, right first root of the buccal ganglion. (From Gillette and Davis, 1976.)

Kennedy and Davis, 1975). The data from the MCG neuron suggest
the need for a more radical departure from current dogma, however,
involving reciprocal connections between classes of neurons in a
motor network. By this view a motor network is conceived as a
horizontal hierarchy with extensive reciprocal interactions between
different levels. Such an arrangement in fact renders the notion
of levels meaningless within the boundaries of the network, although as developed below, differential access to external output
could nevertheless provide the basis for cellular specialization.

The concept of reciprocity implies that one of two interacting
neurons secondarily acquires the functional role of the other, although the latency of a second-order interaction is longer and the
effect probably weaker. For example, if a motoneuron is excitatorily coupled to a command neuron in the same motor system,
excitation of the motoneuron can in principle activate the entire
motor system. Results of this kind have in fact been obtained
from the feeding system of Pleurobranchaea (Unpublished data).
Owing to such an arrangement, a motoneuron secondarily acquires
the functional capacity of command.

The Concept of Distributed Function

As outlined above, the concept of reciprocity implies functional redundancy. Such redundancy is reinforced by the investment of more than one functional role in the same neuron(s). As
shown above, for example, the metacerebral giant neuron apparently
acts not only as a command cell, but also as an oscillator, and
perhaps even as a coordinating, sensory or motor neuron. Similarly,
feeding motoneurons were previously shown to have properties of
neuronal oscillators (Siegler et al., 1974). Finally, the coordinating neurons in the buccal ganglion of Pleurobranchaea (the
corollary discharge neurons) are probably homologous with the
cyberchron neurons that generate the oscillatory feeding rhythm
in the buccal ganglion of Helisoma (Kater, 1974). In other words,
the functions of oscillation and coordination may also reside in
the same neurons. These interpretations are of course inconsistent
with the concept of compartmentalization. They suggest instead the
concept of distributed function, according to which several functional properties reside in the same neuron, and different populations of neurons therefore have overlapping functions.

The Concept of Emergent Properties

The concept of distributed function need not imply that a
given network property is completely represented in any single
neuron. Instead, network properties may reflect the interaction

of cells with different properties, in which case the network property is described as an emergent property. The oscillation of Pleurobranchaea's feeding network provides an example of how such an emergent property might arise. Several neurons within the network exhibit a depolarization-induced increase in membrane conductance, including the metacerebral giant neuron (see previous page) and many feeding motoneurons (Siegler et al., 1974). Assuming that the conductance increase is to potassium, as shown for the parabolic burster neuron (R15) of Aplysia (Junge and Stevens, 1975), a train of action potentials resulting from a tonic input will periodically self-terminate, leading to the formation of repetitive bursts of impulses at a period determined by the time constant of the conductance change. In other words, this membrane property in principle can lead to the oscillation of a motor network even though no single neuron within the network has the capacity of endogenous oscillation.

The Concept of Consensus

The concept of emergent properties can be applied not only to the function of oscillation, but also to the special case of command. As reviewed earlier, data from the lobster swimmeret system suggest that command neurons must normally operate together rather than alone, in which case the command function may be viewed as an emergent network property. Similarly, the metacerebral giant neuron of Pleurobranchaea appears to be but one of several feeding command neurons which may normally operate in consensus. Willows (this volume) suggests that the command of swimming behavior in Tritonia is also an emergent property of the trigger network, although as mentioned above, the trigger network may not be involved in initiating swimming (Getting, 1975). The evidence is incomplete, but nonetheless suggestive that command cells operate normally by consensus rather than by unilateral activity.

GENERALITY AND LIMITATIONS OF THE PROPOSED CONCEPTS

The concepts proposed in this chapter are illustrated in Figure 5. These concepts provide an appealing explanation of select data, but it is not yet certain that these concepts have general application. Indeed, the proposed concepts have been formulated in large part from investigation of a specialized molluskan feeding system, and it is not even clear how fully the proposed concepts apply within this central motor network. As summarized above, however, there are indications that at least some of the proposed concepts may apply to other motor systems, notably the crustacean cardiac ganglion, the crustacean stomatogastric ganglion, and the swimming system of Tritonia.

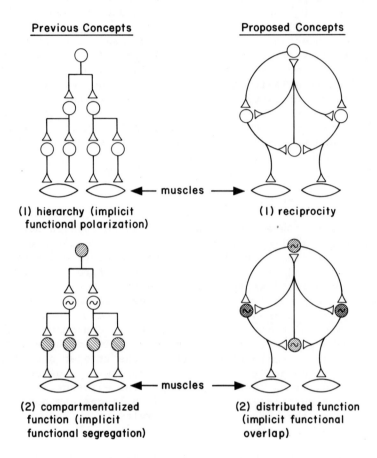

Figure 5. Organizational Concepts in Central Motor Networks. Schematic portrayals of the old (left column) and new (right column) concepts discussed in this chapter. Each open circle represents a separate neuron, with synaptic terminals symbolized as open triangles. Remaining functions are symbolized as follows: ⌀, command; ⊘, oscillation; ⊗, motor. See text for further details. (Figure continued on next page.)

CENTRAL ORGANIZATION OF INVERTEBRATES

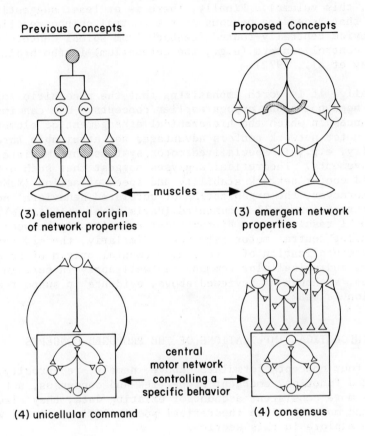

Figure 5. Organizational Concepts in Central Motor Networks.

In addition to their potential application to invertebrate preparations, there is increasing evidence that the proposed concepts may apply to vertebrate locomotor systems as well. Vertebrate locomotion is now viewed as the product of local, autonomous oscillator networks that interact reciprocally with one another by means of coordinating neurons (Grillner, 1975, and this volume). These elements are believed to be activated by descending command neurons whose cooperative action is required for the effective excitation of locomotor sequences (Humphrey et al., 1970; see also Grillner, this volume). Finally, there is at least suggestive evidence that ascending neurons in the ventral spinocerebellar tract provide centrally-routed feedback from spinal centers to putative control centers (e.g., the cerebellum) in the brain (Arshavsky et al., 1972).

Finally, it is worth emphasizing that the recognition of new concepts need not alone oppugn earlier concepts. One can imagine circumstances in which the hierarchical arrangement of elements within a motor network confers advantages not attainable through reciprocity, e.g., in specialized motor systems that mediate rapid escape movements. Theoretical analyses suggest that such organization could confer network stability and speed of action (McMurtrie, 1975). Moreover, the organization of unrelated behavioral acts into hierarchies is well documented (Davis et al., 1974a, b), and may well result from a hierarchical relationship between the corresponding central motor networks. Similarly, the old concepts of compartmentalization of function, elemental origin of network properties and unicellular command may well apply in certain circumstances, although as reviewed above, evidence in support of this proposition is weak.

THEORETICAL IMPLICATIONS OF THE PROPOSED CONCEPTS

The four concepts formulated above, namely, reciprocity, distributed function, emergent properties and consensus, not only provide a more consistent account of existing data; they also offer broader and more flexible theoretical possibilities, six of which I wish to explore in this section.

Redundancy

The concepts of reciprocity and distributed function both imply extensive redundancy within a central motor network. In sensory systems, redundancy of sensory units increases the resolving power of a sense organ, and thereby reduces potential perceptual ambiguity (e.g., Cohen, 1964). Similarly, we may speculate that redundancy in motor systems, both at motor and pre-motor levels, increases the resolution of control over the muscular force that ultimately underlies behavioral acts.

Moreover, in the absence of such redundancy, a motor system would presumably be more vulnerable to the loss of strategic elements from injury or developmental perturbation. In theory, at least, duplication of function protects both against such loss and against the corresponding behavioral deficit.

Positive Feedback

The existence of extensive reciprocal connections within a central network suggests the theoretical possibility of reverberation, a notion that has long been discussed on neuroanatomical grounds, and for which there is some physiological evidence (e.g., Kusano and Grundfest, 1965). The concepts formulated here imply the existence of positive feedback within a motor network, which in turn may be expected to underlie reverberatory activity within the network. Such positive feedback is at least theoretically competent to explain the frequent experimental observation that a motor pattern outlasts the stimulus that elicits it (Davis and Kennedy, 1972a; Dorsett et al., 1973; Willows et al., 1973).

Positive feedback mechanisms are of course self-reinforcing, and therefore subject to uncontrolled runaway. A variety of mechanisms, however, may be envisioned to protect against such instability in central motor networks. Such protective mechanisms include neuronal adaptation, inhibitory interrupts, and the maintenance of the gain of the system at less than unity (Stark, 1968), which could be achieved in central motor networks by suitable modulation of synaptic transfer functions.

Command Flexibility

A well-developed axiom of information theory holds that a system operates most efficiently when the locus of command is also the site of the most important information (e.g., Arbib, 1972). If the source of the most important information varies, then theory dictates that for maximum efficiency, the command locus should also vary. Applying this axiom to the feeding system of Pleurobranchaea, the most important information is provided by the food itself. As the relative position of the food changes during food detection, approach, ingestion and swallowing, the locus of command can change accordingly owing to command redundancy, which results from reciprocity, distributed cellular function and command by consensus. The same principle of variable command locus is extended readily to other motor systems, including those that control locomotion.

Command Specialization

The principles of reciprocity and distributed function imply a breakdown of cellular specialization within a central motor network. In the extreme condition, every cell may be envisioned as connected with every other cell, with a corresponding loss of identity within the network. In such an extreme case, neuronal specialization cannot be based upon differential interconnection within a network. Specialization may nevertheless be retained on the basis of differential access to information <u>outside</u> the network. According to this view, the functional role of command is conferred upon a given population of neurons at least in part by unique convergence upon that population of sensory information that initiates and sustains the behavior in question.

Control Flexibility

Extensive reciprocal interconnection within a motor network increases the potential number of sites at which control of the network can be exercised. Especially when such connections are made in a circuitous series, each synapse or set of synapses represents a single, localized target at which excitation or inhibition of the entire network can be asserted. In theory, at least, all four of the proposed concepts suggest an increase in the number of loci at which such functional gating of a motor network's activity may occur.

Plasticity

All four of the proposed concepts also increase the likelihood that physiological changes associated with learning are distributed widely throughout a central motor network. In a completely equipotential network, such modifications could occur equally in each cell or synapse, and under such conditions the change in any <u>single</u> element may be indetectable. Lashley's unsuccessful effort to localize the memory engram in the rat brain (Lashley, 1929) would seem to signify exactly such a widespread diffusion of the physiological substrates of learning.

On a more optimistic note, we have seen that despite the democratic distribution of cellular function within a motor network, certain cells such as command neurons may occupy privileged positions with respect to access to external sensory information. One may still nourish the hope that the physiological substrates of plastic changes such as learning may be detected in such neurons. Investigations to date on the cellular basis of plasticity in the invertebrates encourage this view (see Davis, 1975, for a review).

Conclusion

In this chapter I have tried to show: 1) that earlier concepts of motor organization were founded unavoidably upon an incomplete factual basis; 2) that such concepts do not account satisfactorily for new data; 3) that new concepts of central motor network organization, emphasizing functional redundancy and neuronal population dynamics, are more capable of explaining both old and new data; and 4) that these new concepts have at least modest theoretical appeal. The concepts proposed here do not yet have sufficient empirical support to be described as principles. Instead they are offered as hypotheses, in the hope that they will be tested on a variety of invertebrate and vertebrate locomotor systems, and in the expectation that the data so obtained will lead to their refinement or will compel their revision.

Acknowledgment

Original research from this laboratory is supported by NIH Research Grants NS 09050 and MH 23254, and NSF Grant GZ 3120. I am indebted to my colleagues, Dr. Mark Kovac, Dr. Jeffrey Ram and especially Dr. Rhanor Gillette, for collaboration and critical discussion of the material presented in this article.

REFERENCES

Alving, B.O., (1968) Spontaneous activity in isolated somata of Aplysia pacemaker neurons. J. Gen. Physiol. 51, 29-45.

Arbib, M.A., (1972) The Metaphorical Brain: An Introduction to Cybernetics as Artificial Intelligence and Brain Theory. Wiley-Interscience, New York.

Arshavsky, Yu.I., Berkinblit, M.B., Fuxon, O.J., Gelfand, J.M. and Orlovsky, G.N., (1972) Origin and modulation in neurons of the ventral spino-cerebellar tract during locomotion. Brain Res. 43, 276-279.

Arvanataki, A. and Chalazonitis, N., (1968) "Electrical properties and temporal organization in oscillatory neurons (Aplysia)," In Symposium on Neurobiology of Invertebrates. (Salanki, J., ed.), Plenum, New York, (169-200).

Atwood, H.L. and Wiersma, C.A.G., (1967) Command neurons in the crayfish central nervous system. J. Exp. Biol. 46, 249-261.

Biology Today, (1972) CRM Publishers, Del Mar, Calif., (554-555).

Bowerman, R.F. and Larimer, J.L., (1974a) Command fibres in the circumoesophageal connectives of crayfish. I. Tonic fibres. J. Exp. Biol. 60, 95-117.

Bowerman, R.F. and Larimer, J.L., (1974b) Command fibres in the circumoesophageal connectives of crayfish. II. Phasic fibres. J. Exp. Biol. 60, 119-134.

Cohen, M.J., (1964) "The peripheral organization of sensory systems," In Neural Theory and Modeling. (Reiss, R.F., ed.), Stanford Univ. Press, Stanford, (273-292).

Connor, J.A., (1969) Burst activity and cellular interaction in the pacemaker ganglion of the lobster heart. J. Exp. Biol. 50, 275-297.

Dagan, D., Vernon, L.H. and Hoyle, G., (1975) Neuromimes: self-exciting alternate firing pattern models. Science. 188, 1035-1036.

Davis, W.J., (1971) Functional significance of motoneuron size and soma position in swimmeret system of the lobster. J. Neurophysiol. 34, 274-288.

Davis, W.J. and Ayers, J.L., (1972) Locomotion: control by positive feedback optokinetic responses. Science. 177, 183-185.

Davis, W.J. and Kennedy, D., (1972a) Command interneurons controlling swimmeret movements in the lobster. I. Types of effects on motoneurons. J. Neurophysiol. 35, 1-12.

Davis, W.J. and Kennedy, D., (1972b) Command interneurons controlling swimmeret movements in the lobster. II. Interaction of effects on motoneurons. J. Neurophysiol. 35, 13-19.

Davis, W.J. and Kennedy, D., (1972c) Command interneurons controlling swimmeret movements in the lobster. III. Temporal relationships among bursts in different motoneurons. J. Neurophysiol. 35, 20-29.

Davis, W.J., (1973) "Neuronal organization and ontogeny in the lobster swimmeret system," In Control of Posture and Locomotion. (Stein, R.B., Pearson, K.G., Smith, R.S. and Redford, J.B., eds.), Plenum, New York, (437-455).

Davis, W.J., Siegler, M.V.S. and Mpitsos, G.J., (1973) Distributed neuronal oscillators and efference copy in the feeding system of Pleurobranchaea. J. Neurophysiol. 36, 258-274.

Davis, W.J., Mpitsos, G.J., Siegler, M.V.S., Pinneo, J.M. and
    Davis, K.B., (1974) Neuronal substrates of behavioral hier-
    archies and associative learning in the mollusk Pleuro-
    branchaea. Amer. Zool. 14, 1037-1050.

Davis, W.J., Mpitsos, G.J. and Pinneo, J.M., (1974a) The behavioral
    hierarchy of the mollusk Pleurobranchaea. I. The dominant
    position of the feeding behavior. J. Comp. Physiol. 90,
    207-224.

Davis, W.J., Mpitsos, G.J. and Pinneo, J.M., (1974b) The behavioral
    hierarchy of the mollusk Pleurobranchaea. II. Hormonal
    suppression of feeding associated with egg-laying. J. Comp.
    Physiol. 90, 225-243.

Davis, W.J., (1975) "Plasticity in the invertebrates," In Neural
    Mechanisms of Learning and Memory. (Bennett, E.L. and
    Rosenzweig, M.R., eds.), MIT Press, Cambridge, Mass.,
    (In press).

Dorsett, D.A., Willows, A.O.D. and Hoyle, G., (1973) Neuronal
    basis of behavior in Tritonia. IV. Central origin of a fixed
    action pattern. J. Neurobiol. 4, 287-300.

Evoy, W.H. and Kennedy, D., (1967) The central nervous organization
    underlying control of antagonistic muscles in the crayfish.
    I. Types of command fibres. J. Exp. Zool. 165, 223-238.

Getting, P.A., (1975) Tritonia swimming: Triggering of a fixed
    action pattern. Brain Research. 96, 128-133.

Gillette, R. and Davis, W.J., (1975) Control of feeding behavior
    by the metacerebral giant neuron of Pleurobranchaèa.
    Neuroscience Abstracts. 1, 571.

Gillette, R. and Davis, W.J., (1976) The role of the metacerebral
    giant neuron in feeding behavior of Pleurobranchaea. (In
    preparation).

Gorman, A.L.F. and Mirolli, M., (1969) The input-output organiza-
    tion of a pair of giant neurones in the mollusc Anisodoris
    nubilis (MacFarland). J. Exp. Biol. 51, 615-634.

Grillner, S., (1975) Locomotion in vertebrates: central mechanisms
    and reflex interaction. Physiol. Rev. 55, 247-304.

Helmholtz, H. Von, (1925) Treatise of Physiological Optics. Vol 3,
    3rd ed., (Southall, P.C., ed. and trans.) Optical Society of
    America, Menasha, Wis.

Hughes, G.M. and Wiersma, C.A.G., (1960) The co-ordination of swimmeret movements in the crayfish, Procambarus clarkii (Girard). J. Exp. Biol. 37, 657-670.

Humphrey, D.K., Schmidt, E.M. and Thompson, W.D., (1970) Predicting measures of motor performance from multiple cortical spike trains. Science. 170, 758-762.

Ikeda, K. and Wiersma, C.A.G., (1964) Autogenic rhythmicity in the abdominal ganglia of the crayfish: The control of swimmeret movements. Comp. Biochem. Physiol. 12, 107-115.

Junge, D. and Stevens, C.L., (1973) Cyclic variation of potassium conductance in a burst-generating neurone in Aplysia. J. Physiol. 235, 155-181.

Kater, S.B., (1974) Feeding in Helisoma trivolvis: The morphological and physiological basis of a fixed action pattern. Amer. Zool. 14, 1017-1036.

Kennedy, D., Evoy, W.J., and Hanawalt, J.T., (1966) Release of coordinated behavior in crayfish by single central neurons. Science. 143, 917-919.

Kennedy, D., Evoy, W.H., Dane, B. and Hanawalt, J.T., (1967) The central nervous organization underlying control of antagonistic muscles in the crayfish. II. Coding of position by command fibers. J. Exp. Zool. 165, 239-248.

Kennedy, D. and Davis, W.J., (1975) "The organization of invertebrate motor systems," In Handbook of Physiology: Neurophysiology, 2nd ed., American Physiological Society, New York, (In press).

Koester, J., Mayeri, E., Liebeswar, G. and Kandel, E.R., (1974) Neural control of circulation in Aplysia. II. Interneurons J. Neurophysiol. 37, 476-496.

Kovac, M., (1974) Abdominal movements during backward walking in crayfish. II. The neuronal basis. J. Comp. Physiol. 95, 79-94.

Kusano, K. and Grundfest, H., (1965) Circus reexcitation as a cause of repetitive activity in crayfish lateral giant axons. J. Cell Comp. Physiol. 65, 325-336.

Larimer, J.L., Eggleston, A.C., Masukawa, L.M. and Kennedy, D., (1971) The different connexions and motor outputs of lateral and medial giant fibres in the crayfish. J. Exp. Biol. 54, 391-402.

Lashley, K.S., (1929) Brain Mechanisms in Intelligence. Univ. of Chicago Press, Chicago.

Livengood, D.R. and Kusano, K., (1972) Evidence for an electrogenic sodium pump in follower cells of the lobster cardiac ganglion. J. Neurophysiol. 35, 170-186.

Maynard, D., (1972) Simpler networks. Ann. N.Y. Acad. Sci. 193, 59-72.

McMurtrie, R.E., (1975) Determinants of stability of large randomly connected systems. J. Theor. Biol. 50, 1-11.

Mendelson, M., (1971) Oscillator neurons in crustacean ganglia. Science. 171, 1170-1173.

Mulloney, B. and Selverston, A.I., (1974) Organization of the stomatogastric ganglion of the spiny lobster. I. Neurons driving the lateral teeth. J. Comp. Physiol. 91, 1-32.

Parnas, I. and Strumwasser, F., (1974) Mechanisms of long-lasting inhibition of a bursting pacemaker neuron. J. Neurophysiol. 37, 609-620.

Pearson, K.G. and Fourtner, C.R., (1975) Nonspiking interneurons in walking system of the cockroach. J. Neurophysiol. 38, 33-52.

Perkel, D.H. and Mulloney, B., (1974) Motor pattern production in reciprocally inhibitory neurons exhibiting post-inhibitory rebound. Science. 185, 181-183.

Selverston, A.I., (1974) Structural and functional basis of motor pattern generation in the stomatogastric ganglion of the lobster. Amer. Zool. 14, 957-972.

Sherrington, C., (1906) The Integrative Action of the Nervous System. Yale Univ. Press, New Haven, Conn.

Siegler, M.V.S., Mpitsos, G.J. and Davis, W.J., (1974) Motor organization and generation of rhythmic feeding output in the buccal ganglion of Pleurobranchaea. J. Neurophysiol. 37, 1173-1196.

Stark, L., (1968) Neurological Control Systems: Studies in Bioengineering. Plenum, New York.

Stein, P.S.G., (1971) Intersegmental coordination of swimmeret motoneuron activity in crayfish. J. Neurophysiol. 34, 310-318.

Strumwasser, F., (1968) "Membrane and intracellular mechanisms governing endogenous activity in neurons," In Physiological and Biochemical Aspects of Nervous Integration. (Carlson, F.D., ed.), Prentice-Hall, Englewood Cliffs, N.J., (329-341).

Tazaki, K., (1971) The effects of tetrodotoxin on the slow potential and spikes in the cardiac ganglion of a crab Eriocheir japonicus. Jap. J. Physiol. 21, 529-536.

Tinbergen, N., (1950) The hierarchical organization of nervous mechanisms underlying instinctive behavior. Symp. Soc. Exp. Biol. 4, 305-312.

Weiss, K.R., Cohen, J. and Kupfermann, I., (1975) Modulatory command function of the metacerebral cell on feeding behavior in Aplysia. Fed. Proc. 34, 418.

Wiersma, C.A.G., (1938) Function of the giant fibers of the central nervous system of the crayfish. Proc. Soc. Exp. Biol. Med. 38, 661-662.

Wiersma, C.A.G. and Novitski, E., (1942) The mechanism of nervous regulation of the crayfish heart. J. Exp. Biol. 19, 255-265.

Wiersma, C.A.G., (1952) Neurons of arthropods. Cold Spring Harb. Symp. Quant. Biol. 17, 155-163.

Wiersma, C.A.G. and Ikeda, K., (1964) Interneurons commanding swimmeret movements in the crayfish Procambarus clarkii (Girard). Comp. Biochem. Physiol. 12, 509-525.

Willows, A.O.D. and Hoyle, G., (1969) Neuronal network triggering a fixed action pattern. Science. 166, 1549-1551.

Willows, A.O.D., Dorsett, D.A. and Hoyle, G., (1973) The neuronal basis of behavior in Tritonia. III. Neuronal mechanisms of a fixed action pattern. J. Neurobiol. 4, 255-285.

Wilson, D.M., (1966) "Central nervous mechanisms for the generation of rhythmic behavior in arthropods," In Nervous and Hormonal Mechanisms of Integration, Symp. Soc. Exp. Biol., vol. XX. Academic Press, New York, (199-228).

Wine, J.J. and Krasne, F.B., (1972) The organization of escape behavior in the crayfish. J. Exp. Biol. 56, 1-18.

COMMAND INTERNEURONS AND LOCOMOTOR BEHAVIOR IN CRUSTACEANS

James L. Larimer

Department of Zoology

University of Texas, Austin, Texas 78712

Wiersma and his colleagues discovered "command fibers" by systematically stimulating small bundles of axons that had been isolated from the interganglionic connectives of crustaceans. These and subsequent studies revealed that these neurons, when driven at relatively low frequencies and in unpatterned trains, are capable of releasing a wide variety of partial or complete behaviors, including the cyclical ones of neural oscillator origins as well as those effecting single movement and positioning. Units of the former type are known which influence swimmeret rhythms, forward and backward walking, swimming, escape and feeding-like movements, while noncyclical behaviors such as tail, abdomen and limb positioning, turning, righting and defense posture are evoked by other units. Additional neurons affect heartbeat, ventilation, stomach and intestinal movements; still others provide inhibition, suppression or even freezing of an ongoing behavior. This list is probably incomplete, since it is doubtful if any study thus far has located all of the commands present in a specific part of the CNS. Since the stimulus train that is adequate to activate most command systems is simple and contains few instructions, command pathways are thought to be more "permissive" than "instructive" in nature. Although the data are not extensive, some command neurons can apparently evoke different behaviors when driven at different frequencies.

Units isolated from all levels of the CNS can exhibit command fiber properties; however, those emerging from the brain apparently evoke behaviors of the greatest complexity. Although "whole" behaviors can be released from single command neurons, others seem to require the participation of several simultaneous commands to be complete.

True redundancy has not been found for the command neurons thus far examined; instead, each appears to release a unique behavior. This, along with the relatively constant cord location of the command neurons and the repeatability of output in different individuals suggests that the associated neural circuitry is genetically determined. When examined at the motor neuron level, command outputs exhibit reciprocity, timing and other characteristics of volitional movements. Many command fiber outputs also appear to be among the normal behavioral repertory of the animals. Finally, command fiber-evoked behaviors are organized in the CNS and can typically proceed with only minor modification without sensory feedback.

A disturbing lack of information exists concerning the role of command systems in normal behavior, their organization and their mechanisms of action. For example, with the single exception of the giant fiber-mediated escape response, the associated circuitry of command elements in crustaceans is not well known. More data are also needed to demonstrate that activity in single or multiple command fibers underlies a particular behavior. The techniques of behavior genetics, surgical ablation, chronic recording, detailed mapping and neural morphology all need to be applied to this system. Efforts are being made in several laboratories to obtain such information. We have chosen to examine the control of locomotor behavior and are attempting to correlate this with known command neurons using several techniques.

In crayfish, two peaks of activity are associated with an imposed light cycle; one is reflexive and occurs at lights on, while the other shows circadian characteristics and appears approximately fifteen minutes after lights off. These two distinct peaks of activity are expressed voluntarily but more importantly, they can be recorded remotely for long periods and are under experimental control via the environmental light cycle. Both of these peaks are abolished by sectioning the circumesophageal connectives although random leg movements persist. If one connective only remains, there is no behavioral deficit. Tests performed on animals which have undergone surgical paring of the single remaining connective revealed that at least three of the four quadrants of the connective contains the descending information necessary to maintain the two peaks (Gordon, Larimer and Page, unpub.). Thus the coupling neurons between the brain and the locomotor oscillators are apparently widely distributed in the connectives. Command neurons for walking are also found in all quadrants, but it is not now known whether these units or others are responsible for the information transfer.

Introduction

The selective stimulation of interneurons in crustacean preparations has yielded numerous examples of units capable of releasing partial or complete behaviors (Wiersma, 1938, 1952; Wiersma and Ikeda, 1964). Since access to these interneurons traditionally was obtained by isolating their axons from the interganglionic connectives, they were termed "command fibers" (Wiersma and Ikeda, 1964). Studies over the past decade have provided an extensive catalogue of movements which can be generated in this way, and the term command fiber has come to mean almost any premotor interneuron which provides a defined motor output. The behaviors that are released range in variety and complexity from movements of single appendages to co-ordinated cyclical behaviors that incorporate nearly all the segments of the organism. Despite the quantity of accumulated information few data are available on their anatomical structure or their activity in freely behaving animals.

Some of the current questions and speculations concerning the role of command fibers in behavior will be discussed under the following three general headings: 1) The evidence that command cells underlie normal voluntary behaviors, 2) Types of locomotion provided by command interneurons, and finally, 3) The influence of impulse pattern and frequency on command fiber output. Only a brief outline of the general characteristics of command interneurons will be given here, since this topic has been reviewed extensively (Kennedy, 1961; Kandel and Kupferman, 1970; DeLong, 1971; Kennedy, 1971; Maynard, 1973; Kennedy, 1973; Kennedy and Davis, 1975; Bowerman and Larimer, 1976).

CHARACTERISTICS OF COMMAND INTERNEURONS IN CRUSTACEANS

Some command interneurons receive direct sensory input, particularly those whose activity releases only simple behavior. In the cases tested, however, the complex commands appear to be distant from sensory input to the extent that activation of multiple sensory modalities often does not affect their activity. Such data suggest that these interneurons occupy several levels in the neural hierarchy.

It is accepted that many behaviors evoked by the stimulation of command interneurons can proceed, at least in their main features, without proprioceptive feedback. This is more a property of the central pattern generator upon which the commands impinge, however, than of the commands themselves.

Evidence exists that some single command interneurons can evoke complete and even complex behaviors (Kennedy et al., 1966; Larimer and Kennedy, 1969b). In other instances the data clearly imply that several commands must act in concert to provide for a complete behavior (Davis and Kennedy, 1972a; Bowerman and Larimer, 1974a, b; Wilkens et al., 1974). For example, an important parameter of the motor program for driving the swimmeret rhythm is the burst period, the time between the onset of sequential bursts in a given motor neuron. A systematic examination of the burst periods provided by each of the swimmeret command interneurons in lobsters revealed that no single one, regardless of the frequency of activation, was able to provide the total range of burst periods observed in intact animals. Instead, each interneuron was able to give only a fraction of the total range. Such "range fractionation" strongly suggests that several commands are at times required to release total behavior (Davis and Kennedy, 1972a).

Behaviors can be produced by the stimulation of command fibers at relatively low frequencies (generally from 2 to 75 Hz). Such rates are well within the range of capability of the crustacean interneurons. Apparently identical command fibers have been located repeatedly in different individuals. Such interneurons not only occupy the same location within the connective, but also exhibit the same voltage thresholds and release the same behaviors. Such data strongly suggest a genetic specification of the cells and their connections.

Premotor interneurons have been isolated from all levels of the central nervous system; however, those providing the most complex outputs appear to emerge from the brain. It has been argued that this organization might be anticipated, since the neural oscillators, coupling systems, etc., required for the expression of many behaviors such as locomotion and swimming are located below the brain (Bowerman and Larimer, 1974b). Such an arrangement suggests a hierarchy of command elements, but this has not been established definitely (Kennedy, 1969; Bowerman and Larimer, 1974a, b, 1975). It is unknown, for example, whether more rostral neurons selectively activate combinations of lower command neurons to organize complex behaviors.

It is the rule that several interneurons releasing similar behaviors can be found within a given level of the central nervous system (CNS). For example, five units located in the circumesophageal connectives (CECs) drove forward walking; four effected backward walking, and the stimulation of five different (non-giant) neurons evoked escape and swimming (Bowerman and Larimer, 1974b). Other interneurons were found to be responsible for several forms of abdominal flexion and extension and for initiating swimmeret activity. Each output appears unique when examined carefully (i.e., motor neuron firing patterns, cinematography of behavior). Thus, little or no redundancy apparently exists. These data also suggest that the elements might be used in groups to provide a wide variation of each behavioral class. Almost nothing, however, is known about the gating process for any of these neurons (except in the case of the lateral and medial giant interneurons; see below).

In addition to those interneurons which appear to initiate behaviors, others appear to inhibit or to suppress ongoing movements in various degrees (Evoy and Kennedy, 1967). The most effective units can freeze the entire animal in a variety of positions (Bowerman and Larimer, 1974a).

Interneuron-driven outputs exhibit reciprocal activation and inhibition of antagonistic muscles characteristic of the voluntary pattern (Kennedy, 1969). This reciprocity is broken in certain outputs where muscles may assume different roles depending on the combination in which they are recruited; e.g., for movements in various planes around a mobile joint (Larimer and Kennedy, 1969b). In addition, command-driven as well as voluntary motor programs either may incorporate or partially disengage proprioceptive reflexes depending upon the requirements of the movement. For example, resistance reflexes must be suppressed in many forms of locomotion (Kennedy, 1969; Evoy and Cohen, 1971; Evoy and Fourtner, 1973; Field, 1974).

It is the general rule that command neurons do not synapse directly upon motor neurons (there are exceptions, however; e.g., the commands to the stomatogastric (Dando and Selverston, 1972) and the giant interneurons discussed below). Instead, various forms of driver interneurons appear to be interposed between the command and the final output. Although examples of these intermediate cells have been located only recently (Kovac, 1974b), considerable indirect evidence points to their existence (Evoy and Kennedy, 1967). In systems providing repetitive behavior, command interneurons almost certainly synapse on a series of central oscillator neurons. However, the information required to ensure the proper phase relationships between appendages is supplied by separate coupling interneurons rather than by the command (Stein, 1971, 1974; Kovac, 1974b). Finally, command interneurons supply other important information. For example, synaptic efficacy of commands can vary significantly in different ganglia, including the preferential distribution of output to one side of the animal.

Each of these properties strongly suggests that command interneurons are important elements in the control hierarchy. Is there evidence that they perform these functions in freely behaving animals?

Normal Behavior and Command Neuron Activity

Direct evidence for the participation of command neurons in the execution of behavior is difficult to obtain because we lack certain information. Most commands in crustaceans are recognized by the isolation and stimulation of their axons in the interganglionic connectives. It is not known, for example, whether the units are being driven orthodromically. Location of the inputs and outputs is often unknown. Indeed, with the exception of the giant interneurons, which may be somewhat atypical command fibers, we do not know the morphology of any of these cells. Although large enough to yield to the various dye filling techniques currently in use, the appropriate experiments have not been done. A more serious deficit is the almost complete lack of chronic recordings from command fibers in freely behaving animals (see below). Furthermore, efforts to stimulate and record from command elements using chronically implanted electrodes in otherwise intact animals were generally unsuccessful in generating clear, reproducible behaviors (T. Page and Larimer, Unpublished data). It was assumed that this stimulation technique failed because the input was artificial and was out of context with other ongoing voluntary commands.

Role of Command Neurons

The data now available, although largely indirect, nevertheless provides substantial evidence for the role of command neurons. As noted above, the catalogue of command-driven movements known to exist in crustaceans includes many that are recognizable as partial or complete behaviors. A summary of these crustacean commands is presented in Table 1. It is apparent from the variety of units present, that their activation in proper sequence or combination could account for many recognized behaviors in crustaceans. These data therefore provide indirect although persuasive evidence that such units must in some way underlie voluntary behavior.

Both direct and indirect evidence exists that activity in the giant neurons and several non-giant cells is responsible for various kinds of swimming and escape behaviors. The lateral and medial

Table 1

Behavior	References
1. Abdominal positioning which achieves varied geometries. Abdominal movements which differ in degree, rate, and direction	Kennedy et al., 1966; Kennedy et al., 1967; Evoy, 1967; Evoy and Kennedy, 1967; Evoy et al., 1967; Fields et al., 1967; Atwood and Wiersma, 1967; Gillary and Kennedy, 1969; Page and Sokolove, 1972; Bowerman and Larimer, 1974a; Page, 1975a, 1975b
2. Movements of the appendages of the terminal segment (uropods and telson), ipsilateral, contralateral, bilateral as well as cyclical movements in all planes	Larimer and Kennedy, 1969a, 1969b; Kovac, 1974a, 1974b
3. Movement of the dactyl of the claws: opening, closing	Smith, 1974
4. Movements of ventilation appendages, acceleration, inhibition, ipsilateral, contralateral, bilateral, and reversal	Wilkens et al., 1974
5. Swimmeret movements. Metachronous beating; CFs evoke different burst periods (beat frequency); recruit different motor neurons; fire MN at different frequencies; segmental effects; power stroke/return stroke effects; ipsilateral/contralateral as well as bilateral output; tonic positioning; inhibition	Hughes and Wiersma, 1960; Wiersma and Ikeda, 1964; Atwood and Wiersma, 1967; Stein, 1971; Davis and Kennedy, 1972a, 1972b, 1972c
6. Escape responses; medial and lateral giant neurons mediate different escape behaviors and exhibit different motor neuron connections	Wiersma, 1938; Roberts, 1968; Schrameck, 1970; Larimer et al., 1971; Zucker et al., 1971; Wine and Krasne, 1972; Zucker, 1972a, 1972b, 1972c; Mittenthal and Wine, 1973

Table 1 (continued)

Behavior	References
7. Escape behavior: a variety are mediated by non-giant interneurons	Schrameck, 1970; Wine and Krasne, 1972; Bowerman and Larimer, 1974b
8. Swimming: units provide short bouts, long bouts, rapid or slow frequencies, all non-giant mediated	Bowerman and Larimer, 1974b; (see also Schrameck, 1970; Wine and Krasne, 1972)
9. Defense responses, partial, complete	Wiersma, 1952; Atwood and Wiersma, 1967; Bowerman and Larimer, 1974a
10. Single or pairs of limbs; positioning or movement	Atwood and Wiersma, 1967; Bowerman and Larimer, 1974a, 1974b
11. Total body positioning, e.g. generalized flexion, ipsilateral flexion; general promotion; ipsilateral promotor	Bowerman and Larimer, 1974a
12. "Turning" behavior, left, right	Bowerman and Larimer, 1974a
13. "Righting" behavior from left and right	Bowerman and Larimer, 1974a
14. "Freezing" behavior, in a variety of limb and body positions	Bowerman and Larimer, 1974a
15. Forward walking: different stepping frequencies, body geometries	Atwood and Wiersma, 1967; Bowerman and Larimer, 1974b
16. Backward walking: different stepping frequencies, body geometries	Bowerman and Larimer, 1974b
17. Heart rate: units produce tachycardia, bradycardia, arrest	Wiersma and Novitski, 1942; Wilkens et al., 1974; Field and Larimer, 1975

Table 1 (continued)

Behavior	References
18. Stomach movements: two units affect both the gastric mill and pyloric filtering apparatus, both excitation and inhibition	Dando and Selverston, 1972
19. Hindgut movements	Wolfe and Larimer, 1971; Winlow and Laverack, 1972; Wolfe, 1973

giant fibers were isolated from the CECs and stimulated separately while the resulting movements were recorded by cinematography at 1000 frames/sec. Analyses of these records revealed the medial giant output caused rapid and complete abdominal flexion which propelled the animal backward. The lateral giant neurons, on the other hand, release a different response, characterized by an incomplete flexion of the more posterior segments such that the escape trajectory is upward, as well as backward (Figure 1A). A comparison of movies of sensory-induced escape responses with films of escape behaviors commanded by direct giant stimulation showed certain behaviors that were apparently identical, although most were different. It was concluded, therefore, that the two sets of giant fibers secure different escape behaviors, but that non-giant cells also exist which trigger other escape responses (Larimer et al., 1971). An analysis of the selective synaptic contacts of the lateral and medial giant neurons on the fast flexor motor neurons explained the major features of the two behaviors generated by these interneurons (Figure 1B), (Larimer et al., 1971; particularly Mittenthal and Wine, 1973).

Chronic recording from the abdominal cord of crayfish undergoing escape and swimming showed that the giant neurons often, but not always, fire in escape maneuvers, but that they seldom fire during swimming (Schrameck, 1970). These conclusions have been extended by several recent studies. The giant neurons tend to react to sudden stimuli. Visual or tactile inputs directed toward the head fire the medial giants whereas caudally directed stimuli release lateral giant activity (Wine and Krasne, 1972). Non-giant escape or swim-evoking interneurons respond to more gradual and complex stimuli. As shown by Schrameck (1970), a bout of swimming is often launched by a giant-mediated tail flick. A non-giant system, however, mediates sustained swimming probably by impinging upon a neural organizing network located in the subesophageal or thoracic ganglia (Wine and Krasne, 1972). Recently, at least five

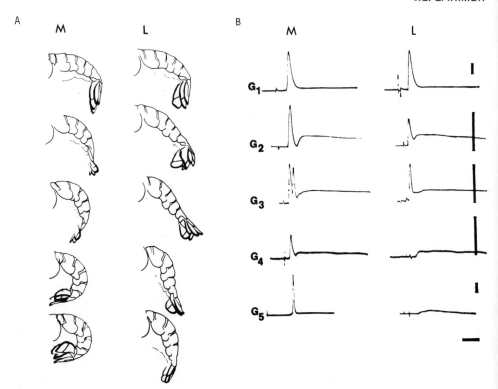

Figure 1. The different behaviors resulting from the stimulation of individual medial and lateral giant interneurons in crayfish and the differential distribution of their synaptic outputs to fast flexor motor neurons in abdominal segments 1-5. (A). Outlines traced from high speed (1000 fps) motion pictures of abdominal movements. Approximately 15 msec separates each position shown. The behavior resulting from the stimulation of the medial giant fiber (M) differs from that obtained by lateral giant stimulation (L) not only in the extent of flexion in each segment but also in the accompanying movements of the terminal appendages. (B). Intracellular records taken from the largest motor neuron emerging from ganglion G_1 through G_5 supplying the fast flexor muscles along the abdomen of crayfish. Selective stimulation of the medial giant interneuron (M) generated EPSPs in the flexor motor neurons in each ganglion, whereas activation of the lateral giant (L) caused EPSPs only in the motor neurons of G_1 - G_3. The smaller depolarizations of the motor neurons are IPSPs received from other interneurons that are also activated by the command fibers. The differential connectivity patterns of the M and L giant interneurons are consistent with the different behaviors they evoke. Calibrations: voltage, 20 mv; time, 10 msec. (A. Reproduced from Larimer et al., 1971, with permission of The Journal of Experimental Biology. B. Modified from Mittenthal and Wine, 1973, with permission of Science).

non-giant command interneurons have been located in the circumesophageal connectives of crayfish that release either escape or swimming behaviors (Bowerman and Larimer, 1974b; Figure 2). Each of these, designated S1-S5, appeared from cinematographic analysis to release a separate form of locomotion. Sustained swimming was always obtained from interneurons located in the vicinity of S1. Others gave single flexions, weak swimming, etc.. Each began with different extensions, reached different flexions or proceeded at different speeds. It seems clear therefore that the CNS of crayfish has at least seven command neurons (two giants and five non-giant cells) which could provide a wide range of escape and swimming locomotion. Fiber counts from the circumesophageal connectives (CECs) show that each connective contains 3,200 axons (Sutherland and Nunnemacher, 1968). It seems reasonable that at least seven of these neurons, if not more, may trigger these important behaviors.

Although swimming utilizes the same fast flexor motor neurons as the escape responses, it is a much more complex behavior. For example, since the fast extensors as well as the flexors are involved, one must assume that a pattern generator with repetitive output and appropriate segmental coupling is needed to coordinate the swimming behavior. Little is known about the generation of the motor program for swimming beyond the suggestions of Wine and Krasne (1972).

The means of activation as well as the detailed synaptic connections of the lateral giant neurons on the fast flexors are well known (Roberts, 1968; Larimer et al., 1971; Zucker et al., 1971; Selverston and Remler, 1972; Zucker, 1972a, 1972b, 1972c; Mittenthal and Wine, 1973), and this description may serve as a useful model of the non-giant escape systems. The lateral giant escape behavior deserves attention too, since it provides the only example of a complete neural circuit of a crustacean command system. Figure 3 illustrates the basic operation of the neural elements (Zucker, 1972a). Phasic stimuli to the hair receptors (TR) of the body surface provide a barrage of input to three or more large tactile interneurons (A), (B) and (C). The major synapses of the primary sensory cells on the tactile interneurons are chemical, but some processes bypass the sensory interneurons to synapse electrically on the command fiber, the lateral giant (LG). The sensory interneurons not only are coupled electrically to each other, but other processes also converge with electrical synapses on the lateral giant fiber. Excitation of the lateral giant interneuron fires the motor giant neuron (MoG) via a rectifying synapse (Furshpan and Potter, 1959) and excites the other fast flexor motor neurons by more conventional electrical junctions.

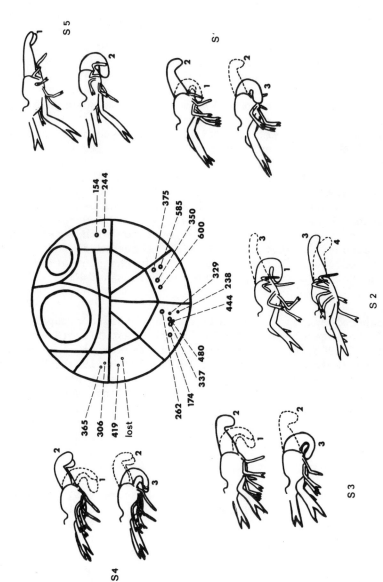

Figure 2. Outlines from moving film records showing various forms of swimming and escape behaviors. These responses were obtained by the stimulation (75 Hz) of small bundles of neurons isolated from the circumesophageal connectives of crayfish. All of the bundles were subdivided as much as possible and contained only small, non-giant axons. During the experiment each command was assigned a number and its location noted on (Legend continued on next page.)

the cross-sectional map as shown. The numbers adjacent to the
outlines are frame numbers from moving film taken at 8 fps. The
cord locations as well as the behaviors appeared to fall into
five classes, designated S1 - S5. Those fibers classed as S3 and
S5 evoked only escape behavior, i.e., one or two fast flexions,
while those belonging to S1, S2 and S4 released swimming. Only S1
was observed to release long bouts of swimming behavior. Fibers in
class S2 are of interest in that the overall behavior did not include streamlining of the body. (Reproduced with permission from
Bowerman and Larimer, 1974, Journal of Experimental Biology.)

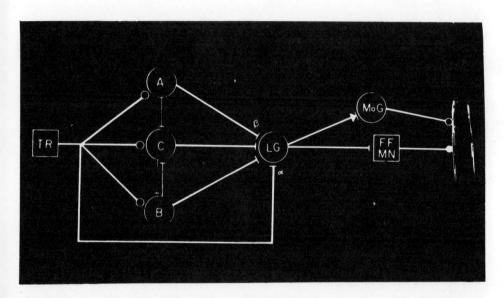

Figure 3. The proposed arrangement of identified cells which
accounts for the escape behavior mediated by the lateral giant
interneurons. (TR) represents the primary sensory input which is
composed of numerous tactile receptors on the carapace of the
abdomen. (A), (B) and (C) are large interneurons: (A) is uni-
segmental, (B) and (C) are multisegmental. (LG) is the lateral
giant interneuron, (MoG) is the largest of the fast flexor motorneurons in each ganglion supplying the fast flexor mass of the
abdomen. Open circles are antifacilitating; closed circles facilitating excitatory chemical synapses, bars are excitatory electrical
synapses, the lateral giant to (MoG) synapse is rectifying and
electrical. The alpha junctions are short latency, temporally
stable synapses while the beta population is of longer latency.
(Reproduced by permission from Zucker et al., 1971, Science.)

It is difficult to predict whether this system is typical of other networks generating simple behaviors, since there are few data for comparison. The two layers of interneurons seen here do not appear to provide the wide latitude of integrative function that certainly characterizes the more complex systems. This network instead meets the criterion of a reflexive system where activation of a restricted sensory modality leads more or less directly to a motor response. Finally, the command cell itself is immediately premotor which also may be atypical. The studies which detailed this network, however, serve as a model for deciphering other restricted systems that generate behavior.

Stimulation of Command Neurons by Electrical and Natural Means

Other direct evidence of the command fiber activity underlying a behavioral output has been obtained from simpler systems by driving the command interneuron with both electrical and natural stimulation to obtain the same motor output. Figure 4 shows how a command neuron (isolated in the first abdominal connective) supplying the slow flexor motor system in three abdominal ganglia (G_2, G_3, and G_4) responded to both electrical (upper) and natural stimulation (lower). The command, driven at 100 Hz, increased the spontaneous activity in all the roots, particularly in the fourth ganglion. Natural stimulation (the flexion of the uropods) drove the interneuron in weak bursts. This produced the same motor program as electrical stimulation including the preferential activation of ganglion four.

This approach also was used to demonstrate command fiber activity in the circumesophageal connectives underlying the movements of the ventilatory appendages and heart rate in crabs (Wilkens et al., 1974). A wide variety of sensory inputs are known in crustaceans that affect the heart and ventilation rates, often concurrently (Larimer, 1964; Ashby and Larimer, 1965; Larimer and Tindel, 1966; Pocock and Larimer, 1968; McMahon and Wilkens, 1972). An equally large assortment of interneurons have been found in the CECs that produce similar responses. In crayfish, Field and Larimer (1975) located some ten interneurons that accelerate, and fourteen that inhibit heart rate. In the crab, an estimated 15 interneurons were mapped that influenced either ventilation or heart rate or both (Wilkens et al., 1974). Thus numerous parallel pathways are present, suggesting perhaps that special sensory inputs are required to gate them and possibly that they are utilized at times in combinations. Records from command interneurons lend support to these suggestions. Tactile stimuli activated some of the interneurons shown by electrical stimulation to be effective in inhibiting the heart and ventilation rhythms (Wilkens et al., 1974). The increased firing frequency of the interneurons in response to

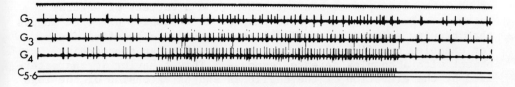

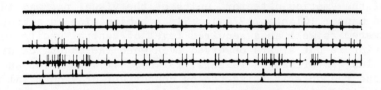

Figure 4. Two records which compare the motor outputs obtained by electrical and natural stimulation of the same command interneuron. The unit here was a typical flexion-evoking command interneuron; however, in this case the axon had been isolated from both the 1-2 and the 5-6 abdominal connectives at different times in the experiment. The motor output was recorded from the small roots supplying the superficial flexor muscles in each segment ($G_2 - G_4$). The axon of the command cell was driven electrically from the connective between G_1 and G_2 (top record) and by natural stimulation to the caudal appendages (bottom record). Proceeding from upper to lower, the traces in both records are as follows: Timing signal 100 Hz, the next three traces are in order, extracellular records from the roots of abdominal ganglion 2 (G_2), ganglion 3 (G_3) and four (G_4). The trace labeled C_{5-6} (upper record) and unlabeled (lower record) shows respectively the spikes in the command fiber axon at the 5-6 connective and at the 1-2 connective before isolation. Finally, the bottom trace in both records shows the stimulus: electrical, 100 Hz (upper record) and natural (lower record). Both methods of activation provided the same output including the preferential activation of the G_4 motor neurons. (Reproduced in modified form from Kennedy et al., 1966, with permission of the authors and Science.)

natural stimulation, however, was too weak to account for the relatively strong behavioral response. This suggests that activity in other unmonitored but parallel commands was largely responsible for the output.

Direct evidence of underlying command activity is difficult to obtain for interneurons with more distributed or complex outputs. These interneurons are apparently remote from direct sensory effects as shown by attempts to activate them from a variety of natural pathways. They seem instead to respond to ensembles of mixed inputs that require sequencing or simultaneous gating to be effective. For example, crayfish can be made to respond to an approaching object with a complex defense response. This behavior is characterized by the elevation of the chelipeds and an extension of the legs and abdomen which lifts the cephalothorax (Wiersma, 1961). A major sensory input which initiates this defense reflex requires the activation of the "motion detector" units of the visual pathway in a specified manner (Glantz, 1974a). Since habituation is a property of the system as well as a required level of central excitation, the behavior is not always observed even when the major features of the stimulus are presented (Glantz, 1974b). If a behavior such as the defense response is triggered by concurrent or sequential commands the difficulty of securing recording and stimulating access to the appropriate group of interneurons at one time is compounded. Apparently, success in recording underlying command fiber activity is not assured, even if the responsible interneurons are identified.

Comparisons of Motor Programs

Several studies in crustaceans have contributed substantial indirect evidence that command fiber activity is responsible for triggering various behaviors. The approach often used is a comparison of the motor program produced by interneuron stimulation with that of a freely behaving animal. This comparison has been accomplished successfully for several simple behavioral routines (Evoy, 1967; Larimer and Eggleston, 1971; Sokolove, 1973; Smith, 1974; Sokolove and Tatton, 1975) but only recently for a complex behavior (Kovac, 1974a, 1974b). If the responsible motor neurons and the command interneuron for a complex behavior are accessible, and if the behavior also can be evoked by natural stimulation in the intact animal, then one can judge, by several criteria, whether both behaviors are identical. Conventional electrophysiological methods are used in these studies to assess the command-driven motor program while chronic recording from motor roots, EMG's and cinematography are usually employed to examine the voluntary motor program. One such behavior that has been analyzed in this manner is a unique form of slow backward walking accompanied by movements of the abdomen and uropods which aid the locomotion. The latter consists of a cyclical motor program of slow flexions and extensions of the abdomen and telson as well as promotion and remotion of the uropods (Kovac, 1974a). This behavior is brought about in the intact animal by threatening or mildly noxious stimuli directed anteriorly. The command interneuron responsible for releasing the cyclical abdominal movements is located in area 85 (Wiersma and Hughes, 1961) of the abdominal cord (Larimer and Kennedy, 1969b; Kovac, 1974a). The motor neurons which take part in the motor program innervate the slow flexor and extensor muscles of the abdomen (Kennedy and Takeda, 1965a, b) and the muscles of the telson and uropods (Larimer and Kennedy, 1969a). It is uncertain whether one of the backward walking commands that can be isolated from the circumesophageal connectives is directly responsible for the cyclical abdominal movements or whether the complete behavior requires the concurrent activation of several commands.

A detailed comparison was made, however, of the command-driven and voluntary motor programs which give rise to the cyclical movements of the abdomen (Kovac, 1974a). Not only were the active motor units the same in both cases, but their phase relationships were identical. The periods of the oscillations were similar although not identical. These differences were largely accounted for by the lack of complete sensory input in the experiments involving electrically-driven behavior. These criteria, as well as the great similarity of overall movement as evidenced from motion picture analysis, give indirect but convincing evidence that the area 85 command interneuron is responsible for the movement.

Other experiments on this cyclical pattern generating system led to identification of the first driver interneuron (Kovac, 1974b). These units, present bilaterally in each abdominal connective, were found to furnish simultaneous excitation to the flexor motor neurons and inhibition to the extensors, as was predicted for flexor drivers (Evoy and Kennedy, 1967). Upon prolonged stimulation of the command interneuron, the driver responded with a series of 600-800 msec bursts each of which preceded the flexor activity in adjacent ganglia by about 100 msec. Direct stimulation of the driver provided a single complete cycle of motor output in lower ganglia that had the same phase and burst period as produced by the command itself. Other data indicated that the burst production is endogenous to the flexor driver system providing both initiation and termination of the flexor burst. The initiation of the extensor firing at the termination of the flexor burst suggests that the extensors may be activated by inhibitory rebound. Thus the behavior of the flexor driver network can account for most aspects of pattern generation in this system. Finally, the observed coupling between flexor driver neurons in the various segments also accounts for much of the intersegmental coordination required for the behavior (Kovac, 1974b).

It is of interest to note that several theoretical studies using analog and digital computer stimulation techniques have shown that simple neural networks composed of only a few cells can be arranged to produce stable, alternating output patterns in two or more channels not unlike those required for the cyclical abdominal movements as well as other behaviors (Reiss, R.F., ed., 1964; Perkel and Mulloney, 1974; Dagan et al., 1975). These models incorporate well known properties and interactions of nerve cells, e.g., postinhibitory rebound and mutual inhibition or excitation. Significantly, in some network configurations, a sustained alternating pattern can be achieved with a single initiating spike in one unit! Finally, either class of pattern generators, those originating from membrane oscillator units or those utilizing networks without pacemakers, can be modeled to achieve the desired alternating output. Such models demonstrate that a command can be extremely simple in theory, and that very few units actually are needed to generate what might otherwise be considered a complex output.

Another way of examining the proposed role of command interneurons is one of combining behavior genetics with more traditional mapping techniques. If a particular behavioral routine is missing in a species, does it also lack the appropriate command interneuron(s) to trigger that behavior? A recent study has shown that this is in fact the case for abdominal positioning movements of some species of crayfish (Page, 1975a, b). In the species most often used by neurophysiologists, Procambarus clarkii, numerous command fibers from both the circumesophageal connectives (Atwood and Wiersma, 1967; Bowerman and Larimer, 1974a) and the abdominal cord (Evoy and Kennedy, 1967) are known to release various abdominal extension movements. Furthermore, a form of abdominal extension behavior also can be produced reliably in Procambarus clarkii by removal of support from beneath the legs; i.e., by a platform drop (Larimer and Eggleston, 1971; Sokolove, 1973). The crayfish, Orconectes rusticus and Orconectes virilis, were examined along with P. clarkii for the expression of the abdominal extension reflex and for the presence of extension command neurons in the circumesophageal connectives (Page, 1975a, b). In agreement with earlier studies, P. clarkii reliably showed the extension behavior (greater than 90% of trials) and was found to possess a variety of extension commands. In some 80 trials of the extension reflex using Orconectes, the majority extended slightly (from an initial flexed position), fewer than 10% extended above the horizontal, while the remainder gave no extension when suspended for 4-5 seconds. Sampling of interneurons in the circumesophageal connectives of Orconectes revealed only two axon bundles with extension command activity, and one of these upon subdivision was found to contain two commands (Page, 1975a). That is, only when both subdivisions of this bundle were stimulated simultaneously was extension behavior observed. The other bundle could not be subdivided and thus was judged to contain a conventional, albeit very weak, extension command interneuron. Yet other command interneurons, e.g., those providing flexion, were present in their normal complement in the CECs of Orconectes. Such correlation strongly suggests that this behavior requires activity in extension command neurons located in the circumesophageal connectives.

Several additional questions remain, however, concerning the role of command interneurons in the release of the abdominal extension behavior. Why does the command interneuron present in the CECs of Orconectes fail to trigger the extension behavior in response to the platform drop? Does this interneuron require some separate and specific input that is not contained in the stimulus, or are multiple command lines required in Orconectes in order to release the complete behavior? In this regard, a search for extension evoking interneurons in the abdominal connectives of Orconectes showed them to be present at that level in the CNS (Page, 1975a). Do these lower interneurons not receive an input from the platform drop, or is there in Orconectes but not in Procambarus, a simultaneous and competing or inhibiting behavior released by the stimuli associated with the platform drop? There is some evidence for this since the command interneuron of the CEC activated the extensor inhibitor neuron along with the extensor motor neuron. Finally, if one rendered inoperative the CEC command interneurons for extension in Procambarus by cutting the two connectives, will these animals (also possessing lower level extension command fibers) continue to respond to platform drops with hyperextension behavior? This test has been performed with Procambarus clarkii in the course of other experiments, and some were found to respond, although weakly, when deprived of their neural connections to the brain (Larimer and Gordon, Unpublished data). Interpretation of these observations would be more conclusive if the overall morphology of the extension command system were known. It is known that other interneurons in Procambarus degenerate very slowly even when the cell bodies are separated from the axons (Bittner et al., 1974). If the axons were continuous through the cord, and if they received sensory input along their length, then severing them above the thorax would have left surviving segments that could continue to mediate the response. These results still point to the conclusion that the extension command systems of the two species differ in some fundamental way that presently is not understood.

Surgical Intervention

Recently the role of interneurons in the normal expression of behavior has been examined by surgically removing segments of the circumesophageal connectives and by assaying the animals (crayfish) for behavioral deficits (Gordon, Larimer and Page, Unpublished data). The activity associated with changes in daily illumination was studied and two distinct responses were seen in intact animals. One is a reflexive, non-circadian "lights-on" response in which the overall activity, forward, backward locomotion, heart rate, etc., increases for about one hour after the onset of the light phase of a 12:12 LD cycle. The other response is, by several criteria, a circadian rhythm of similar activity which begins about one half-hour after the onset of the dark phase of the light cycle and continues typically for several hours (Page and Larimer, 1972; 1975). The latter behavior is entrained by an extraretinal photoreceptor and exhibits free-running activity in total darkness, but is otherwise similar to the reflexive behavior (Page and Larimer, 1975). Both forms of behavior are abolished by severing the circumesophageal connectives, suggesting that neural rather than hormonal signals are required for their expression. Our initial question was whether the coupling is carried in one area of the cord or on many parallel pathways. The approach was to sever one connective completely and to remove all but one quadrant of the remaining connective. When assayed, animals with any single quadrant except the dorso-medial one retained both behaviors. From these data, it is assumed that multiple coupling interneurons, widely distributed in the cord, are mediating these behaviors. It is of interest to note that the dorso-medial quadrant is not totally devoid of locomotor commands, but does contain fewer than any of the other three quadrants. Although this correlation is suggestive, further experimentation is needed to determine whether command interneurons or neurons of some other type are responsible for the coupling.

Locomotion and Command Interneurons

In addition to the various types of escape and swimming behaviors described earlier, other forms of locomotion, including forward walking, backward walking and swimmeret rhythms are also obtained from the stimulation of specific interneurons in crustaceans. Of these, the swimmeret control system is the most fully understood and provides a model for the study of other forms of multiappendage locomotion. The neural elements contributing to swimmeret control are distributed conveniently through the abdominal cord allowing experimental accessibility. The system contains all the major ingredients to provide for multisegmental locomotion. These include:

(1) the presence of central oscillators and associated interneurons in each of the abdominal ganglia for pattern generation;

(2) the presence of interneurons to ensure intersegmental coordination of the phase of each limb within the rhythm; and

(3) a variety of command neurons that not only set the system into activity but modulate the basic rhythm in different ways.

Since each of these important aspects of the swimmeret system will be dealt with independently in this volume, the present discussion will be confined to an examination of the forward and backward walking commands in crustaceans.

A survey of the command interneurons in the CECs of crayfish revealed the presence of about a dozen different interneurons that effected walking behaviors (Bowerman and Larimer, 1974b). The major points of interest concerning the walking command data are as follows: 1) It was necessary to provide the animals with a circular treadmill, and ensure that the limbs touched the surface, in order to obtain what appeared to be normal interneuron-driven walking behavior. 2) On the basis of cord locations, and from the similarities of associated body geometry, there appear to be at least 5 interneurons which produce forward walking behaviors. In addition, as many as four of these were located in a single connective of a single individual. Figure 5A shows the map locations for the axons of these interneurons as well as outlines of the assumed body positions. In many cases, forward walking appeared normal and was accompanied by cheliped elevation, abdominal extension and swimmeret activity. 3) The walking commands that were examined more thoroughly showed a frequency threshold and some frequency-dependent fractionation of behavior. Low frequency stimulation gave rise to some limb and abdominal movement but all parts of the behavior were intact when the threshold frequency for walking itself was reached. At higher stimulation frequencies of CF activation (to 75 Hz) there was a diminished latency of onset of the behavior, but the other features of the motor output apparently were not affected. 4) A similar analysis revealed four interneurons which appeared to release distinct forms of backward walking (Figure 5B). These outputs were somewhat more variable than the forward walking programs with respect to the presence or absence of some accompanying behavior; cheliped lifting, abdominal flexion, etc.

Detailed analyses of normal walking behaviors are available for only a few species of crustaceans; e.g., for lobsters (McMillan, 1975), crayfish (Parrack, 1964), and crabs (Barnes et al., 1972; Evoy and Fourtner, 1973; Barnes, 1975). Conclusions concerning the relationships between the central triggering interneurons and the various forms of locomotion cannot be made among these species since command neuron data for walking are available only for the crayfish. In addition, the cine films of the electrically-evoked walking behaviors as with other behaviors in the initial survey were taken at 8 fps and are not adequate for the analysis of stepping patterns (Bowerman and Larimer, 1974b). All three species examined thus far

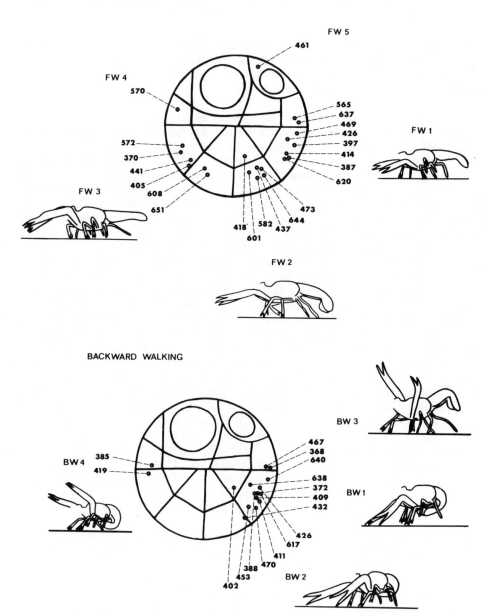

Figure 5. Types of forward and backward walking behaviors evoked by the stimulation (at 50–100 Hz) of interneurons isolated from the circumesophageal connectives of crayfish. Outline tracings of body geometries were taken from filmed records (8 fps) of the behavior. The command fiber   (Legend continued on next page.)

exhibit multiple gaits, omnidirectional movements, varying speeds and numerous compensating adjustments for variation in load and terrain. Thus because of the complexity of the system, data on crayfish and lobster locomotion were obtained only for forward walking. Both lobsters and crayfish exhibit six gaits (assuming each limb steps only once per cycle), and shifts between the different gaits are frequent. The dominant gait for both animals is characterized by a stepping sequence of (four limbs, one side) 1-4-2-3. Frequencies of the use of the remaining five gaits are also established for both species (Parrack, 1964; McMillan, 1975).

Could the different command fibers furnish not only excitation to the ganglionic oscillators but also select for certain gaits? If so, is the selection process for a particular gait due to the recruitment of a special segmental coordinating interneuron which orders the new sequence, or is some other mechanism operating? Clearly, these questions must be approached by a systematic analysis of the voluntary gaits followed by a comparison with the gait or gaits evoked by interneuron stimulation. More attention also should be paid to the problem of possible frequency and pattern dependence within the system. These comparisons can now be made and more can be learned about the triggering mechanism itself. There are anticipated technical difficulties in searching for oscillators and the interganglionic coordinating neurons in the thoracic region, however, since the anatomical complexity is great and the CNS here is encased in skeletal material. It is encouraging to note that ganglionic oscillator neurons similar to those that might underlie these behaviors have been located recently in both crustacean (Mendelson, 1971) and insect ganglia (Pearson and Fourtner, 1975).

Frequency and Pattern Influences on Command Fiber Output

In the earlier studies discussed here, most commands are set into activity by complex inputs, probably originating from an array of sensory modalities and requiring subtle central states of excitability for activation. They require therefore something conceptually equivalent, in ethological terms, to a "releaser" in order to be gated (Konishi, 1971). There is no question that the

---

number and approximate location of each is given on the cross-sectional map of the connective. From these locations and the corresponding behaviors there appeared to be five types of forward walking and four types of backward walking behaviors; however, these categories are somewhat artificial. (Reproduced with permission from Bowerman and Larimer, 1974b, Journal of Experimental Biology.)

higher integrative interneurons that impinge upon command cells contain a great deal of information. Earlier considerations regarding the frequency and pattern dependence of command interneuron output, however, suggest that they may be more important as trigger elements than as carriers of complex instructions, i.e., they are permissive rather than instructive (DeLong, 1971).

This concept, however, may not be entirely correct for all of these interneurons. An important question is how the neural elements (the driver networks, pattern generators and motor neurons), that are postsynaptic to command interneurons decode the information content of their input. This has been tested to some degree by examining the output of command systems driven at different frequencies and patterns. Although the available data are somewhat limited, the emerging evidence suggests that there are gradations of responses in different command systems with respect to the amount of information they can convey. These range from networks relatively independent of the complexity of the input to others that exhibit extraordinary frequency and pattern sensitivity.

Even those outputs which respond weakly to frequency variations show some decoding. For example, in the abdominal positioning system in crayfish, Evoy and Kennedy (1967) showed that flexion commands driven at increasing frequency tended to diminish the interspike interval and to recruit new units gradually. Thus command fiber frequency could be of considerable importance in grading tension in tonic systems of this type. Gillary and Kennedy (1969) examined the effects of command input frequency on the output pattern of the largest of the tonic flexor motor neurons, a cell that inherently fires in bursts. Increasing frequency of command fiber discharge did not appreciably affect the latency of onset of the motor neuron discharge in this case, but did increase both the number of impulses per burst as well as burst frequency. The command, driven in short repetitive bursts, produced greater output than one stimulated at the same frequency but with a constant interval train.

The swimmeret system responds to increased frequency in some command interneurons with a diminished latency to onset (Wiersma and Ikeda, 1964) and with an increased frequency of the rhythm. Otherwise the output remains basically the same (Davis and Kennedy, 1972a). Both the burst discharge frequency of the motor neurons and the number of motor neurons recruited into each burst were also observed to increase with increasing frequency in the command. Although the recruitment generally appeared to follow the size principle, there were notable exceptions (Davis and Kennedy, 1972a). Where there is relative independence of the output from the complexity of command stimulation as was shown in the cases above, one must conclude that the command fiber lines carry relatively little complex information but serve either to trigger driver interneurons for positioning behavior or to activate a neural oscillator for releasing repetitive behavior. Here, the details of the motor program reflect the properties of the neural "center" rather than the information in the command.

In contrast, the outputs of other interneurons are strongly frequency-dependent, even to the extent that different behaviors emerge at different frequencies. Thus the decoding capabilities of some neural networks apparently are sufficient to permit the incoming commands themselves to carry significant instructions. In the most striking example, multiple behaviors were found to be generated from a single command fiber over a relatively narrow range of frequencies (Atwood and Wiersma, 1967). In this study, command fiber CM_4 from the CEC of the crayfish was located some 40 times, offering an unusual opportunity for examining its output under different conditions. When stimulated at 2-4 Hz there was a remotion of the fifth periopods. When the stimulus was increased to about 4 Hz the abdomen was flexed anteriorly but was extended posteriorly along with the telson. Finally, there were often swimmeret movements as well. At 10 Hz the claws were extended and raised, while the legs were positioned laterally, i.e., the preparation assumed a "defense posture." Further study of CM_4 showed that its output was pattern-sensitive. Pairs of stimuli were more effective than unpatterned trains of the same average frequency (Atwood and Wiersma, 1967).

In another example, taken from the swimmeret system, a strong response was observed to an abrupt change in command activity (Wiersma and Ikeda, 1964). It was noted that the swimmeret rhythm could be initiated readily by driving a selected interneuron at 30 Hz but not at 12-16 Hz. However, once the output was triggered, it could be maintained for long periods by an input of 12-16 Hz. This frequency sensitivity resembles that observed for tension development in many muscles (Blaschko et al., 1931; Wiersma and Ripley, 1952; Wilson and Larimer, 1968).

A remarkable frequency dependence of a different kind was observed in the swimmeret control system of lobsters. In this case, occasional commands were encountered that were effective in generating rhythmic output but only when driven in a narrow frequency band (Davis and Kennedy, 1972a). Command activation at frequencies outside this range produced other responses in the motor neurons including changes in their rate of firing and order of recruitment. Since units of this kind would ordinarily go undetected in most experimental procedures, they may be more common than now supposed.

Two command neurons emerge from the brain and synapse directly on several motor neurons that control the striated muscles of the grinding teeth (gastric mill) and the filtering apparatus (pyloric region) of the lobster stomach. These interneurons are of interest for two reasons. They synapse directly on a mixed network of interneurons <u>and</u> motor neurons responsible for generating rhythmic motor patterns for the two sections of the stomach. In addition, the two groups, the pyloric and gastric mill neurons, respond strongly and differentially to increasing frequency of the command inputs. At 2 Hz the pyloric cycle frequency increases while the motor neurons of the gastric mill are unaffected. At 5, 10 and 20 Hz the gastric mill group receives increasing inhibition while the pyloric cycle becomes first disrupted then completely inhibited (Dando and Selverston, 1972).

Finally, an example of a frequency-dependent output was taken from the simple motor system of the first limb of the crayfish. Here most commands stimulated in the CEC activate both the excitor and inhibitor neurons to the dactyl opener muscle. However, the ratio of output in the two motor neurons was found to be strongly dependent upon the frequency of command fiber stimulation. This caused opening at times and closing at others depending upon the concurrent activity to the dactyl closer muscle (Smith, 1974).

It appears likely that a systematic study of frequency and pattern dependence will reveal other examples. However, the present sample testifies to their existence.

In order to realize the full behavioral output, the frequency-sensitive neural centers must receive the appropriate codes at their inputs. It would be of interest to know whether command interneurons normally fire in complex ways, and whether they utilize frequency, microstructure, probability, burst duration or some other type of code (Perkel and Bullock, 1968; Bullock, 1968). Unfortunately, since we have not been successful in recording from command interneurons in freely behaving animals the codes that might be utilized are unknown. Clearly, the decoding capability exists in some output pathways, which can execute instructions more complex than a simple "on" or "off." In summary, the current evidence suggests that a gradation of commands exists. Some exhibit little frequency dependence of their output on the one hand, while others exhibit strong frequency and pattern dependence.

It can be argued that the representation of neurons within this distribution should be skewed toward the less frequency dependent classes. For example, outputs which vary strongly in rate of command activation would tend to introduce certain serious constraints on the motor control system. More specifically, a neural organization in which a single premotor interneuron releases a series of different behaviors when driven at different frequencies might introduce the following requirements:

(1) Each interneuron in this class should be responsible for only related and non-conflicting behaviors. (Evolution might provide a system such as this.)

(2) A very precise mechanism would be needed to activate the interneuron to the desired level from moment to moment to prevent the accidental release of a non-adaptive behavior.

(3) Precision of the encoding of command frequency as well as stability of the decoding of its signal with time would be required to prevent an inadvertent shift from one behavior to another.

(4) Additional and equally specialized mechanisms would be required to select one behavior, or a pair, from the total array available. An even more serious difficulty might arise if it became necessary to re-sequence the behaviors within the array.

(5) Control of the command interneuron by the activating or stabilizing circuits could require the dedication of many other neurons. This would tend to cancel any numerical advantage achieved by allowing a few command neurons to release many different behaviors. One might predict too, that the various command interneurons should not be extensively cross-connected with others, particularly at the upper levels, since this would severely limit the range of available behavioral output. Thus the arrangement of separate elements, each releasing a single behavior, would appear to offer the most options. The separate elements then could be sequenced or combined freely for extension of the repertoire to that of a freely behaving animal.

Acknowledgment

The author wishes to thank Dr. William Gordon, Benjamin Williams, Trevor Pollard and Rowland Aertker for reading and criticizing the manuscript. The original research from this laboratory was supported by NIH Grant NS05423.

REFERENCES

Ashby, E.A. and Larimer, J.L., (1965) Modification of cardiac and respiratory rhythms in crayfish following carbohydrate chemoreception. J. Cell. and Comp. Physiol. 65, 373-380.

Atwood, H.L. and Wiersma, C.A.G., (1967) Command interneurons in the crayfish central nervous system. J. Exp. Biol. 46, 249-261.

Barnes, W.J.P., Spirito, C.P. and Evoy, W.H., (1972) Nervous control of walking in the crab, Cardisoma guanhumi. II. Role of resistance reflexes in walking. A. verg. Physiol. 76, 16-31.

Barnes, W.J.P., (1975) Leg co-ordination during walking in the crab, Uca pugnax. J. Comp. Physiol. 96, 237-256.

Bittner, G.D., Ballinger, M.L. and Larimer, J.L., (1974) Crayfish CNS: Minimal degenerative-regenerative changes after lesioning. J. Exp. Zool. 189, 13-36.

Blaschko, H., Cattell, M. and Kahn, J.L., (1931) On the nature of the two types of response in the neuromuscular system of the crustacean claw. J. Physiol. 73, 25-35.

Bowerman, R.F. and Larimer, J.L., (1974a) Command fibers in the circumesophageal connectives of crayfish. I. Tonic fibers. J. Exp. Biol. 60, 95-117.

Bowerman, R.F. and Larimer, J.L. (1974b) Command fibers in the circumesophageal connectives of crayfish. II. Phasic fibers. J. Exp. Biol. 60, 119-134.

Bowerman, R.F. and Larimer, J.L., (1976) Command neurons in Crustaceans. Life Science (In press).

Bullock, T.H., (1968) Representation of information in neurons and sites for molecular participation. Proc. Nat. Acad. Sci. 60, 1058-1068.

Dagan, D., Vernon, L.H. and Hoyle, G., (1975) Neuromimes: Self-exciting alternate firing pattern models. Science. 188, 1035-1036.

Dando, M.R. and Selverston, A.I., (1972) Command fibers from the supraoesophageal ganglion to the stomatogastric ganglion in Panulirus argus. J. Comp. Physiol. 78, 138-175.

Davis, W.J. and Kennedy, D., (1972a) Command interneurons controlling swimmeret movements in the lobster. I. Types of effects on motor neurons. J. Neurophysiol. 35, 1-12.

Davis, W.J. and Kennedy, D., (1972b) Command interneurons controlling swimmeret movements in the lobster. II. Interactions of effects on motor neurons. J. Neurophysiol. 35, 13-19.

Davis, W.J. and Kennedy, D., (1972c) Command interneurons controlling swimmeret movements in the lobster. III. Temporal relationships among bursts in different motor neurons. J. Neurophysiol. 35, 20-29.

DeLong, M., (1971) Central patterning of movement. Neurosciences Res. Prog. Bull. 9, 10-30.

Devoy, W.H., (1967) "Central commands for postural control in the crayfish abdomen," In Invertebrate Nervous Systems. (Wiersma, C.A.G., ed.), University of Chicago Press, Chicago, (213-217).

Evoy, W.H. and Kennedy, D., (1967) The central nervous organization underlying control of antagonistic muscles in crayfish. I. Types of command fibers. J. Exp. Zool. 165, 223-238.

Evoy, W.H., Kennedy, D. and Wilson, D., (1967) Discharge patterns of neurons supplying tonic abdominal flexor muscles in the crayfish. J. Exp. Biol. 46, 393-411.

Evoy, W.H. and Cohen, M.J., (1971) "Central and peripheral control of arthropod movements," In Advances in Comparative Physiology and Biochemistry, 4. (Lowenstein, O., ed.), Academic Press, New York, (225-266).

Evoy, W.H. and Fourtner, C.R., (1973) "Crustacean walking," In Control of Posture and Locomotion. (Stein, R.B., Pearson, K.G., Smith, R.S. and Redford, J.B., ed.), Plenum Press, New York, (477-493).

Field, L.H., (1974) Sensory and reflex physiology underlying cheliped flexion behavior in hermit crabs. J. Comp. Physiol. 92, 394-414.

Field, L.H. and Larimer, J.L., (1975) The cardioregulatory system of crayfish: The role of circumesophageal interneurons. J. Exp. Biol. 62, 531-543.

Fields, H.L., Evoy, W.H. and Kennedy, D., (1967) Reflex role played by efferent control of an invertebrate stretch receptor. J. Neurophysiol. 30, 859-874.

Furshpan, E.J. and Potter, D.D., (1959) Transmission at the giant motor synapses of the crayfish. J. Physiol. 145, 289-325.

Gillary, H.L. and Kennedy, D., (1969) Pattern generation in a crustacean motor neuron. J. Neurophysiol. 32, 595-606.

Glantz, R.M., (1974a) Defense reflex and motion detector responsiveness to approaching targets: The motion detector trigger to the defense reflex pathway. J. Comp. Physiol. 95, 297-314.

Glantz, R.M., (1974b) The visually evoked defense reflex of the crayfish: habituation, facilitation and the influence of picrotoxin. J. Neurobiol. 5, 263-280.

Hughes, G.M. and Wiersma, C.A.G., (1960) The co-ordination of swimmeret movements in the crayfish, Procambarus clarkii (Girard). J. Exp. Biol. 37, 657-670.

Kandel, E.R. and Kupferman, I., (1970) The functional organization of invertebrate ganglia. Ann. Rev. Physiol. 32, 193-258.

Kennedy, D. and Takeda, K., (1965a) Reflex control of abdominal flexor muscles in the crayfish. I. The twitch system. J. Exp. Biol. 43, 211-227.

Kennedy, D. and Takeda, K., (1965b) The reflex control of abdominal flexor muscles in the crayfish. II. The tonic system. J. Exp. Biol. 43, 229-246.

Kennedy, E., Evoy, W.H. and Hanawalt, J.T., (1966) Release of co-ordinated behavior in crayfish by single central neurons. Science. 154, 917-919.

Kennedy, D., Evoy, W.H., Dane, B. and Hanawalt, J.T., (1967) The central nervous control underlying control of antagonistic muscles in the crayfish. II. Coding of position by command fibers. J. Exp. Zool. 165, 239-248.

Kennedy, D., (1969) "The control of output by central neurons," In The Interneuron. (Brazier, M.A.B., ed.), U. Cal. Press, Berkeley, (21-36).

Kennedy, D., Selverston, A.I. and Remler, M.P., (1969) Analysis of restricted neural networks. Science. 164, 1488-1496.

Kennedy, D., (1971) Crayfish interneurons. The Physiologist. 14, 5-30.

Kennedy, D., (1973) "Control of motor output," In Control of Posture and Locomotion. (Stein, R.B. et al., eds.), Plenum Press, New York, (429-436).

Kennedy, D. and Davis, W.J., (1975) "The organization of invertebrate motor systems," In Handbook of Physiology: Neurophysiology. (In press).

Konishi, M., (1971) Ethology and Neurobiology. Amer. Sci. 59, 56-63.

Kovac, M., (1974a) Abdominal movements during backward walking in crayfish. I. Properties of the motor program. J. Comp. Physiol. 95, 61-78.

Kovac, M., (1974b) Abdominal movements during backward walking in crayfish. II. The neuronal basis. J. Comp. Physiol. 95, 79-94.

Larimer, J.L., (1964) Sensory-induced modification of ventilation and heart rate in crayfish. Comp. Biochem. Physiol. 12, 25-36.

Larimer, J.L. and Tindel, J.R., (1966) Sensory modification of heart rate in crayfish. Anim. Behav. 14, 239-245.

Larimer, J.L. and Kennedy, D., (1969a) Innervation patterns of fast and slow muscles in the uropods of crayfish. J. Exp. Biol. 51, 119-133.

Larimer, J.L. and Kennedy, D., (1969b) The central nervous control of complex movements in the uropods of crayfish. J. Exp. Biol. 51, 135-150.

Larimer, J.L. and Eggleston, A.C., (1971) Motor programs for abdominal positioning in crayfish. Z. vergl. Physiol. 74, 388-402.

Larimer, J.L., Eggleston, A.C., Masukawa, L.M. and Kennedy, D., (1971) The different connections and motor outputs of lateral and medial giant fibers in the crayfish. J. Exp. Biol. 54, 391-402.

Maynard, D.M., (1973) The command neuron. B.I.S. Conference Report #30. June, 1973. 6th Annual Winter Conference on Brain Research.

McMahon, B.R. and Wilkens, J.L., (1972) Simultaneous apnoea and bradycardia in the lobster Homarus americanus. Canad. J. Zool. 50, 165-170.

McMillan, D.L., (1975) A physiological analysis of walking in the American lobster Homarus americanus. Phil. Trans. R. Soc., (London) B270, 1-59.

Mendelson, M., (1971) Oscillator neurons in crustacean ganglia. Science. 171, 1170-1173.

Mittenthal, J.E. and Wine, J.J., (1973) Connectivity patterns of crayfish giant interneurons: Visualization of synaptic regions with cobalt dye. Science. 179, 182-184.

Page, C.H. and Sokolove, P.G., (1972) Crayfish muscle receptor organ: Role in regulation of postural flexion. Science. 175, 647-650.

Page, C.H., (1975a) Control of postural movements of the crayfish abdomen by circumesophageal connective fibers: MRO and extensor motor neuron activities in Orconectes and Procambarus. J. Comp. Physiol. 102, 65-76.

Page, C.H., (1975b) Generic differences in the abdominal extension reflexes of the crayfish Orconectes and Procambarus. J. Comp. Physiol. 102, 77-84.

Page, T.L. and Larimer, J.L., (1972) Entrainment of the circadian locomotor activity rhythm in crayfish. J. Comp. Physiol. 78, 107-120.

Page, T.L. and Larimer, J.L., (1975) Neural control of circadian rhythmicity in the crayfish. I. The locomotor activity rhythm. J. Comp. Physiol. 97, 59-80.

Parrack, D.W., (1964) Stepping sequences in the crayfish. Ph.D. Dissertation, U. Michigan, Ann Arbor.

Pearson, K.G. and Fourtner, C.R., (1975) Nonspiking interneurons in the walking system of the cockroach. J. Neurophysiol. 38, 33-52.

Perkel, D.H. and Bullock, T.H., (1968) Neural coding. Neurosciences Res. Prog. Bull. 6, 221-348.

Perkel, D.H. and Mulloney, B., (1974) Motor pattern production in reciprocally inhibitory neurons exhibiting postinhibitory rebound. Science. 185, 181-183.

Pocock, M.W. and Larimer, J.L., (1968) Behavioral responses to electrolytes in the crayfish Procambarus clarkii. Physiol. Zool. 41, 332-340.

Reiss, F., ed., (1964) Neural Theory and Modeling. Stanford University Press, Stanford, California.

Roberts, A., (1968) Some features of the central coordination of a fast movement in the crayfish. J. Exp. Biol. 49, 645-656.

Schrameck, N.E., (1970) Crayfish swimming: Alternating motor output and giant fiber activity. Science. 169, 698-700.

Selverston, A.I. and Remler, M.P., (1972) Neural geometry and activation of crayfish fast flexor motor neurons. J. Neurophysiol. 35, 797-814.

Smith, D.O., (1974) Central nervous control of excitatory and inhibitory neurons of opener muscle of the crayfish claw. J. Neurophysiol. 37, 108-118.

Sokolove, P.G., (1973) Crayfish stretch receptor and motor unit behavior during abdominal extensions. J. Comp. Physiol. 84, 251-266.

Sokolove, P.G. and Tatton, W.G., (1975) Analysis of postural motor neuron activity in crayfish abdomen. I. Coordination by premotor neuron connections. J. Neurophysiol. 38, 313-331.

Stein, P.S.G., (1971) Intersegmental coordination of swimmeret motor neuron activity in crayfish. J. Neurophysiol. 34, 310-318.

Stein, P.S.G., (1974) Neural control of interappendage phase during locomotion. Am. Zool. 14, 1003-1016.

Sutherland, R.M. and Nunnemacher, R.F., (1968) Microanatomy of crayfish thoracic cord and roots. J. Comp. Neur. 132, 499-518.

Wiersma, C.A.G., (1938) Function of the giant fibers of the central nervous system of the crayfish. Proc. Soc. Exp. Biol. Med. N.Y. 38, 661-662.

Wiersma, C.A.G. and Novitski, E., (1942) The mechanisms of the nervous regulation of the crayfish heart. J. Exp. Biol. 19, 225-265.

Wiersma, C.A.G., (1952) Neurons of arthropods. Cold Spring Harbor Symp. Quant. Biol. 17, 155-163.

Wiersma, C.A.G. and Ripley, S.H., (1952) Innervation patterns of crustacean limbs. Physiol. Comparata Oecol. 2, 391-405.

Wiersma, C.A.G., (1961) "Reflexes and the central nervous system," In The Physiology of Crustacea, Vol. II, (Waterman, T.H., ed.), Academic Press, New York, (241-279).

Wiersma, C.A.G. and Hughes, G.M., (1961) On the functional anatomy of neuronal units in the abdominal cord of the crayfish Procambarus clarkii (Girard). J. Comp. Neur. 116, 209-228.

Wiersma, C.A.G. and Ikeda, K., (1964) Interneurons commanding swimmeret movements in the crayfish Procambarus clarkii (Girard). Comp. Biochem. Physiol. 12, 509-525.

Wilson, D.M. and Larimer, J.L., (1968) The catch property of ordinary muscle. Proc. Nat. Acad. Sci. 61, 909-916.

Wilkens, J.L., Wilkens, L.A. and McMahon, B.R., (1974) Central control of cardiac and scaphognathite pacemakers in the crab, Cancer magister. J. Comp. Physiol. 90, 89-104.

Wine, J.J. and Krasne, F.B., (1972) The organization of escape behavior in the crayfish. J. Exp. Biol. 56, 1-18.

Winlow, W. and Laverack, M.S., (1972) The control of hindgut motility in the lobster Homarus gammarus (L.) 3. Structure of the sixth ganglion (6 A.G.) and associated ablation and microelectrode studies. Mar. Behav. Physiol. 1, 93-121.

Wolfe, G.E., (1973) Neuronal circuits involved in the control of the crustacean intestine. Ph.D. Dissertation, University of Texas at Austin.

Wolfe, G.E. and Larimer, J.L.,(1971) The intestinal control system in the crayfish P. clarkii. Am. Zool. 11, 666.

Zucker, R.S., Kennedy, D. and Selverston, A.I., (1971) Neuronal circuit mediating escape responses in crayfish. Science. 173, 645-650.

Zucker, R.S., (1972a) Crayfish escape behavior and central synapses. I. Neural circuit exciting lateral giant fiber. J. Neurophysiol. 35, 599-620.

Zucker, R.S., (1972b) Crayfish escape behavior and central synapses. II. Physiological mechanisms underlying behavioral habituation. J. Neurophysiol. 35, 621-637.

Zucker, R.S., (1972c) Crayfish escape behavior and central synapses. III. Electrical junctions and dendrite spikes in fast flexor motor neurons. J. Neurophysiol. 35, 638-651.

TRIGGER NEURONS IN THE MOLLUSK TRITONIA

A.O.D. Willows

University of Washington, Friday Harbor Laboratories

Friday Harbor, Washington 98250

Many acts of behavior across a diverse range of animal species are elicited in their fully coordinated form by a particular "sign" stimulus. Commonly, the critical releasing capabilities of this stimulus are related to qualitative features other than its duration. Put another way, duration and other details of the response elicited often do not depend upon the duration of the stimulation. Instead, the "sign" stimulus acts as a trigger to cause the full unfolding or "playback" or a response, which is apparently pre-programmed neurally.

Neurophysiological experiments on three species of nudibranch mollusk suggest that the neural bases for sign stimulus "recognition" and the subsequent "release" of an appropriate behavioral response are vested in a well-defined neural network. Several properties of these neurons emerged: (i) The network is electrically active, producing a short train of impulses and rapid bursts, whenever the sign stimulus is presented and a response occurs. (ii) With extracellular stimulation of the network for a brief period of time relative to the duration of the response, the normal playback of the response occurs. (iii) The network consists of several neurons in well-defined topographical groups on both sides of the mid-line of the central nervous system. (iv) All members of the group on both sides of the midline are closely electrically coupled to one another and, (v) The network has an inherent burst generating mechanism which gives it a characteristic discrete, patterned output.

Such a system has several emergent properties. It tends to be insensitive to random inputs (noise) and to produce a vigorous and regenerative response when stimulated by a number of parallel inputs. In this regard it can act as a stimulus filter. Its endogenous burst generating mechanism is effective in converting a maintained depolarizing input (from either sensory or artificial sources) into a train of short intense impulse bursts. The intensity of these bursts could be used to overcome a high threshold in the motor pattern generator, a threshold which would be important in preventing the inappropriate activation of the response by "noise" or other weak inputs.

Such interactive neural systems may be present in the central nervous systems of other animals. If so, they would be recognizable by several criteria, including their electrical coupling, their differential sensitivity to stimuli of different types, and the detailed structure of the burst patterns that they produce.

Introduction

All behavioral actions, particularly the stereotyped act released by a particular stimulus, in principle are controlled by neural machinery having two fundamental components. The first of these two is that diverse stimuli must be recognized and separated. Sensory and perhaps interneuronal elements which encounter a range of incoming sensations must discern the appropriate stimuli necessary for a certain response. Secondly, a motor output pattern must be produced. This requires the production of nerve impulses in spatially and temporally coordinated patterns. These two components are the neurophysiological correlates of stimulus filtering or recognition and pattern generation. This chapter deals with a system of neurons that has some of the characteristics and physiological properties mentioned above.

Whether peripherally or centrally located, systems of neurons that encounter and selectively respond to particular stimuli and subsequently elicit a particular response, must accomplish several levels of integrative activity. For instance, sensory inputs irrelevant to the response are disregarded. Random or noise activity in the system is suppressed. Furthermore, the system's output must be the correct transformation of the input. Thus, if the response must move in some phase or duration relationship to stimulus parameters, then the neurons must receive this phase and duration information. If the response intensity, duration or phase bear little or no relation to the stimulus parameters, then the

transformation is of a different sort. In this instance, the
stimulus only need be recognized as appropriate for transformation
into an all-or-none output with an amplitude sufficient to exceed
the threshold of the pattern generator. These integrative functions are sought in analyses of neurophysiological substrates of
stimulus filtering or discrimination networks.

Re-identifiable neurons were studied in earlier experiments
using intracellular and extracellular techniques to determine how
they control the release of an escape-swimming response in the
nudibranch mollusk, Tritonia. Although the cells' specific action
for controlling release of swimming remains uncertain, a possible
triggering role has emerged. In addition, the input-output transformation function of these neurons was examined for their intrinsic
burst generating mechanism.

Materials and Methods

Experiments were conducted on neurons in the pleural ganglia
of the nudibranch mollusk genus, Tritonia. Three species were
used, two of which have a well developed escape swimming response
(T. diomedia, T. hombergi) and one which does not (T. tetraquetra).
All except one species were obtained in the waters of Puget Sound,
and studied at the Friday Harbor Laboratories on San Juan Island.
T. hombergi was obtained in Liverpool Bay and studied at Menai
Bridge, North Wales. All the nervous systems were studied while
perfused with sea water at the local ambient temperature. Both
isolated ganglia and whole animal preparations were employed. The
techniques employed have been described elsewhere (Willows and
Hoyle, 1969; Willows et al., 1973a, b; Willows and Dorsett, 1975).
Electrophysiological techniques were conventional and included 3M
KCl-filled micropipettes for intracellular studies and sea-water
filled suction electrodes for extracellular stimulating and
recording.

Results: Role of Trigger Group Neurons in Elicitation of Escape-Swimming Behavior

When the nudibranch mollusks, Tritonia diomedia or T. hombergi
come into contact with any of a number of echinoderm species anywhere on their epithelial surface, they execute a stereotyped
swimming response (Willows et al., 1973a). The basic form of the
responses in both species is similar although differences in readiness to respond particularly over a series of stimulus presentations
are apparent (Willows and Dorsett, 1975). The North American
species, T. diomedia, responds more readily, is more likely to
respond over a series of trials, and swims for a longer duration on

average, than does T. hombergi. Thus, for instance, the likelihood of a response on each of 5 successive trials in T. diomedia is 100% whereas after 5 trials the likelihood of a response in T. hombergi is nearly zero. Similarly, the average swim duration in T. hombergi on any trial is about 50% of what it is for T. diomedia.

There are surfactants present in the mucus of the tube-feet of echinoderms which apparently act as a specific stimulus (sign stimulus or releaser). The natural releaser in this instance can be imitated satisfactorily by the placement of salt crystals on the animal's skin but most other forms of stimulation (mechanical, chemical, visual) are not effective. In many other animals there are examples of stimuli which are similarly specific releasers. For example, any object brought suddenly into the visual field of many decapod crustacea elicit a defensive posture involving widespread coordinated movements. Quivering movements of male stickleback fish produce egg laying in a receptive female. Small rodents elicit a particularly stereotyped posture and pounce in domestic dogs. Similar examples can be drawn from all animal groups that display organized behavior.

One other North Pacific species, T. tetraquetra, which resembles T. diomedia in several ways, does not respond to physical contact with echinoderms in a vigorous manner. Nor does it respond by swimming to skin contact with salt or other substances.

The gross neuroanatomy of the brain (cerebral, pedal, pleural ganglia) of all three species is similar. There are pigmented neurons (diameter, 50-800 microns) present on the surfaces of the ganglia. In particular, there are quite large neurons (300-800 microns) on the posterior surfaces of the pleural ganglia in all three species. In the region between the central commissure and the above mentioned giant neuron somata, there is on each pleural ganglion, at least one cluster of smaller neurons (50-100 microns) ranging in number from 5 to 35 depending upon the species selected. Neurons in these latter clusters in T. diomedia have been called trigger group neurons (TGN's). These cell groups play the role of the innate releasing mechanism in regard to the escape swimming response. Evidence for this follows. First, in those intact animal preparations stimulated and/or recorded intracellularly, penetration of one of the TGN's at the outset of an experiment occasionally elicited the entire escape-swimming sequence. Such events were rare however, occurring only 5 times out of over 200 experiments in T. diomedia. It has also been observed once in 32 cases in T. hombergi. In no case in many more penetrations (perhaps 1000) of other brain cells in both species, did a similar event occur. It is noteworthy that in every case out of the 6, when swimming was elicited upon initial penetration early in the experiment, the cell produced a short burst of impulses (caused most likely by penetration damage). Furthermore, the response could not be elicited again simply by directly depolarizing the cell to produce a similar burst of impulses.

More frequently, it was noted that mechanical pressure and vibration caused by attempts to drive an electrode through the epineurium and into one of the neurons in this region, elicited swimming behavior. This was seen in both T. diomedia (many times) and in T. hombergi (twice), and could in some cases be repeated in the same preparation, especially if the electrode pressure was applied near the midline border of the group.

Attempts were made using T. diomedia to localize the region of the pleural ganglia where the response could be elicited by placing a 200 micron diameter suction electrode on the ganglion surface and delivering short trains (0.25-20 sec.) of 10 Hz pulses. As before, the region over the TGN group was most sensitive with particular sensitivity along the midline border of the group. By this method, escape-swimming could be elicited in most preparations and responses could be triggered repeatedly by brief stimulation.

In all cases described above, the intracellular mechanical and extracellular electrical stimulation that released the escape swimming response was effective at durations shorter than the duration of the behavior. Typically, excitation lasting for less than 1 sec. produced responses that were on the order of 30-100 sec. long. Nor was there any noticeable correlation between the duration of the stimulus and the amplitude or duration of the response. In general, the response was released in an all or nothing manner with an apparent threshold in the stimulus level.

In experiments with T. diomedia, T. hombergi, and T. tetraquetra, it was possible to record from neurons in the central pleural ganglion region while presenting the animal with normal stimuli. A clearly defined continuum existed in the responses of neurons in these three species to the same stimuli (Figure 1). In T. diomedia a few crystals of NaCl placed directly on the center of the oral veil invariably produced a volley of epsp's and several nerve impulses. These impulses sometimes occurred in a few short, high frequency pairs or triplets with interposed epsp's. Other forms of stimulation produced either no response or a much reduced, synaptically driven volley. If several (2-5) neurons in the region were monitored simultaneously, it was observed that all produced spike bursts (pairs, triplets) nearly synchronously when the animal was stimulated with salt crystals and the underlying psp's likewise occurred nearly simultaneously in all. It was determined that all neurons in the groups on both sides of the central commissure responded as if they were receiving inputs from sign stimulus receptors in parallel and indicating that they were closely coupled to one another. There were 15-25 neurons visible in the central

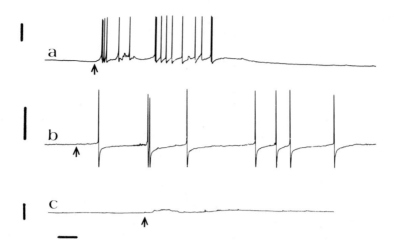

Figure 1. Response of neurons in TGN region to skin stimulation (at arrow) by one drop of saturated salt solution in three species of Tritonia. (a) Tritonia diomedia responds with vigorous volley of synaptic potentials and spikes, including several bursts of 2-3 impulses. Similar responses are seen in all neurons in this region of the pleural ganglia. (b) T. hombergi produces a weaker response, and only one burst of 2 spikes. Some neurons in this species (not shown) produce no response to such stimulation. (c) In T. tetraquetra, salt stimulation produces only a weak synaptic volley and no spikes. No neurons were found that produced more than one or two spikes to such stimulation. Calibration: 20 mV, 2 s.

region of each pleural ganglion, all of which shared these response properties.

It was determined also that these neurons were electrically coupled to one another both within each group and across the midline commissure. The coupling factor measured as the ratio of post-junctional neuron polarization to pre-junctional neuron polarization for D.C. signals ranged from 0.05-0.40 depending upon the neuron pair chosen. This ratio decreased as the frequency of the impressed currents was increased above 1 Hz., and was reduced below 1% at and above 10 Hz in most cells (Getting, 1974). Thus, the depolarizing up-swing of an action potential is transmitted less efficiently through these junctions than is the much longer duration, after-hyperpolarization.

The possibility of a measurable aspect of the TGN response as correlated with the duration of the escape-swimming response was tested in the following manner. The response of TGN's to skin application of a constant amount of salt over 7 trials at 5 minute inter-trials was measured in two animals. Also, the response of several other neurons, including flexion neurons that drive different phases of the swimming cycle were monitored. The number of swimming flexion cycles made by the animal also was recorded. The number of swim cycles, the number of flexion neuron (FN) bursts, the number of spikes in the TGN response, the duration of the TGN response, the average frequency of the TGN response and its frequency during the first 5 sec. are recorded in Table 1. It is clear from these data that there is no close correlation between response duration (measured as number of swim cycles) and any other parameter except possibly the overall duration of the TGN response. There is a parallel decline in both these measures in both animals. This observation may not be indicative of a causal relation since very short periods (less than 1 sec.) of TGN stimulation reliably can produce very much longer (about 1 min.) responses. Furthermore, in the course of these experiments it often has been observed that the TGN system may respond only very briefly (1-5 sec.) to normal stimuli and yet produce fully developed escape-swimming. Thus, overall, there is little basis for asserting any connection between the duration or other aspect of TGN activity and the duration of swimming behavior.

The situation in T. hombergi, the European species that swims only less readily and for a shorter duration, was somewhat different. The large, and giant neurons were present in the same locations on the pedal and pleural ganglion. In several cases, the electrophysiological activity and functional roles of these neurons were observed directly and comparisons were made to the suspected homologs in T. diomedia. For instance, the 3 neurons found to generate bursts spontaneously in T. diomedia (specifically LPl 2, LPl 3, and RPl 3) were present and active in precisely the expected locations in T. hombergi. The functional roles of many pedal and pleural cells (e.g., LPl 1, RPl 1, LPd 1, RPd 1) were also determined in T. hombergi. It was found that these neurons control regional and local branchial tuft withdrawal as they also do in T. diomedia. Similarly, dorsal and ventral flexion movements were elicited by certain pedal ganglion neurons. Despite these homologies however, significant differences in the neurons in the central pleural ganglion (TGN) region were found. It was apparent at the outset that the number of neurons visible in this area was considerably less (approximately 10/ganglion as compared with 25 in T. diomedia). Furthermore, the spontaneous firing and response properties of many of them differed substantially. About one-half of the neurons did not produce accelerating impulse bursts when depolarized directly, nor did they do so when excited via normal sensory pathways in whole animal preparations. Some of these in fact did not respond vigorously to skin applications of salt, but instead were especially sensitive to mechanical vibrations nearby

Table 1

Animal	Trial	Number of Swim Cycles and (FN Bursts)	Spikes in TGN Response	Duration of TGN Response in sec.	Average Frequency impulses/sec.	Frequency of TGN Impulses During Initial 5 sec.
1	1	5 (5)	15	41	0.37	0.8
1	2	5 (5)	11	43	0.26	1.2
1	3	4 (4)	9	33	0.27	1.4
1	4	4 (4)	14	33	0.42	1.4
1	5	3 (3)	15	24	0.62	1.4
1	6	3 (3)	12	24	0.50	1.2
1	7	3 (3)	11	24	0.46	1.2
2	1	4 (4)	34	30	1.1	1.4
2	2	4 (4)	32	32	1.0	2.0
2	3	3 (3)	27	24	1.1	2.6
2	4	2 (2)	20	15	1.3	2.0
2	5	3 (3)	26	29	0.9	1.8
2	6	2 (2)	24	28	0.9	2.0
2	7	1 (1)	13	11	1.2	1.6

in the water. Nor were these latter neurons electrically coupled to one another.

Quite unlike T. diomedia, no more than 4 neurons in any single ganglion were found in 9 preparations which responded selectively to salt applied to skin areas, produced recurrent bursts of impulses when directly depolarized, and were electrically coupled to any of their neighbors. By these criteria, the TGN group was considerably less well developed in T. hombergi by comparison with T. diomedia (Willows and Dorsett, 1975).

Six experiments were done on T. tetraquetra, a species that is apparently unable to swim in response to skin contact with starfish, salts, and other applied substances. The physiological results for all neurons excepting those in the TGN region were similar to those in both other species. The large and giant neurons on pedal and pleural ganglia present in the expected locations, had comparable pigmentation and relative sizes. The bursting neurons (LPl 2, LPl 3, RPl 2) were present and active. The giant pleural (LPl 1, RPl 2) and pedal neurons (LPd 1, RPd 1) were found in the anticipated locations and when depolarized directly in intact animal prepraratrions, caused branchial tuft withdrawal in the expected regions.

The distribution of neurons in the central TGN region of each pleural ganglion was even more sparse in T. tetraquetra than in T. hombergi. Only 4-7 neurons could be seen in these otherwise clear, translucent regions. It is unclear whether the space was filled with neuropil, glia, or microscopic neurons. However, it is certain that the number of visible (larger than 20µ) pigmented neurons was much reduced from the number seen in T. diomedia. Thirty-eight of these neurons were penetrated with microelectrodes and stable records were obtained for at least 5 minutes from each. In no case did any of these neurons show a tendency to produce accelerating bursts when directly depolarized for several seconds. Instead, each produced a gradually adapting spike train or in a few cases, just 1 spike followed by damped membrane oscillations (Figure 2). In no case were neurons found in this region that produced impulse bursts when starfish or salt crystals were placed on the skin of the animal. Instead the characteristic response of cells in this region to several salt crystals placed on the oral veil was a short and subthreshold volley of epsp's. Furthermore, only 4 pairs of neurons out of 15 tested were found in these areas that were electrically coupled to one another. The coupling factor in each case was less than 0.10. Tapping in the region with a microelectrode (over 50 instances) produced no evident behavioral response, nor did electrical stimulation of any of the above mentioned neurons.

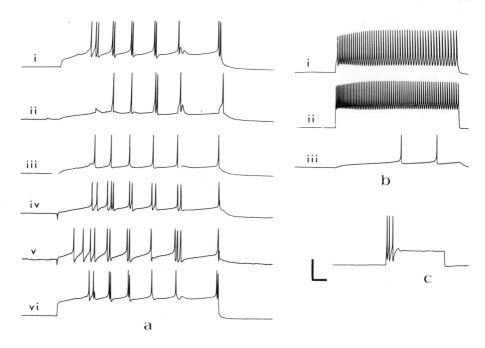

Figure 2. Response of TGN neurons to maintained depolarizing currents. (a) Six neurons in <u>Tritonia diomedia</u> recorded simultaneously. All neurons except channel 3 are depolarized through recording electrode. Entire group fires in loose synchrony with 6 impulse bursts in each. After-hyperpolarization on last spike of each burst is both larger in amplitude and longer in duration than those preceding. The same response was seen in several neurons of <u>T. hombergi</u>. (b) In <u>T. tetraquetra</u>, 3 electrically coupled neurons recorded simultaneously. Depolarization of upper 2 produces rhythmic firing in both, with 2 impulses in third neuron. Interimpulse intervals are constant, and no bursts are produced. Only 4 such coupled neurons were found in <u>T. tetraquetra</u> out of 15 pairs tested. (c) Most common response in <u>T. tetraquetra</u> to maintained depolarization was 1-3 impulses followed by damped membrane oscillations. This response was seen often in <u>T. hombergi</u> also. Calibration: 30 mV, 0.4 s.

The results of experiments in <u>T. diomedia</u> and <u>T. hombergi</u> in which direct stimulation of a TGN neuron occasionally produced a complete swimming response, and the subsequent observations that several nearby neurons were electrically coupled to one another, suggested that it would be important to understand the interactions and input-output relationships for the group. The fact that excitation of a single neuron occasionally elicited swimming, yet would not do so repeatedly, indicated that the TGN neurons were not individually responsible for eliciting the behavior. The activity of the TGN group in T. diomedia was then examined to determine what aspects of the activity of the group as a whole conferred the capacity to trigger the release of escape-swimming.

The Mechanism of Burst Generation

Recordings from single TGN's made during normally elicited escape responses indicated that the burst of impulses occurring at the beginning of swimming lasted 0.5-15 seconds. The impulse train at the beginning of swimming was in the form of 1-10 rapidly accelerating, short impulse bursts, with a similar number of impulses interspersed. The bursts were often composed of 3-5 spikes and were clearly demarcated by the appearance of relatively large negative-going after-hyperpolarization following the last spike of each burst. This after-hyperpolarization was 1-10 mv more negative than that following individual spikes and had a duration of 100-600 msec, compared with a duration of 25-100 msec. after an individual spike (Figures 2, 3). The burst structure therefore was composed of an accelerating spike train that was terminated, and then separated from ensuing spikes or bursts by a larger and longer than usual after-hyperpolarization (Getting and Willows, 1973, 1974). These bursts were seen preceding swimming activity in both T. diomedia and T. hombergi. They were never observed in any neurons of T. tetraquetra.

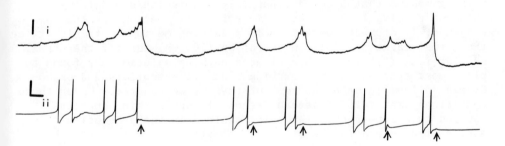

Figure 3. Response of 2 neurons in T. hombergi to maintained depolarization. Last impulse in each group has longer after-hyperpolarization than any preceding. In upper trace, a non-firing neuron at higher gain shows time and voltage course of subthreshold activity, with larger, prolonged hyperpolarizing wave invading cell at end of each burst. Evidence of the effect of this wave, which originated in a third neuron (not shown), can be seen as a diphasic post-synaptic potential in bottom trace (arrows). Calibration: 0.5 mV (upper), 20 mV (lower) and 0.4 s.

If more than one TGN was recorded simultaneously under these conditions, it was apparent that all were excited synchronously and that some individual spikes and all impulse bursts occurred synchronously in all. In particular, the last spike in each burst, with its larger and longer after-hyperpolarization occurred synchronously throughout the network.

Neurons showing such activity also responded in a characteristic way when depolarized for several seconds by currents injected through a microelectrode. Individual neurons in either T. diomedia or T. hombergi stimulated in this way produced recurrent, accelerating impulse bursts for as long as the stimulating currents continued (Figure 3). If more than one neuron was recorded simultaneously during such stimulation, it was apparent that both the depolarizing current and the impulse bursts were occurring synchronously in all the recorded cells. On the assumption that the bursts recorded under conditions of direct stimulation were produced by the same interactive mechanism as those seen when the animal was stimulated through normal sensory channels, we tried to establish the underlying causes. There were two aspects of the burst generation that required consideration separately; the accelerating, excitation phase and the burst termination and establishment of an interburst interval.

In regard to the first, it was apparent that a regenerative feedback mechanism was likely to be involved. To determine whether excitation within the system of neurons was made regenerative by synaptic pathways either intrinsic to the TGN network or involving other neurons, the capability of the system to generate bursts in a bathing medium that blocked synaptic transmission was determined (Willows et al., 1973b). Recordings from two neurons in the TGN system of T. diomedia were made in a bathing medium containing twice the normal magnesium ion concentration (100 mM) and 1/2 the normal calcium ions (2.5 mM). This altered bathing solution was found to block most spontaneously occurring synaptic potentials and orthodromic volleys in other neurons of these same ganglia. Under these conditions, maintained (several seconds) depolarizing currents applied to one neuron in the pair resulted in the production of a series of accelerating bursts in both cells.

Also, it was shown that the burst generating machinery was not sensitive to reduced temperature. Normally formed bursts were produced in neurons that were cooled to 5°C, indicating that temperature dependent processes such as electrogenic pumps probably were not involved. This same conclusion was indicated by the observation that bursting persisted in sea water in which Li^+ was substituted for Na^+.

On the other hand, one neuron could be prevented from participating in the group activity by simply injecting sufficient hyperpolarizing current to stop action potential generation. Under these hyperpolarized conditions, a depolarizing wave and superimposed shorter duration depolarizing potentials could be seen in the neuron in phase with the burst generation in other TGN's. Thus, whatever processes underlie the regenerative excitatory build-up in the system, they are unaffected by hyperpolarization of one cell, and they result in membrane depolarizations of that cell.

A single impulse in one neuron of the system tends to produce another (Willows and Hoyle, 1969; Getting and Willows, 1973). If one records from several TGN's simultaneously, both spontaneously occurring and driven impulses are produced which can be seen to re-excite additional impulses. Based upon the evidence described above, the regenerative feed-back processes within the TGN system appear not to depend upon chemically mediated synaptic pathways. It is more likely that the close electrical coupling that exists between all cells serves to transmit excitation throughout the system. Thus, any depolarizing activity in the system, whether from artificially applied currents, from synaptic input volleys, or from spikes in other neurons in the system, depolarizes many other neurons in the system to varying degrees with a short latency. The process of regeneration is also enhanced by facilitation of the post-junctional psp's. When a pre-junctional neuron produces a train of spikes due to either applied or natural stimulation, successive spikes have longer durations (Getting and Willows, 1974). The lower frequency components in these slower spikes are transmitted more efficiently through the electrical junctions, causing the production of larger psp's in other TGN's. Firing tends to be regenerative on both these counts.

The second aspect of the burst formation mechanism was the enlarged and prolonged after-hyperpolarization which generates the interburst interval. This could not be accounted for on the basis of a large and prolonged IPSP impinging on TGN's from within or from outside the system because, as before, when chemically mediated synaptic transmission was interrupted by suitable manipulations of the bathing sea water (high Mg^{++}, low Ca^{++}) the large hyperpolarizing wave persisted (Getting and Willows, 1974). Furthermore, the amplitude and synchrony of the hyperpolarizing wave was unaffected by altering the membrane potential (Getting, 1974), suggesting that an ionic conductance change was not involved.

On the other hand, the generation of the wave was altered and even blocked by manipulating the firing activity of other TGN's. For instance, a hyperpolarization applied directly to two TGN's that were known to be coupled to a third, could prevent burst formation in the latter. Instead of the usual series of bursts in response to maintained depolarization, the third showed only a smoothly adapting spike train (Getting and Willows, 1974).

It was noticed also that spikes in individual TGN's that were unaccompanied by synchronous ones in others, were always characterized by smaller after-hyperpolarizing waves than were spikes that occurred synchronously in many neurons. Furthermore, when one or more TGN's were voltage clamped to follow the potential of another TGN, the after-hyperpolarization of the latter was increased in amplitude and duration.

These observations and the nature of the after-hyperpolarization can be understood in terms of loading of individual TGN's by the others in the system. A single isolated spike in one TGN is the result of currents across its membrane. These currents are shunted partly through the electrical junctions into other TGN's. The after-hyperpolarization is affected more strongly by this shunting than the depolarizing phase of the spike because of the low pass filter properties of the junctions. When several neurons are voltage clamped to a common potential, no potential difference can exist between them and accordingly, no currents will flow between them. Under these conditions, the neurons involved are effectively uncoupled electrically and the shunting effect of one upon the other is greatly reduced. This provides an explanation for the observation that in voltage clamping experiments, the amplitude of the after-hyperpolarization is increased.

Under normal conditions during the build up of firing in TGN's the frequency of impulses tends to increase in all units. The close electrical coupling in the system tends to assure that both the build up of excitation and the firing of individual impulses is synchronous. After only a few impulses, the synchrony in the group becomes complete and the membrane potentials of many neurons in effect are clamped together. Since little potential difference now exists between the cells, even during the spike, little current flows between them and the shunting of each neuron by the others is greatly diminished. Under these conditions, the fully developed spike after-hyperpolarization of each neuron is no longer shunted and appears much larger. This unshunted wave is the large, prolonged hyper-polarization which is responsible for the inter-burst interval.

From the above it is apparent that both aspects, i.e., acceleration and termination of burst generation in TGN's when depolarized by applied currents, are the direct result of factors that are intrinsic to the system. In particular, close electrical coupling through non-rectifying junctions seems to be the essential property which underlies this input-output transformation in the TGN system.

The question remains as to what, if any, specific aspect of
TGN activity is responsible for the trigger function of the system.
It is implicit, however, in the above arguments that the fast
bursts of impulses generated when the system is depolarized by
brief or prolonged sensory inputs are somehow crucial in the innate
releasing mechanism for swimming.

One way to test this idea is to manipulate TGN's to produce or
prevent such bursts in intact animal preparations and to observe
simultaneously the effect that these manipulations have upon the
swimming response of the animal. This was done, in part, in the
experiments of Willows and Hoyle (1969), when direct depolarizations
of single TGN's were applied in an unsuccessful effort to elicit
escape swimming. One may not safely conclude that TGN's are
uninvolved from such an experiment, however. It is possible, indeed
likely, that currents injected into single TGN's will have only a
local effect in the system, i.e., cause impulses and bursts in only
one or a few neurons. It has been suggested (Getting, Personal
communication) that a comprehensive TGN system response could be
produced by injecting depolarizing currents simultaneously into
several (3-5) such neurons, and that the response of the system
might be monitored faithfully in another unstimulated cell. This
procedure would have the advantage of permitting the injection of a
larger amount of current into the system on a widespread basis
thereby stimulating the production of bursts throughout the system.
Experiments were performed to test this idea. Several (2-6) electrodes were placed in different TGN's using all but one to inject
current and one presumably to monitor the response of the system as
a whole (Figure 4). It was found that very similar results were
obtained whether any number from 2 to 6 electrodes were used.
Direct, salt crystal stimulation applied to the skin elicited the
usual bursts of impulses. Typically, the number of impulses concentrated into the last 100 msec. of such bursts were 3-4. These
bursts recurred at irregular intervals over a period of 2-15 sec.
In all of the 6 preparations and 35 instances where it was
attempted, it proved to be impossible to drive bursts which resembled the naturally occurring ones in the monitored neuron. The
results were the same whether depolarizing currents were strong
enough to produce maximal spiking rate, complete depolarizing
blockade or irreversible damage. In all but 3 cases, depolarization of the TGN system by this means produced a few individual
spikes that appeared to be synchronous with others (because of the
large after-hyperpolarization) at long inter-spike intervals (over
100 msec). In only 3 instances out of 35 attempts, 2 spiked bursts
were produced. It was apparent from observing the activity in the
neurons into which current was injected (through a bridge circuit)
that strong depolarizations (over 50 mV) were ineffective at producing high frequency bursts of spikes in these neurons.
Instead, the spike generating mechanism was blocked, apparently due

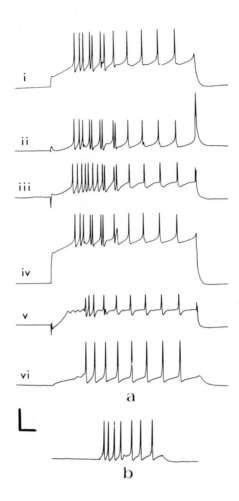

Figure 4. Strong depolarization of 5 TGN's in T. diomedia does not produce bursting in a sixth. (a) Channels 1-5 depolarized through recording electrodes sufficiently to produce maximal firing frequency. Stronger depolarization (not shown) produces block of impulse generating mechanism. Response of monitored neuron (Channel 6) is series of single impulses which appear to be synchronous with others in other neurons. (b) Using stronger depolarization sufficient to irreversibly damage 5 neurons, response of sixth neuron shown is a similar train of single impulses, with no apparent bursting. Calibration: 30 mV, 0.4 s.

to a failure of sodium channels to inactivate. Nor did weaker depolarizations of the cell somas produce burst firing frequencies that were comparable to the natural situation.

Experiments also were carried out to determine if bursting and swimming could be blocked by strong hyperpolarization of several neurons in the system (Figure 5). With 5 neurons in the system hyperpolarized sufficient to produce 20-40 mV hyperpolarization in the monitored cells, swimming could still be elicited by the usual forms of natural stimulation. However, under these conditions, many neurons in the system apparently produced normal looking bursts. Indeed, it was not possible to block bursting in any of the monitored neurons tested in the 6 cases that were studied by these means.

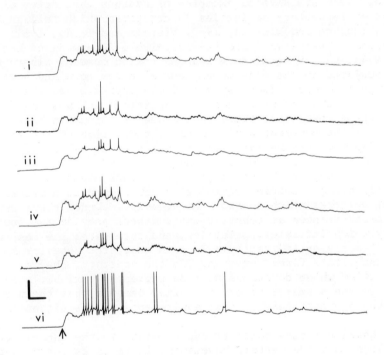

Figure 5. Strong hyperpolarization of 5 TGN's in T. diomedia does not block bursting in a sixth. Neurons on channels 1-5 were hyperpolarized sufficiently to maintain channel 6 26 mV. below resting level throughout record. One drop of salt solution was applied to oral veil of animal (arrow). Large depolarizing volley elicits some spikes in hyperpolarized neurons and several impulses and bursts in a sixth monitor cell. Animal swam 3 cycles. Same results were obtained in other experiments using hyperpolarization of up to 40 mV. Depolarizing volleys caused by salt stimulation often exceeded 60 mV. Calibration: 30 mV, 2 s.

These experiments, like the observations involving stimulation of single neurons, were not useful in showing whether TGN bursting per se is critical for the innate releasing mechanism for swimming. They do show however that it is unlikely that single spikes either individually or occurring synchronously in the system in low frequency trains are adequate releasing activity for escape swimming.

Discussion

A clear distinction should be made between the driving and the triggering functions of the central nervous system in regard to patterned or stereotyped behavioral responses. Most instances of central nervous system control over locomotory responses that have been studied involve the driving function of neurons. For instance, there are several examples of neurons that drive either postural or locomotory activities in crustacea which fall in this category (Hughes and Wiersma, 1960; Wiersma and Ikeda, 1964; Ikeda and Wiersma, 1964; Atwood and Wiersma, 1967; Bowerman and Larimer, 1972, 1974a, b). Several such neurons called command fibers have been stimulated in the crustacean ventral nerve cord and have been found to drive either aspects of the rhythmic beating of swimmerets or the postural changes associated with walking. In each of these instances, there is a close relationship between the duration of the movement produced and the duration of the stimulus train. The movement begins after a variable but relatively short latency following the onset of stimulation, and continues until the stimulation ends, or shortly thereafter. Thus, command fibers apparently produce driving stimulation for the pattern generating or other networks which control the behavior. Under these experimental circumstances at least, command fibers provide the neural activity which initiates, maintains, and terminates the response either directly or indirectly. In each case, the behavior involved is not one with a fixed duration. On the contrary, they are part of activities whose durations must be extremely variable since their utility in the normal life of the animal demands that their duration be matched appropriately to the requirements of the situation.

Although evidence quite strongly points to the capability of such fibers to drive certain movements, there is as yet no direct evidence that they are in fact used in this way under normal circumstances. Proof of such involvement requires monitoring the activity of these fibers during the execution of normal behavioral activities, an experimental situation which has not yet been accomplished in crustacea.

The neurons described in these studies of molluskan swimming play a substantially different role. At the outset, the swimming response itself is fundamentally different in that its duration on the first trial of a series is relatively fixed (about 1 min.) and independent of the stimulus strength over a wide range. The duration of the electrical activity in the TGN system similarly bears no simple relationship to the duration of the response, with TGN activity varying from 5-50 sec. eliciting responses of about the same duration, i.e., approximately 1 m. Thus the response is, in a sense, ballistic and not driven by TGN's. Instead, it is apparently triggered or released by this activity. The TGN system may be supplying the triggering burst of excitation to the motor pattern generating networks that subsequently drive the response.

The evidence on the specific role of the TGN's in eliciting the response is not unambiguously clear however. There are several observations which are consistent with a triggering function:

(1) Occasionally, brief intracellular stimulation of single trigger cells in T. diomedia or T. hombergi produces escape-swimming.

(2) Direct, gross stimulation of the entire brain region near the TGN's reliably causes escape-swimming.

(3) Whenever swimming occurs as a result of either normal sensory stimulation or electrical excitation to appropriate nerve trunks, a burst or bursts of impulse activity can be recorded in any TGN.

(4) The physiological development of the TGN system varies in three Tritonia species in which swimming is evolved and used in differing degrees. There are more such neurons having the appropriate interactions and sensitivities in the most actively swimming species (T. diomedia), less in the reluctant swimming species (T. hombergi) and few, or none in the non-swimmer (T. tetraquetra).

On the other hand, it is puzzling that direct excitation of single TGN's or even small groups is not an adequate stimulation to elicit escape swimming. One might expect that stimulation of one or a few TGN's would cause sufficient excitation in many others to mimic the effect of physiological inputs. The close electrical coupling between many TGN's, with a coupling factor to D.C. currents of as much as 0.40 would suggest that close co-ordination might occur system-wide to even local inputs. However, this is apparently not the case, since strong depolarizing currents applied to as many as 5 TGN's produce only synchronous spikes in other TGN's and do not generate the high frequency bursts that characterize the response of these cells to appropriate sensory inputs. Instead, one observes only single spikes that appear to be occurring synchronously in many TGN's. This result suggests that a substantial number of TGN's other than those stimulated are only weakly influenced by local inputs to the system and act as a distributed, electrical load or shunt on the group, preventing the development of a vigorous regenerative burst. The observation that bursts occur widely in the system when it is excited via normal sensory pathways implies that these inputs must impinge upon the system at many different points in parallel.

Similarly, the observation that it is apparently not possible to block the swimming response by hyperpolarizing 1-5 neurons to prevent firing is somewhat puzzling. However, the relatively large number of the coupled neurons in the system (in T. diomedia) may, once again, offer an explanation. As has been observed here, although the imposed currents may be made large enough to produce as much as 40 mV of hyperpolarization in unstimulated neurons (through the electrical junctions), this manipulation is nevertheless inadequate to prevent normal sensory inputs from driving bursting in many neurons of the TGN group, including those that are strongly hyperpolarized through the junctions.

These two observations together encourage the view that the TGN system probably does not trigger swimming by single impulses occurring in individual neurons. Neither does the innate releasing mechanisms depend upon single impulses that occur synchronously in many TGN's. If the TGN system is indeed directly responsible for releasing the swimming behavior, then it may do so through the development of one or a few synchronous, relatively high frequency bursts of impulses. This possibility requires further testing.

It is quite possible, despite these findings, that the TGN's are not directly involved in eliciting escape swimming. They may, for instance, act in parallel with the actual innate releasing mechanism and be responsible for driving some other discrete or general aspect of the response. They might, for instance, be responsible for increasing general body wall tonus. This is a necessary component of the preparation for swimming since in this animal, all movements are made against the hydrostatic resistance of the hemocoelic fluids.

Conclusion

The TGN system has a number of emergent physiological properties which, in principle, match it well to the role of the innate releasing mechanism. It can be tuned sharply to appropriate sensory inputs and made insensitive to other signals. If the receptors that are sensitive to components of the mucus secreted in starfish tube-feet are connected in parallel into many different neurons of the TGN system (Willows et al., 1973a) then inputs of this type would produce depolarizations throughout the system. These inputs would be effective in initiating the regenerative firing described above. Furthermore, such inputs could not be ignored or overridden by random inhibitory inputs impinging upon the system or by such inputs occurring in some small subset of the system. The observation that strong direct hyperpolarization of 5 neurons is inadequate to prevent bursting and swimming is confirmation of this idea.

In addition, random or inappropriate inputs to a few cells in the system are likely to be suppressed and not cause bursting, or swimming. Once again, it is the number of neurons involved and their coupling to one another which confers this property. Signals which are local to a few neurons in the group are effectively shunted into the several other, unstimulated cells, and a regenerative response prevented. The observation that strong, but local inputs (from 5 neurons) to the system do not elicit bursting is confirmation in this regard.

The TGN system effectively converts maintained depolarizations from either sensory or experimental depolarizations into recurrent, accelerating bursts, rather than a maintained or adapting spike train. This is contrary to what one would expect of a single, large neuron which might be considered to be formally analogous to the TGN system. If the key for the lock of the innate releasing mechanism is a burst, or bursts, or impulses with high intra-burst frequencies, then this system is ideally designed to convert brief or prolonged starfish contact into the initiation of a response. Other signals, whether random noise or accidental, will be suppressed and ignored. The lock and key analogy is made even more reasonable, if one considers that the bursts produced certainly would be substantially more effective in producing a response in a post-synaptic neuron or neurons particularly if that synapse required facilitation for effective transmission. This synapse would then act as a barrier to any and all inputs excepting those from the appropriate triggering source (which would be dependent upon suitable neural connections into the TGN's) and having the necessary impulse frequency (which would be dependent upon both the connectivity and, to a lesser extent, the spatial and temporal qualities of the sensory stimulus).

Finally, it should be useful to mention that neural systems having these properties should be recognizable in other animals, including many others in which intracellular recording from multiple units may not be feasible. One expects that a system which is as fundamentally simple as this, including only a small group of neurons coupled by non-rectifying electrical junctions might well be found in numerous other animals. The bases for recognition of a neuron that is part of such an interactive group would be minimally, three. If intracellular recording were possible, then units should show accelerating impulse bursts to maintained depolarizing currents and the last impulse in each train should develop a larger and longer after-hyperpolarization. The sign stimulus should evoke in such neurons a similar bursty response. If one were limited to extracellular recording, then it might not be possible to recognize the larger after-hyperpolarization. However, the other features should still be evident and characteristic.

REFERENCES

Atwood, H.L. and Wiersma, C.A.G., (1967) Command interneurons in the crayfish central nervous system. J. Exp. Biol. 46, 249-261.

Bowerman, R.F. and Larimer, J.L., (1972) Command fibers in the circumesophageal connections of crayfish. Amer. Zool. 12, 692.

Bowerman, R.F. and Larimer, J.L., (1974a) Command fibers in the circumesophageal connectives of crayfish. I. Tonic fibers. J. Exp. Biol. 60, 95-117.

Bowerman, R.F. and Larimer, J.L., (1974b) Command fibers in the circumesophageal connectives of crayfish. II. Phasic fibers. J. Exp. Biol. 60, 119-134.

Getting, P.A. and Willows, A.O.D., (1973) Burst formation in electrically coupled neurons. Brain Research. 61, 13-18.

Getting, P.A., (1974) Modification of neuron properties by electrotonic synapses. I. Input resistance, time constant, and integration. J. Neurophysiol. 37, 846-857.

Getting, P.A. and Willows, A.O.D., (1974) Modification of neuron properties by electrotonic synapses. II. Burst formation by electrotonic synapses. J. Neurophysiol. 37, 858-868.

Hughes, G.M. and Wiersma, C.A.G., (1960) The co-ordination of swimmeret movements in the crayfish, Procambarus clarkii (Girard). J. Exp. Biol. 37, 657-670.

Ikeda, K. and Wiersma, C.A.G., (1964) Autogenic rhythmicity in the abdominal ganglia of the crayfish: The control of swimmeret movements. Comp. Biochem. Physiol. 12, 107-115.

Wiersma, C.A.G. and Ikeda, K., (1964) Interneurons commanding swimmeret movements in the crayfish, Procambarus clarkii (Girard). Comp. Biochem. Physiol. 12, 509-525.

Willows, A.O.D. and Hoyle, G., (1969) Neuronal network triggering a fixed action pattern. Science. 166, 1549-1551.

Willows, A.O.D., Dorsett, D.A. and Hoyle, G., (1973a) The neuronal basis of behavior in Tritonia. III. Neuronal mechanism of a fixed action pattern. J. Neurobiol. 4(3), 287-300.

Willows, A.O.D., Getting, P.A. and Thompson, S., (1973b) "Bursting mechanisms in Molluskan locomotion," In Control of Posture and Locomotion. (Stein, R.B., Pearson, K.G., Smith, R.S. and Redford, J.B., eds.), Plenum Press, N.Y., (457-475).

Willows, A.O.D. and Dorsett, D.A., (1975) Evolution of swimming behavior in Tritonia and its neurophysiological correlates J Comp. Physiol. 100, 117-133.

Note: An article recently published by Getting (Brain Research 96, 1975, 128-138) using Tritonia esculans, reports that 30 mV. hyperpolarization of 4 TGN's prevents firing in these cells and in a fifth monitor cell. Under these circumstances, the swimming impulse pattern could still be elicited in the pattern generation. As well, direct stimulation of 4 cells sufficient to produce an impulse pattern in an additional monitor cell (which resembled that seen when swimming was elicited earlier) failed to elicit the swimming motor output. These results add to the suggestion above that TGN activity is not necessarily the proximate cause of swimming. Since alternative explanations of all these results are possible and since there are substantial differences between Getting's results and those described in Figures 4 and 5, further work must be done to clarify these points. Joint experiments are planned.

SOME ASPECTS ON THE DESCENDING CONTROL OF THE SPINAL CIRCUITS GENERATING LOCOMOTOR MOVEMENTS

S. Grillner

Department of Physiology

GIH, Stockholm, Sweden

The neural organization of the vertebrate locomotor system is compared to the current picture of the corresponding invertebrate networks. The descending control is given particular attention and the experimental data available is reviewed.

Introduction

Do vertebrates have "command fibers" as defined in invertebrate motor systems? This term, command fibers, has been used to classify neurones, which if stimulated repetitively elicit a certain type of motor behaviour such as locomotor movements (Wiersma and Ikeda, 1964). Such neurones are assumed to control a pattern generating network (generator/oscillator) that activates the appropriate motoneurones in proper sequence (see Figure 1A). If one assumes a single generator for each limb, the interlimb coordination is achieved by interaction between the different generators by "coordinating neurones" (see Stein, 1974). A similar type of organization has been considered in relation to vertebrate locomotion (Brown, 1911; von Holst, 1939; Arshavsky et al., 1965; Gelfand and Tsetlin, 1966; Grillner, 1975).

Although invertebrate command neurones have been shown to drive many different behaviours including different types of locomotion, it has not yet been possible to record their activity during naturally occurring locomotion. This record would confirm the actual command is tonic in nature as generally assumed (see Figure 1A). It is now realized (cf. Davis, this volume) that, as separate functional entities, these terms (command generator/ oscillator, coordinating neurones and motoneurones) are not without ambiguity particularly if a generalization is applied to all movements. For instance, motoneurones are part of the oscillator network in the feeding systems of lobsters and snails (Selverston, 1974 and this volume; Davis, this volume) and the command fibers act directly on the motoneurones in the tail flip escape reaction (see Larimer, this volume). This might just imply that the nervous system has "chosen" the simplest solution in each case. If motoneurones are used for only one type of movement, they might as well be part of the oscillator network. On the other hand, if they participate in different types of movements, interneuronal pattern generators would seem more purposeful. Not only do some systems lack one of the different functional entities but, not unexpectedly, they demonstrate interaction between the different operational levels (Figure 1B, C; see Davis, this volume). Therefore, the simplest hierarchical view must be abandoned.

Since phasic afferent input can be completely prevented without destroying the motor pattern (e.g., Wilson, 1964; Grillner and Zangger, 1975; Grillner et al., 1976) it is clear that the scheme of Figure 1A can be proposed (invertebrates and vertebrates) with the addition of internal feedback loops (Figure 1B; cf. Arshavsky et al., 1972a, b; Perret, 1973; Szentágothai and Arbib, 1974; Davis, this volume). However, although it is realized that afferent input does not decide the normal pattern of activity, these signals, when present, can be very important in influencing both motoneurones and the generators and in modifying the step cycle according to the instantaneously required readjustments (e.g., the duration of the stance phase) in the natural environment (see Grillner, 1975; Rossignol et al., 1975; Pearson and Duysens, this volume).

By stimulating _individual_ command fibers a "generator" can often be switched on. However, with increasing frequency of stimulation the network is rarely driven within its entire natural frequency range but is usually active only within a certain part (Larimer, this volume). Since there are usually several command fibers it is generally believed that the natural frequency range is obtained through a combined activity of some or of all command fibers. Although the evidence for the functional role of these neurones is compelling, it should be noted that their activity has never been recorded during naturally elicited locomotion. A somewhat different type of neurones, thought to elicit movements by triggering rather than by driving them, have been called the "trigger neurones" (Willows, this volume). These neurones fire when escape swimming is elicited in _Tritonia_, but stimulation of individual neurones has not been shown to elicit swimming although gross stimulation in the general area where these neurones are

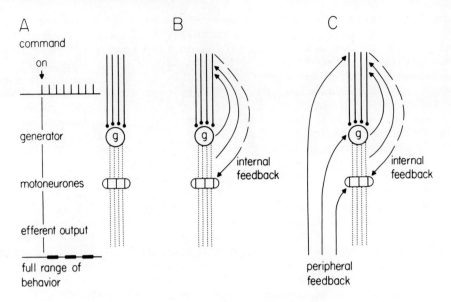

Figure 1. Schematic outline of the subdivisions of the locomotor network currently used particularly for invertebrates. The simplest hierarchical scheme (A) with added internal (B) and peripheral (C) feedback. Several generators (g) can be coordinated by "coordinating neurones" (not indicated).

located elicits escape swimming (Willows, this volume). Although these experiments are suggestive they do not exclude the possibility that the "trigger neurones" studied could be part of some completely different aspect of the escape response (such as alerting the animal).

I. "TONIC" CONTROL OF LOCOMOTION IN SPINAL PREPARATIONS

In walking vertebrates there is strong evidence concerning the presence of individual pattern generators for single limb coordination, for interaction between generators, for interlimb coordination and for the separation of motoneurones from pattern generators (Grillner, 1975; Edgerton et al., this volume; Miller and van der Meché, this volume). We will now present the evidence both for and against the similar "command function" in the context of vertebrate locomotion. In 1910 Roaf and Sherrington described that alternating locomotor movements of the hindlimbs could be elicited in the cat by stimulation of the cut surface of the cervical spinal cord near the lateral funiculus (Figure 2). Lennard and Stein could similarly elicit swimming movements in the turtle by local stimulation within the lateral funiculus but not in the dorsal columns (Stein, this volume). Furthermore, stimulation of the entire cut spinal cord of the dogfish can elicit or enhance the frequency of coordinated swimming movements (Lissman, 1946; Grillner et al., 1975) even under curarized conditions which prevent all phasic afferent feedback (Grillner et al., 1975). These effects are probably due to stimulation of descending "command fibers" driving the "generator network" but effects due to ascending fibers activating the same network antidromically via collaterals cannot be excluded.

A completely different set of experiments in spinal preparations is also relevant. These are based on the finding that there are noradrenergic fibers descending from the brainstem to the spinal cord but that there are no intraspinal neurones containing noradrenaline (Carlson et al., 1964; Dahlstrom and Fuxe, 1965; Maeda et al., 1973). To mimic a physiological activation of these fibers one may either (1) inject the noradrenergic precursor DOPA i.v. which is taken up into the spinal cord and very likely acts through a release of noradrenaline from the terminals of the noradrenergic fibers (Andén et al., 1966b) or (2) inject Clonidine, which is a direct stimulator of "α-receptors" also within the spinal cord. An injection of DOPA i.v. or Clonidine in a proportion of cats with their spinal cords transected at a lower thoracic level, can cause a release of stepping movements (Grillner, 1969a; Forssberg and Grillner, 1973). If such cats are put on a treadmill they can perform walking movements with a "normal" EMG pattern and can adapt to various treadmill speeds (Budakova, 1973; Forssberg and Grillner, 1973; Grillner, 1975). Rhythmic activity can be obtained in such spinal preparations even when afferent feedback is removed entirely by transecting all dorsal roots and/or curarizing the animals (Grillner and Zangger, 1974, Unpublished data), i.e., under conditions similar to Figure 1A or B. When spontaneous efferent locomotor bursts occur, their frequency can be influenced by tonic stimulation of dorsal roots or of the dorsal columns antidromically. In both cases an acceleration of the activity is generally obtained. This can be related to the general finding that many different locomotor networks (when operating) tend to be driven by nonspecific afferent stimuli (see Grillner, 1975).

A tonic activation of noradrenergic receptors may well be equated to a release of noradrenaline from tonically active noradrenergic fibers (Andén et al., 1966a, b). Therefore these results strongly suggest that an activation of certain descending noradrenergic fibers will influence the spinal network for generating walking movements in the cat, i.e., it would act as one type of command system. Even if this interpretation is correct, it does not follow that noradrenergic fibers are the only command fibers for locomotion. In fact the fibers activated by electrical stimulation of the cut spinal cord might not be noradrenergic since they are unmyelinated. One would expect rather that it would be difficult to activate such fibers by electrical stimulation at a strength that would not also influence a large number of other fibers in the white matter (see text under (7) Noradrenergic).

II. "TONIC" DESCENDING CONTROL FROM THE BRAINSTEM

Cats with the neuraxis transected above the subthalamic nuclei or with chronic intercollicular decerebration can initiate and maintain spontaneous locomotor movements for some periods of time (Hinsey et al., 1930; Bard and Macht, 1958; Orlovsky, 1969; see Grillner, 1975). These experiments indicate that nervous structures located in mesencephalon are of importance for the initiation of spontaneous locomotor movements, but so far this has not been observed in preparations with only "lower" parts of the nervous system intact.

It is therefore of particular interest in this respect that Shik et al., (1966a, b, 1967) demonstrated the importance of an area below the inferior colliculus (the nucleus cuneiform or according to Steeves et al., (1976) nucleus coeruleus). A weak tonic stimulation of this region (50-60 Hz) elicits walking of all four limbs on the treadmill provided that the belt is moved at a suitable speed (cf. Figure 2). If the strength of stimulation was increased somewhat the cat would tend to increase both the speed of the movement and the horizontal propulsive force (impulse), i.e., it would change to a faster walk or a trot, and increase further to gallop. Therefore the mode of interlimb coordination is changed. If such a stimulation is performed on a decerebrate cat standing on the floor, it will start walking (for fish see Kashin et al., 1974). An intact cat with a chronically implanted electrode in this region will start walking when stimulated, but it will avoid obstacles and walk around in a seemingly normal manner (Shik, 1971; Sirota and Shik, 1973).

Thus, with an increase of the strength of stimulation (i.e., the number of cells/fibers activated by each stimulus), the speed increased from slow walk to gallop. We therefore can conclude that the number of cells activated is directly related to the energy-output of the limbs, i.e., the speed of locomotion. The frequency of stimulation does not appear to be critical (Shik et al., 1966b). The region from which locomotor movements can be elicited is usually very circumscribed (Shik et al., 1966a, 1967; Grillner and Shik, 1973). Steeves et al., (1976) have reported noradrenergic cell bodies (nucleus coeruleus) could be demonstrated less than 200 μm from each effective stimulation point. This again suggests that noradrenergic neurones might be involved in the generation of these movements (see text under (7) Noradrenergic).

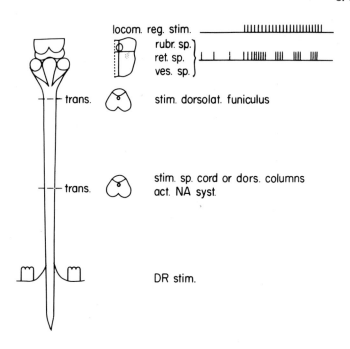

Figure 2. Schematic representation of different cat preparations in which locomotion can be induced (see text). Tonic stimulation of the locomotor region is effective in the high decerebrate cat, and stimulation of the dorsolateral funiculi in the high spinal preparation or dorsal columns and dorsal roots (DR), when DOPA has been injected in the low spinal preparation (see text). During locomotion rhythmic efferent activity occurs in rubro-, reticulo- and vestibulospinal neurones (i.e., rubr. sp; ret. sp; ves. spin.).

The effect of the brainstem stimulation is apparently what would be anticipated if "command neurones" in the brainstem were stimulated directly or via a relay. The more neurones activated the more effective are the spinal centers driven. Low strength of stimulation alone does not evoke locomotion but when coupled with the moving treadmill belt under the cat, locomotion occurs. These and other data show that many different peripheral and central stimuli will summate to initiate locomotor activity (see above; Shik et al., 1967; see Grillner, 1975). This summation can be due to either (1) all systems acting directly on the generator network or (2) all systems converging on a local set of "lower level" spinal command interneurones that exert the final control on the locomotor network (cf. Figure 3). It is pertinent to consider what changes are induced in the spinal cord by the brainstem stimulation under these threshold conditions. These changes are

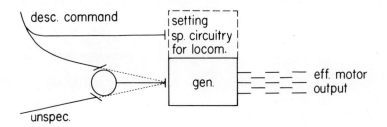

Figure 3. Schematic representation of possible actions of the descending command. The command is assumed to drive the generator in addition to "set" the spinal circuitry for locomotion. The generator network is also driven by unspecific peripheral stimuli (unspec.). The possibility that the descending command and the unspecific signal converge on a lower level "command neurone" is indicated.

required for the initiation of locomotor movements but in addition, e.g., treadmill action is required.

The effects of such subthreshold stimulation were tested in curarized cats in which the relevant stimulation parameters of the brainstem had been tested earlier during walking (Grillner and Shik, 1973). By comparing monosynaptic test reflexes before and during brainstem stimulation, it was found that no direct excitability changes were induced in the α-motoneurone pools tested, whereas an increased level of activity occurs in the γ-efferents (see Table 1; recall that the animals were curarized and therefore the increased γ-activity could not influence the excitability of α-motoneurones). During brainstem stimulation the short latency reflex transmission (central part < 4 ms) to extensor motoneurones from ipsilateral electrically stimulated cutaneous and high threshold muscle afferents was depressed, and instead a long-latency (100-250 ms) reflex discharge could be elicited. These discharges are very similar to the ones obtained after DOPA in spinal walking cats (Jankowska et al., 1967a, b; Grillner, 1973) and are definitely related to the spinal generators for locomotion (Edgerton et al., this volume). It thus appears that the brainstem induces an increased excitability in the locomotor network itself or a facilitation of the transmission from these nonspecific afferents to the network. There is no rhythmicity however.

These effects are elicited from both the "mesencephalic" locomotor region and a spinal preparation following injection with DOPA. In both cases, the stimulus has modified the preparations so that locomotor movements can be induced. Therefore it is

Table 1

	Mesencephalic preparation locom. reg. stim.	Spinal preparations		
		DOPA	Clonidine	Chronic
Walk on treadmill	$+^1$	$+^2$	$+^3$	$+^4$
Tonic effects				
a) depression of short latency effects from FRA	$+^5$	$+^6$	$+^7$	?
b) effect on alpha-motoneurone excitability	$\sim 0^5$	$\sim 0^6$	$\sim 0^7$	?
c) effect on discharge rate of gamma	$+^8$	$+^9$	?	?
Phasic alternate activity in flexor and extensor				
a) alpha motoneurones	$+^{10}$	$+^{11}$	$+^7$	$+^4$
b) static gamma	$+^8$	$+^9$	?	?
c) dynamic gamma	$+^{12}$	$+^9$	?	?
d) "Ia interneurones"	$+^{13}$	$+^{14}$	?	?
e) depression of Renshaw inhibition during phasic activity	$+^{15}$	$+^{16}$	?	?

Comparison between the "mesencephalic walking cat" and the low spinal cat (Th 12) injected with DOPA i.v. The different statements in the table are based on the following papers: (1) Shik et al., 1966; (2) Grillner, 1969a; Budakova, 1971; Forssberg and Grillner, 1973; (3) Forssberg and Grillner, 1973; (4) Grillner, Forssberg and Sjöström, Unpubl., cited in Grillner, 1973;
(5) Grillner and Shik, 1973; (6) Andén et al., 1966a; (7) Grillner, 1973; (8) Severin et al., 1967a; Severin, 1970; (9) Grillner, 1969a, b; Bergmans and Grillner, 1969. The data for the DOPA preparation refer to participation   (Legend continued on next page.)

reasonable to believe that these changes are required for the generation of locomotor movements to occur. From the fact that the transmission in one pathway is modified "under these preparatory conditions" it does not follow however, that the same conditions will prevail when the actual locomotor network is operating (it would have been active if a stronger stimulation had been used or if the treadmill had been moved under the animal). During walking but not under the preparatory conditions (cf. Grillner, 1975) recurrent inhibition from the ventral roots appears to be depressed due to inhibition of the "Renshaw cells" (Severin et al., 1968; Bergmans et al., 1969). Moreover, at least specific cutaneous inputs from the dorsum of the foot cause different effects in different phases of the step cycle (Forssberg et al., this volume).

It is therefore likely that the "command signal" elicited by the "brainstem stimulation" causes much more complex effects than just releasing a certain group of nerve cells that constitute the pattern generator for locomotion (Figure 3). It is likely that many parts of the spinal circuitry are modified; for example certain reflex pathways are depressed because it is not meaningful for them to operate during locomotion, while others might be facilitated; others again might be put under the command of the "generator" to phasically open and close during different parts of the step cycle (see Forssberg et al., this volume). Our data about these effects are still fragmentary but more data can easily be obtained.

III. REQUIREMENTS FOR AN EFFECTIVE CONTROL OF LOCOMOTION

In the preceding paragraphs it has been shown that a tonic stimulation of the brainstem or the spinal cord (electrical or "biochemical") can elicit stereotype locomotor movements. Most probably they are due to orthodromic activation of descending fibers which cause both a release of a spinal interneuronal network generating the locomotor movements and other modifications of the spinal neuronal circuitry. In this respect there is a similarity to the invertebrate "command fiber" concept. However, the experiments neither decide the number of different descending

---

of these neurones in late discharge, but it should be noted that α-γ-linkage exists in spinal locomotor activity (Sjöström and Zangger, 1975); (10) Severin et al., 1967b; (11) Grillner and Zangger, 1974. (12) In the decorticate preparation, dynamic gamma activity appears to take part (Perret and Buser, 1972), whereas it is unknown for the mesencephalic preparation; (13) Feldman and Orlovsky, 1975; (14) Fu et al., 1975; Edgerton et al., this volume; (15) Severin et al., 1968; (16) Bergmans et al., 1969.

systems (and neurons) involved, nor how these systems precisely influence the spinal generator and interact with other stimuli.

Before discussing these topics further, it might be useful to consider the requirements for effective control of locomotion.

(1) It should be possible to <u>initiate</u> locomotion, if necessary rapidly (e.g., escape running); to maintain a certain speed and when required, to modify the steady-state velocity and finally to <u>arrest</u> the locomotion, again rapidly if necessary.

(2) The locomotor movements or each step should be <u>adapted to the environment</u> so that the foot is always placed in the proper location, and that minute or large changes in direction will take place when needed.

(3) The "locomotor posture" should be modifiable to match different situations such as crouching locomotion, swimming, running in snow, carrying or pulling loads. This is also related to the different needs for the equilibrium control during locomotion (see Grillner, 1975).

There is evidence to suggest that the noradrenergic descending system can be of importance (see above and below). This is, however, a slow unmyelinated system. The need for a rapid induction of locomotion as well as other experimental data, (Grillner, 1975; Steeves et al., 1976; Jordan and Steeves, this volume) suggest that other descending systems also are involved, some of which would be faster. In the mesencephalic cat, however, the induction of locomotion takes usually more than 0.5 s from the onset of brainstem stimulation, but no serious attempt has been made to measure the absolute minimum. The data would thus be compatible with the activation of a very slow system. The participation of faster systems seems desirable but there is no direct evidence for any particular type of neurone (see below). The role of slow systems could be to provide a background excitation on the locomotor network that would summate with signals from several other systems for initiation of the stepping cycle and for modification or maintenance of steady-state locomotion.

The nervous system should not only be able to initiate the locomotor movements quickly but also it seems equally important to quickly stop them. This cannot be achieved merely by an inhibition of the motoneurons (this would cause collapse) but rather by rapid deceleration followed by assumption of a certain posture. In this context it is thought-provoking that Bowerman and Larimer, (1974a, b) describe "neurons" that when stimulated, freeze a movement, i.e., cause the animal to assume a certain posture. The mechanism of this action is not clear.

The rapid adaptations of locomotion to the environment are largely based on anticipation of the required adjustments, e.g., exactly when and where it is suitable to place the foot on ground during each step. This requires very fast conducting pathways which, at least in part, can be expected to act directly on motoneurones and on last order inhibitory interneurones rather than on the generator itself. These descending pathways by adding excitation to a certain combination of motoneurones and by inhibition (or disfacilitation) to others could direct the foot to a somewhat different location in accordance with specified functional needs (such as beside a pool of water rather than in the middle). It is noteworthy that several descending pathways influence α-motoneurones as well as inhibitory interneurones innervating only specific target motor nuclei (i.e., "Ia-interneurones" see Figure 4). It might be relevant in this context to mention that fast rubro-, vestibulo- and reticulo-spinal neurones are active during particular phases of each step cycle, and, that for the majority, their activity is dependent on an intact cerebellum (in the mesencephalic or decorticate cat, Orlovsky, 1970; 1972a, b; Perret, 1973; see below). These and other fast descending neurones could also be used for commands related to anticipatory corrections. It should be mentioned that activity in these rapidly conducting pathways does not appear to be necessary for the generation of locomotion. There is no evidence (see below) that these pathways are part of the intrinsic generator network but rather, they may be an adjuvant providing phasic corrections acting on the generator output level (cf. Grillner, 1975). However, during each step cycle, phasic activity in these different descending pathways by mere direct action can add to the excitation or inhibition of the motoneurones, thereby contributing further to the final level off activity in these neurones. It is interesting to note that the same type of phasic "descending" activity occurs in the lobster stomatogastric 12 neurone system, which controls the "gastric mill." It can generate rhythmic activity independently of the other ganglia and also it can be influenced from the commissural ganglion. In this latter ganglion, there are neurones with phasic activity (see Selverston, 1974, and this volume; Russel, 1976) such as the one discussed above for rubral neurones.

Phasic activity in ascending neurones occurs with and without afferent feedback (Figure 4); phasic descending activity without afferent feedback may be due to internal feedback, e.g., via the ventral spinocerebellar pathway which causes this phasic activity (Arshavsky et al., 1972a, b; Perret, 1973; cf. Grillner, 1975). Peripheral phasic signals are transmitted via the dorsal spinocerebellar tract (DSCT) for example.

Another type of adaptation is that of changing the "locomotor posture" of normal walk to that of crawling, swimming, climbing

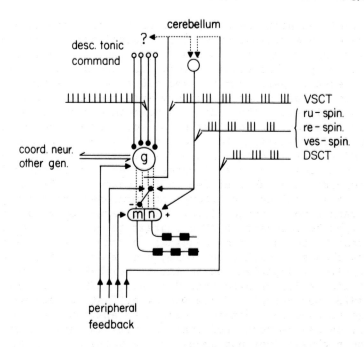

Figure 4. Some patterns of activity in the walking cat. Peripheral signals influence the "generator" specific last order inhibitory interneurones ("Ia interneurones" driven also from the generator), motoneurones and cerebellum via the dorsal spinocerebellar pathway (DSCT). Internal "feedback" is provided by the ventral spinocerebellar pathway (VSCT). Phasic activity occurs in several descending pathways (rubrospinal, ru-spin; reticulospinal, re-spin; vestibulospinal, ves-spin). A descending tonic command is assumed to act on the generator(s) for each limb and the coordination to be achieved by their interaction through coordinating neurones.

uphill or downhill or to that used when pulling, pushing or carrying a load. For humans one can presumably add the changes adopted when skiing, bicycling or when walking on very different terrains. During these adaptations it is clear that the degree of activity in different muscles must be markedly changed since joint motion can be varied and under certain conditions the movements can be completely suppressed in particular joints. This emphasizes that the motor output during locomotion even at a given speed must be expected to vary considerably with the "locomotor posture" adopted.

This may imply that it is possible to control the output from the locomotor generator so that certain muscles can be blocked and others enhanced. Further it is conceivable to gate effects according to alternate pathways so that muscles not "normally" active in a certain phase of a movement may become activated. It should also be mentioned that the "generator" for a limb may be subdivided in smaller units controlling the muscles at one joint. During normal locomotion these different generators would be tightly coupled by mutual excitation (Grillner, 1973; Edgerton et al., this volume). The advantage is that, in certain motor behaviours, the flexion-extension movements in one joint could be driven in isolation and, under other conditions, the unit generators could be recombined for example by permitting the hip to be out of phase with ankle and knee resulting in backwards locomotion (see Figure 12 of Edgerton et al., this volume). If this is true, descending systems could exert their control by isolating a certain part of the limb generator thereby driving alternate movement of one joint without movements of other joints (see Edgerton et al., this volume).

After this survey of "descending locomotor effects" and of the requirements for this control, it may be appropriate to consider some of the known physiological effects of different descending systems.

A. The pyramidal tract (PT) is very heterogenous with only a small component of fast myelinated fibers which in higher vertebrates are known to mono- and poly-synaptically activate α- and γ-motoneurones of different motor nuclei and to influence the transmission in reflex arcs facilitating both "Ia inhibitory interneurones" as well as interneurones of other reflex arcs (Phillips, 1966; Koeze et al., 1968; Lundberg, 1970; Hultborn, 1972; Jankowska et al., 1975). The involvement of different PT-neurones in voluntary movements has been studied extensively (Evarts, 1972). A transection of the pyramidal tract at the bulbar level leads to a particular motor deficit; it namely abolishes independent finger movements in baboons. Otherwise, the animals have a normal motor repertoire (Lawrence and Kuypers, 1968a).

Stimulation of the pyramids at the mesencephalic level is known to elicit locomotion in high decerebrate cats, provided that the pyramids are transected in the lower brainstem. Thus, only if the direct corticospinal fibers are transected, can pyramidal fibers influence the brainstem and cause locomotor movements (Shik et al., 1968). These effects might be due to synaptic activation of the "mesencephalic locomotor region," since a lesion including this and more ventral regions of mesencephalon abolishes the response. As attractive as this possibility appears it should be realized that the effect of the lesion might also be due to transection of corticobulbar fibers passing through the region damaged.

B. Neurones in the rubrospinal tract are known to be rhythmically active in phase with the locomotor movements (Orlovsky, 1972a) provided the cerebellum is intact at least for a large proportion of the cells (see also Perret, this volume). But neither a removal of the cerebellum nor a lesion of the red nucleus abolishes locomotion. Correspondingly, a longlasting stimulation of the red nucleus does not elicit locomotor movements and shortlasting trains delivered in different parts of the step cycle only facilitate the ongoing flexor activity without influencing the rate of stepping or the structure of the step cycle (Orlovsky, 1972a, c).

In primates the rubrospinal tract has monosynaptic connections to α-motoneurones (Shapovalov, 1972) but in the cat it has almost exclusively di- or poly-synaptic effects on motoneurones facilitating different excitatory and inhibitory reflex arcs including excitatory effects on "Ia inhibitory interneurones" (Hongo et al., 1969a, b; Hultborn, 1972). A combined lesion including the pyramidal and the rubrospinal tracts abolishes not only independent finger movements but also independent hand movements. During locomotion and other large movement synergies the entire limb is used including the fingers and the hand (Lawrence and Kuypers, 1968b).

C. (1) Similar to the rubrospinal neurones, the vestibulospinal neurones originating in Deiters' nucleus are active "in phase" during locomotion; however, they fire in a different part of the step cycle (i.e., mainly during the extension phase). Lesion and stimulation experiments also show the corresponding results (Orlovsky, 1972b, c). The fast fibers in this tract have mainly mono- and disynaptic excitatory effects to extensor α- and γ-motoneurones (Grillner et al., 1969, 1970; Wilson and Yoshida, 1969a; Akaike et al., 1973), as well as facilitating "Ia inhibitory interneurones" to antagonists (Grillner et al., 1966; Grillner and Hongo, 1972) and crossed excitatory and inhibitory reflexes from high threshold cutaneous and muscular afferents (ten Bruggencate et al., 1969).

(2) Vestibulospinal fibers originating in the medial vestibular nucleus descend to the cervical and thoracic spinal cord and can have both monosynaptic excitatory and inhibitory effects (Wilson and Yoshida, 1969b; Akaike et al., 1973). Their possible role during locomotion has not been explored.

D. Other systems descending from the brainstem.

(1) Fibers from N. reticularis pontis caudalis descend in the medial longitudinal fasciculus and the ventral part of the spinal cord to make monosynaptic connections mainly with flexor motoneurones (Grillner and Lund, 1968; Wilson and Yoshida, 1969a). It is likely that most of the reticulospinal neurones studied by Orlovsky (1970) belong to this tract. These neurones are active in the flexor phase of the step cycle and their rhythmic activity depends on an intact cerebellum.

(2) Fibers from <u>nucleus reticularis gigantocellularis</u> descend in the ventral part of the lateral fasciculus (Nyberg-Hansen, 1966). The function of these neurones is not clear although Shapovalov (1972) has suggested that they may have monosynaptic connections to spinal motoneurones.

(3) The <u>dorsolateral reticulospinal system</u> has been defined physiologically in decerebrate cats. Lesions at the obex or of the lateral funiculus at different levels along the cord cause a release of the transmission in the shortlatency reflex arcs mediating the flexor reflex (flexor excitation, extensor inhibition), (Eccles and Lundberg, 1959; Holmqvist and Lundberg, 1959, 1961; Engberg et al., 1968a). Whether these effects are due to direct action of reticulospinal fibers or to relay in the spinal cord is not clear (Engberg et al., 1968a). It should be noted that the tonic depression of shortlatency reflexes by this system is not concomitant with a release of the late spinal reflex discharges related to the locomotor network (see (7)).

(4) A lesion at the <u>lower pontine level</u> in decerebate cats has revealed an important finding; there is a release of <u>short-latency inhibitory</u> reflexes to <u>flexor</u> α-motoneurones (e.g., trisynaptic) from cutaneous and high threshold muscle afferents (Holmqvist and Lundberg, 1961). The tonic depression of the "ordinary" shortlatency flexor reflex remains. This demonstrates that a powerful reflex pathway that is completely concealed under "ordinary" experimental conditions, can be selectively controlled from the brainstem. The significance of this system during locomotion is unknown.

(5) Fibers descending in the ventral funiculus cause postsynaptic inhibition in extensor and flexor α-motoneurones. Such effects can be elicited from the ventral part of the medullary reticular formation probably corresponding to Magoun's inhibitory center (Jankowska et al., 1968).

(6) 5-HT fibers originate in the raphe nuclei and descend throughout the spinal cord (Dahlström and Fuxe, 1965). The physiological effect of these fibers has been studied by injecting the 5-HT precursor 5HTP or receptor stimulators. The effect of such an activation of these descending fibers is a release of "late reflex discharges," a depression of "shortlatency flexor reflexes" and an enhanced γ-activity (Andén et al., 1964; Engberg et al., 1968b; Ahlman et al., 1971). These effects are similar to those obtained with DOPA but in addition, 5HTP causes an increase in direct excitability of the α-motoneurones (Andén et al., 1964). From these data it could be suspected that 5-HT fibers would also release locomotor movements. A few preliminary experiments in acute spinal animals (cat or rat) with 5HTP administered i.v. do

not support this supposition (Grillner and Shik, Unpublished data; Grillner, Rossignol and Trolin, Unpublished data). However, some days after a chronic spinal transection, spontaneous but feeble spinal locomotor movements are markedly enhanced after injection of a 5-HT receptor stimulator. This is also observed when the injection is combined with peripheral 5-HT receptor blockers (Grillner, Rossignol and Trolin, Unpublished data).

(7) <u>Noradrenergic</u> cell bodies located in locus coeruleus and in the lower brainstem send their axons down to the lumbosacral spinal cord (Maeda et al., 1973; cf. Jones et al., 1974; cf. Chu and Bloom, 1974; Kuypers and Maisky, 1975). It thus appears that there is a noradrenergic "coeruleospinal" pathway in the cat. It is particularly interesting that the effective stimulation points in the mesencephalon are within 200 μm distance from noradrenergic cell bodies (Steeves et al., 1976). This is interesting in relation to previous data on spinal cats in which descending noradrenergic fibers presumably are activated by injecting DOPA i.v. (see above) and in which coordinated locomotor movements can be obtained (Grillner, 1969a; Budakova, 1973; Forssberg and Grillner, 1973). There is a strong similarity between the effect of stimulation of the mesencephalic locomotor region and the intravenous injection of DOPA in the spinal cat (see above). Based on this and other evidence it has been suggested that the noradrenergic system could be one "command fiber system" for locomotion, activated from the "mesencephalic locomotor region" (Grillner and Shik, 1973). However, it is unlikely that this system alone is responsible for locomotor behaviour since a lesion (by 6 OH-dopamine-injection) of significant degree to noradrenergic descending fibers (marked reduction in NA-content below the lesion) in the thoracic spinal cord neither prevents locomotion in otherwise intact cats (Jordan and Steeves, this volume; Grillner, Rossignol and Trolin, Unpublished data) nor induction of locomotion from the mesencephalic locomotor region (Jordan and Steeves, this volume).

Among all these different descending motor systems there is none whose mode of operation and significance in movement control is satisfactorily known. Concerning the large fiber systems we have some interesting information about their pattern of activity during various movements, including locomotion and how this activity is generated. For the slow fiber systems, i.e., the majority of the fibers, we have a very limited knowledge, because they are technically much more difficult to analyze. Moreover, the detailed knowledge of the effects of different descending systems on the spinal cord has almost exclusively been obtained on anesthetized or decerebrate preparations that are not engaged in active movement. Admittedly it is important to know the effects that can be revealed under such conditions. It must be realized, however, that when an active movement is carried out or planned, these pathways might exert very specific and important types of controls which are simply impossible to deduce from experiments on motionless or inactive preparations.

In conclusion we know that tonic stimulation can drive locomotor movements, indicating some similarity in function between vertebrate neurones and invertebrate "command neurones." Such tonic neuronal commands have not yet been recorded during naturally occurring movements. Neurones in several fast descending systems can be phasically active in different parts of the step cycle, but their activity does not appear to be related to driving the generators, but rather to phasic corrections of the motor output.

Acknowledgment

This work was supported by the Swedish Medical Research Council (project no. 3026) and Magnus Bergwalls stiftelse.

REFERENCES

Ahlman, H., Grillner, S. and Udo, M., (1971) The effect of 5HTP on the static fusimotor activity and the tonic stretch reflex of an extensor muscle. Brain Res. 27, 393-396.

Akaike, T., Fanardjian, V.V., Ito, M. and Ohno, T., (1973) Electrophysiological analysis of the vestibulospinal reflex pathway of rabbit. II. Synaptic actions upon spinal neurones. Exp. Brain Res. 17, 497-515.

Andén, N.-E., Jukes, M.G.M., Lundberg, A. and Vyklický, A., (1964) A new spinal flexor reflex. Nature. 202, 1344-1345.

Andén, N.-E., Jukes, M.G.M., Lundberg, A. and Vyklický, L., (1966a) The effect of DOPA on the spinal cord. 1. Influence on transmission from primary afferents. Acta Physiol. Scand. 67, 373-386.

Andén, N.-E., Jukes, M.G.M., Lundberg, A. and Vyklický, L., (1966b) The effect of DOPA on the spinal cord. 2. A pharmacological analysis. Acta Physiol. Scand. 67, 387-397.

Arshavsky, Yu.I., Kots, Ya.M., Orlovsky, G.N., Rodionov, I.M. and Shik, M.L., (1965) Investigation of the biomechanics of running by the dog. Biophysics. 10, 737-746.

Arshavsky, Yu.I., Berkinblit, M.B., Fukson, O.I., Gelfand, I.M. and Orlovsky, G.N., (1972a) Activity of neurons of the VSCT during locomotion. Biofizika. 17, 883-890.

Arshavsky, Yu.I., Berkinblit, M.B., Fukson, O.I., Gelfand, I.M. and Orlovsky, G.N., (1972b) Origin of modulation in neurones of the ventral spinocerebellar tract during locomotion. Brain Res. 43, 276-279.

Bard, P. and Macht, M.B., (1958) "The behaviour of chronically decerebrate cats," In Neurological Basis of Behaviour. Ciba Found. Symp. (Wolstenholme, G.E.W. and O'Connor, C.M., eds.), J. and A. Churchill Ltd. London, (55-75).

Bergmans, J. and Grillner, S., (1969) Reciprocal control of spontaneous activity and reflex effects in static and dynamic γ-motoneurones revealed by an injection of DOPA. Acta Physiol. Scand. 77, 106-124.

Bergmans, J., Burke, R. and Lundberg, A., (1969) Inhibition of transmission in the recurrent inhibitory pathway to motoneurones. Brain Res. 13, 600-602.

Bowerman, R.F. and Larimer, J.L., (1974a) Command fibres in the circumoesophageal connectives of crayfish. I. Tonic fibres. J. Exp. Biol. 60, 95-117.

Bowerman, R.F. and Larimer, J.L., (1974b) Command fibres in the circumoesophageal connectives of crayfish. II. Phasic fibres. J. Exp. Biol. 60, 119-134.

Brown, T.G., (1911) The intrinsic factors in the act of progression in the mammal. Proc. R. Soc. B. 84, 308-319.

Bruggencate, G. ten, Burke, R., Lundberg, A. and Udo, M., (1969) Interaction between the vestibulospinal tract, contralateral flexor reflex afferents and Ia afferents. Brain Res. 14, 529-532.

Budakova, N.N., (1973) Stepping movements in the spinal cat due to DOPA administration. Fiziol. Zh. USSR. 59, 1190-1198.

Carlsson, A., Falck, B., Fuxe, K. and Hillarp, N.-Å., (1964) Cellular localization of monoamines in the spinal cord. Acta Physiol. Scand. 60, 112-119.

Chu, N.-S. and Bloom, F.E., (1974) The catecholamine-containing neurons in the cat dorsolateral pontine tegmentum: distribution of the cell bodies and some axonal projections. Brain Res. 66, 1-21.

Dahlström, A. and Fuxe, K., (1965) Evidence for the existence of monoamine neurons in the central nervous system. II. Experimentally induced changes in the intraneuronal amine levels of bulbospinal neuron systems. Acta Physiol. Scand. 64, (Suppl. 247), 1-36.

Eccles, R.M. and Lundberg, A., (1959) Supraspinal control of interneurones mediating spinal reflexes. J. Physiol. (Lond.). 147, 565-584.

Engberg, I., Lundberg, A. and Ryall, R.W., (1968a) Reticulospinal inhibition of transmission in reflex pathways. J. Physiol. (Lond.). 194, 201-223.

Engberg, I., Lundberg, A. and Ryall, R.W., (1968b) The effect of reserpine on transmission in the spinal cord. Acta Physiol. Scand. 72, 115-122.

Evarts, E.V., (1972) Contrasts between activity of precentral and postcentral neurons of cerebral cortex during movement in the monkey. Brain Res. 40, 25-31.

Feldman, A.G. and Orlovsky, G.N., (1975) Activity of interneurones mediating reciprocal Ia inhibition during locomotion. Brain Res. 84, 181-194.

Forssberg, H. and Grillner, S., (1973) The locomotion of the acute spinal cat injected with Clonidine i.v. Brain Res. 50, 184-186.

Fu, T.-C., Jankowska, E. and Lundberg, A., (1975) Reciprocal Ia inhibition during the late reflexes evoked from the flexor reflex afferents after DOPA. Brain Res. 85, 99-102.

Gelfand, I.M. and Tsetlin, M.L., (1971) "Mathematical modeling of mechanisms of the central nervous system," In Models of the Structural-Functional Organization of Certain Biological Systems. (Gelfand, I.M., Gurfinkel, V.S., Fomin, S.V. and Tsetlin, M.L., eds.), MIT Press. Cambridge, Massachusetts, (1-22). (Translation from Russian, original publication in 1966).

Grillner, S., Hongo, T. and Lund, S., (1966) Interaction between the inhibitory pathways from the Deiters' nucleus and Ia afferents to flexor motoneurones. Acta Physiol. Scand. 68, (Suppl. 277), 61.

Grillner, S. and Lund, S., (1968) The origin of a descending pathway with monosynaptic action on flexor motoneurones. Acta Physiol. Scand. 74, 274-284.

Grillner, S., (1969a) Supraspinal and segmental control of static and dynamic γ-motoneurones in the cat. Acta Physiol. Scand. 6, Suppl. 327, 1-34.

Grillner, S., (1969b) The influence of DOPA on the static and the dynamic fusimotor activity to the triceps surae of the spinal cat. Acta Physiol. Scand. 77, 490-509.

Grillner, S., Hongo, T. and Lund, S., (1969) Descending monosynaptic and reflex control of γ-motoneurones. Acta Physiol. Scand. 75, 592-613.

Grillner, S., Hongo, T. and Lund, S., (1970) The vestibulospinal tract. Effects on alpha-motoneurones in the lumbosacral spinal cord in the cat. Exp. Brain Res. 10, 94-120.

Grillner, S. and Hongo, T., (1972) "Vestibulospinal effects on motoneurones and interneurones in the lumbosacral cord," In Basic Aspects of Central Vestibular Mechanisms. (Brodal, A. and Pompeiano, O., eds.), Elsevier Publ. Co., Progr. in Brain Res. 37, 243-262.

Grillner, S., (1973) "Locomotion in the spinal cat," In Control of Posture and Locomotion. (Stein, R.B., Pearson, K.G., Smith, R.S. and Redford, J.B., eds.), Plenum Press, New York, (515-535).

Grillner, S. and Shik, M.L., (1973) On the descending control of the lumbosacral spinal cord from the "mesencephalic locomotor region. Acta Physiol. Scand. 87, 320-333.

Grillner, S. and Zangger, P., (1974) Locomotor movements generated by the deafferented spinal cord. Acta Physiol. Scand. 91, 38A-39A.

Grillner, S., (1975) Locomotion in vertebrates: central mechanisms and reflex interaction. Physiol. Rev. 55, 247-304.

Grillner, S. and Zangger, P., (1975) How detailed is the central pattern generator for locomotion? Brain Res. 88, 367-371.

Grillner, S., Perret, C. and Zangger, P., (1976) Central generation of locomotion in the spinal dogfish. Brain Res. (In press).

Hinsey, J.C., Ranson, S.W. and McNattin, R.F., (1930) The role of the hypothalamus and mesencephalon in locomotion. Arch. Neurol. Psychiat. (Chic.). 23, 1-43.

Holmqvist, B. and Lundberg, A., (1959) On the organization of the supraspinal inhibitory control of interneurones of various spinal reflex arcs. Arch. ital. Biol. 97, 340-356.

Holmqvist, B. and Lundberg, A., (1961) Differential supraspinal control of synaptic actions evoked by volleys in the flexion reflex afferents in alpha motoneurones. Acta Physiol. Scand. 54, (Suppl. 186), 1-51.

Holst, E., von, (1939) Die relative Koordination. Ergebn. Physiol. 42, 228-306.

Hongo, T., Jankowska, E. and Lundberg, A., (1969a) The rubrospinal tract. I. Effects on alphamotoneurones innervating hindlimb muscles in cats. Exp. Brain Res. 7, 344-364.

Hongo, T., Jankowska, E. and Lundberg, A., (1969b) The rubrospinal tract. II. Facilitation of interneuronal transmission in reflex paths to motoneurones. Exp. Brain Res. 7, 365-391.

Hultborn, H., (1972) Convergence on interneurones in the reciprocal Ia inhibitory pathway to motoneurones. Acta Physiol. Scand. (Suppl. 375), 1-42.

Jankowska, E., Jukes, M.G.M., Lund, S. and Lundberg, A., (1967a) The effect of DOPA on the spinal cord. 5. Reciprocal organization of pathways transmitting excitatory action to alpha motoneurones of flexors and extensors. Acta Physiol. Scand. 70, 369-388.

Jankowska, E., Jukes, M.G.M., Lund, S. and Lundberg, A., (1967b) The effect of DOPA on the spinal cord. 6. Half-centre organization of interneurones transmitting effects from the flexor reflex afferents. Acta Physiol. Scand. 70, 389-402.

Jankowska, E., Lund, S., Lundberg, A. and Pompeiano, O., (1968) Inhibitory effects evoked through ventral reticulospinal pathways. Arch. ital. Biol. 106, 124-140.

Jankowska, E., Padel, Y. and Tanaka, R., (1975) Projections of pyramidal tract cells to α-motoneurones innervating hind-limb muscles in the monkey. J. Physiol. (Lond.). 249, 637-667.

Jones, B.E. and Moore, R.Y., (1974) Catecholamine-containing neurons of the nucleus locus coeruleus in the cat. J. comp. Neurol. 157, 43-52.

Kashin, S.H., Feldman, A.G. and Orlovsky, G.N., (1974) Locomotion of fish evoked by electrical stimulation of the brain. Brain Res. 82, 41-47.

Koeze, T.H., (1968) The independence of corticomotoneuronal and fusimotor pathways in the production of muscle contraction by motor cortex stimulation. J. Physiol. (Lond.). 197, 87-105.

Kuypers, H.G.J.M. and Maisky, V.A., (1975) Retrograde axonal transport of horseradish peroxidase from spinal cord to brain stem cell groups in the cat. Neurosci. Letters. 1, 9-14.

Lawrence, D.G. and Kuypers, H.G.J.M., (1968a) The functional organization of the motor system in the monkey. I. The effects of bilateral pyramidal lesions. Brain. 91, 1-14.

Lawrence, D.G. and Kuypers, H.G.J.M., (1968b) The functional organization of the motor system in the monkey. II. The effects of lesions of the descending brainstem pathways. Brain. 91, 15-36.

Lissman, H.W., (1946) The neurological basis of the locomotory rhythm in the spinal dogfish (Scyllium canicula, Acanthias vulgaris). II. The effect of deafferentation. J. Exp. Biol. 23, 162-176.

Lundberg, A., (1970) "The excitatory control of the Ia inhibitory pathway," In Excitatory Synaptic Mechanisms. (Andersen, P. and Jansen, J.K.S., eds.), Universitets-forlaget. Oslo, (333-340).

Maeda, T., Pin, C., Salvert, D., Ligier, M. and Jouvet, M., (1973) Les neurones contenant des catecholamines du tegmentum pontique et leurs voies de projection chez le chat. Brain Res. 57, 119-152.

Nyberg-Hansen, R., (1966) "Functional organization of descending supraspinal fibre systems to the spinal cord. Anatomical observations and physiological correlations," In Reviews of Anatomy, Embryology and Cellbiology. Springer-Verlag, (39, Heft 2).

Orlovsky, G.N., (1969) Spontaneous and induced locomotion of the thalamic cat. Biophysics. 14, 1154-1162.

Orlovsky, G.N., (1970) Activity of reticulospinal neurones during locomotion. Biophysics. 15, 761-771.

Orlovsky, G.N., (1972a) Activity of rubrospinal neurons during locomotion. Brain Res. 46, 99-112.

Orlovsky, G.N., (1972b) Activity of vestibulospinal neurons during locomotion. Brain Res. 46, 85-98.

Orlovsky, G.N., (1972c) The effect of different descending systems on flexor and extensor activity during locomotion. Brain Res. 40, 359-371.

Perret, C. and Buser, P., (1972) Static and dynamic fusimotor activity during locomotor movements in the cat. Brain Res. 40, 165-169.

Perret, C., (1973) Analyse des méchanismes d'une activité de type locomoteur chez le chat. These de doct., Université de Paris VI. Paris.

Phillips, C.G., (1966) "Changing concepts of the precentral motor area," In Brain and Conscious Experience. (Eccles, J.C., ed.), Springer-Verlag, New York, (389-421).

Roaf, H.E. and Sherrington, C.S., (1910) Further remarks on the mammalian spinal preparation. Quart. J. exp. Physiol. 3, 209-211.

Rossignol, S., Grillner, S. and Forssberg, H., (1975) Factors of importance for the initiation of flexion during walking. Neuroscience Abstracts. 1, 181.

Russel, D.F., (1976) Rhythmic excitatory inputs to the lobster stomatogastric ganglion. Brain Res. 101, 582-588.

Selverston, A.I., (1974) Structural and functional basis of motor pattern generation in the stomatogastric ganglion of the lobster. Am. Zool. 14, 957-972.

Severin, F.V., Orlovsky, G.N. and Shik, M.L., (1967a) Work of the muscle receptors during controlled locomotion. Biofizika. 12, 575-586. (Eng. transl.).

Severin, F.V., Shik, M.L. and Orlovsky, G.N., (1967b) Work of the muscles and single motoneurones during controlled locomotion. Biofizika. 12, 762-772. (Eng. transl.).

Severin, F.V., Orlovsky, G.N. and Shik, M.L., (1968) Recurrent inhibitory effects on single motoneurones during an electrically evoked locomotion. Bull. exp. Biol. Med. 66, 3-9. (In Russian).

Severin, F.V., (1970) The role of the gamma motor system in the activation of the extensor alpha motor neurones during controlled locomotion. Biofizika. 15, 1138-1145, (Eng. transl.).

Shapovalov, A.I., (1972) Extrapyramidal monosynaptic and disynaptic control of mammalian alpha-motoneurones. Brain Res. 40, 105-115.

Shik, M.L., Severin, F.V. and Orlovsky, G.N., (1966a) Control of walking and running by means of electrical stimulation of the mid-brain. Biophysics. 11, 756-765.

Shik, M.L., Orlovsky, G.N. and Severin, F.V., (1966b) Organization of locomotor synergism. Biophysics. 11, 1011-1019.

Shik, M.L., Severin, F.V. and Orlovsky, G.N., (1967) Structures of the brain stem responsible for evoked locomotion. Fiziol. Zh. USSR. 12, 660-668. (In Russian).

Shik, M.L., Orlovsky, G.N. and Severin, F.V., (1968) Locomotion of the mesencephalic cat elicited by stimulation of the pyramids. Biophysics. 13, 143-152.

Shik, M.L., (1971) The controlled locomotion of the mesencephalic cat. Proc. Intern. un. Physiol. Sci. 8, 104-105.

Sirota, M.G. and Shik, M.L., (1973) Locomotion of the cat during stimulation of the midbrain. Fiziol. Zh. USSR. 59, 1314-1320. (In Russian).

Sjöström, A. and Zangger, P., (1975) α-γ-linkage in the spinal generator for locomotion in the cat. Acta Physiol. Scand. 94, 130-132.

Steeves, J.D., Jordan, L.M. and Lake, N., (1976) The close proximity of catecholamine containing cells to the "mesencephalic locomotor region," (MLR). Brain Res. (In press).

Stein, P.S.G., (1974) Neural control of interappendage phase during locomotion. Am. Zool. 14, 1003-1016.

Szentágothai, J. and Arbib, M.A., (1974) Conceptual models of neural organization. Neurosci. Res. Prog. Bull. 12, 313-510.

Wiersma, C.A.G. and Ikeda, K., (1964) Interneurons commanding swimmeret movements in the crayfish, Procambarus Clarki (Girard). Comp. Biochem. Physiol. 12, 509-525.

Wilson, D.M., (1964) "The origin of the flight-motor command in grasshoppers," In Neuronal Theory and Modeling. (Reiss, R.F., ed.), Stanford University Press, (331-345).

Wilson, V.J. and Yoshida, M., (1969a) Monosynaptic inhibition of neck motoneurones by the medial vestibular nucleus. Exp. Brain. Res. 9, 365-380.

Wilson, V.J. and Yoshida, M., (1969b) Bilateral connections between labyrinths and neck motoneurones. Brain Res. 13, 603-607.

NEURONAL MECHANISMS FOR RHYTHMIC MOTOR PATTERN GENERATION IN A SIMPLE SYSTEM

Allen I. Selverston

Department of Biology, University of California

San Diego, La Jolla, California   92093

The lobster stomatogastric ganglion is an ideal system with which to study the genesis of rhythmic patterned motor activity. This ganglion operates the striated musculature of the lobster stomach. All muscles operate under direct neural control and the stomach itself can be considered an internalized appendage whose movements are similar to those observed in crustacean locomotory systems. The isolated, deafferented ganglion produces two rhythms, the gastric mill rhythm and the pyloric rhythm, each with a different range of movements and a different pattern of bursts. When the ganglion is connected to the rest of the central nervous system, it receives modulating input over these rhythms from sensory receptors in the wall of the stomach and from higher centers in the central nervous system. The ganglion contains about 30 cells, 23 of which are motorneurons. The synaptic connectivity between the motorneurons has been determined for both the gastric and pyloric rhythms. Subthreshold activity can be seen in all of the cell bodies and all motorneuron axons can be traced to their muscles and isolated for recording or antidromic stimulation. The mechanism of burst generation for the <u>pyloric</u> rhythm appears to consist of endogenous bursts generated by three electrotonically coupled cells. When the synaptic input to these cells is blocked, the cells continue to oscillate and fire whereas other cells are converted to a free running mode. Similar effects can be observed by hyperpolarizing these three cells to shut off their activity and observing the effects on follower cells. When ramp currents are passed into the bursting cells, they begin to burst at threshold and the burst number and inter-burst interval goes up with depolarization. The mechanism underlying the <u>gastric</u> bursts appears to be the result of the cooperative activity of many non-bursting cells connected primarily by inhibitory synapses. No single cell in the gastric mill network has been found to be capable of endogenous burst generation. When depolarized they fire continuously with some accommodation. Occasionally some single cells fire a spontaneous burst without synaptic input. The burst generation mechanism for the pyloric rhythm, therefore, appears to reside within the membranes of the endogenous bursting cells whereas the burst generation mechanism for the gastric mill network appears to be an emergent property of the network itself. It appears to be entirely reasonable, on the basis of preliminary modeling studies, to suggest that the pyloric rhythm burst generation mechanism can be supplemented by properties emerging from the pyloric network as a whole. Similarly, although the gastric system appears to be predominantly a rhythm generated by cooperative synaptic effects, bursting in some of the cells may also play a part in the overall generation of the gastric mill pattern.

Introduction

Locomotion in all animals involves the generation of rhythmic motor patterns. It is of fundamental interest to understand what kinds of neural networks are utilized in the production of such patterns.

One approach to the investigation of the basic circuitry involved in locomotory systems is to examine the numerically simpler networks of invertebrates. Such networks provide considerable information toward an understanding of invertebrate behavior and toward the development of conceptual frameworks for more complex systems.

Stomatogastric System

There are considerable similarities between the nervous mechanisms which operate a lobster stomach and those which are responsible for vertebrate limb movements. The stomach muscles are derived ectodermally, striated, and operate under extrinsic nervous control. Both rely heavily on cyclical patterns generated by central processes. The patterns in both must be switched on

PATTERN GENERATION IN SIMPLE SYSTEM

and off easily and must be modulated by sensory feedback for adjustment to variable environmental conditions.

As with locomotion, different groups of muscles must be brought into activity in proper sequence. These groups usually are arranged in antagonistic pairs so that the motor patterns in their simplest form consist of alternating bursts of activity to each muscle.

The behavior exhibited by a lobster's stomach is of smaller range than that of a vertebrate limb. Several important features make it an extremely useful preparation.

(1) Its behavior is not perfectly rigid but has a broad dynamic range. Once turned on it does not run to completion blindly but is modulated continuously by higher centers.

(2) Only about thirty neurons are involved in the generation of the central motor patterns.

(3) Synaptic relationships between most neurons can be experimentally determined.

This chapter will describe the two rhythms generated by the stomatogastric ganglion in terms of the neural circuitry underlying them. Also the mechanisms involved and their experimental verification will be discussed. And finally, a comparison of the features this system has in common with other known pattern generating networks will be made.

In the stomatogastric ganglion the pattern generating mechanism comprises primarily motorneurons. Two interneurons are present, only one of which plays an important functional role. There is no other large pool of interneurons which must definitely contribute to delimiting the total behavioral range. Only one nerve carries afferent information to the ganglion, the stomatogastric nerve (stn). When it is cut, the ganglion is deafferented completely, (i.e., both sensory and CNS inputs cut off). In this condition, the ganglion generates two rhythms easily monitored by placing electrodes on any of the motor nerves. The ganglion functions as a neuronal machine. Preparations are made by removing the identified nerves from the wall of the stomach after tracing them to the muscles they innervate (Figure 1). The nerves and the ganglia are pinned out in a Sylgard lined Petri dish and covered with saline. Pin electrodes are placed on all the important nerves for extracellular monitoring or stimulating. The stomatogastric ganglion is desheathed and transilluminated. For experiments in which the influences of higher centers, particularly the

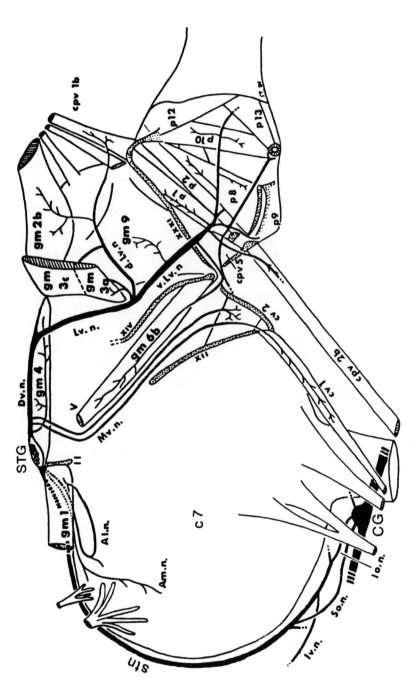

Figure 1. Diagram showing a side view of the lobster stomach with the principal muscles and nerves of the stomatogastric system indicated. The stomatogastric ganglion (STG) lies in the opthalmic artery on the dorsal surface. The paired commissural ganglia, CG, can be seen near the oesophagus and the single oesophageal ganglion (not labeled) (Legend continued on next page.)

is at the junction of the two ion nerves and the stomatogastric nerve (stn). Details of the muscle and never anatomy can be found in Selverston et al., (1976).

commissural (CG) and oesophageal (OG) ganglia are studied, a combined preparation is made which leaves these three ganglia connected to the stomatogastric ganglion (STG) via the superior (son) and inferior (ion) oesophageal nerves.

Pyloric System

The behavior of the pyloric region of the stomach consists basically of alternate dilation and contraction. The phase relationships between the various muscles are more complicated however, resembling those seen in hip muscles during walking. Phase relationships for the fourteen pyloric neurons are shown for a deafferented ganglion in Figure 2. In the combined preparation, the cycle period is increased from one or two seconds between bursts to about two bursts/second.

Intracellular recordings made from pyloric neuron cell bodies show three different forms of electrical activity; oscillations in the resting potentials, attenuated spike potentials and subthreshold postsynaptic potentials. By recording spontaneous activity from two or three pyloric neurons simultaneously, it has been possible to determine the synaptic connections between individual elements in the system. For example Figure 3 shows that when a PD cell bursts, the LP, VD and PY cells are all inhibited. Generally it is possible to see fixed latency monosynaptic potentials in follower cells. When a burst in one cell produces a smooth hyperpolarization in a follower cell, we call the interaction functional (f). Electrical coupling between neurons can be demonstrated by the passage of current from one cell to the other.

A synthesis of such data has produced the circuit shown in Figure 2C. All of the synapses are inhibitory. Connections can be both unidirectional or reciprocal. In one case, the PD not only inhibits the VD, but is also electrically coupled to it.

Gastric System

There is a separate group of cells in the stomatogastric ganglion responsible for producing a second rhythm. This second or gastric rhythm operates the three "teeth" of the gastric mill in a behavior closely resembling chewing. There are four movements:

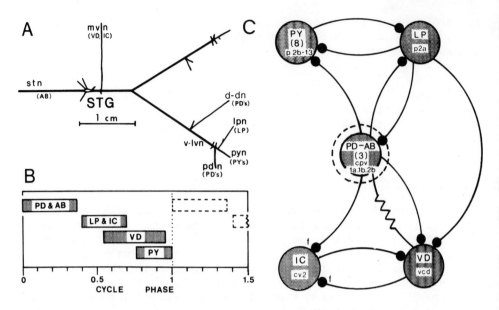

Figure 2. (A), the nerves containing pyloric rhythm axons as they would appear pinned out in a Petri dish. (B), phase relationships of pyloric motor units. (C), pyloric system circuitry. Dashed circle indicates the PD-AB units are electrically coupled. The resistor represents an electrical synapse between the PD-AB group and the VD. Black circles are chemical inhibitory synapses. A small f by some synapses refers to the fact that the synapse is functional but has not been shown rigorously to be monosynaptic. The number of neurons of each type and the muscles they innervate are indicated.

(1) The two lateral ossicles or teeth are pulled together.

(2) The single medial tooth is pulled down and forward over them.

(3) The lateral teeth are pulled apart.

(4) The medial tooth is reset upwards and backwards.

These four basic movements use two pairs of antagonistically arranged muscles. The neurons which are used by this system are shown in Figure 4A. The muscles closing the lateral teeth are operated by the LG and MG neurons. Those opening them, by the two LPG neurons. The medial tooth muscles which provide the powerful downward and forward "power stroke" are controlled by four GM motor neurons. The neuron responsible for resetting the tooth is the DG,

PATTERN GENERATION IN SIMPLE SYSTEM

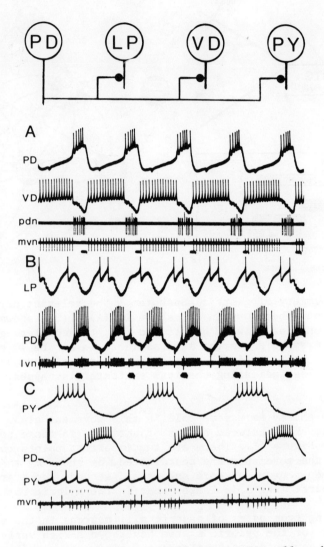

Figure 3. Spontaneous firing pattern of the PD cell and its follower neurons in a deafferented preparation. The PD is inhibitory to the LP, VD and PY neurons and intracellular recordings from pairs of these cells show the periodic interruption of firing which occurs. (A), shows inhibition of the VD by the PD; (B), shows inhibition of the LP and (C), inhibition of two PYs. Spikes in the axons of these motorneurons can be seen in the extracellular traces. Vertical calibration (in C) is 10 mV. Time marks are one/second in A and B and sixty/second in C.

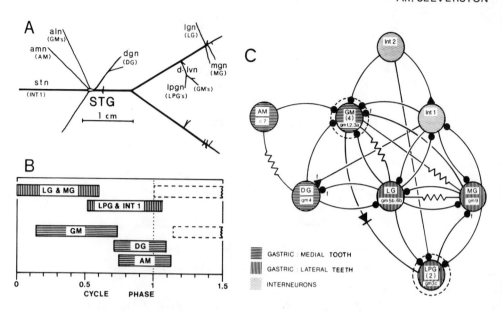

Figure 4. Summary of gastric subsystem. The relevant nerves and the motor axons they contain are shown in (A). (B), shows the phase relationships between bursts and (C), the complete gastric circuit. The symbols are the same as those in Figure 2C except for the diode representing a rectifying electrical synapse and closed triangles indicating chemical excitatory synapses.

although the AM may have some role in this movement also. The phase relationships between bursts of activity in these neurons are shown in Figure 4B. There are two important points to make here, first the movements are coordinated, as shown by the fixed phases between neurons of each subset and second, the overall cycle period and burst length is longer than in the pyloric rhythm.

Recording from pairs of gastric mill neurons in the same way as described for the pyloric system has shown a rich variety of monosynaptic contacts and spontaneous firing patterns (Figures 5, 6). In addition two interneurons were found, one of which plays an important role in the operation of the gastric rhythm. The circuit for the gastric mill subset is shown in Figure 4C.

Although the gastric and pyloric rhythms can be generated in the completely deafferented ganglion, a combined preparation leaving the inputs from the oesophageal and commissural ganglia intact has a strong influence on the reliability and period of both outputs. In particular, the gastric mill motor output works reliably in all combined preparations.

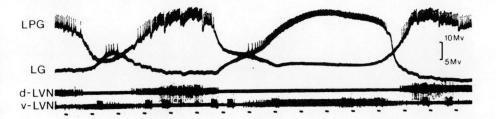

Figure 5. Spontaneous cyclic activity for two gastric mill motor-neurons recorded intracellularly from a deafferented preparation. The antagonistic bursts of activity can be seen extracellularly in the d-LVN and v-LVN traces. Time marks are one/second.

The inputs to the stomatogastric ganglion are just beginning to be worked out, but one pair of afferent fibers (CNS) should be mentioned here. Russell has found two neurons in the commissural ganglia which provide excitatory drive to the gastric mill circuit (Russell, 1976). The unique feature of these inputs, referred to as E neurons, is that they provide phasic excitation synchronized with the gastric rhythm. Functionally, they are periodically inhibited by interneuron 1. They fire in bursts to the GM, LPG, LG and possibly the MG motorneurons. When interneuron 1 is shut off by hyperpolarizing current, the E neurons fire continuously and cause tonic firing of the GM cells as a result of their excitatory synapses onto these cells. The LG and MG neurons are strongly electronically coupled so it is difficult to know whether the excitation from the E fibers is to one or to both.

The LG and the MG both inhibit interneuron 1 and one would predict that if these cells were "removed" from the gastric circuit by simultaneous hyperpolarization of each, interneuron 1 would be free to fire continuously. Also since interneuron 1 inhibits the E neurons, one would expect little input to the stomatogastric ganglion from this source. When Russell performed this experiment (Russell, 1976) the E fiber bursts did terminate as did the firing of the GM neurons (Figure 7).

Since the E neurons receive continuous inhibitory modulation from the gastric rhythm and provide periodic bursts of excitation to components of this rhythm, they must be considered part of the network itself. Although the gastric rhythm can be observed in the deafferented preparation (sensory and CNS), in reality it is controlled by a network distributed between the commissural and the stomatogastric ganglia.

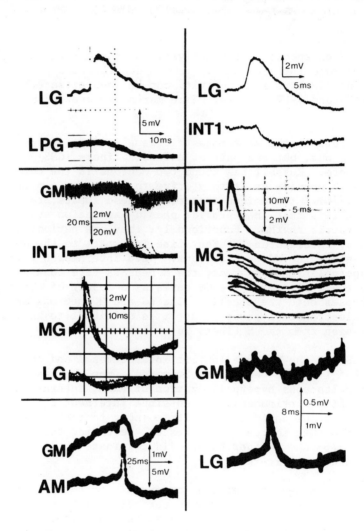

Figure 6. Fast sweeps of representative monosynaptic connections between pairs of neurons in the gastric subsystem.

PATTERN GENERATION IN SIMPLE SYSTEM

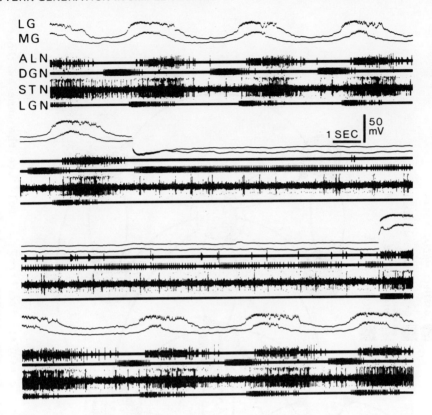

Figure 7. Shut off of LG and MG neurons by hyperpolarization of their cell bodies removes inhibition from interneuron 1. This allows interneuron 1 to fire tonically and inhibit the E neurons continuously. This stops firing of the GMs and the LG but turns on the DGN (see circuit diagram, Figure 8). Normal bursts resume when the cells are released from hyperpolarization. (Russell, Unpublished data.)

The complete circuit incorporating the two E neurons is shown in Figure 8.

Burst Generating Mechanism for Pyloric Cycle

Cells with endogenous bursting capability have been known for a long time. It was not surprising to find them in the stomatogastric ganglion. What was unique however was the possibility of understanding their functional role in a neural network. Maynard (1972) suggested, on the basis of indirect evidence, that the PD

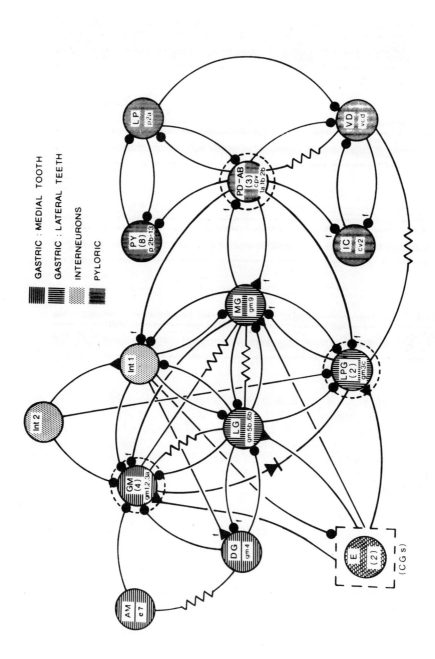

Figure 8. Summary diagram of the stomatogastric neural circuitry. Symbols are the same as those in Figures 2 and 4. The two E neurons are in the commissural ganglia (CG).

cells were endogenous bursters. Unlike such bursting cells in molluscs, the direct evidence that these cells are true endogenous bursters has been hard to obtain.

Perhaps the most direct experiment is physical isolation of the burster neuron cell body. The cell bodies of most crustacean neurons do not have electrogenic membranes. The site of burst generation is somewhere within the neuropil. Isolation of the soma does not isolate the burst generating site.

Another approach is to uncouple synaptic inputs to the PD-AB group chemically or pharmacologically. If the PD-AB group continued to burst in the absence of any synaptic input, this would constitute excellent proof that they were endogenous bursters. Experiments in which the ganglion was bathed in various ratios of low Ca^{++} high Mg^{++} known to block synaptic transmission in many invertebrates, have been inconclusive. PD-AB bursting does not continue indefinitely when controls show synapses have been blocked. Rather than interpreting this as negative evidence for endogenous bursting, it probably only suggests that Ca^{++} is necessary for the burst generating mechanism itself.

More suggestive evidence derives from the behavior of the PD-AB group within the pyloric network as a whole. Altering the firing rate of other cells in the pyloric subset has very little effect on the PD-AB group. However increasing or decreasing the firing rate of the PD-AB group, by passing current into their cell bodies, alters the firing rates of the other cells markedly. In fact if one completely shuts off the PD-AB group by hyperpolarization, the follower cells fire continuously and not in bursts.

The commissural ganglia also play a role in the maintenance of the pyloric rhythm. With these ganglia intact, the pyloric cycle rate is 2-4 times faster and the whole system operates reliably and consistently. As with the gastric network, the pyloric system is distributed into the commissural ganglia. Russell has found neurons in the commissural ganglia, termed P neurons, and which fire in bursts with the pyloric rhythm (Russell, 1976). P neurons appear to be inhibited by the AB interneuron; when the PD-AB group is hyperpolarized this periodic inhibition is removed and the P cells fire continuously (Figure 9).

Another indication of the bursty nature of the PD-AB cells can be demonstrated by passage of ramp currents into the cell bodies. When a depolarizing current reaches threshold for a PD cell, it begins to fire in bursts. As the cell becomes more depolarized, the number of spikes per burst increases and the interburst interval decreases (Figure 10). The reverse occurs during the hyperpolarizing phase of the ramp current. Such experiments of course

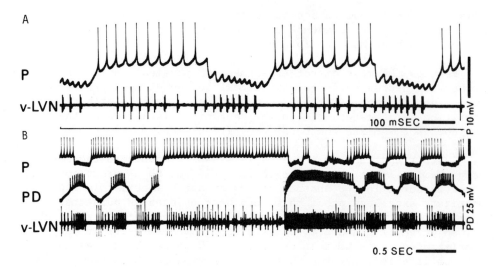

Figure 9. Experiment to demonstrate the inhibition of a commissural ganglion P cell by a PD cell. (A), shows periodic inhibition occurring at the same time the PD cell fires (small units in v-LVN trace). (B), shows tonic firing of P cell which results from lack of inhibition from PD when the PD is hyperpolarized. (Russell, Unpublished data.)

reveal the intramembrane mechanisms responsible, but demonstrate they are voltage dependent as is the mechanism for the action potential.

If indeed the PD-AB group represent a tightly coupled group of endogenous bursters, then their functional role in the pyloric network is not difficult to understand. However, the evidence for such endogenous capability is not yet complete. Consider the other cells in this network are spontaneously active and fire tonically rather than in bursts. Then the PD-AB group can be considered as the source of periodic interruptions of these continuous firing patterns; inhibitory synapses between other cells as necessary for producing the correct phase relationships between them (Maynard and Selverston, 1975). Since it is the motorneurons themselves which provide the cyclic bursts, a pool of premotor interneurons is not required.

Network Oscillators

An examination of all of the gastric mill cells has shown that none is capable of sustained endogenous burst activity. When

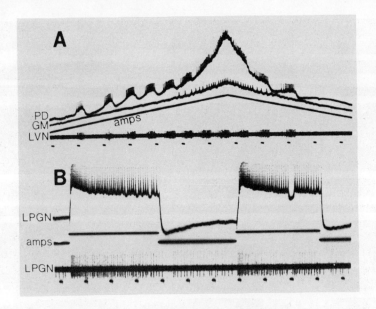

Figure 10. The response of a gastric mill (GM) and a pyloric cell (PD) to a ramp current is different. In (A), simultaneous depolarization and repolarization shows that the PD responds by bursting at a rate proportional to the amount of current while the GM fires steadily at a frequency also proportional to the amount of current. In (B), a square wave current pulse applied to the soma of an LPG cell causes steady firing with some accommodation.

these cells are depolarized with ramp or step currents (Figure 10) they fire steadily, not in bursts, with a firing rate proportional to the amount of depolarization.

Synaptic uncoupling experiments are more definitive for the neurons in the gastric mill subset also. Low Ca^{++} solutions cause cells which are bursting either to fire steadily or to become silent (Figure 11). These results suggest that unlike the pyloric rhythm, there is no single cell or group of cells which provide the main driving force for the network. Instead the synaptic connectivity itself produces properly phased, alternating bursts - a so-called emergent property of the network. The role of the E fibers appears to distribute the network outside of the ganglion and although the deafferented preparation can burst, the phasic excitatory drive produced by these cells insures reliable activity. Trying to understand the mechanism of burst production by a complex network is not a trivial problem. Not only do synaptic strengths differ, the total variety of nonlinear interactions (facilitation, depression, postinhibitory rebound, etc.) must act cooperatively to

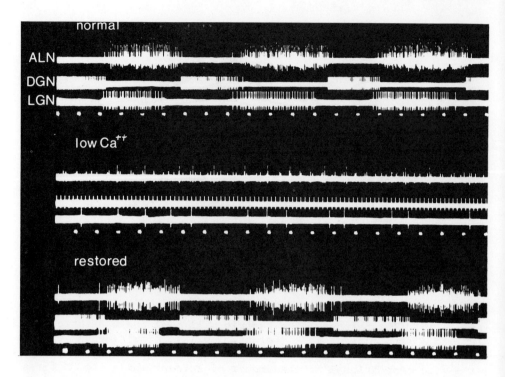

Figure 11. Effect of low calcium saline on gastric mill bursting. The top three traces show normal gastric rhythm recorded extracellularly from GM, DG and LG neurons (in ALN, DGN and LGN traces respectively). The low calcium saline blocks synaptic transmission and abolishes the rhythmic activity. Normal activity is restored when calcium is restored. Time marks one/second. (Mulloney and Selverston, Unpublished data.)

produce the observed patterns. We can make an attempt at explaining the output by examining the network qualitatively. Interneuron 1 and the lateral teeth neurons LG and MG form a reciprocal inhibitory unit if one considers the electrical coupling between the latter two. On the basis of modeling studies, such reciprocally inhibitory networks have been proposed as possibilities for the production of alternate bursting in antagonists (Wilson and Waldron, 1968). Once such bursting is initiated, the LG-MG antagonists, the LPGs, would have their tonic firing interrupted periodically. Similarly, the tonically firing GM neurons would be made to burst by periodic inhibition from interneuron 1. The only group of cells which cannot fit easily into such a scheme are the DG-AM pair, the antagonists to the four GMs. They are normally silent when the gastric mill is not operating so that it would

appear they need some excitatory inputs or particularly strong
inhibitory ones (so that they could fire from postinhibitory rebound) to activate them. They do get weak, possibly indirect,
excitation from interneuron 1 as well as some inhibition from the
LG cell. It is possible for the DG-AM pair to burst alternately
with the GMs even when interneuron 1 is not bursting (Selverston
et al., 1976). The precise nature of the medial tooth mechanism
still is obscure.

Cooperativity Between Network and Cellular Oscillators

From the complete wiring diagram it is known the pyloric and
gastric mill groups are connected by about five functional connections so that activity in one group has some influence on the
other. However, the functional role of these interactions is not
known. A more basic question is whether or not a central pattern
generator could utilize both mechanisms: cooperativity and endogenous bursting. Two observations relate to this. First a computer
simulation of the pyloric rhythm which did not include PD-AB
bursting as a parameter, was able to produce properly patterned
bursts between PD, LP and PY cells (Warshaw and Hartline, 1975).
This suggests that the pyloric network may have the inherent
capability of generating bursts and that if burster cells are
present, their role may be more one of timing than of burst production. In a similar vein, it is not uncommon to see a quiescent
gastric system cells suddenly burst for a cycle or two without any
apparent synaptic input (Selverston et al., 1976). Such bursting
may be caused by some membrane instability. This may not be the
same as that which causes bursting in true endogenously active
cells. It may add, however, to the total burst generating mechanism of a synaptic network and indeed may assume a primary role.
Computer simulations in which such cellular parameters as endogenous bursting could be assigned to or removed from cells within
a network would go a long way toward determining which, if either,
has the more important role.

There are several important cellular properties which have
not been mentioned thus far, and which may play a role in generating and controlling the motor rhythms.

(1) Nonspiking inhibition is the ability of current spread into
the terminal from sources other than spikes, to cause the graded
release of transmitter (Maynard and Walton, 1975). Thus far such
release has only been achieved by passing large currents into the
presynaptic cell. Whether or not such release occurs under physiological conditions is not known.

(2) Gating of presynaptic terminals occurs when there are multiple spike initiating zones for one cell. A main axon may fire without causing the release of transmitters from terminals on branches within the neuropil. Similarly, terminals on these branches may release transmitters as a result of inputs which are ineffective in firing the main axon.

(3) Synaptic-electrotonic interactions may occur which dynamically alter the values of these parameters. The strength of a chemical synapse may be altered for example if the resistance of the postsynaptic membrane changes as a result of current flowing through an adjacent electrical junction. Similarly, the coupling coefficient between two electrically connected cells may change as a result of current flow through adjacent chemical synapses (Spira and Bennett, 1972). There is no evidence yet that such interactions occur in the stomatogastric system, but with so many neurons connected with both kinds of synapses, the possibility certainly exists.

Common Features of Known Pattern Generating Networks

In terms of neural circuitry, repetitive behaviors are now the best understood. It may be useful to compare the stomatogastric system with some of the other better known invertebrate systems. Such comparisons can be made with reference to three important points.

(1) What is the source of the repetition?

(2) What is the source of the total motor pattern underlying the behavior?

(3) What is the role of sensory input in motor pattern generation?

In some cases, these points are related closely so as to be inseparable. In other instances, the role of each is quite distinct. For example, in the stomatogastric system the basic source of repetition is different for each of the two rhythms. The pyloric rhythm appears to depend on a group of cells (the PD-AB group) each of which is an endogenous oscillator. Due to the electrical coupling between each they tend to fire synchronously. The firing pattern for other pyloric cells derives from inhibitory synaptic connections, many of which are reciprocal, between the other motorneurons. The repetitive nature of the gastric rhythm however appears not only to emerge as a network property, but also to use the same network as the mechanism by which motorneurons are made to fire in the correct sequence. It is clear that both rhythms

can be generated in the completely deafferented stomatogastric
ganglion, but that a distributed network between it and the
commissural ganglia greatly enhances the effectiveness of both
rhythms. The stomach wall contains many sensory receptors. However it is not known how strongly they influence either rhythm.
Since the system can operate without them, their role in the
overall pattern generating mechanism can be modulatory only.

A good place to begin a comparison of the stomatogastric
ganglion with other networks is the insect flight motor system,
since this is an entirely centrally generated pattern (Wilson,
1966). On the basis of extracellular recordings and modeling
studies Wilson and his colleagues suggested as with the stomatogastric gastric rhythm, both the cyclic bursting and the proper
phasing were a network property. Sensory input was thought to have
an excitatory effect. One that was averaged over time, and not one
that supplied strong phasic information to the system. More recent
work strongly challenges both of these early hypotheses. On the
basis of intracellular recordings made from flight motorneurons in
the metathoracic ganglion, no evidence has been found for reciprocal inhibition between motorneurons operating antagonistic muscles
(Hoyle and Burrows, 1973). In addition Wendler (1974) has shown
that proprioceptive feedback can have more than just a general
excitatory effect. By artificially moving the wings, he has shown
that the entire motor rhythm can be entrained to a new frequency.
These results have been extended by Burrows (1975). The proprioceptive input coming from stretch receptors in the wings form a
monosynaptic feedback loop which inhibits the elevator motorneurons
and excites the depressors. This sensory modulation is, in
Burrows' view, important for the cyclic pattern generating
mechanism.

The dual excitatory-inhibitory effect of a proprioceptor on a
motor network appears to be quite common and may be an important
generalizable phenomenon. This work does not imply that the basic
pattern is no longer centrally generated, but that the role of
sensory reflexes may be more important than heretofore imagined.

There appears to be only one good example of a reflexively
generated oscillatory motor pattern in the literature and that is
for the swimming behavior of the scallop (Mellon, 1969). Swimming
in the leech appears to utilize circuitry similar to that found in
the locust flight motor system. Once again proprioceptive feedback
from stretch receptors in the dorsal and ventral muscles have a
dual excitatory-inhibitory role (Kristan, 1974). In this case the
afference excites the antagonistic inhibitor while simultaneously
inhibiting the inhibitor to the agonist. Since the peripheral
inhibitors also inhibit agonist excitor motorneurons, the receptor
feedback acts indirectly on these motorneurons also. Since stretch

would produce inhibition of the excitors to the contracting muscles and removal of inhibition from the relaxed (stretched) muscles, an oscillatory mechanism easily could be generated.

If one does not consider the rhythm as centrally generated, four types of underlying mechanisms can be posited.

(1) Bursts arise as a network property.

(2) Bursts are driven by nonspiking oscillator cells.

(3) Bursts are driven by spiking endogenous bursters.

(4) Bursts are driven by groups of electrically coupled interneurons.

The gastric mill rhythm of the lobster falls into the first category although elements distributed outside of the stomatogastric ganglion proper have a significant function. Similarly the feeding behavior of the snail Heliosoma appears to be driven by a network of cells called a cyberchron (Kater, 1974). This system also obtains feedback from a stretch receptor which produces simultaneous excitation and inhibition of antagonist motorneurons. The oscillator also appears to have a dual output, exciting and inhibiting antagonists simultaneously. One additional point about the snail feeding system is that agonist motorneurons are electrically coupled. Although they do not fire one to one, they do tend to discharge at about the same time. This feature is found in many of the networks which have been described (Getting and Willows, 1974; Kater, 1974).

Nonspiking interneurons capable of providing oscillatory drive to motorneurons have been found in crustaceans and insects (Mendelson, 1971; Pearson and Fourtner, 1975). A pair of oscillatory interneurons occur in the lobster subesophageal ganglion which at depolarization or hyperpolarization produce spiking in antagonist scagphagnothite motorneurons without spiking themselves. A similar nonspiking interneuron in the cockroach shows membrane oscillations in phase with motorneuron bursts during walking movements of the legs. When depolarized this interneuron excites levator motorneurons. However, unlike the crustacean oscillator, it had no effect on antagonistic motorneurons when it was hyperpolarized.

Endogenous burster cells are found commonly in invertebrate preparations. While their action is in many cases unknown, they probably are responsible for the pyloric motor rhythm in the stomatogastric ganglion. Modulation of such endogenous bursters by sensory feedback or central command fibers will likely be found in many other instances.

Finally burst production by electrically coupled interneurons has been described for the trigger-group cells in Tritonia (Getting and Willows, 1974). Sensory input appears to set off a burst which accelerates due to the positive feedback between the cells. The bursts are terminated by a hyperpolarizing wave which results from an accentuation and prolongation of the spike after-hyperpolarization. This reduces the functional shunting of spike current and is in effect electrically mediated inhibition. The bursts of activity in these cells drives the dorsal flexor motorneurons so that they burst periodically (Willows et al., 1973). The dorsal flexor motorneurons are connected to the ventral flexor motorneurons by reciprocal inhibition. Such a network apparently can produce alternating bursts between antagonistic muscles and are the basis for the swimming behavior of this animal.

Conclusions

Work on central pattern generating mechanisms at the unit level is now delineating the range of mechanisms which can be used for the generation of oscillatory motor patterns. The stomatogastric system represents a good model for such network analysis even though much remains to be learned about it. Questions remain. How are the patterns controlled? Why is inhibition used so extensively for phase control? Still to be determined is whether modulation is by command fibers or by distributed networks.

REFERENCES

Burrows, M., (1975) Monosynaptic connections between wing stretch receptors and flight motorneurons of the locust. J. Exp. Biol. 62, 189-219.

Getting, P.A. and Willows, A.O.D., (1974) Modification of neuron properties by electrotonic synapses. II. Burst formation by electrotonic synapses. J. Neurophysiol. 37, 858-868.

Hoyle, G. and Burrows, M., (1973) Neural mechanisms underlying behavior in the locust Schistocerca gregaria. II. Integrative activity in metathoracic neurons. J. Neurobiol. 4, 43-68.

Kater, S.B., (1974) Feeding in Helisoma trivolvis: the morphological and physiological basis of a fixed action pattern. Am. Zool. 14, 1017-1036.

Kristan, W.B., (1974) Neural control of swimming in the leech. Am. Zool. 14, 991-1001.

Maynard, D.M., (1972) Simpler networks. Ann. N.Y. Acad. Sci. 193, 59-72.

Maynard, D.M. and Selverston, A.I., (1975) Organization of the stomatogastric ganglion of the spiny lobster. IV. The pyloric system. J. Comp. Physiol. 100, 161-182.

Maynard, D.M. and Walton, K.D., (1975) Effects of maintained depolarization of presynaptic neurons on inhibitory transmission in lobster neuropil. J. Comp. Physiol. 97, 215-243.

Mellon, D.F., Jr., (1969) The reflex control of rhythmic motor output during swimming in the scallop. Z. vergl. Physiol. 62, 318-336.

Mendelson, M., (1971) Oscillator neurons in crustacean ganglia. Science. 171, 1170-1173

Pearson, K.G. and Fourtner, C.R., (1975) Nonspiking interneurons in the walking system of the cockroach. J. Neurophysiol. 38, 33-52.

Russell, D.F., (1976) Rhythmic excitatory inputs to the lobster stomatogastric ganglion. Brain Research. (In press).

Russell, D.F., (1976) Ph.D. Thesis, University of California, San Diego, (In preparation).

Selverston, A.I., King, D.G., Russell, D.F. and Miller, J.P., (1976) The stomatogastric nervous system: structure and function of a small neural network. Progress in Neurobiology. (In press).

Spira, M.E. and Bennett, M.U.L., (1972) Synaptic control of electrotonic coupling between neurons. Brain Research. 37, 294-300.

Warshaw, H.S. and Hartline, D.K., (1975) Simulation of network activity in stomatogastric ganglion of the spiny lobster, Panulirus. Brain Research. (In press).

Wendler, G., (1974) The influence of proprioceptive feedback in locust flight coordination. J. Comp. Physiol. 88, 173-200.

Willows, A.O.D., Dorsett, D.A. and Hoyle, G., (1973) The neuronal basis of behavior in Tritonia. III. Neuronal mechanisms of a fixed action pattern. J. Neurobiol. 4, 255-285.

Wilson, D.M., (1966) "Central nervous mechanisms for the generation of rhythmic behavior in arthropods," In Nervous and Hormonal Mechanisms of Integration. Academic Press, N.Y., (199-228).

Wilson, D.M. and Waldron, I., (1968) Models for the generation of the motor output pattern in flying locusts. Proc. I.E.E.E., June, (1058-1064).

CENTRAL NERVOUS CONTROL OF COCKROACH WALKING

Charles R. Fourtner

Department of Biology, State University of New York at Buffalo, Buffalo, New York 14214

During the past few years there has been a concerted effort to determine the neuronal circuit of and the neurophysiological basis for rhythmic leg movements in walking insects. In the cockroach there is strong evidence that rhythmic motor patterns, similar to those occurring during normal walking, spontaneously occur after complete deafferentation of the segmental ganglion. This led to the hypothesis that there are segmental rhythm generators that operate independently from the segmental sensory input. Subsequent morphological and physiological investigations of the metathoracic ganglion of the cockroach demonstrated the presence of interneurons whose membrane potentials oscillate during rhythmic leg movements. One interneuron, Interneuron I, oscillates such that its cyclic depolarizations are synchronized with the flexion phase (return stroke) of the rhythmic leg movements. When experimentally depolarized, Interneuron I drives and specifically recruits the flexion motoneurons in the same sequence as in normal walking. Interneuron I also appears to be an integral part of the neuronal circuit generating the rhythmic leg movements since short depolarizing pulses in Interneuron I can reset the cycle time of the rhythmic leg movements. An unusual phenomenon of Interneuron I and indeed of most interneurons within the metathoracic ganglion of the cockroach is that they are not excitable, that is they do not produce nerve spikes. These interneurons modulate activity in motoneurons by minute changes in membrane potentials. This is the first locomotory system in which identified nonspiking neurons have been demonstrated to play an important role in generating a specific motor pattern.

Introduction

During the early and middle part of this century there were several studies on insects, in which a variety of ablation and lesion techniques were utilized to describe the possible coordination and control of insect locomotion (for literature review see Hughes, 1965 and Wilson, 1966). In the past decade the interest in insect locomotion has increased tremendously with the advent of high-speed cinematography, electromyography, intracellular recording and intracellular staining techniques. High speed cinematography has permitted a greater degree of accuracy in the measurements of phase relationships between various walking appendages and a more detailed account of the movement of leg segments with respect to one another. Electromyographic recordings, using small diameter insulated wires, are useful in the study of muscle and motoneuron activity in both free-walking organisms and in restrained organisms producing rhythmic leg movements. Microelectrode techniques have allowed single cell analysis of synaptic events in motoneurons and interneurons (for literature review see Pearson et al., 1973). Single cell staining techniques have greatly advanced the understanding of the neuronal mechanisms underlying locomotion since neurons which previously could be identified only physiologically now can be identified morphologically (Kater and Nicholson, 1973). This is of particular interest in studies of invertebrate central nervous systems in which there is a relatively small number of neurons within the nervous system.

Results from experiments utilizing the above techniques have led to a number of hypothetical models of the neural control of insect walking (Pearson and Iles, 1970, 1973; Delcomyn, 1971, 1973; Graham, 1972; Pearson et al., 1973). It is clear from these studies that insect walking is controlled by a central pattern generator which is sensitive to a variety of peripheral sensory inputs. It is also clear that the peripheral sensory inputs have a greater effect on the central pattern generator at lower walking speeds than at higher speeds. This conceptual and experimental framework of a central pattern generator with a variety of peripheral and central modifiers is also a well-established paradigm for vertebrate locomotion (Grillner, 1975).

This chapter reviews the evidence for a central pattern generator controlling rhythmic leg movements of the cockroach metathoracic legs (third pair of walking legs). Also it describes in some detail a few of the more recent results obtained with intracellular recordings from the metathoracic ganglion of the cockroach.

For further information on peripheral inputs and the mechanisms by which these inputs affect the walking central pattern generator in the cockroach, refer to Pearson's paper in this volume; for some of the more general characteristics of arthropod walking, please refer to Hoyle, this volume.

Behavioral Analysis of Leg Movements

Cinematographic analyses of locomotory movements of the metathoracic leg demonstrate two distinct phases during the stepping cycle: the protraction phase, which is comparable to the swing phase in vertebrate leg movements; and the retraction or stance phase. According to these analyses, increases in the duration of the step cycle result in increases in the duration of the stance phase. The swing phase does not increase as rapidly. For example, in both the cockroach and the locust, plotting swing duration versus step duration at walking speeds greater than 2 Hz yields a slope of less than 0.09. The slope of stance versus step duration over the same stepping range is approximately 0.7 to 0.8 (data from Delcomyn and Usherwood, 1973 and Burns, 1973). This relationship has also been described in other walking systems (Evoy and Fourtner, 1973; Goslow et al., 1973).

The metathoracic leg plays a role in a number of different behaviors; walking, flight, righting and cercal grooming. Of these various behaviors the rhythmic leg movements produced during walking and cercal grooming are the most similar and easily confused. However there are a number of characteristic differences between the two behaviors. First, the arcs which the leg joints traverse are different. The coxa-femur joint (considered as a single joint although there is a very small segment, the trochanter between the coxa and the femur) and the femur-tibia joint traverse a greater arc during walking than they do in cercal grooming (Figure 1). Secondly, during cercal grooming the abdomen is arched about $15°$ to $20°$ toward the grooming metathoracic leg, whereas in locomotion the abdomen is not drawn toward either side. Thirdly, the highest frequency of rhythmic leg movements during cercal grooming is much lower than the fastest walking frequency (approximately 7 Hz for cercal grooming as opposed to 24 Hz for walking), (Reingold, 1975).

Although these differences between walking and grooming exist, it is possible that one central pattern generator could produce the rhythmic leg movements observed in both behaviors. It may be that one central pattern generator could be activated by different command inputs or combinations of command inputs (Davis and Kennedy, 1972). The different command fibers could alter one or more of the components of the central pattern generator and thereby produce minor changes in the motor output. In addition Barnes et al.,

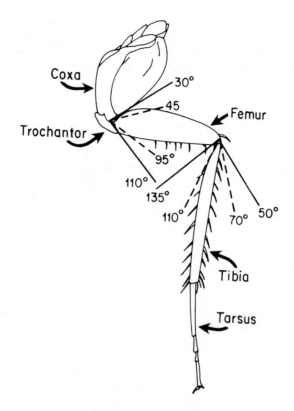

Figure 1. Drawing of the metathoracic leg of a cockroach illustrating the maximum flexion and extension of the coxa-femur joint and the femur-tibia joint during walking (solid line) and cercal grooming (broken line). (Data from Reingold, 1975).

(1972) have demonstrated that reflexes normally observed in quiescent crabs are not evoked during walking. Again it may be that different command inputs can differentially select sensory input and thereby produce minor changes in the motor output.

Patterned Activity in Motoneurons

Movement at the coxa-femur joint is produced by two groups of antagonist muscles, the femur flexor (levator) muscles and the femur extensor (depressor) muscles. The extensor muscles are innervated by several motor axons; however at slow walking speeds (1-10 Hz) only one slow motor axon is active. This motor axon is designated D_s (Pearson and Iles, 1970). At higher walking speeds at least one additional larger axon is recruited, axon D_f. The flexor muscles also are innervated by a large number of motor axons, but there are only two slow motor axons active at slow walking speeds. These are designated as axons 5 and 6 according to axon diameter. Axon 6, the larger of the two, is recruited for walking speeds greater than 2 Hz and at least one larger axon is recruited at speeds about 10 Hz. These data are consistent with the size principle which holds that smaller diameter motor axons are recruited prior to larger diameter axons for a given motor program (Henneman et al., 1965; Davis, 1971; Hinkle and Camhi, 1972). EMG recordings from the extensor and flexor muscles of the femur of the cockroach have demonstrated a very strong reciprocal relationship between the antagonists in both free walking and restrained roaches (Figure 2). The EMG data also supports the behavioral data. The duration of the flexor burst (swing phase) of the stepping cycle remains fairly constant, whereas the duration of the extensor burst (stance phase) increases as the duration of the stepping cycle increases.

In addition to recording EMGs, it is possible also to record directly from the nerves containing motor axons 5 and 6 (nerve 6Br4) and motor axon D_s (nerve 5r1). This experimental approach allows all segmental afferents to the metathoracic ganglion and the anterior connectives of the metathoracic ganglion to be severed. Using this deafferented preparation Pearson and Iles (1970) and Pearson (1972) demonstrated that the metathoracic ganglion could produce a rhythmic motor pattern in which there was a strong reciprocity

Figure 2. Reciprocal activity during spontaneous rhythmic leg movements. Upper trace, recording from nerve 6Br4 containing the flexor motoneurons; axon 5 (smaller spike) and 6 are active during the bursts. Lower trace, EMG from the extensor muscle innervated by the slow extensor motoneuron, D_s. The first burst in the lower trace is the initial burst in this sequence of rhythmical activity. In five sequences of rhythmic leg movements in each of three different preparations, the initial burst always was observed in the extensor muscle.

between the extensor and flexor motoneurons (as recorded from the motor nerves 5r1 and 6Br4 respectively). Again as in the free walking and the intact restrained animal, there was no overlap between the extensor and flexor bursts. This pattern was observed in 50% of the deafferented preparations. The spiking patterns of motoneurons 5, 6 and D_s during the bursts in a deafferented preparation were similar to the spiking patterns occurring during free walking with one notable exception. The spike frequency of D_S was highest at the beginning of the burst in deafferented preparations and on the other hand was highest near the termination of the bursts in the free walking situation. Since the tonic firing rate of D_S was much higher (about 300 Hz) in the deafferented preparation, presumably due to a release of either central or sensory inhibition, Pearson and Iles (1970) suggested that the initial high firing rate of D_S in the rhythmic deafferented preparation was due to postinhibitory rebound from the inhibitory input during the flexor bursts.

From their experiments on the cockroach Pearson and Iles developed a model of the rhythmic pattern generator which produces walking in the cockroach. This model has been further amplified in later papers (Pearson, 1972; Pearson et al., 1973). The Pearson and Iles model is an asymmetric model which (Figure 3) consists of a command input which drives a bursting interneuron and tonically excites the extensor motoneuron, D_s. The bursting interneuron converts the command input into a bursting output; this output then drives the flexor motoneurons 5 and 6 and inhibits the extensor motoneuron D_s. The experimental bases for this hypothetical model are as follows: 1) rhythmic bursting patterns persisted even following complete deafferentation which implies a central pattern generator; 2) the duration of the levator bursts was relatively constant which suggests a pacemaker neuron with a constant depolarizing phase and a variable hyperpolarizing phase; 3) there was a strong reciprocity between the flexor motoneurons and the extensor motoneurons during rhythmic leg movements; 4) D_s had a tendency to fire at a higher frequency at the beginning of the burst rather than later in the burst which suggests D_S is rebounding from inhibition produced by the bursting interneuron, and 5) in a number of cases bursting occurred in the flexor motoneurons with no reciprocal activity in the extensor motoneurons and the opposite situation (bursting in extensor motoneurons without activity in flexor motoneurons) never occurred. Therefore Pearson and Iles suggested that 5 and 6 were driven by the bursting interneuron and the tonic activity in D_S would be inhibited by the interneuron.

It is clear from the studies of Pearson and Iles, (1970) and Pearson (1972) that any further information concerning the pattern generator in the walking system of the cockroach would have to be gleaned from experiments with intracellular recordings from the

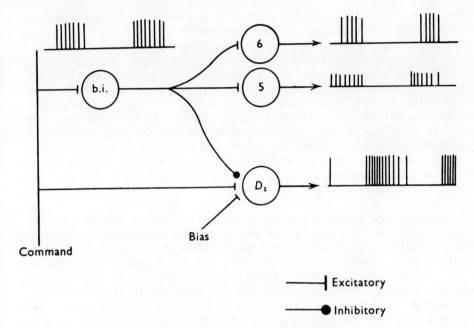

Figure 3. Hypothetical model proposed by Pearson and Iles (1970) for the genesis of reciprocal activity in the flexor and extensor motoneurons in the metathoracic leg of the cockroach. (From Pearson and Iles, J. Exp. Biol. 52, 1970).

metathoracic ganglion. During the past two years studies of this type have been pursued (Pearson et al., 1973 and Pearson and Fourtner, 1975).

Intracellular Recordings from the Metathoracic Ganglion

Cockroaches were mounted on a pedestal with the dorsal surface up so that the metathoracic legs moved freely in air. The gut and some flight muscles were removed to expose the ganglion. Staple pins around the anterior and posterior connectives held the ganglion on a stainless steel platform. Intracellular recordings were obtained from motoneurons D_s, 5 and 6. The recording sites were directly from the neuropile region of the ganglion rather than the somata region. Since the recordings were from neurites of cells, the records should give a more complete survey of the true synaptic activity occurring in the neuron. During rhythmic leg movement the spiking output of motoneurons was produced by a large depolarization of the neuritic membrane, and in some cases the depolarization was terminated by IPSPs. The depolarization was not produced by large EPSPs but appeared to result from either a summation of many minute

EPSPs or a graded input. This type of activity in motoneurons has been seen in other insect motoneurons which are driven by a postulated pacemaker or interneuron pattern generator (Bentley, 1969; Hoyle and Burrows, 1973; Burrows, 1974). In other studies on the metathoracic ganglia Pearson and Fourtner (1975) have described a class of interneurons termed nonspiking interneurons. These are cells which do not produce detectable spike activity to large depolarizing pulses yet produce intense changes in motoneuronal output with only one or two millivolt alterations in membrane level. They identified four different physiological types of nonspiking interneurons that could affect either the extensor motoneurons or the flexor motoneurons. Two of these, Interneurons I and II, were investigated extensively.

Interneuron I has a number of characteristics suited to the study of the central control of rhythmic leg movements. 1) The membrane potential of Interneuron I oscillates at the same frequency as observed with rhythmic leg movements. 2) The depolarizing phase of the Interneuron I oscillation is synchronized with the burst activity in the flexor motoneurons; the hyperpolarizing phase corresponds with the burst activity in extensor motor neurons (Figure 4A). 3) In a quiescent animal an intracellularly applied depolarization of Interneuron I activates flexor motoneurons according to axon size. For example, it activates axon 5, then axon 6, and at higher levels will bring in larger axons that also innervated the flexor muscles. This is consistent with the order of recruitment occurring as the frequency of leg movement increases in normal walking animals. Not only did the depolarization activate the flexor motoneurons of the coxa-femur joint but it also produced flexion of the femur-tibia joint. Thus, this interneuron is capable of exciting the motoneuronal pool active during normal walking. 4) In cases in which motoneuron D_s remained active following penetration of Interneuron I, depolarizing pulses produced a weak inhibition of D_s (see Pearson and Fourtner, 1975 for more details). 5) In most cases penetration of Interneuron I abolished spontaneous rhythmic leg movements. Obviously Interneuron I was involved in the neural circuitry producing rhythmic leg movements. To determine if Interneuron I played a role specifically in the pattern generator itself or whether it was simply a premotor interneuron situated between the pattern generator and the motoneurons, resetting experiments were attempted. Two experiments were done to test if short depolarizing pulses applied during a bout of rhythmic leg movement could reset the timing of the rhythmic leg movements. This would imply that the interneuron itself was part of the pattern generating system. Figure 5 gives the results of two such experiments which demonstrate that when a short depolarizing pulse was applied during an expected hyperpolarization of interneuron I, the cycle time was reset; the next normal size burst in 6Br4 was delayed. A depolarizing pulse applied during the depolarizing

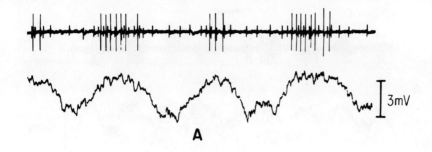

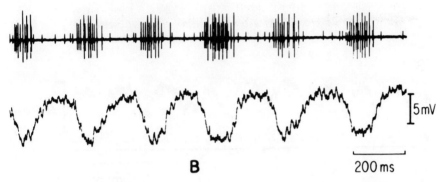

Figure 4. Intracellular recordings from nonspiking interneurons during rhythmic leg movements. (A). Upper trace, extracellular recording from the flexor motor nerve 6Br4; lower trace, intracellular recording from Interneuron I corresponds to the bursts in 6Br4. (B). Upper trace, extracellular recording from 6Br4; lower trace, intracellular recording from Interneuron II. Note that the hyperpolarizations in Interneuron II correspond to the bursts in 6Br4. (From Pearson and Fourtner, J. Neurophysiol. 38, 1975.)

phase of the oscillations in Interneuron I produced no quantitative change in the cycle timing (Pearson and Fourtner, 1975). Unfortunately, the neurons could not be held long enough to test if hyperpolarizing pulses would indeed reset the cycle time. However, the above data indicate that Interneuron I is a part of the neuronal circuitry of the central pattern generator which produces rhythmic leg movements in the cockroach. In eight preparations iontophoretic injection of Co^{++} into interneurons physiologically identified as Interneuron I revealed a neuron with a highly branching neuritic region along the lateral border of the ipsilateral region of the neuropil. This region seems to overlap the

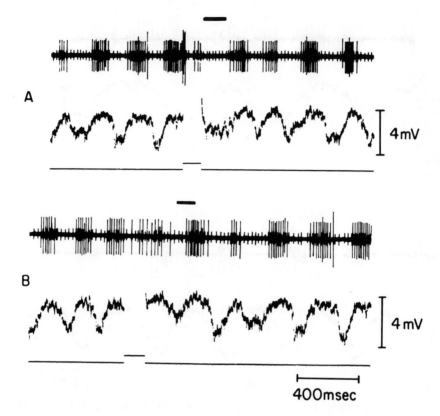

Figure 5. Resetting the flexor burst activity (cycle) with short depolarizing pulses. Two examples demonstrating that short depolarizing pulses (25 nA) applied during the hyperpolarizing phase of Interneuron I oscillations can reset the timing of subsequent flexor bursts. The heavy bars above the top trace indicate the time of the expected burst if the pulse had not been applied. Top traces, extracellular recordings from 6Br4. Middle traces, intracellular recordings from Interneuron I. Bottom traces, current recording. Note the strong inhibition of activity in 6Br4 in A even though Interneuron I is depolarized; the inhibition is also evident in B. Therefore, even strong depolarization of Interneuron I during its hyperpolarizing phase does not excite the flexor motoneurons. (5A from Pearson and Fourtner, J. Neurophysiol. 38, 1975.)

neurites of the flexor motoneurons (Pearson and Fourtner, 1975). However, the soma was located on the ventral contralateral surface of the ganglion. Interneuron I had the same location and major branching patterns in all preparations. No axonal processes leaving the ganglion via the anterior or posterior connectives or via any of the segmental nerves were observed in the above experiment. Therefore, it appears as though Interneuron I is an intraganglionic interneuron.

Interneuron II also oscillated during rhythmic leg movements in such a manner that the hyperpolarizing phase coincided with the flexor burst; the depolarizing phase with the extensor burst (see Figure 4B). In quiescent organisms, a depolarizing pulse would inhibit ongoing activity in the flexor motoneuron whereas a hyperpolarizing pulse would initiate activity. So Interneuron II displays characteristics physiologically antagonistic to those of Interneuron I. However, depolarization of Interneuron II does not appear to excite the extensor motoneuron D_s. Unfortunately, there were no experiments in which rhythmic leg movements occurred and in which short pulses could be injected. Therefore, it is impossible to state specifically whether Interneuron II is a part of the rhythm generating system. Furthermore, morphologically, identification of Interneuron II has not been accomplished. In fact Interneuron II might be a small class of intraganglionic nonspiking neurons (Pearson and Fourtner, 1975).

As stated above one of the interesting characteristics of the nonspiking interneurons is that the genesis of action potentials is not necessary for their proper functioning. The nonspiking interneurons exert their effect via very small changes in their membrane potential level which apparently modulate the amount of transmitter released at the synaptic terminals. Perhaps this system has evolved due to the close proximity of the input and output regions of these neurons.

This type of slow graded (nonspiking) activity is found in a variety of receptors as well as in some second order sensory neurons in the invertebrate and vertebrate visual systems (for review see Pearson, 1975). A nonspiking neuron has been physiologically identified in the crustacean respiratory system (Mendelson, 1971). A single nonspiking neuron functions as an oscillator and generates the total respiratory motor output. Recently, in a study of the stomatogastric ganglion (a ganglion associated with the digestive system in crustaceans) of the lobster, Maynard and Walton (1975) suggested that slow graded potentials in spiking neurons play an important role in synaptic transmission. Maynard (1972) further has suggested that the origin of the pyloric rhythm is the result of these slow potentials rather than a function of the spiking output. These studies indicate that a considerable amount of information is transmitted between neurons by graded potentials rather than by the classically accepted mechanism of a neural code produced by trains and volleys of action potentials.

Command Integrating Interneuron

In other experiments, Pearson and Fourtner (Unpublished observations) have recorded from another type of nonspiking element within the metathoracic ganglion of the cockroach. This element appears to oscillate very weakly (total amplitude less than 2 mV) in phase with Interneuron I. That is, its depolarizing phase is synchronous with the flexor motoneuron burst. However, depolarization of the neuronal element does not activate the flexor motoneurons in the typical fashion displayed by Interneuron I. This neural element instead produces reciprocal bursts in the extensor motoneurons and flexor motoneurons. Figure 6 is an example of one such sequence. A long duration pulse of 15 nA produced a sequence of rhythmic leg movements, and in each of the three different experiments the depolarization produced an initial activation of the extensor motoneuron. Clearly this neural element has quite different physiological characteristics from those described for Interneuron I. Unfortunately, penetration of these cells has not been maintained long enough to use different pulse amplitudes. This method would test if the level of membrane potential of this neuronal element could control the frequency of rhythmic leg movement. Morphological identification of this neuron by iontophoresis of cobalt has not been done. It is of interest, however, that a similar neuronal element has been found in the central pattern generator of the respiratory rhythm in the crab. Mendelson (1971) in his work on decapod crustaceans penetrated a neuron which displayed weak oscillations during rhythmic bursts of the respiratory motoneurons. When depolarized, this cell would produce rhythmic activity similar to that seen in the normal respiratory cycle. Mendelson demonstrated also that there seems to be a single abortive spike or in some cases, a single action potential following depolarization of this neuronal element. He assumed that he had penetrated the axon terminal of a command unit receiving a feedback from the oscillator. The neuronal element illustrated in Figure 6 could be a terminal of a command fiber, but there was no evidence of spontaneous action potentials in the cell and no action potentials were evoked by direct stimulation of the cell. Therefore, this neural element may be an intraganglionic interneuron similar to others described by Pearson and Fourtner (1975). This interneuron could receive input from the command system as well as tonic and phasic sensory input from segmental regions. It may convert the various spiking inputs into graded outputs which directly activate the rhythm pattern generator circuit producing rhythmic leg movements.

Model for the Walking Rhythm Pattern Generator

Figure 7 is a hypothetical model which incorporates several new findings concerning the central control of rhythmic leg

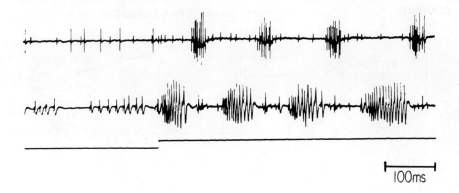

Figure 6. Recording of reciprocal burst activity produced by depolarization of a nonspiking neuron in the metathoracic ganglion of the cockroach. Top trace, extracellular recording from 6Br4. Middle trace, EMG of extensor muscle activity. Bottom trace, recording of current applied to nonspiking neuron. Note that the initial burst in the sequence of reciprocal activity occurs in the extensor muscle. This phenomenon is similar to the initiation of spontaneous rhythmic leg movements in intact beasts (cf. Figure 2). The sequence of activity continued only while the neuron was depolarized. Current 15 nA.

movements in the cockroach. This particular model is based upon the original model proposed by Pearson and Iles, 1970 (see Figure 3) in which they suggested that a bursting interneuron produced the bursts recorded in the flexor motoneurons. From intracellular recordings of several neurons within the metathoracic ganglion of the roach, it appears that the central pattern generator of rhythmic leg movements consists of a small network of at least two nonspiking interneurons rather than a single pacemaker neuron (Pearson and Fourtner, 1975).

The following experimental evidence strongly suggests that Interneuron I plays an important role in the genesis of rhythmic leg movements for the following reasons: 1) strong oscillations of the membrane potential occur during rhythmic leg movements; 2) there is selective excitation of the motoneuron pool producing flexion of the coxa-femur and the femur-tibia joints; 3) and, most importantly, applied membrane depolarizations reset the timing of rhythmic leg movements (Pearson and Fourtner, 1975). Other evidence leads to the conclusion that Interneuron I is not the only neuron in the central pattern generator. First of all, lengthy depolarizations of Interneuron I never produced oscillations in Interneuron I nor rhythmic activity in the flexor and extensor motoneurons. Secondly, there is a very strong inhibition of the

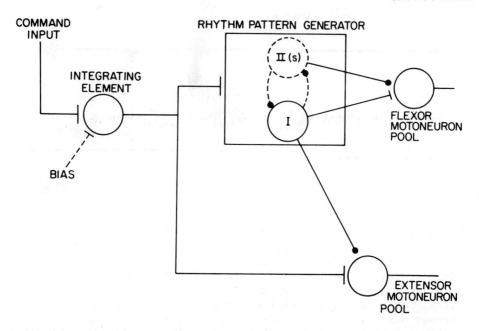

Figure 7. Hypothetical model of the central mechanism producing rhythmic leg movements in the cockroach. The broken lines suggest very tentative proposals since there is no direct evidence that Interneuron II is part of the rhythm pattern generator, nor is there evidence that Interneurons I and II are linked via reciprocal inhibition. The bias input in the integrating element is input from segmental afferents and other central neurons, for example, coordinating fibers (Stein, 1974). Lines ———⊣ designate excitatory connections and ———● designate inhibitor connections.

flexor motoneurons during the hyperpolarizing phase of the oscillations in Interneuron I; large depolarizing currents applied to Interneuron I during this hyperpolarizing phase could not excite the flexor motoneurons. Thirdly, the depolarizing phase of Interneuron I apparently is terminated by a number of large IPSPs or by some type of active hyperpolarization. It may be that depolarization in another nonspiking interneuron (perhaps Interneuron II) directly inhibits Interneuron I. In their original model, Pearson and Iles argued that D_s is active tonically and is inhibited strongly by the pattern generator; the results of Pearson and Fourtner, (1975) indicate that Interneuron I produces at least a weak inhibition of the extensor motoneuron D_s.

Interneuron II tentatively has been placed within the pattern generator for the following reasons. 1) it oscillates at the same frequency as observed rhythmic leg movements; 2) its oscillations are 180° out of phase with those of Interneuron I and 3) it strongly inhibits the flexor motoneurons. This inhibition may be direct onto the flexors or it may be indirect by Interneuron I. Definitive resetting experiments have yet to be attempted in order to discern if Interneuron II is part of the pattern rhythm generator or simply an inhibitory premotor neuron. The reciprocity between Interneurons I and II suggests a reciprocal inhibitory network between Interneuron I and Interneuron II similar to models of central pattern generators hypothesized for other systems (Wilson and Waldron, 1968; Maynard, 1972; Selverston, this volume).

The synaptic connections between nonspiking interneurons and those between nonspiking interneurons and motoneurons have not been morphologically and physiologically identified. It must be emphasized that the model of the rhythm pattern generator illustrated in Figure 7 is simply a working hypothesis.

The preliminary results on the command integrating element shown in Figure 6 also are incorporated in the model. Pearson and Iles (1970) proposed that command input directly activated the interneuronal pattern generator and tonically excited the extensor motoneuron D_s. In the model proposed here a neural element is placed between the command input and the rhythm generator and D_s. The element suggested is a nonspiking interneuron which would convert and integrate the spiking input from the command fibers as well as other neural input into a graded potential. The graded output from the integrating interneuron would then activate the rhythm generator producing the flexor bursts as well as tonically activate the extensor motoneurons. Membrane depolarization experiments indicate that long term depolarizations of the integrating element lead to initial excitation of the extensor motoneurons before excitation of the flexor motoneurons. The initial excitation of the extensor motoneuron can be accomplished by a direct pathway to the extensor motoneuron whereas excitation of the flexor motoneurons would be retarded due to the initial delay in activating the rhythm pattern generator.

Without morphological identification, it is difficult to state whether this integrating element is a single nonspiking interneuron or a region of Interneuron I located some distance from the output region to the flexor motoneurons (the integrating element does weakly oscillate in a mode similar to Interneuron I). However, the physiological evidence, that depolarization of the integrating element produces an initial excitation of D_s, indicates that the integrating element and Interneuron I are different neurons, and that some element of the rhythm pattern generator feeds back onto the integrating element.

Conclusion

The motor pattern producing cockroach walking is generated by an asymmetric rhythm pattern generator which drives the flexor (swing phase) motoneuron pool and inhibits tonic activity in the extensor (stance phase) motoneuron pool.

A hypothetical model is proposed to describe some of the more recent results obtained from intracellular studies on insect central nervous systems.

Acknowledgment

I should thank E. Golder for comments on the manuscript. Supported by N.S.F. Grant #BMS 74-22818.

REFERENCES

Barnes, W.J.P., Spirito, C.P. and Evoy, W.H., (1972) Nervous control of walking in the crab, Cardisoma guanhumi. II. Role of resistance reflexes in walking. Z. vergl. Physiol. 76, 16-31.

Bentley, D.R., (1969) Intracellular activity in cricket neurons during the generation of behavior patterns. J. Insect. Physiol. 15, 677-700.

Burns, M.D., (1973) The control of walking in orthoptera. I. Leg movements in normal walking. J. Exp. Biol. 58, 45-58.

Burrows, M., (1974) Modes of activation of motoneurons controlling ventilatory movements of the locust abdomen. Trans. R. Soc., Lond. B. 269, 29-48.

Davis, W.J., (1971) Functional significance of motoneuron size and soma position in the swimmeret system of the lobster. J. Neurophysiol. 34, 274-288.

Davis, W.J. and Kennedy, D., (1972) Command interneurons controlling swimmeret movements in the lobster. I. Types of effects on motoneurons. J. Neurophysiol. 35, 1-12.

Delcomyn, F., (1971) The locomotion of the cockroach, Periplaneta americana. J. Exp. Biol. 54, 443-352.

Delcomyn, F., (1973) Motor activity during walking in the cockroach Periplaneta americana. II. Tethered walking. J. Exp. Biol. 59, 629-642.

Delcomyn, F. and Usherwood, P.N.R., (1973) Motor activity during walking in the cockroach Periplaneta americana. I. Free walking. J. Exp. Biol. 59, 643-654.

Evoy, W.H. and Fourtner, C.R., (1973) Nervous control of walking in the crab, Cardisoma guanhumi. III. Proprioceptive influences on intra- and intersegmental coordination. J. comp. Physiol. 83, 303-318.

Graham, D., (1972) A behavioral analysis of the temporal organization of walking movements in the 1st instar and adult stick insect (Carasius morosus). J. comp. Physiol. 81, 23-52.

Grillner, S., (1975) Locomotion in vertebrates: central mechanisms and reflex interactions. Physiol. Rev. 55, 247-304.

Goslow, G.E., Reinking, R.M. and Stuart, D.G., (1973) The cat step cycle: hind limb joints, angles and muscle lengths during unrestrained locomotion. J. Morph. 141, 1-41

Henneman, E., Somjen, G. and Carpenter, D.O., (1965) Functional significance of cell size in spinal motoneurons. J. Neurophysiol. 28, 559-620.

Hinkle, M. and Camhi, J.M., (1972) Locust motoneurons: bursting activity correlated with axon diameter. Science. 175, 553-556.

Hoyle, G. and Burrows, M., (1973) Neural mechanisms underlying behavior in the locust Schistocerca gregaria. II. Integrative activity in metathoracic neurons. J. Neurobiol. 4, 43-67.

Hughes, G.M., (1965) "Locomotion: terrestrial," In The Physiology of the Insecta. (Rockstein, M., ed.), Academic Press, New York, (1-94).

Kater, S.B. and Nicholson, C., (1973) Intracellular Staining in Neurobiology. Springer-Verlag, New York.

Maynard, D.M., (1972) Simpler networks. Ann. N.Y. Acad. Sci. 193, 59-72.

Maynard, D.M. and Walton, K.D., (1975) Effects of maintained depolarization of presynaptic neurons on inhibitory transmission in lobster neuropil. J. comp. Physiol. 97, 215-243.

Mendelson, M., (1971) Oscillator neurons in crustacean ganglia. Science. 171, 1170-1173.

Pearson, K.G. and Iles, J.F., (1970) Discharge patterns of coxal levator and depressor motoneurones of the cockroach *Periplaneta*. J. Exp. Biol. 52, 139-165.

Pearson, K.G., (1972) Central programming and reflex control of walking in the cockroach. J. Exp. Biol. 56, 173-193.

Pearson, K.G. and Iles, J.F., (1973) Nervous mechanisms underlying intersegmental coordination of leg movements during walking in the cockroach. J. Exp. Biol. 58, 725-744.

Pearson, K.G., Fourtner, C.R. and Wong, R.K., (1973) "Nervous control of locomotion in the cockroach," In Control of Posture and Locomotion, (Stein, R.B. et al., ed.), Plenum Press, New York, (495-514).

Pearson, K.G., (1975) Nerve cells without action potentials. (In press).

Pearson, K.G. and Fourtner, C.R., (1975) Nonspiking interneurons in the walking system of the cockroach. J. Neurophysiol. 38, 33-52.

Reingold, S.C., (1975) Developmental and short term behavioral plasticity in the cockroach, *Periplaneta americana*. Ph.D. Dissertation, Cornell University.

Stein, P.S.G., (1974) Neural control of intra-appendage phase during locomotion. Amer. Zool. 14, 1003-1016.

Wilson, D.M., (1966) Insect walking. Ann. Rev. Enton. 11, 103-123.

Wilson, D.M. and Waldron, I., (1968) Models for the generation of the motor output pattern in flying locusts. Proc. I.E.E.E. 56, 1058-1064.

NEURAL CONTROL OF FLIGHT IN THE LOCUST

M. Burrows

Department of Zoology, University of Cambridge

England

The wings of locusts are moved by two sets of synchronous muscles, elevators and depressors, innervated by about 80 identifiable motoneurons with somata in the thoracic ganglia. Within these ganglia is the basic pattern generator for flight. In isolation they can be made to produce a pattern of alternating elevator and depressor motoneuron spikes. Attention is focused upon the elements of the pattern generator and upon the effects of sensory feedback. Interconnections between the motoneurons alone are not responsible for the generation of the flight pattern. To get information about the interneurons which are responsible, intracellular recordings are made from several motoneurons at once to allow the identification of any common synaptic inputs. Two interneurons synapse upon most elevator and depressor motoneurons of both sets of wings and upon spiracular closer and opener motoneurons of each thoracic segment. These interneurons carry information about two rhythms, a slow rhythm of 0.1-0.5 Hz which is in time with ventilation, upon which is superimposed a faster one of 20 Hz, a frequency similar to that of the wingbeat in flight. In spiracular motoneurons, the synaptic input is suprathreshold so that the spiracular motoneurons may express both rhythms in opening and closing the spiracles. In the flight motoneurons it is subthreshold but sums with unpatterned inputs to evoke motoneuron spikes with a periodicity similar to that in flight.

Movements of the wing are monitored by a variety of sense organs. Of these the single celled stretch receptor which responds to elevation of the wing, can be identified. It makes connections with ipsilateral flight motoneurons, exciting depressors and inhibiting elevators so forming a negative feedback loop. A second feedback loop activated by depression of the wing causes excitation of ipsilateral and contralateral elevators and inhibition of depressors. The effects of both loops are usually subthreshold but sum with other inputs to evoke motoneuron spikes. In flight they would be expected to stabilise the amplitude of the wing movements.

Introduction

Locusts fly by flapping two pairs of wings at a frequency of about 20 Hz. Each wing is moved by about 10 power producing muscles arranged in two sets, elevators and depressors. There are about 20 motoneurons to the muscles of each wing of which some 15 can already be recognised as individuals, either by recording their spikes in the muscles, or by recording their synaptic inputs and spike outputs in their somata. The somata are 40-80 μm in diameter and lie superficially in the ventral cortex of each of the three thoracic ganglia. The ganglia themselves are linked by paired connectives. The movements of the wings are monitored by a variety of sense organs both upon the surface of the wings and at their hinge with the thorax. Some of these sense organs can be identified as individuals in recordings made from the main wing nerve and can be manipulated independently. Within the thoracic ganglia is a pattern generating mechanism for flight. When isolated from afferents and the cerebral ganglia, this mechanism produces a repetitive and alternating output in elevator and depressor motoneurons in response to a random input (Wilson, 1961; Wilson and Wyman, 1965).

Flight of the locust therefore provides an opportunity to study the mechanisms of a central pattern generator, the co-ordination of different appendages and the role of afferents when both the central and peripheral neural elements can be observed individually. Although stated as separate mechanisms it is unlikely they can be divorced either from each other, or from the mechanisms involved in the co-ordination of other movements which the animal must perform. There are two ways to approach these problems both of which require extrapolations of data. First, the locust's behavior is allowed to proceed almost without impediment while recordings are made peripherally of motor spike sequences or muscle potentials. Particular peripheral elements then are ablated and any resulting modifications of the patterns noted.

From the observed patterns, inferences are made about the components and their interconnections within the central nervous system that might have generated these patterns. This has been the preferred method since the inaugural work of Wilson (1961). This method describes well the patterns of motor and some sensory spikes during flight but does not explain the mechanisms. The second method is to record from elements within the central nervous system with microelectrodes but in a preparation whose behavioural repertoire is more restricted. Particular central components then can be manipulated separately and any pathways between them demonstrated. Assumptions necessary to this method are: 1) that the observed pathways in fact are used during normal behaviour, and 2) that their effect in the normal is identical to the experimental situation. This is the method favored at present by this author because it has the potential for describing the actual mechanisms. The limitations to observing normal behaviour gradually will be reduced. Presently it is possible to record from insects while patterns of behaviour such as ventilation, kicking and even sequences of flight-like activity proceed. It would seem sensible to define the components involved in the co-ordination of flight from recordings within the central nervous system. These definitions then would serve as a framework for understanding the results obtained in the periphery.

Sensory Feedback Loops

The role of afferents in the co-ordination of flight was not investigated intensively until recently. This is because a narrow limit is imposed upon experimental perturbations of the normal pattern in order to reveal the effects of afferents which are present normally. Other types of experiments may demonstrate different effects of afferents but do not indicate the role in normal behaviour. For example, each wing base contains a one-celled sense organ, the stretch receptor which responds to elevation of the wing and in flight spikes a few times towards the end of each upstroke. The spike of the stretch receptor can be readily recorded in the wing nerve (N1) because of the large (6 µm) diameter of its axon. Removal of the sense organs from the base of three or four wings caused a 1/2 reduction in wing beat frequency and it was assumed the stretch receptor alone was responsible for the effect (Wilson and Gettrup, 1963). Stimulation of the nerve containing the stretch receptor, a procedure which probably activates many other afferents, raises the wingbeat frequency of a partially isolated central nervous system with a time constant of 1-2 s. This stimulation does not entrain the motor output (Wilson and Wyman, 1965). The stretch receptor was never manipulated on its own. It was suggested that it did not have a cycle by cycle influence on the flight pattern, but merely raised the

overall frequency. Therefore the phasic information provided by
the receptor was ignored by the central nervous system according
to Wilson and Wyman (1965). In fact, the phasic information
provided by the array of receptors of the wing is not ignored.
The effect on the central nervous system can be observed by moving
one wing of a flying locust at a frequency ± 13 percent of the
natural frequency (Wendler, 1974). The other wings adopt the
imposed frequency within one or two cycles. Other indications of
the effects of afferents follow. First, removal of sense organs
from the wings is associated with an increased variance in the
intervals between the spikes of antagonistic motoneurons (Kutsch,
1974). Secondly, the flight pattern of newly moulted adults that
are allowed to move their wings is less regular than in those with
fixed wings (Altman, 1975). Possibly the soft wings do not move
as expected and therefore provide an afference which is in conflict
with the centrally generated motor pattern. Therefore one must
consider the connections made between afferents and neurons in the
central nervous system. The wing stretch receptor is of use in
determining what connections an identified sensory neuron makes
with known motoneurons.

Stretch Receptors

The stretch receptor has a complex anatomical projection with
a profusion of axonal branches in the ipsilateral halves of all
three thoracic ganglia (Burrows, 1975a). Stimulation of the nerve
containing the axon of a forewing stretch receptor evokes an excitatory post-synaptic potential (EPSP) in the first basalar, a depressor motoneuron, and an inhibitory post-synaptic potential
(IPSP) in the tergosternal, an elevator motoneuron of the same wing
(Figure 1a). These potentials are associated with the spike of a
large axon, attributable to the stretch receptor (Figure 1b). The
EPSP in the depressor motoneuron follows each spike of the stretch
receptor with a constant latency of 1.0 ms and at frequencies of
up to 125 Hz. The connection is probably monosynaptic, and also
chemically mediated (Burrows, 1975a). The IPSP in the elevator
follows with a delay of 4-5 ms so that the connection may or may
not be monosynaptic. Upon elevation of the wing the frequency of
spikes of the stretch receptor increases, the depressor motoneuron
becomes depolarized (Figure 1c) and the elevator motoneuron becomes
hyperpolarized (Figure 1e). The EPSP's in the depressor decrease
in amplitude probably for a variety of reasons, one of which may
be fatigue. The first EPSP's of a group elicited by electrical
stimulation are always the largest (Figure 1d). This could provide
an amplification mechanism at the start of flight when the initial
elevation evokes large EPSP's in the depressor and could help
ensure that the alternating pattern is established quickly.

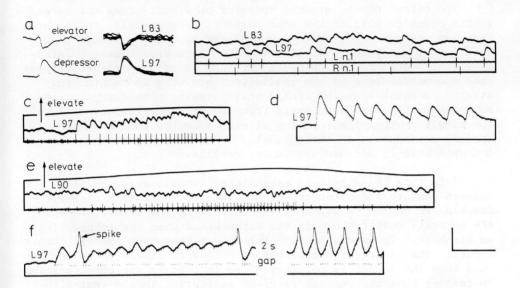

Figure 1. The negative feedback loop formed by the stretch receptor of a wing. (a) An electrical stimulus to a forewing nerve (N1) evokes an IPSP in an elevator (1st tergosternal, 83) and an EPSP in a depressor motoneuron (1st basalar, 97) of that wing. (b) These potentials are caused by the spike of the axon of the stretch receptor. The spike of the contralateral stretch receptor apparently evokes no potentials. (c) Elevation of the wing (upwards on first trace) causes an increase in the frequency of stretch receptor spikes. The EPSP's in a depressor summate and decrease in amplitude. (d) The first EPSP's evoked by a sequence of electrical stimuli are always of larger amplitude. (e) Upon elevation of the wing an elevator motoneuron (1st posterior tergocoxal) receives a barrage of IPSP's from the stretch receptor which hyperpolarize it. (f) Simulation of the pattern of spikes of the stretch receptor during flight; four stimuli at 100 Hz are repeated at intervals of 60 ms, the approximate wingbeat period. Waves of depolarization are evoked in a depressor motoneuron which can lead to spikes on each group of stimuli. Calibration: vertical (a,c,d) 3 mV; (b,e,f) 10 mV; wing movement 80°; horizontal (a) 130 ms; (b-f) 200 ms.

Similar connections are made by the stretch receptor of each wing with motoneurons innervating muscles of the appropriate wing. Elevators are inhibited and depressors are excited. Elevation of a wing activates a negative feedback loop which then inhibits the power producing elevators and excites the depressors, thereby terminating elevation.

The effect of the stretch receptor apparently does not spread to the opposite wing of the same segment. For example, the right forewing stretch receptor does not appear to evoke potentials in left forewing motoneurons (Figure 1b). There are, however, interganglionic connections whereby a forewing stretch receptor synapses upon the motoneurons of the ipsilateral hindwing and a hindwing stretch receptor upon the ipsilateral forewing motoneurons. Some motoneurons thus receive inputs from two stretch receptors. In all the single neuron of a forewing stretch receptor synapses upon some 17 motoneurons in three ganglia and a hindwing stretch receptor upon some 15 motoneurons in two ganglia.

This type of experiment establishes functional connections between sensory and motoneurons but gives no indication of their function in normal behaviour. The effects of the stretch receptor are normally subthreshold in the motoneurons when the locust lies on its back. To have a behavioural effect they must sum with other inputs. The amount of other sensory inflow is likely to be much less when the locust is not flying, so that it is not unreasonable to suppose that the general level of excitation in the central nervous system will be lower. A loop of this type might be expected to have a cycle-by-cycle influence on the timing of the motor spikes in flight. That it could have such an effect can be shown by applying a d.c. depolarization to a depressor motoneuron, which will simulate an unpatterned input from other neurons and superimposing upon this a pattern of stretch receptor spikes simulating their pattern in flight. The EPSP's from the stretch receptor can then evoke spikes in the depressor motoneuron and entrain those spikes at frequencies similar to the wingbeat in flight (Figure 1f).

Depression Receptors

Although the stretch receptor is the only sensory neuron of the wing identified with certainty, it is not the only receptor to have an effect upon flight motoneurons (Figure 2). Stimulation of a wing nerve (N1) with increasing intensity first activates the stretch receptor. Then it evokes potentials of opposite sign to those which have been caused by the stretch receptor (Figure 2a). The axons of the receptors mediating these effects can be stimulated selectively in a branch of the wing nerve, N1C (Figure 2b). Most axons in this nerve spike in response to a depression of the wing. It is probable that depression sensitive receptors mediate the effects observed in the flight motoneurons. Many axons contribute to the effect as the amplitude of the EPSP in the elevator increases with the increase in stimulus intensity. Eventually a spike may be evoked (Figure 2c). Since many unknown sensory neurons seem to be involved, the pathways which evoke the potentials

PATTERN GENERATION IN LOCUST

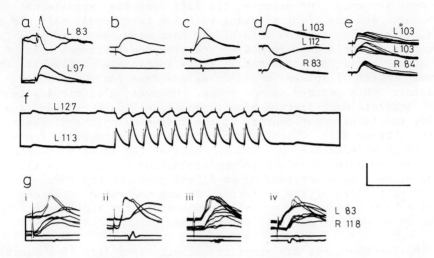

Figure 2. The feedback loop formed by wing receptors with axons in Nlc. (a) A single stimulus is applied to the whole left wing nerve (N1). At low strengths the large axon of the stretch receptor is stimulated evoking an EPSP in a left depressor (1st basalar, 97) and an IPSP in a left elevator (1st tergosternal, 83). As the stimulus intensity is increased the amplitude of the EPSP in the depressor decreases and a spike occurs in the elevator. (b) The axons mediating these new effects can be stimulated on their own in Nlc; they evoke an EPSP in elevator and an IPSP in depressor motoneurons. (c) The EPSP may lead to a spike in the elevator which can be monitored in the muscle. (d,e) Three motoneurons simultaneously are penetrated to show the widespread effects of these afferents. There are contralateral effects. (f) Single stimuli repeated at 60 ms intervals and delivered to Nlc of the left hindwing evoke IPSPs in a left depressor (1st basalar, 127) and spikes in a left elevator (tergosternal, 113). (g) The left forewing tergosternal (L83) and the right hindwing anterior tergocoxal (R118) motoneurons are penetrated and each of the four wing nerves is stimulated in turn: (i) right hindwing, (ii) left hindwing, (iii) right forewing, (iv) left forewing. Afferents of each wing affect both motoneurons. Myograms are on the lower two traces. Calibration: vertical (a,d,e) 3 mV; (b,c,f,g) 10 mV; horizontal (a,d,e) 65 ms; (b,c) 80 ms; (f) 200 ms; (g) 40 ms.

in the motoneurons cannot be determined precisely. Present data seem to indicate the involvement of interneurons. The effects of these sensory neurons are widespread. They evoke EPSP's in most ipsilateral elevators; IPSP's in most depressors (Figures 2d, e). They also affect contralateral flight motoneurons of the same or of the next segment. For example, the left forewing tergosternal motoneuron and the right hindwing anterior tergocoxal, both elevators, receive inputs from each of the four wing nerves (Figure 2g). Therefore the afferents in N1C of one wing can influence most of the locust's flight motoneurons. It is impossible to simulate the normal pattern of spikes in flight as was done for the stretch receptor. This pattern is not known. However, electrical stimuli which activate many neurons simultaneously will evoke repetitive spikes in elevator motoneurons at the frequency of the wingbeat in flight (Figure 2f). There is therefore a second negative feedback loop whose effects are opposite to those of the stretch receptor and which is able to influence the production of spikes in flight motoneurons. An understanding of flight must include these loops as integral parts of the pattern producing mechanism. These and possibly other loops are probably responsible for the phasic effect revealed by Wendler's (1974) experiments.

Moving the wings may sometimes elicit other less reproducible effects. Variability may occur because the effects are mediated by interneurons whose thresholds are only occasionally exceeded in the type of preparation used. Although examining them is difficult, their effects should not be ignored. Elevation of a forewing may evoke spikes in the ipsilateral hindwing subalar muscle so that the hindwing is depressed (Figure 3a). This response is not caused by the forewing stretch receptor although it is activated by this movement. It apparently does not synapse upon the hindwing subalar motoneuron. There are thus effects which spread from fore- to hindwing. Elevation of the right forewing evokes reciprocal potentials in the left hindwing motoneurons so that additional effects spread from one side of the body to the other (Figure 3b). EPSP's occur in depressor motoneurons which correspond with IPSP's in elevators. Sometimes spikes may be evoked in both an elevator and a depressor motoneuron and these alternate as they would in flight (Figure 3c, d). Still more rarely a sequence of alternating elevator and depressor spikes may be evoked at the same frequency and with similar phasing as that which occurs in normal flight (Figure 3e). In this way afferents from the wing co-ordinate or perhaps trigger a patterned output by the central nervous system. There is therefore a considerable repertoire of sensory effects which could influence flight. This is not surprising because efficient flight requires that the motor output be matched both to the vagaries of the environment and to the changes in the configuration of the body.

Common Synaptic Driving of Motoneurons

Experiments performed in the periphery have given scanty information about the neurons of the central nervous system which

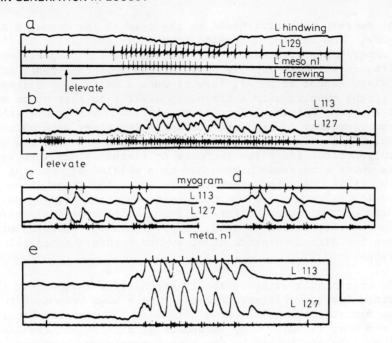

Figure 3. Other effects of afferents. (a) Elevation of the left forewing evokes spikes in the left hindwing subalar muscle (129, second trace) which depresses the hindwing. The forewing stretch receptor (third trace) does not appear to directly mediate the effect. (b-e) Elevation of the right forewing evokes reciprocal responses in elevator and depressor motoneurons of the left hindwing. (b) There are EPSP's in the depressor (1st basalar, 127) which coincide with IPSP's in the elevator (tergosternal, 113). (c,d) Spikes may occur alternately in both motoneurons. (e) Elevation of the forewing evokes a sequence of alternating spikes in the elevator and depressor motoneurons which resembles their pattern of activity during flight. Calibration: vertical (b) 8 mV; (c-e) tergosternal 113 15 mV, 1st basalar, 127 10 mV; wing movement 80°; horizontal (a) 200 ms, (b-e) 100 ms.

comprise the pattern generating mechanism. It is clear that interneurons are involved because appropriate connections between the motoneurons themselves do not exist. Peripheral experiments unfortunately have led to confusion about the properties of the central pattern generator. First, it has become an assumption (cf. Altman, 1975) that the frequency of the central pattern generator is only half the normal flight frequency. This is the frequency of the repetitive output produced by an isolated thoracic nervous system or of a locust devoid of most of its sensory input from the wings. It is worth considering that the frequency of the central pattern generator could be the same as the normal wingbeat frequency. The lack of afference or general excitation in the above experimental procedures may prevent the full expression of this frequency. When wind is blown at the head of nymphal locusts, whose wings cannot be moved, they produce a pattern of spikes in their flight muscles with a frequency similar to that of an adult flight pattern (Kutsch, 1974; Altman, 1975). Furthermore, the wingbeat frequency of an adult gradually increases with maturation (Kutsch, 1971). This was attributed to a maturation of the central pattern generator since the increase in frequency was observed in locusts whose wing sense organs had been ablated at moulting (Kutsch, 1974). Another experiment in which the wings of newly moulted adults were fixed and then released after varying periods indicated that different afference from the progressively hardening cuticle was responsible (Altman, 1975). This confusion points to the need for direct evidence about events within the central nervous system.

To obtain such evidence about interneurons involved in flight, recordings from several motoneurons at once have been made to observe synaptic potentials common to each (Burrows and Horridge, 1974; Burrows, 1975b). Long sequences of such potentials lead one to assume they are evoked by the same interneuron. This procedure allows a description of the pattern of connections of the inferred interneurons and reveals functional properties of the interneurons which probably would not otherwise have come to light.

The left and the right tergosternal (elevator) motoneurons of the forewing undergo a slow rhythmic depolarization in time with the expiratory phase of the ventilatory cycle (Figure 4a). The depolarization is subthreshold. Only during flight, in a dissected preparation or in a normal animal, do the flight muscles contribute force to ventilation. The synaptic potentials in the two motoneurons are matched during the expiratory depolarization but there are synaptic potentials which occur independently. There is no evidence for electrical coupling between the two motoneurons. The common synaptic potentials occur also in ipsilateral fore- and hindwing tergosternal motoneurons (Figure 4b) and in the left and right hindwing tergosternal motoneurons (Figure 4c). A recording made simultaneously from three of these motoneurons shows that their major depolarizing synaptic inputs are the same (Figure 4d). During the expiratory depolarization the membrane potential of the motoneurons undergoes rapid oscillations. These ripples consist of groups of EPSP's and may occur with a regular periodicity of about 50 ms. This is similar to the frequency of wingbeats in flight. There are thus two rhythms of synaptic potentials in these flight motoneurons; a slow rhythm in time with ventilation and a fast rhythm whose period is similar to that of flight. The slow rhythm is affected by those stimuli which affect ventilation and never

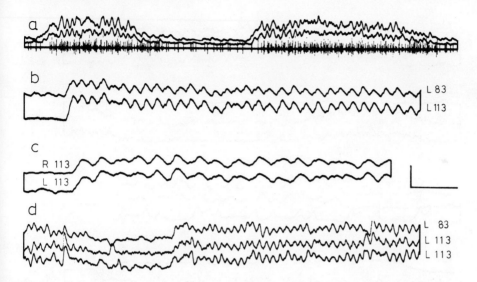

Figure 4. Patterns of synaptic potentials in flight motoneurons evoked by interneurons. (a) The left (first trace) and the right (second trace) forewing tergosternal motoneurons are depolarized rhythmically in time with ventilation (third trace shows burst of spikes in prothoracic N6 to muscles of the neck involved in head pumping). (b) The depolarization during expiration is subdivided into ripples with a period similar to that of the wingbeats in flight. The synaptic potentials in the left forewing (L83) and left hindwing (L113) tergosternal motoneurons are matched. (c) The left and right hindwing tergosternal motoneurons also have synaptic potentials in common. (d) A simultaneous recording from three elevator motoneurons of different wings to show that they receive inputs from the same interneurons. Calibration: vertical (a) 4 mV; (b,c) 5 mV; (d) 3 mV; horizontal (a,d) 400 ms; (b) 200 ms; (c) 100 ms.

occurs in the absence of ventilation. Slow and ventilatory rhythms are one and the same rhythm.

The distribution of these rhythms in other flight motoneurons can be assessed by recording from groups of motoneurons, one of which is known to receive the inputs. The rhythms are found in motoneurons of each of the three thoracic ganglia (Figure 5a). They occur in some depressor motoneurons, for example the forewing dorsal longitudinals (Figure 5a) but apparently not in the hindwing basalars (Figure 5b). In all some 30 flight motoneurons share these synaptic potentials but they do not occur in leg motoneurons within the same ganglia (Figure 5c). The rhythms are thought of as

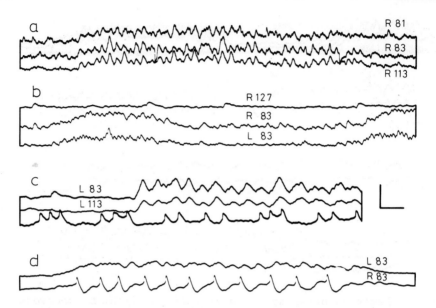

Figure 5. Sphere of influence of the co-ordinating interneurons.
(a) Flight motoneurons in each of the three thoracic ganglia
receive inputs; prothoracic, forewing dorsal longitudinal (depressor, R81); mesothoracic, forewing tergosternal (elevator, R83);
metathoracic, hindwing tergosternal (elevator, R113). (b) Some
depressors for example, the hindwing basalar (R127, first trace)
do not apparently receive inputs. (c) Leg motoneurons, for
example the slow depressor tarsus (third trace) also do not receive
inputs. (c) The inputs to elevator motoneurons can evoke spikes.
The right forewing tergosternal (second trace, R83) is depolarized
by an applied d.c. current and spikes during the expiratory depolarization caused by the interneurons. Calibration: vertical
(a,b) 3 mV; (c, traces 1 and 2, d) 5 mV; (c, trace 3) 10 mV;
horizontal: (a) 200 ms; (b-d) 100 ms.

real potentials caused by interneurons for the following reasons.
They are not movement artifacts caused by abdominal movements or
by air moving through the ganglionic tracheae. They persist when
the abdomen and tracheae are removed. They are not field potentials because their amplitude can be altered by changing the
membrane potential of a motoneuron. Moreover, they do not occur
in all motoneurons of one ganglion. The rhythms are not caused
by a rhythmic afference set up by the ventilatory movements
because they persist in an isolated thoracic nervous system. The
simplest explanation is that the potentials are caused by a pair
of symmetrically arranged interneurons which originate in the

metathoracic ganglion, each of which makes widespread and bilateral connections with the flight motoneurons. Each flight motoneuron thus receives an input from both interneurons. An EPSP in a motoneuron is presumed to be caused by a spike in one of the interneurons. Each interneuron conveys information about both rhythms. The slow ventilatory rhythm is coded in the overall burst of spikes during expiration and the fast rhythm in the interval between those spikes.

Do the Rhythms Have a Behavioural Role?

The rhythmic depolarizations of the flight motoneurons are usually subthreshold. If air is blown at the head the motoneurons are more likely to spike during expiration. Histograms of the intervals of the spikes during expiration then show peaks at 50 ms and multiples thereof. Similarly if a tergosternal motoneuron is depolarized with a d.c. current, spikes occur on each or on alternate oscillations of the membrane potential during expiration (Figure 5d). Therefore both slow and fast rhythms can be expressed as motoneuron spikes and must have a behavioural consequence under certain conditions.

The Slow Rhythm

In a locust which is not flying the flight muscles are not known to contribute force to ventilation although this does occur in some beetles (Miller, 1971). However, in addition to the flight motoneurons the interneurons also synapse upon many ventilatory motoneurons (Burrows, 1975c). Air enters the network of tracheae through spiracles whose apertures are typically controlled by a closer and an opener muscle. Air is then circulated around the body by the pumping movements of the abdomen. An intracellular recording from any of the closer motoneurons of the thoracic spiracles shows that they produce bursts of spikes during expiration and at the same time as the slow depolarization of the flight motoneurons (Figure 6a). The synaptic potentials which underly these spikes are matched with those in a tergosternal flight motoneuron so that they must be caused by the same interneurons (Figure 6b). The spikes in the spiracular closer motoneurons are governed by both rhythms; the overall burst of spikes by the slow rhythm and the grouping of the spikes within that burst by the period of the fast rhythm. The opener motoneurons of the thoracic spiracles spike during inspiration and receive a barrage of IPSP's during expiration (Figure 6c). These IPSP's match the EPSP's recorded in a tergosternal flight motoneuron (Figure 6d, e). Therefore the interneurons have reciprocal effects on antagonistic spiracular motoneurons, inhibiting the openers and exciting the closers. The

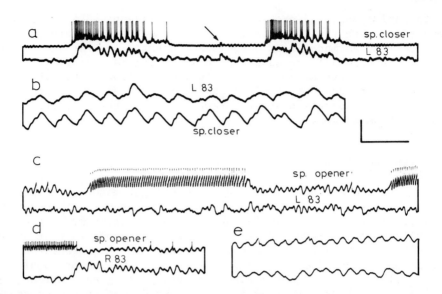

Figure 6. Co-ordination of spiracular motoneurons. (a) A <u>closer</u> motoneuron of the mesothoracic spiracle spikes during expiration and during the wave of depolarization that occurs in a flight motoneuron. The EPSP's in the two motoneurons are matched during expiration and one common EPSP occurs during inspiration (arrow). (b) The spikes in a metathoracic spiracular closer motoneuron are abolished by an applied d.c. hyperpolarization to reveal that its EPSP's match those in a flight motoneuron (forewing tergosternal, L83). (c) A metathoracic <u>opener</u> motoneuron spikes during inspiration and is inhibited by IPSP's during expiration. (d,e) The IPSP's in the opener motoneuron match the EPSP's in the flight motoneuron. Calibration: vertical (a) trace 1, 25 mV; trace 2, 5 mV; (b) trace 1, 3 mV; trace 2, 10 mV; (c-e) trace 1, 10 mV; trace 2, 5 mV; horizontal: (a,c,d) 400 ms; (b) 100 ms; (e) 200 ms.

responsibility for the co-ordination of the movements of the spiracles throughout the thorax must rest largely with these interneurons. In addition these interneurons synapse upon abdominal motoneurons responsible for normal pumping movements and upon motoneurons of neck muscles which participate in an auxiliary form of ventilation, head pumping. So far 20 ventilatory motoneurons have been shown to receive inputs from these interneurons. Together with the flight motoneurons, some 50 motoneurons each receive common synaptic potentials from these interneurons.

One function of the interneurons clearly is to assist in the co-ordination and patterning of the ventilatory movements. A function for the slow rhythm thus is established.

The Fast Rhythm

The similarity between the period of the fast rhythm and the wingbeat period is striking and suggests the possibility the two may be related causally. There is no direct evidence for this and the chance of finding the answer seems rather remote. It should be stressed that the rhythm merely may be the fortuitous consequence of the need for the interneurons to spike at some frequency. Nevertheless, the fast rhythm does occur in flight motoneurons and can under conditions which are likely to be encountered in flight, cause the motoneurons to spike at intervals similar to the wingbeat period. There are problems in accepting a causal relationship between the two rhythms. First, both elevators and some depressors receive common inputs though typically the amplitude of the potential in the latter is smaller. In flight their spikes obviously must alternate. Secondly, the input from the interneurons is intermittent; flight is continuous. Sometimes, however, the common EPSP's occur continuously, increasing in frequency during each expiration. This is precisely the pattern required during flight. Ventilation could proceed normally and yet there would still be a continuous input of the correct frequency to the flight motoneurons. It would seem a reasonable working hypothesis that these interneurons function in flight.

Interaction of Pattern Generators

Having established that the interneurons have a ventilatory function and a possible role in flight, there is another unanswered question. Why are the flight motoneurons depolarized in the ventilatory rhythm and the ventilatory motoneurons depolarized in the fast rhythm? Perhaps one set of neurons perform two functions for the sake of parsimony, although the implications appear more profound.

The grouping of spikes in the fast rhythm in spiracular closer motoneurons causes uneven movements of the spiracular valves but the effect on the movement of air is not known. The pattern of synaptic potentials in the closer motoneurons is an adequate explanation for their pattern of spikes. Therefore the interneurons must play a major role in ensuring that the spiracles close and open. During flight it is necessary that ventilation continues and that the spiracles open and close to some extent. Therefore it is a reasonable inference that these interneurons must be active during flight. If they are it might be expected that the flight rhythm would be modulated by the ventilatory rhythm because the flight motoneurons receive information about both rhythms. In tethered flight an adult female locust had a flight frequency of 17.8 Hz and its abdomen vibrated in time with the movements of the wing (Figures 7a, b). Some of the movement no doubt is due to mechanically transmitted vibrations from the thorax but some is due to the patterning of the spikes of the abdominal motoneurons in the flight rhythm (Camhi and Hinkle, 1972). The abdomen also moves with a slower rhythm, the ventilatory rhythm, upon which are superimposed the more rapid vibrations of the flight rhythm (Figures 7c, d). A sequential plot of the intervals between successive spikes of a hindwing elevator muscle shows that the intervals decrease at the start of inspiration and lengthen during expiration (Figures 7c, d). There is thus a clear rhythmic modulation of the

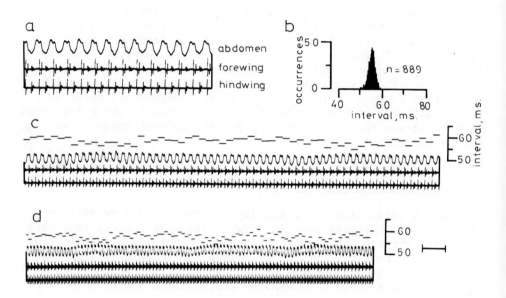

Figure 7. Modulation of flight by the ventilatory rhythm. (a) Vertical movements of the abdomen, and myograms of fore and hindwing elevator (large spike), and depressor muscles are recorded during tethered flight. (b) Histogram of the intervals between hindwing elevator spikes. 889 intervals were recorded in 50 s so that the flight frequency was 17.8 Hz. (c,d) Segments from the middle portion of a tethered flight which lasted for 10 mins. The traces are: first, sequential plot of the intervals between hindwing elevator spikes; second, movements of the abdomen; third and fourth, fore and hindwing myograms. The abdomen moves in the flight rhythm and more slowly in the ventilatory rhythm. During inspiration (upward deflection of the second trace) the interval between elevator spike decreases. Calibration (a) 100 ms; (c) 200 ms; (d) 400 ms.

PATTERN GENERATION IN LOCUST

flight frequency in time with the ventilatory rhythm. The change in interval is quite small, about $\pm4\%$ of the mean (a variation in frequency of just over 1 Hz) and its effect during free flight has yet to be assessed. Nevertheless the effect is present. The obvious candidates to mediate the effect are the interneurons which make widespread connections with flight and ventilatory motoneurons.

Interactions between flight and ventilation are also indicated by experiments in which a forewing is moved (Figure 8). Elevation of a forewing is associated with a ventral movement of the abdomen (Figure 8a). The response is tonic and is maintained for as long as the wing is held elevated. There is also an effect on the rate of ventilation; typically this increases during a maintained elevation and falls upon depression (Figure 8b). The movement of the abdomen involves all the segments and is mediated in part by motoneurons which are used in normal ventilatory movements (Figures 8b, c). The mechanism of this effect has yet to be explored, but any change in the ventilatory rate must affect the pattern of activity in the interneurons described above.

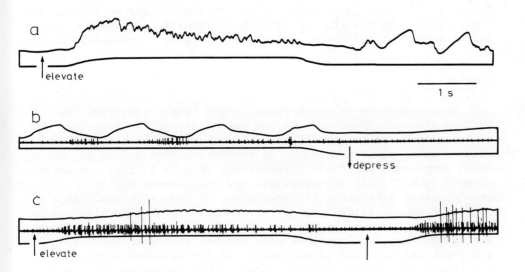

Figure 8. Elevation of a forewing evokes movements of the abdomen. (a) Upon elevation of the forewing the abdomen is moved ventrally (upward movement of first trace) and is held there until the wing is depressed. (b) The wing is held elevated and then depressed whereupon the ventilatory rate is reduced. (c) A continuation of (b) to indicate that the abdominal motoneurons which are affected by elevation of the wing are those which contributed towards normal ventilatory movements.

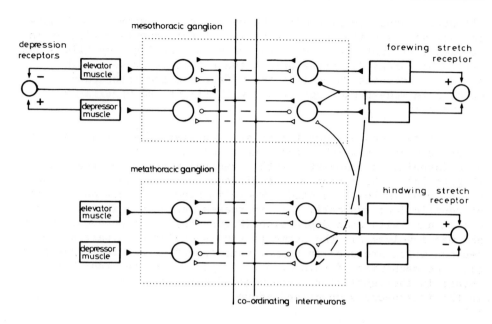

Figure 9. Pathways between the components of flight so far identified. Each wing has its own stretch receptor and depression receptor loops. Δ represent excitatory synapses, O inhibitory ones.

Conclusion

What presently is known about flight barely scratches the surface although insect flight has been analysed for more than thirty years. An adequate listing of components with their interactions seems far from sight. It is obvious from a list of the known components and their interactions (Figure 9) that the network already is difficult to understand and the outputs difficult to predict under particular circumstances. A look at a small aspect of flight necessitates further consideration of the interaction of different patterns of behaviour. So far this only involves flight and ventilation but doubtless will include mouthpart movements, heartbeat and in other grasshoppers and crickets the production and reception of song (Kutsch, 1969). This emphasizes that referring to "systems" within the nervous system may be an oversimplification for the sake of convenience. There may be considerable sharing of components. The method of analysis of central components by intracellular recording still contributes useful information. For example, the fact that a ventilatory modulation of flight was not noted before (despite the hours of myograms previously taken from flying locusts) is probably because there was no reason to expect such an effect. The discovery of central components with unexpected effects or connections therefore can lead to more precise and to more purposefully directed behavioural experiments.

Acknowledgment

This work was supported by a grant from the Nuffield Foundation.

REFERENCES

Altman, J.S., (1975) Changes in the flight motor pattern during development of the Australian plague locust, Chortoicetes terminifera. J. comp. Physiol. 97, 127-142.

Burrows, M. and Horridge, G.A., (1974) The organization of inputs to motoneurons of the locust metathoracic leg. Phil. Trans. R. Soc. Lond. B. 269, 49-94.

Burrows, M., (1975a) Monosynaptic connexions between wing stretch receptors and flight motoneurons of the locust. J. Exp. Biol. 62, 189-219.

Burrows, M., (1975b) Co-ordinating interneurons of the locust which convey two patterns of motor commands: their connections with flight motoneurons. J. Exp. Biol. 63, 713-733.

Burrows, M., (1975c) Co-ordinating interneurons of the locust which convey two patterns of motor commands: their connections with ventilatory motoneurons. J. Exp. Biol. 63, 735-753.

Camhi, J.M. and Hinkle, M., (1972) Attentiveness to sensory stimuli: central control in locusts. Science. 175, 550-553.

Kutsch, W., (1969) Neuromuskulare Aktivitat bei verschiedenen Verhaltensweisen von drei Grillenarten. Z. vergl. Physiol. 63, 335-378.

Kutsch, W., (1971) The development of the flight pattern in the desert locust, Schistocerca gregaria. Z. verg. Physiol. 74, 156-168.

Kutsch, W., (1974) The influence of wing sense organs on the flight motor pattern in maturing adult locusts. J. comp. Physiol. 88, 413-424.

Miller, P.L., (1971) Rhythmic activity in the insect nervous system: thoracic ventilation in non-flying beetles. J. Insect. Physiol. 17, 395-405.

Wendler, G., (1974) The influence of proprioceptive feedback on locust flight co-ordination. J. comp. Physiol. 88, 173-200.

Wilson, D.M., (1961) The central nervous control of flight in a locust. J. Exp. Biol. 38, 471-490.

Wilson, D.M. and Gettrup, E., (1963) A stretch reflex controlling wingbeat frequency in grasshoppers. J. Exp. Biol. 40, 171-185.

Wilson, D.M. and Wyman, R.J., (1965) Motor output patterns during random and rhythmic stimulation of locust thoracic ganglia. Biophys. J. 5, 121-143.

CENTRAL GENERATION OF LOCOMOTION IN VERTEBRATES

V.R. Edgerton[x], S. Grillner[xx], A. Sjöström[xx] and P. Zangger[xxx][o]

Different evidence is presented showing that the detailed pattern of locomotion is generated centrally by spinal α-γ-linked circuits. Data concerning rhythmic interneuronal activity related to the spinal generator for locomotion will be discussed.

Introduction

This chapter presents a summary of different types of experiments performed to show how stereotype locomotor movements, mainly of one single limb are generated by the central nervous system. Different approaches and problems are discussed under separate headings.

---

[x]Department of Kinesiology, U.C.L.A., Los Angeles, California 90024, U.S.A.
[xx]From the Department of Physiology, University of Göteborg, Göteborg, Sweden
[xxx]Institut de Physiologie, Université de Fribourg, Pérolles, 1700 Fribourg, Swiss
[o]Names published in alphabetical order (Presenter: S. Grillner)

I. THE EFFECT OF DORSAL ROOT TRANSECTION ON THE ELECTRO-MYOGRAPHICAL ACTIVITY DURING LOCOMOTION

In the beginning of the last century (for references see Grillner, 1975) the question was posed as to whether the generation of movements requires intact dorsal roots in the moving limb. It was found that basic features of the movements during locomotion can remain even after extensive deafferentation (Taub, this volume; for references see Grillner, 1975) but no or few attempts were made to compare the movements or the electromyographical activity before and after deafferentation in otherwise intact animals. In the newt, Székely et al., (1969) showed that the very complex electromyographical pattern of activity exhibited by the forelimbs during locomotion was not changed after bilateral transection of the dorsal roots supplying the forelimbs (cf. also Harcombe-Smith and Wyman, 1970). Similar experiments (Grillner and Zangger, 1975) were made on high decerebrate cats that can be made to walk, trot or gallop on a treadmill by stimulating in the mesencephalic "locomotor region" with increasing strength (Shik et al., 1966, 1967). In such preparations the EMG-pattern in one hindlimb was compared before and after transecting the ipsi- and the contralateral dorsal roots supplying the limb girdle investigated (Figure 1). The timing of the onset and termination of EMG in different muscles remain unchanged. Note particularly that the short burst in the knee flexor in relation to the long flexor bursts in other muscles persists after deafferentation. The temporal pattern of activity of the different muscles under normal undisturbed locomotion can thus remain after deafferentation of the limb girdle (Grillner and Zangger, 1975) and is therefore not dependent on phasic afferent signals from the limb as suggested previously (see Lundberg, 1969).

Although this pattern can be generated without afferent signals it sometimes occurs that after deafferentation the pattern at times becomes modified. For instance the knee flexor, semitendinosus, which is also a hip extensor can for some cycles show a large extensor activity during stance but not during swing and some cycles later revert to the normal flexor pattern (Grillner and Zangger, 1975 and Perret, Personal communication).

II. CAN THE SPINAL CORD ITSELF GENERATE LOCOMOTOR MOVEMENTS?

The next area to be investigated as to generation of movements is the spinal cord. Some fish can perform continuous swimming movements after a spinal transection (e.g., Gray and Sand, 1936), with proper movements of the body and a phase coupling between

PATTERN GENERATION IN VERTEBRATES

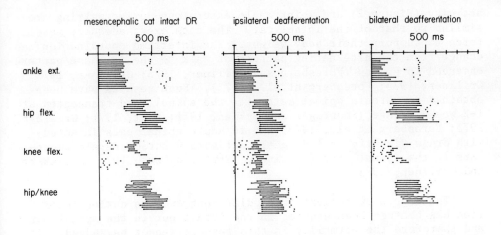

Figure 1. The effect of dorsal root transection on the EMG pattern during walking in the mesencephalic cat. The periods of EMG activity in 12 to 15 consecutive step cycles are compared before and after ipsilateral and then contralateral transection of all dorsal roots (L3-S4). The bars indicate the main period of EMG activity during successive cycles. Each cycle starts with the onset of activity in the ankle extensor (lateral gastrocnemius) and the dot to the right shows the termination of the cycle. The hip flexor is iliopsoas, the knee flexor is semitendinosus. This muscle has in addition to its main activity in flexion also a smaller and more variable burst during the extension phase which is indicated by two dots. The lowermost muscle is lateral sartorius (hip/knee) which is a flexor of the hip but has also an extensor function at the knee. The data were obtained in a high mesencephalic cat stimulated in the locomotor region (modified from data in Grillner and Zangger, 1975). The treadmill speed was 1.6 m/s in the left panel and 2.0 m/s in the other panels. Note that the characteristic short period of activity of semitendinosus remains after the transections as well as the extensor bursts as in the intact cat (Engberg and Lundberg, 1969).

adjacent muscular segments as in intact fish (Grillner, 1974 and Grillner and Kashin, this volume). For the different classes of tetrapods a great deal of evidence has accumulated that the hindlimbs can perform alternating movements after a spinal transection (for ref. see Grillner, 1975). Recently it has been shown that spinal cats can be made to walk on the treadmill with a speed determined by that of the treadmill belt. The pattern of EMG activity (Figure 2) and joint angle excursions during walking are similar to that of the intact cat. The cats exert adequate muscle force to support their body weight. The interlimb coordination can change from the alternating pattern of walk to the in phase pattern observed in gallop (Forssberg and Grillner, 1973; Grillner, 1973; Grillner, 1975; Forssberg et al., 1975). Such results have been obtained in chronic spinal cats with the spinal cord transected at 1-2 weeks of age (Shurrager and Dykman, 1951; Hart, 1971; Grillner, 1973; Forssberg et al., 1975) and in acute spinal cats injected with drugs directly or indirectly activating noradrenergic receptors in the spinal cord (Grillner, 1969; Budakova, 1973; Forssberg and Grillner, 1973).

The degree of activation of different muscles during locomotion has neither been studied in the intact nor in the spinal cat and therefore the normality in this respect cannot be judged. It should further be noted that these spinal walking cats have severe equilibrium deficits and tend to fall over (cf. Grillner, 1973, 1975) and do of course lack the adaptability of the intact cat but they are clearly capable of stereotype locomotor movements.

III. LOCOMOTOR ACTIVITY PRODUCED BY THE SPINAL CORD ITSELF DEPRIVED OF PHASIC AFFERENT INPUT THROUGH DORSAL ROOT TRANSECTION OR CURARIZATION

Brown (1911, 1913, 1914) has shown that alternating activity occurred in ankle muscles in spinal deafferented animals for a short period just after transecting the cord (see also Perret, 1973) and also during "narcosis progression" (Viala and Buser, 1969; under curare in Perret, 1973). Further it has been shown that after injection of drugs acting either directly or indirectly on noradrenergic receptors, a spinal network is released that generates alternating activation of flexors and extensors reflexly (Andén et al., 1966a, b; Jankowska et al., 1967a, b; Grillner, 1973 and see below Section VI). Under some conditions such alternating activity can occur spontaneously particularly after potentiation of the noradrenergic effects with Nialamide (Jankowska et al., 1967b; Viala and Buser, 1971; Grillner and Zangger, 1974).

Figure 3 shows the alternating EMG activity in flexors and extensors in a deafferented acute spinal (Th 12) preparation. This activity occurs in all joints of the limb. The limb is actually brought backward and forward as during locomotion, and the two limbs usually are alternating as during walk or trot but occasionally they can be active in phase as during gallop. The activity in this network can be influenced by tonic stimulation. A bilateral stimulation of the dorsal roots with e.g., 50 Hz (see legend Figure 3) can increase the frequency of alternation from e.g., 0.5

PATTERN GENERATION IN VERTEBRATES

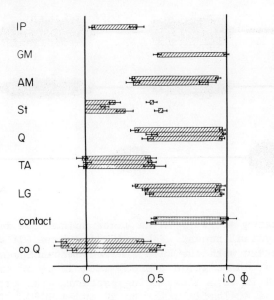

Figure 2. Diagrammatical representation of the electromyographical activity of different hindlimb muscles during locomotion in chronic spinal cats. The average onset and termination of different hindlimb muscles ($\pm$ S.D.) is plotted in relation to a normalized cycle starting with onset of the ipsilateral knee flexor semitendinosus. One to three different cats are shown for each individual muscle. The following muscles were recorded bipolarly (see Forssberg and Grillner, 1973): hip flexor (IP, iliopsoas), hip extensor (GM, gluteus medius and AM, adductor magnus), knee flexor (St, semitendinosus), knee extensor (Q, quadriceps, normally vastus lateralis), ankle flexor (TA, tibialis anterior), ankle extensor (LG, lateral gastrocnemius) and also the contralateral quadriceps (coQ). Foot contact is indicated between LG and coQ. Note that the extensor activity generally precedes foot contact and the shorter duration of St in relation to the other flexors and its short burst of activity during the extension phase. The different cats had their spinal cords transected between 1 and 2 weeks after birth and were then kept for 4 and 15 months. For the data on the figure, the cats were walking with their hindlimbs on a treadmill belt (see Grillner, 1973). (From Forssberg, Grillner and Sjöström, Unpublished data).

Hz to 1 Hz. With increasing strength of stimulation the frequency of the efferent output can increase even more (Grillner and Zangger, 1974). These results apply also to acutely operated chronic spinal cats (Grillner and Zangger, 1974). The above evidence suggest that the network generating the alternating activity

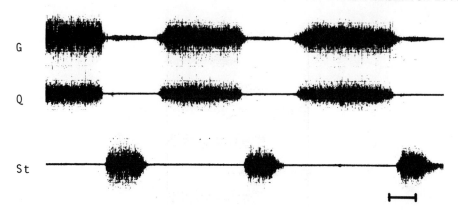

Figure 3. Alternating activity in an acute spinal cat during stimulation of L6 dorsal roots. Ankle (G) and knee (Q) extensors were recorded together with the knee flexor semitendinosus (St). Bilateral repetitive stimulation (70 Hz) with 20 µA of the dorsal roots (L6) resulted in this alternating pattern. The cat had received 50 mg/kg of Nialamide prior to 30 mg/kg DOPA. Time calibration for 1 s (Grillner and Zangger, Unpublished data).

after deafferentation is closely related to the generation of locomotion (see also preceding section on spinal locomotion).

In addition, experiments with curarization were performed to exclude the possibility that afferents entering via the ventral roots could give phasic information involved in the generation of the alternating activity. After blocking the neuromuscular transmission it is apparent that no movements occur and therefore also no phasic information related to movements. Under such conditions efferent alternating activity can be recorded in different flexor and extensor filaments (see Figures 4-8) and the frequency of burst can be similarly increased by tonic stimulation of cut dorsal roots with different frequencies (e.g., 50 Hz) (Grillner and Zangger, 1974 and Unpublished data).

In the same type of curarized preparation (see legend) motoneurons were recorded intracellularly (Edgerton, Grillner and Sjöström, Unpublished data). Flexor and extensor motoneurons were active at the same time as the corresponding flexor or extensor efferents (Figure 4). In many α-motoneurons the membrane-potential oscillated in phase with the activity in the corresponding filaments without action potentials (the neurons readily discharged spike trains when being depolarized). If the locomotor activity was enhanced so that the efferent burst frequency (and amplitude) increased, such silent motoneurons often could be made

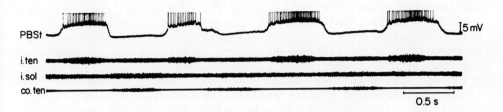

Figure 4. Rhythmic activity in a flexor motoneuron. The upper intracellular recordings are from a flexor motoneuron (posterior biceps semitendinosus identified antidromically from the muscle nerve) in an acute spinal curarized cat injected with Nialamide (50 mg/kg) and DOPA (30 mg/kg). Voltage calibration applies to the intracellular recording, time calibration to all records. The simultaneous activity in filaments to the ipsilateral tenuissimus (i.ten), and soleus (i.sol) and the contralateral tenuissimus (co.ten). Note the bursts of activity in the motoneuron; the first two spikes occur at very short intervals (not visible with this time resolution) and are followed by a comparatively long pause in bursts 1, 2 and 4 (From Edgerton, Grillner and Sjöström, Unpublished data).

to discharge in each "locomotor burst." It is of considerable interest that many α-motoneurons hence can show fluctuations of the membrane potential in phase with the locomotion although without inducing any spikes. This implies that the generating network acts on most motoneurons but that the quantitative effect varies.

The period of hyperpolarization between each burst or depolarization could be due simply to a disfaciliation or it could be combined with postsynaptic inhibition. In some motoneurons postsynaptic inhibition has been demonstrated by using chloride injections and hyperpolarizing currents (Edgerton, Grillner and Sjöström, Unpublished data). This also could be predicted from the findings of rhythmically active "Ia interneurons" (cf. Feldman and Orlovsky, 1975; Fu et al., 1975 and see below).

Hence, it may be concluded that the spinal locomotor network can operate without phasic afferent signals and that in the different motoneurons, periods of excitation and inhibition alternate (for fish see also von Holst, 1935; Grillner et al., 1975).

IV. SPINAL α-γ-LINKAGE IN LOCOMOTION

In the preceding section it was shown that alternating "locomotor activity" can be induced in α-motoneurons in the spinal deafferented state. It may be asked if these spinal commands are α-γ-linked or not. Sjöström and Zangger (1975 and Unpublished data; see also Perret, 1973) in the same type of preparation have recorded single identified γ-efferents to flexors and extensors together with α-efferents and thereby have shown that there is an α-γ-coactivation in both cases. They have further documented this linkage by recording primary endings from muscle spindles. Their data do not allow further differentiation into static and dynamic γ-motoneurons (Matthews, 1962). In studies of the late reflex discharges that can be elicited after DOPA i.v. (Andén et al., 1966a; Jankowska et al., 1967a) it was shown that there is a coactivation of α- and both static and dynamic γ-motoneurons to the same muscle (Bergmans and Grillner, 1968, 1969; Grillner, 1969a, b). Since there are good reasons to believe that the activity during these late reflex discharges is related to the "locomotion network" (see below Section VI, and Figure 6), these data together all strongly suggest that there is a spinal $α-γ_S-γ_D$-linkage to both flexors and extensors. The advantage of such a linkage in locomotion has been discussed previously by Grillner (1969b).

It is often suggested that the γ-motoneurons should be activated from the brainstem. The previous demonstration of α-γ-linkage in locomotion (Severin et al., 1967; Severin, 1970; Perret and Buser, 1972; Perret and Berthoz, 1974) could have been due to descending α-γ-coactivation. However, it is now clear that the spinal generator itself provides an α-γ-linkage.

V. DOES THE SPINAL GENERATOR OPEN AND CLOSE REFLEX PATHWAYS IN PHASE WITH THE DIFFERENT PARTS OF THE STEP CYCLE?

It has been shown that an identical stimulus can give opposite effects during the stance and the swing phase, i.e., a phase dependent reflex reversal (Forssberg et al., 1975; Forssberg et al., this volume). The evidence to suggest that this switching is performed centrally by the generator is summarized by Forssberg et al., (this volume). Recent intracellular motoneuronal recordings have shown that during "spinal curarized locomotion" (preparation described in Section III and VI) a stimulation applied to the dorsum of the foot or to a peripheral nerve can give very different synaptic effects during the periods of depolarization and hyperpolarization, suggesting a phasic change of the transmission in the short latency reflex arcs to the α-motoneurons (Edgerton, Grillner, Rossignol and Sjöström, Unpublished data).

VI. INTERNEURONAL ACTIVITY RELATED TO THE SPINAL GENERATOR

Orlovsky and Feldman (1972) recorded from lumbosacral neurons in decerebrate walking cats. They showed that many neurons were rhythmically active but with their main activity located in different parts of the "step cycle." Thus they were not distributed strictly in a flexor and an extensor group. Recently it was shown that the interneurons known to transmit the reciprocal Ia inhibition from muscle spindles to α-motoneurons were rhythmically active even after removing all rhythmic afferent inflow (Feldman and Orlovsky, 1975). Hence it could be concluded that the phasic activity in these "classical" relay interneurons was due to a central activation. In their experimental situation it is not clear if the activation of any given neuron is due to the spinal generator or to phasic activity in the different descending pathways, which Orlovsky (1972a, b,c) earlier has shown to be phasically active and which converge onto e.g., Ia interneurons (see Lundberg, 1970; Hultborn, 1972).

Another relevant set of experiments have been performed in acute spinal cats injected with DOPA i.v. (Jankowska et al., 1967b). In this preparation it had been shown that an entirely new type of reflex discharge occurred with a long central latency (0.1-0.2 s) and of long duration (0.5 s) when a short train of pulses to ipsi- or contralateral nerves at high strength was given (Andén et al., 1966a; Jankowska et al., 1967a). Such a stimulation resulted in ipsilateral flexor and contralateral extensor discharges which occur in all preparations that can generate walking movements (Grillner, 1973). It was suggested that the neuronal network giving rise to these late discharges would be related to the spinal generator of locomotion (Jankowska et al., 1967b). When recording lumbosacral interneurons many were found to be activated in relation to these "later" efferent discharges (1) by either ipsilateral or (2) by contralateral stimulation or (3) by stimulation of both sides (Jankowska et al., 1967b). These results were interpreted as giving evidence for an interneuronal halfcenter organization (see below). In a later investigation (Fu et al., 1975) it was found that the Ia interneurons giving rise to disynaptic inhibition of α-motoneurons were activated in such late discharges. It was therefore inferred that they were activated from the central generator.

In an ongoing series of experiments (Edgerton, Grillner and Sjöström, Unpublished data) lumbar (L7 interneurons (n = 89) have been recorded in the spinal curarized cat injected with DOPA and Nialamide (see Grillner and Zangger, 1974 and Section III) during spontaneous and induced rhythmic alternating activity. The interneurons were recorded simultaneously with the efferent activity in ipsi- and contralateral flexor and extensor filaments. In this way the neuronal activity could be correlated to the efferent activity. Figure 5 shows an interneuron with an activity strictly alternating with the ipsilateral flexor and related to the activity in the contralateral flexor and ipsilateral extensor. The activity is linked tightly to the efferents. Note in Figure 6 how the interneuronal activity is tightly related to that of the ipsilateral flexor and how the variability in one is also reflected in the other. Figure 7 shows a neuron that is active during the flexion phase but with the main burst limited to the last third of this phase (note the logarithmic frequency scale). It is striking the termination of the ipsilateral flexor and of the neuronal activity is very closely related in contrast to the onset (compare right and left diagrams in Figure 7). Figure 8 shows schematically the activity of two simultaneously recorded neurons together with six different efferents. These graphs show the constancy of the different efferent bursts in relation to the two interneurons. To estimate the relation between the activity in interneurons and that of different efferents, correlation ratios were calculated to compare both the onset and the termination of flexor or extensor bursts with that of the interneurons. So far 61 interneurons have been analyzed statistically: 31 were related to the activity of the ipsilateral flexor with a mean correlation ratio (r) of 0.92 ($\pm$0.10 S.D.); 20 to the extensors (r = 0.93 $\pm$ 0.06 S.D.) and 10 to the contralateral flexor (r = 0.95 $\pm$ 0.06 S.D.). All neurons with rhythmic activity have been classified in these three groups.

All the interneurons thus are related tightly to some efferent discharge. It might therefore be relevant to ask during which part of the efferent burst different interneurons are active. The histograms of Figure 9B and C show the occurrence of the midpoint

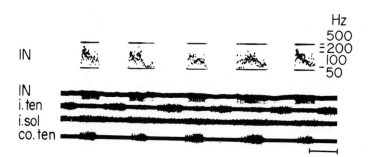

Figure 5. Rhythmically active interneuron. The firing pattern of the interneuron (IN) is shown as direct recordings and as instantaneous frequency (Hz) on a logarithmic scale. The simultaneous recording from filaments to the ipsilateral tenuissimus (i.ten) and soleus (i.sol) and the contralateral tenuissimus (co.ten). Preparation as in Figure 4 (From Edgerton, Grillner and Sjöström, Unpublished data).

PATTERN GENERATION IN VERTEBRATES

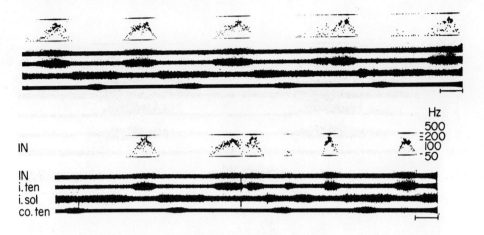

Figure 6. Rhythmically active interneuron and the response to a nerve stimulation. The display of data as in Figure 5. This interneuron has its main activity together with the ipsilateral flexor. The two series are a continuation of each other and show the variability of the discharge in relation to the variations in the flexor discharge. The ipsilateral stimulation (ipsilateral peroneal nerve, short trains of 3 pulses at 20 times threshold) in the fourth last burst elicits a "late discharge" in the flexor and a concomitant increase in the activity in the interneuron. Note the modification in the spontaneous bursts occurring after the stimulation (From Edgerton, Grillner and Sjöström, Unpublished data).

of interneuron discharge (halfway between onset and termination) for different neurons in relation to the flexor or the extensor burst. It is apparent that the majority of the interneurons have a midpoint coinciding with that of either the flexor or the extensor burst. However, apparently there are also some neurons with their midpoints distributed throughout the step cycle. On the basis of such data one should be cautious in trying to advocate two homogeneous main populations of neurons particularly when considering neurons such as illustrated in Figure 7. Here the level of activity is very low at the midpoint and the discharge pattern is skewed with a large peak occurring much later. For such cases a histogram of the peak frequencies would be more informative. This also can be ambiguous since many of the neurons have a steady or slightly variable discharge throughout its period of activity in each cycle (Figure 5) with no clear peak or sometimes even with two peaks.

At this point, it may be concluded that many neurons have a more or less steady discharge coinciding largely with the flexor

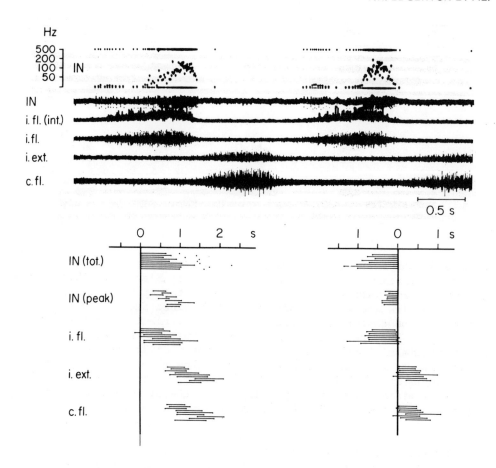

Figure 7. Interneuron with asymmetric activity. Representation as for interneuron in Figure 5. Note that the activity is related to that of the ipsilateral flexor but the large increase in activity occurs only in the last part of the burst. Below is showed in a schematic form (as in Figure 1) the periods of activity in several successive cycles of the filaments and of the interneuron. In the left graph the cycle starts with the onset of activity in the interneuron and in the right graph, the zeropoint is moved to the end of the interneuron burst. It can be seen that the termination of activity in the interneuron is related tightly to the termination of the flexor burst. For comparison both the total duration of the interneuronal activity in each cycle and the duration of the main burst are given (From Edgerton, Grillner and Sjöström, Unpublished data).

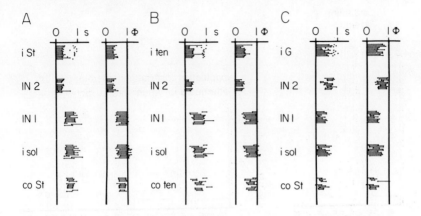

Figure 8. The activity of two simultaneously recorded interneurons in relation to the efferent activity in six different filaments. The results are displayed as in Figures 1 and 7. The filaments are the ipsilateral semitendinosus (i.St), soleus (i.sol) tenuissimus (i.ten) lateral gastrocnemius (i.G) and contralateral semitendinosus (co.St) and tenuissimus (co.ten). Note that one interneuron has an activity related tightly to the ipsilateral St and whereas the other occurs during the ipsilateral extensor discharge. Left graph in A, B and C shows the events in real time whereas to the right the cycle duration is normalized (Edgerton, Grillner and Sjöström, Unpublished data).

or the extensor burst. Others may have a definite peak distributed at any point within the step cycle and still have a high correlation ratio for the onset or termination of a flexor or an extensor burst (cf. Figure 7). It should be stressed that under the given conditions each neuron often had a very consistent discharge pattern with regard to the shape of the frequency curve.

Although no strict quantitative correlation between the level of activity in the efferents and the interneurons has yet been performed, the activity sometimes does appear very strongly related. Thus in some interneurons an increased frequency occurs together with an increased activity in the efferents and a pause in the interneuron coincides with a decreased activity in the efferents. Other neurons active, through one burst, may show a reversed relationship in that each increase in interneuronal discharge coincides with a decrease in efferent activity and vice versa. Other neurons start their activity only for example, at the peak of the efferent activity and with their main burst related to the cessation of efferent activity.

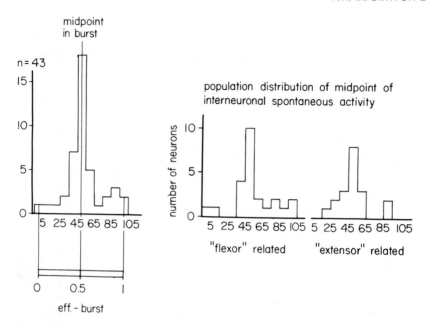

Figure 9. Histogram showing the relation between the midpoint of the interneuron burst and the efferent nerve activity. A. The midpoints (halfway between onset and termination) of 43 interneurons were calculated in relation to the efferent burst to which it was related (extensor or flexor). The period of activity of the different efferents was normalized (100%). The midpoints for the different neurons are plotted as a histogram in relation to the efferent burst. B. and C. The same procedure as in A but the midpoints of "extensor" neurons and "flexor" interneurons are plotted separately. Note that most interneurons have a midpoint that coincides with that of the efferent and that there appears to be no difference between flexor and extensor related neurons (From Edgerton, Grillner and Sjöström, 1975).

When trying to relate the recorded interneurons to previous interneuron studies it is possible to conclude that all interneurons could be activated in a certain relation to the "late discharges" discussed above (Jankowska et al., 1967a, b). The interneuronal activity had the same relation to the efferent activity after stimulation as during the spontaneous efferent activity. Furthermore, when giving such stimuli during spontaneous "locomotor" activity the cycle is perturbed and the subsequent rhythmic activity is related to the induced "late discharge" (Figure 6). These findings support the previous suggestion that the interneuronal data from "late discharges" are related to the "generating network" (Jankowska et al., 1967a, b; Grillner, 1973). Most of the interneurons were not activated with short latency (below approx. 15 ms) from any peripheral nerve but some were activated from cutaneous nerves and five at "gr. I strength" from different hindlimb nerves. These group I neurons were active at the same time as the muscle from which they had their group I input and they also were antidromically inhibited and can therefore be assumed to be the so called group I neurons mediating inhibition to the antagonist motoneurons. It can thus be concluded that these "Ia interneurons" (as indicated by Feldman and Orlovsky, 1975; Jankowska et al., 1975) are part of the central spinal network which generates the rhythmic activity. The "Ia interneurons" active, presumably inhibit the antagonists thereby assuring their silence. Three neurons were antidromically activated from the contralateral side (Th 12) (recorded in L7) and had rhythmic activity, which was closely related to the efferents. Such neurons could serve as coordinating neurons for e.g., the hind-forelimb coordination.

VII. CONCLUDING REMARKS AND COMMENTS ON POSSIBLE GENERATOR MODELS

To summarize the preceding sections, we know that the spinal generator can (see Figure 10):

(1) alternately excite and inhibit α-motoneurons (Section III),

(2) exert this inhibition at least partly via the same interneurons that mediate the reciprocal inhibition from Ia afferents (Section III, VI),

(3) coactivate γ-motoneurons with the α-motoneurons to the same muscle (Section IV),

(4) and very likely phasically open and close reflex pathways e.g., from the dorsum of the foot to motoneurons resulting in reflex reversals (Section V).

It is outside the scope of this article to discuss how the generators are driven from supraspinal structures (see e.g., Grillner, this volume) or influenced from afferents related to the position of the hip (Rossignol et al., 1975) or other types of afferent input (Pearson and Duysens, this volume).

The schematical organization in Figure 10 implies that when one muscle is active the antagonist will be inhibited. Under conditions when there is a co-contraction of antagonist muscles (which does occur during locomotion), an increased activity in one of the muscles would be expected to coincide with a decrease in the antagonist. Indeed this can often be observed during locomotion under deafferented conditions (Grillner and Zangger, Unpublished data).

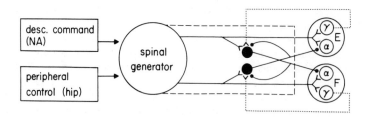

Figure 10. Diagrammatical representation of part of the generator network for one limb. The spinal generator is shown to the left with its linked output on α- and γ-motoneurons to the flexors (F) and extensors (E) respectively and the activation of the reciprocal inhibitory interneurons to the antagonists. On this interneuron there also will be an indirect activation via the γ-Ia afferent and also it is shown that there is a reciprocal inhibition between the two groups of interneurons.

These findings can thus be explained by the reciprocal organization at the output stage and need not in itself imply anything about the intrinsic organization of the generator in terms of the different models discussed below. The fact that "Ia interneurons" to antagonists have reciprocal inhibitory connections (Hultborn and Illert, 1974) is, however, of some possible interest in relation to the generation of alternating activity (see below).

It is noteworthy that the descending pathways (rubro-, vestibulo-, and reticulospinal tracts) known to be phasically active during locomotion in mesencephalic and decorticate cats (Orlovsky, 1972a, b, c; Perret, 1973) can influence directly and indirectly not only α-motoneurons but also γ-motoneurons and "Ia interneurons." For example, the vestibulospinal pathway does excite mono- and disynaptically α- and also γ-extensor motoneurons and disynaptically does inhibit flexors via the "Ia inhibitory pathway" (Grillner and Hongo, 1972). There is indeed a very extensive convergence pattern onto these "Ia interneurons" from many descending pathways as well as different types of afferent pathways and the stepping generator (Hongo et al., 1969; Lundberg, 1970; Hultborn, 1972 and see above). Although quantitatively the Ia input appears strong it must, in view of all these different inputs, be asked if it is misleading to call them "Ia interneurons" (Grillner and Hongo, 1972) because this links the name to only one input. Each group of "Ia interneurons" inhibits only one or a few synergistic motor nuclei. These interneurons are utilized by the spinal generator for locomotion as well as the large muscle spindle afferents and various descending tracts to elicit inhibition in specific target motor nuclei. We therefore raise the possibility that this class of interneurons are one of the main relays for mediating inhibition to specific spinal motor nuclei rather than related specifically to one particular type of afferent input.

Although these facts give some information about the actual generating network they shed very little light on the basic working principle of the generator itself. Several different types of generators have often been discussed, and we will consider three of them below (Figure 11; cf. Grillner, 1975).

Type 1. Székely (1968) and also Gurfinkel and Shik (1972) have considered a model with a closed chain of neurons. Some parts of the chain are connected with one group of motoneurons and other parts with other groups of motoneurons (Figure 11C). In this way it is very easy to generate rhythmic coordinated activity in different motoneuronal groups and it is simple to model a very complex efferent activity with co-contraction in different muscle groups. It is noteworthy that this model was developed from studies on the newt forelimb in which a very complex pattern of muscle activity exists (Székely, 1968; Székely et al., 1969).

Type 2. A model with reciprocal inhibition has been discussed for the cat (Brown, 1911, 1913, 1914; Jankowska et al., 1967a, b) and for different invertebrates (Wilson, 1964; Wilson and Waldron, 1968; Selverston, 1974; Mulloney and Selverston, 1974a, b; Perkel and Mulloney, 1974). This model (Figure 11A) is founded basically on two neurons or groups of neurons which inhibit each other and in addition have high enough excitability to discharge when not inhibited. Such a system can be made to oscillate (i.e., so that the periods of activity alternate between the two neurons) if fatigue is introduced as a crucial factor. The fatigue could operate as an accumulating afterhyperpolarization in the discharging neuron(s) which would stop the spiking or it could operate by a successively decreasing transmitter release in the inhibitory synapse possibly combined with postinhibitory rebound. In this way each neuron (e.g., X) could be active for one period during which it is effectively inhibiting the other cell (Y) and when cell X is fatiguing cell Y is disinhibited and will start firing and thereby inhibit effectively cell X and so forth. Such a network has not been shown to generate locomotor movements but appears to be responsible for generating similar alternating movements of the "gastric mill" of the lobster by a 12-neuron system of the stomatogastric ganglion (Selverston and Mulloney, 1974; Mulloney and Selverston, 1974a, b).

Type 3. The third model (Figure 11B) can be described as asymmetric, driving e.g., the flexor motoneurons while reciprocally inhibiting the extensors. This could be achieved in several ways, for example: (a), By one cell with oscillating membrane potential (pacemaker) or a population of such cells synchronized by electric coupling. Nonspiking "pacemaker" interneurons appear to be responsible for the stepping movements in the cockroach leg by exciting the flexors during each period of depolarization and at the same time inhibiting the extensors via an interneuron (cf. Mendelson, 1971; Pearson et al., 1973; Pearson and Fourtner, 1975; Fourtner, this volume). A similar network was described for the lobster pyloric system (14 neurons) driven by three electrically coupled spiking motoneurons with oscillating membrane potential (Maynard, 1972; Selverston, 1974). (b), A network of "ordinary" electrically coupled neurons if driven can generate rhythmic bursts of activity presumably on the basis of mutual excitation and accumulating afterhyperpolarization being particularly important due to the low pass filter characteristics of these electrical junctions as described for "trigger" interneurons (Willows, this volume). Such a network could substitute for the pacemaker neuron(s) in the asymmetric model drawn in Figure 11, but otherwise the organization for alternation could be analogous.

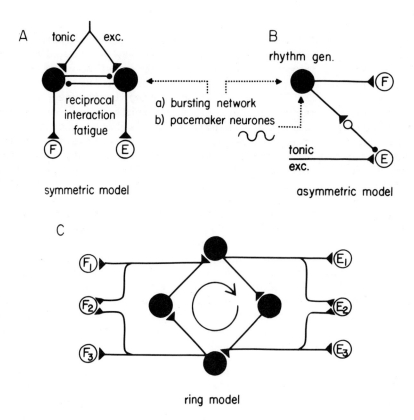

Figure 11. Schematical representation of different generator models (modified from Grillner, 1975).

PATTERN GENERATION IN VERTEBRATES

In the vertebrate locomotor system there is no direct evidence for either type of model. The ring model (1) would appear to be rigid and it would be difficult to explain sudden changes in the output of the generator as exemplified in the previous discussion of the deafferentation experiments (Section I), in which e.g., semitendinosus which is usually mainly active with the flexors can change to the pattern of full extensor for some period of time (see above Section I). Such and other indirect evidence, however, does not disprove this possibility, and the interneuronal activity recorded could be used to support such a network (1) as well as model 2. The other models (2 and 3) have been demonstrated to operate in crustaceans, insects or molluscs and might well be applicable to vertebrates. It is possibly of interest that "normal" nerve cells in Aplysia with a stable membrane potential can be modified to "spiking pacemaker cells" by different pharmacological agents. One type of neuroendocrinological cell becomes a "burster" when monoamines are applied (Boisson and Chalazonitis, 1973). Such "induced pacemaker cells" could be reverted to ordinary cells by other agents. Some cells instead stopped bursting when monoamines were applied (Chalazonitis and Ducreux, 1971; Chalazonitis and Morales, 1974 and Personal communication). These results raise the interesting possibility that transmitter substances in addition to their well known action of inhibiting and exciting nerve cells, could also modify their firing behaviour from steady firing to bursting behaviour and back again. It is perhaps noteworthy that monoamines have been involved in the effects discussed on invertebrate neurons and that noradrenergic substances can contribute to the generation of locomotion in the spinal cord of the cat (see Section II).

Generally it is assumed that the coordination between limbs occurs via interaction between the generators of each limb of a pair and that they can be coordinated in alternation or in phase (Miller et al., 1973; Stein, 1974; cf. Grillner, 1975). If there is reciprocal inhibition of equal weight between the two generators an alternating pattern would occur, if on the other hand, there is mutual excitation between the two generators they would tend to be in phase (see Figure 12). Thus if there are two alternate pathways between the two generators, a switch from one pathway mediating reciprocal inhibition to one with mutual excitation would result in a switch from alternation between the limbs to in phase movements as during gallop.

It may now be asked if the generator for one limb, is one undividable entity or not. Some evidence for the latter possibility can be obtained from experiments in which several muscles at different joints are recorded simultaneously. Sometimes flexors at one joint can be tonically active with no modulation at all, whereas muscles at other joints are clearly rhythmically active.

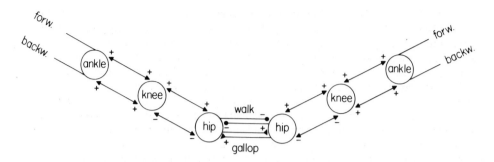

Figure 12. Schematical and speculative representation of possible interactions between generators.

This and other evidence (Grillner and Zangger, Unpublished results) would suggest the possibility that muscles controlling one joint for example, could be controlled by one "unit generator." Normally the different generators for the joints would be tightly coupled to produce the walking movements, i.e., that there is mutual excitation between the different "unit generators." Under other conditions a possible generator for the hip could be reciprocally coupled to that of ankle and knee and then backwards walking would result (cf. Grillner, 1973). Under other conditions again e.g., supraspinal pathways could utilize only one "unit generator" without activating the rest of the network thereby producing alternating movements in only one joint. It is interesting to note Lawrence's and Kuypers' (1968a, b) findings showing that one deficit of a pyramidal tract lesion (monkey) is an inability to make independent finger movements; if such a lesion is combined with a transection of mainly rubrospinal fibers the animals can no longer perform independent hand movements but still use the arm for taking food with flexion-extension movements of the entire arm as well as using the arm during climbing, walking, etc. Their results would be compatible with the view suggested here that one way for a supraspinal control of movements could be to brake larger spinally programmed movement synergies into pieces and control independently different "unit generators." Whether this is correct remains to be shown, but such an organization would certainly increase the versatility of the spinal motor network.

Acknowledgment

This investigation has been supported by the Swedish Medical Research Council (proj. no. 3026) and the Medical Faculty in Göteborg. P.Z. was supported from Swiss National Foundation. The skillful assistance of Mrs. M. Svanberg is gratefully acknowledged.

REFERENCES

Andén, N.-E., Jukes, M.G.M., Lundberg, A. and Vyklicky, L., (1966a) The effect of DOPA on the spinal cord. 1. Influence on transmission from primary afferents. Acta Physiol. Scand. 67, 373-386.

Andén, N.-E., Jukes, M.G.M. and Lundberg, A., (1966b) The effect of DOPA on the spinal cord. 2. A pharmacological analysis. Acta Physiol. Scand. 67, 387-397.

Bergmans, J. and Grillner, S., (1968) Changes in dynamic sensitivity of muscle spindle primary endings induced by DOPA. Acta Physiol. Scand. 74, 629-636.

Bergmans, J. and Grillner, S., (1969) Reciprocal control of spontaneous activity and reflex effects in static and dynamic γ-motoneurones revealed by an injection of DOPA. Acta Physiol. Scand. 77, 106-124.

Boisson, M. and Chalazonitis, N., (1973) Réactivites bioélectriques propres d'un neurone géant sécrétoire (Aplysia depilans). C.R. Acad. Sci. Paris, 276, 1025-1028.

Brown, T.G., (1911) The intrinsic factors in the act of progression in the mammal. Proc. R. Soc. B. 84, 308-319.

Brown, T.G., (1913) The phenomenon of "narcosis progression" in mammals. Proc. R. Soc. B. 86, 140-164.

Brown, T.G., (1914) On the nature of the fundamental activity of the nervous centres; together with an analysis of the conditioning of rhythmic activity in progression, and a theory of the evolution of function in the nervous system. J. Physiol. 48, 18-46.

Budakova, N.N., (1973) Stepping movements in the spinal cat due to DOPA administration. Fiziol. Zh. (USSR). 59, 1190-1198.

Chalazonitis, N. and Ducreux, C., (1971) Stabilization par la dopamine de l'oscillabilité normale d'une neuromembrane type (Soma neuronigue à ondessalves: Aplysia et Helix). C.R. Soc. Biol. Paris. 165, 1350-1353.

Chalazonitis, N. and Morales, T., (1971) Dépolarisation par l'oubaine des motoneurones normalement au artificiellement oscillant (Aplysia). C.R. Soc. Biol. Paris. 165, 1923-1928.

Engberg, I. and Lundberg, A., (1969) An electromyographic analysis of muscular activity in the hindlimb of the cat during unrestrained locomotion. Acta Physiol. Scand. 75. 614-630.

Feldman, A.G. and Orlovsky, G.N., (1975) Activity of interneurons mediating reciprocal Ia inhibition during locomotion. Brain Res. 84, 181-194.

Forssberg, H. and Grillner, S., (1973) The locomotion of the acute spinal cat injected with Clonidine i.v. Brain Res. 50, 184-186.

Forssberg, H., Grillner, S. and Rossignol, S., (1975) Phase dependent reflex reversal during walking in chronic spinal cats. Brain Res. 85, 103-107.

Forssberg, H., Grillner, S. and Sjöström, A., (1975) The locomotor capacity of chronic spinal cat (In preparation).

Fu, T.-C., Jankowska, E. and Lundberg, A., (1975) Reciprocal Ia inhibition during the late reflexes evoked from the flexor reflex afferents after DOPA. Brain Res. 85, 99-102.

Gray, J. and Sand, A., (1936) The locomotory rhythm of the dogfish (Scyllium canicula). J. Exp. Biol. 13, 200-209.

Grillner, S., (1969a) Supraspinal and segmental control of static and dynamic γ-motoneurones in the cat. Acta Physiol. Scand. Suppl. 327, 1-34.

Grillner, S., (1969b) The influence of DOPA on the static and the dynamic fusimotor activity to the triceps surae of the spinal cat. Acta Physiol. Scand. 77, 490-509.

Grillner, S. and Hongo, T., (1972) "Vestibulospinal effects on motoneurones and interneurones in the lumbosacral cord," In Basic Aspects of Central Vestibular Mechanisms. (Brodal, A. and Pompeiano, O., eds.), Elsevier Publ. Co. Progr. in Brain Res. 37, 243-262.

Grillner, S., (1973) "Locomotion in the spinal cat," In Control of Posture and Locomotion. (Stein, R.B., Pearson, K.G., Smith, R.S. and Redford, J.B., eds.), Plenum Press, New York, (515-535).

Grillner, S., (1974) On the generation of locomotion in the spinal dogfish. Exp. Brain Res. 20, 459-470.

Grillner, S. and Zangger, P., (1974) Locomotor movements generated by the deafferented spinal cord. Acta Physiol. Scand. 91, 38-39A.

Grillner, S., (1975) Locomotion in vertebrates - central mechanisms and reflex interaction. Physiol. Rev. 55, 247-304.

Grillner, S., Perret, C. and Zangger, P., (1975) Central generation of locomotion of the spinal dogfish. Brain Research. (In press).

Grillner, S. and Zangger, P., (1975) How detailed is the central pattern generation for locomotion? Brain Res. 88, 367-371.

Gurfinkel, V.S. and Shik, M.L., (1973) "The control of posture and locomotion," In Motor Control. (Gydikov, A.A., Tankov, N.T., Kosarov, D.S., eds.), Plenum Press, New York, (217-234).

Harcombe-Smith, E. and Wyman, R.J., (1970) Diagonal locomotion in de-afferented toads. J. Exp. Biol. 53, 255-263.

Hart, B.L., (1971) Facilitation by Strychnine of reflex walking in spinal dogs. Physiology and Behaviour. 6, 627-628.

Holst, E. von, (1935) Erregungsbildung und Erregungsleitung im Fischrückenmark. Pflügers Arch. ges. Physiol. 235, 345-359.

Hongo, T., Jankowska, E. and Lundberg, A., (1969) The rubrospinal tract, II. Facilitation of interneuronal transmission in reflex paths to motoneurones. Exp. Brain Res. 7, 365.

Hultborn, H., (1972) Convergence on interneurones in the reciprocal Ia inhibitory pathway to motoneurones. Acta Physiol. Scand. Suppl. 375, 1-42.

Hultborn, H., Illert, M. and Santini, M., (1974) Disynaptic Ia inhibition of the interneurones mediating the reciprocal Ia inhibition of motoneurones. Acta Physiol. Scand. 91, 14-16A.

Jankowska, E., Jukes, M.G.M., Lund, S. and Lundberg, A., (1967a) The effect of DOPA on the spinal cord. 5. Reciprocal organization of pathways transmitting excitatory action to alpha motoneurones of flexors and extensors. Acta Physiol. Scand. 70, 369-388.

Jankowska, E., Jukes, M.G.M., Lund, S. and Lundberg, A., (1967b) The effect of DOPA on the spinal cord. 6. Halfcentre organization of interneurones transmitting effects from the flexor reflex afferents. Acta Physiol. Scand. 70, 389-402.

Lawrence, D.G. and Kuypers, H.G.J.M., (1968a) The functional organization of the motor system in the monkey. I. The effects of bilateral pyramidal lesions. Brain. 91, 1-14.

Lawrence, D.G. and Kuypers, H.G.J.M., (1968b) The functional organization of the motor system in the monkey. II. The effects of lesions of the descending brainstem pathways. Brain. 91, 15-36.

Lundberg, A., (1969) Reflex control of stepping. The Nansen Memorial Lecture V. Universitetsforlaget, Oslo.,(1-42).

Lundberg, A., (1970) "The excitatory control of the Ia inhibitory pathway," In Excitatory Synaptic Mechanisms. (Andersen, P. and Jansen, J.K.S., eds.), Universitetsforlaget, Oslo, (333-340).

Matthews, P.B.C., (1962) The differentiation of two types of fusimotor fibre by their effects on the dynamic response of muscle spindle primary endings. Quart. J. Exp. Physiol. 47, 324-333.

Maynard, D.M., (1972) Simpler networks. Ann. N.Y. Acad. Sci. 193, 59-72.

Mendelson, M., (1971) Oscillator neurons in crustacean ganglia. Science. 171, 1170-1173.

Miller, S. and Van der Burg, J., (1973) "The function of long propriospinal pathways in the co-ordination of quadrupedal stepping in the cat," In Control of Posture and Locomotion. (Stein, R.B. et al., eds.), Plenum Press, New York, (561-577).

Mulloney, B. and Selverston, A.I., (1974a) Organization of the stomatogastric ganglion of the spiny lobster. I. Neurons driving the lateral teeth. J. Comp. Physiol. 91, 1-32.

Mulloney, B. and Selverston, A.I., (1974b) Organization of the stomatogastric ganglion of the spiny lobster. III. Coordination of the two subsets of the gastric system. J. Comp. Physiol. 91, 53-78.

Orlovsky, G.N., (1972a) The effect of different descending systems on flexor and extensor activity during locomotion. Brain Res. 40, 359-371.

Orlovsky, G.N., (1972b) Activity of rubrospinal neurons during locomotion. Brain Res. 46, 99-112.

Orlovsky, G.N., (1972c) Activity of vestibulospinal neurons during locomotion. Brain Res. 46, 85-98.

Orlovsky, G.N. and Feldman, A.G., (1972) Classification of lumbo-sacral neurons according to their discharge patterns during evoked locomotion. Nejrcfisiologia. 4, 410-417 (In Russian). (English version of same journal, 311-317).

Pearson, K.G., Fourtner, C.R., and Wong, R.K., (1973) "Nervous control of walking in the cockroach," In Control of Posture and Locomotion. (Stein, R.B. et al., eds.), Plenum Press, New York, (495-514).

Pearson, K.G. and Fourtner, C.R., (1975) Nonspiking interneurones in walking system of cockroach. J. Neurophysiol. 38, 33-52.

Perkel, D.H. and Mulloney, B., (1974) Motor pattern production in reciprocally inhibitory neurons exhibiting post-inhibitory rebound. Science. 185, 181-183.

Perret, C. and Buser, P., (1972) Static and dynamic fusimotor activity during locomotor movements in the cat. Brain Res. 40, 165-169.

Perret, C., (1973) Analyse des méchanismes d'une activité de type locomoteur chez le chat. Thèse de doct., Université de Paris VI, Paris.

Perret, C. and Berthoz, A., (1974) Evidence of static and dynamic fusimotor action on the spindle response to sinusoidal stretch during locomotor activity in the cat. Exp. Brain Res. 18, 178-188.

Rossignol, S., Grillner, S. and Forssberg, H., (1975) Factors of importance for the initiation of flexion during walking. Neuroscience Abstracts. 1, 181.

Selverston, A.I., (1974) Structural and functional basis of motor pattern generation in the somatogastric ganglion of the lobster. Am. Zool. 14, 957-972.

Selverston, A.I. and Mulloney, B., (1974) Organization of the stomatogastric ganglion of the spiny lobster. II. Neurons driving the medial tooth. J. Comp. Physiol. 91, 33-51.

Severin, F.V., Orlovsky, G.N. and Shik, M.L., (1967) Work of the muscle receptors during controlled locomotion. Biofizika. 12, 575-586. (Eng. transl.).

Severin, F.V., (1970) The role of the gamma motor system in the activation of the extensor alpha motor neurones during controlled locomotion. Biofizika. 15, 1138-1145. (Eng. transl.).

Shik, M.L., Severin, F.V. and Orlovsky, G.N., (1966) Control of walking and running by means of electrical stimulation of the mid-brain. Biofizika. 11, 659-666. (English version of same journal), (756-765).

Shik, M.L., Severin, F.V. and Orlovsky, G.N., (1967) Structures of the brain stem responsible for evoked locomotion. Fiziol. Zh. USSR. 12, 660-668.

Shurrager, P.S. and Dykman, R.A., (1951) Walking spinal carnivores. J. Comp. Physiol. Psychol. 44, 252-262.

Sjöström, A. and Zangger, P., (1975) α-γ-linkage in the spinal generator for locomotion in the cat. Acta Physiol. Scand. 94, 130-132.

Stein, P.S.G., (1974) Neural control of interappendage phase during locomotion. Am. Zool. 14, 1003-1016.

Székely, G., (1968) "Development of limb movements: Embryological, physiological and model studies," In Ciba Foundation Symposium on Growth of the Nervous System. (Wolstenholme, G.E.W. and O'Connor, M., eds.), J. & A. Churchill Ltd. London, (77-93).

Székely, G., Czéh, G. and Voros, G., (1969) The activity pattern of limb muscles in freely moving normal and deafferented newts. Exp. Brain Res. 9, 53-62.

Viala, G. and Buser, P., (1969) Activités locomotrices rythmique stéréotypées chez le lapin sous anesthésie légère. Exp. Brain Res. 8, 346-363.

Viala, D. and Buser, P., (1971) Modalités d'obtention de locomoteurs chez le lapin spinal par traitements pharmacologiques (DOPA, 5-HTP, d'amphétamine). Brain Res. 35, 151-165.

Wilson, D.M., (1964) "The origin of the flight-motor command in grasshoppers," In Neuronal Theory and Modeling. (Reiss, R.F., ed.), Stanford Univ. Press, Stanford, (331-345).

Wilson, D.M. and Waldron, I., (1968) Models for the generation of the motor output pattern in flying locust. Proc. IEEE. 56, 1058-1064.

MECHANISMS OF INTERLIMB PHASE CONTROL

Paul S.G. Stein

Department of Biology, Washington University

St. Louis, Missouri 63130

During locomotion there is a phase lag in the movement of one limb when compared to the movement of another limb. The neurons responsible for the regulation of interlimb phase have been described in the swimmeret system of the crayfish. These neurons are termed coordinating neurons. Their axons originate in one limb-bearing ganglion of the ventral nerve cord and terminate in neighboring ganglia of the nerve cord. When the axons of coordinating neurons are cut, interlimb phase regulation is destroyed. These interneurons are active during a specific fraction of the movement cycle of a limb (the "modulator" limb). This activity pattern persists in the absence of sensory discharge from the "modulator" limb. The discharge of coordinating neurons can alter the timing of locomotory movements in another limb (the "modulated" limb). Such an alteration is termed a phase shift. Both the magnitude and direction of the phase shift depend upon the arrival time of the coordinating neuron discharge in the movement cycle of the modulated limb. A systematic plot of these phase shifts is termed the "phase response curve" (PRC). The PRC can be obtained utilizing a "cut command neuron" preparation. Such a preparation leaves the coordinating neurons intact and permits separate electrical control of each limb. The magnitude of interlimb phase can be predicted if the values of the following are known:
(1) the intrinsic cycle period of the modulator limb,
(2) the intrinsic cycle period of the modulated limb,
and (3) the PRC.

It is likely that interlimb phase is regulated by coordinating neurons in other multilimbed organisms. This chapter will present an experimental strategy based on the measurement of the interlimb PRC. Such a strategy is designed to reveal the common features of interlimb phase control which are shared by both the arthropods and the vertebrates.

Introduction

During forward locomotion of a limbed animal, the movements of the limbs are rhythmic and coordinated with one another (Gray, 1968). This chapter will examine the neural mechanisms responsible for interlimb coordination in the arthropods and the lower vertebrates. In these organisms similar principles of central nervous system organization are found. For example, each limb is innervated by a local control center within the central nervous system. Each local control center when isolated from the remainder of the central nervous system can still produce coordinated rhythmic output to agonist and antagonist limb muscles (Ikeda and Wiersma, 1964; Pearson, 1972; Stein, 1974; Grillner, 1975). The local control center is therefore a biological oscillator. Moreover since each limb has its own local control center, then it follows that the locomotory pattern generator includes a set of biological oscillators. The approximate frequency of each control center is regulated in part by descending tonic neural activity. Such activity does not contain the precise timing information necessary to regulate the interlimb phase lag. Such regulation is produced by coordinating neurons which transmit phase information among the control centers. These coordinating neurons therefore serve to couple individual oscillators. It follows that interlimb coordination can be understood by examining the behavior of coupled biological oscillators.

There has been extensive mathematical work examining the coupling of biological oscillators (Pavlidis, 1973). For the optimal application of this work, knowledge of the differential equations describing the important variables of each oscillator is required. Such information is simply not available in locomotory systems. An alternate approach can be utilized when the oscillator equations are not known (Moore et al., 1963; Perkel et al., 1964; Pavlidis, 1973). In this approach, the interoscillator coupling signal is considered a critical variable. This coupling signal has two characteristics. First, the signal carries phase information about the state of one oscillator. Second, the signal can phase modulate the rhythm of a second oscillator. A measure of this modulation, the Phase Response Curve (PRC; Pavlidis, 1973; Pinsker, 1976), can be experimentally determined.

The PRC is important in several ways. First it can be utilized to predict the magnitude of the interoscillator phase lag and the range of stable phase-locking under a wide variety of conditions. Second the PRCs of several organisms can be compared and similarities or differences readily observed. Third the PRC is a basic measure of the biological oscillator. Any biophysical analysis of an oscillator must be able to predict what the PRC will be for a given coupling signal. The PRC therefore represents an important conceptual level intermediate between the biophysical and the behavioral levels of analyses.

For precise mathematical definitions of phase, see Pavlidis (1973); for this discussion, note that interlimb phase equals interlimb latency divided by cycle period. Cycle period is the time between two successive occurrences of a standard event. Interlimb latency is the time between the occurrence of a standard event in the cycle of one limb and the next occurrence of the standard event in another limb. A typical standard event is the onset of retraction in a limb.

In some behaviors the coupling signal among oscillators can be clearly demonstrated. For example circadian rhythms can be controlled by a repetitive regime of light and dark cycles (Aschoff, 1965). In these experiments the coupling signal is external to the organism and can be readily manipulated by the experimenter. In the study of interlimb coordination during locomotion, the major coupling signal among the several limbs resides within the central nervous system. Early work on interlimb control (e.g., von Holst, 1939) demonstrated effects attributable to a central coupling signal, but did not identify the signal. In later work CNS recording techniques were utilized and the interlimb coupling signal could be identified (e.g., Pearson and Iles, 1973; Stein, 1971, 1974). This chapter will discuss both the early and the late experiments.

Selected papers on arthropod and lower vertebrate nervous systems will be reviewed. Much important work in other groups will not be discussed such as work in annelids (e.g., Gray et al., 1938; Kristan et al., 1974), molluscs (e.g., Willows et al., 1973) and mammals (e.g., Kulagin and Shik, 1970; Miller and van der Méche, this volume). Instead three experimental systems will be discussed in detail. These systems are (1) fish fins (von Holst, 1939, 1973), (2) insect legs (Pearson and Iles, 1973) and (3) crustacean swimmerets (Stein, 1971, 1974). A common set of principles characterize all three systems. Moreover it is apparent to this author that in order to understand the nature of interlimb phase coupling in many other organisms, the mathematics of coupled oscillators must be utilized in the design and analysis of future experiments. A brief summary of some of the relevant mathematics will be presented at the end of this chapter.

Fish Fins (von Holst, 1939)

The studies of Erich von Holst during the 1930's are an excellent example of the early work on interlimb phase control (von Holst, 1939; English translation, 1973, Chapter 2). The following discussion is based upon the terminology utilized in the English translation.

von Holst (1973) classified interlimb coordination states into two categories, absolute and relative coordination. In order to discuss these two coordination states, limb behavior must be examined during conditions when the movement frequency of each limb remains constant when averaged over many cycles. Absolute coordination between a pair of limbs applies to situations in which both limbs are moving at the same frequency. Such a state is also termed 1:1 entrainment. During this state interlimb phase is stable. Absolute coordination is the most common locomotory behavior. Relative coordination refers to situations in which the two limbs are moving at different frequencies. A preferred phase or set of phases is observed, but the phase does not remain constant from cycle to cycle. Relative coordination can be observed in naturally-occurring behavior but is rare (von Holst, 1939; Wendler, 1966; Macmillan, 1975). Some types of relative coordination are described as a "gliding coordination" (Wendler, 1966, 1974) in which there will be a temporary phase lock at a preferred phase, then a phase drift, which is followed by phase lock, then drift, and so on. Another form of relative coordination is observed when the ratio of frequencies of two limbs can be expressed as a fraction of two small integers, such as 2:1, 3:2, etc. (Wilson, 1967; von Holst, 1973; Pavlidis, 1973). During such "integral fraction coordination," several preferred phases may be observed. At least one of these preferred phases usually corresponds to the preferred phase observed during an absolute coordination state.

von Holst recognized the theoretical importance of the relative coordination state and devised experimental strategies to elicit such states. He was able to reveal fundamental properties of the central nervous system in these experiments. A wide variety of organisms, including man (von Holst, 1973), were investigated. The most quantitative work examined the coordination of fish fins. In one example of this work, the medulla of the fish, Labrus, was sectioned thus producing an animal displaying spontaneous fin movements with both absolute and relative coordination (Figure 1). The right and left pectoral fins were in absolute coordination with each other. Under certain circumstances, e.g., after + in Figure 1B, the dorsal fin was in absolute coordination with the pectoral fins. In other circumstances, relative coordination was observed, e.g., during the initial portions of Figure 1A and 1B when the

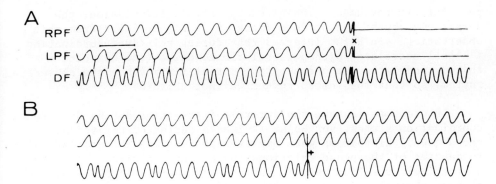

Figure 1. Relative and absolute coordination of fish fin movements. Recordings of the movements of the right pectoral fin (RPF), the left pectoral fin (LPF), and the dorsal fin (DF) of Labrus. Upward deflection of the mechanogram tracing indicates pectoral fin movement in the anteriad direction and dorsal fin movement to the left. In A, the recordings were halted at x while tactile pressure was applied to the anteriad portion of the trunk. Once the stimulus inhibited the pectoral fin movements, then recording of fin movements commenced. Note that in the initial portions of A and B (prior to the x and +, respectively) the dorsal fin is moving at a higher frequency than the pectoral fins. In these records, there is a tendency for the upward portion of the dorsal fin trace to coincide with the downward portion of the left pectoral fin trace; this tendency is characteristic of the relative coordination state. In this state, there is considerable cycle-by-cycle variation in the cycle duration of the dorsal fin movement. In this example the dorsal fin movement is the modulated or dependent rhythm. Note that when the pectoral fins are inhibited (after the x in A) or when the pectoral fins move at the same frequency as the dorsal fin (after the + in B), the duration of the dorsal fin cycle remains the same for several cycles. The coupling observed after the + in B is characteristic of the absolute coordination state. In this example the transition from relative to absolute coordination occurred spontaneously. The horizontal bar is a two second time mark. (Modified from von Holst, 1939.)

dorsal fins moved at a higher frequency than the pectoral fins. When the system was in absolute coordination, all of the rhythms were "uniform," i.e., they were of constant frequency. When the system was in relative coordination, the rhythm of the dorsal fin displayed "periodicity" (ibid; p. 50 = "departure from uniformity of appearance"). The cycle-by-cycle variation of the dorsal fin movement did not occur randomly, but it was directly related to the current phase of the pectoral fin movement cycle. When the pectoral fin rhythm was inhibited by tactile stimulation, the dorsal fin rhythm was uniform (after x in Figure 1A). From this observation, von Holst inferred that during relative coordination a signal correlated with pectoral fin movement (the dominant rhythm) modulated the phase of the dorsal fin rhythm (the dependent rhythm) so that the latter rhythm was not uniform.

In order to measure the interfin interaction during different coordination states, von Holst devised a series of analytical techniques. Histograms were employed to describe the statistical tendencies of the phase relationships of one fin with respect to another. If the fins were in an absolute coordination state, then a phase histogram (ibid., p. 52) would display a sharp peak at the preferred phase, which was also termed the "coaction position." If a "gliding" relative coordination (see also Wendler, 1966) was observed, then the phase histogram would display a broad peak at the "coaction position." The fact that the maximum of the peak observed during both absolute and relative coordination was similar is consistent with the hypothesis that the interaction in the relative coordination state is basically the same as that observed during an absolute coordination state. On the other hand, if the relative coordination is such that there is an integral-fraction relationship between the rhythms, then a multiple-peaked relationship is observed (von Holst, 1973, p. 53).

A slightly more detailed analysis of the relative coordination state is provided by the method of "continuous time and speed registration" (ibid., p. 55). In this procedure the "action time" (= duration of movement in one direction) of either the "phase" (= upward movement of the trace) or of the "counter-phase" (= downward movement of the trace) is plotted as a function of elapsed time (ibid., p. 56). Note that the use of the word "phase" in this context (von Holst, 1973) is different than the conventional meaning of phase and quotation marks will be used when "phase" refers to a portion of the cycle. This procedure reveals that there is a smooth oscillation of the action time plot. In order to understand the systemic basis for this oscillation, a third procedure is necessary.

This method, that of "time tables," yields further insight into the underlying processes of the relative coordination state. In this procedure the mean value of the action time is calculated. The percent deviation of each action time from the mean action time is also calculated. The percent deviation (DELTA %, Figure 2) may be plotted as a function of the relative phase of the two rhythms. Several observations can be made from such a table. First there is a fluctuation of the value of the percent deviations of the action time which depends upon the phase relationship between the two rhythms. Several observations can be made from such a table.

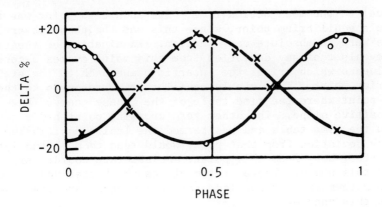

Figure 2. Analysis of the cycle-by-cycle variation of the modulated (dependent) rhythm during the relative coordination state: the "time table" of von Holst. The abscissa (PHASE) indicates the phase relationships of the dorsal fin movement (the dependent rhythm) with respect to the pectoral fin movement (the dominant rhythm). The ordinate (DELTA %) indicates the percentage deviation of the "action time" (AT; the duration of fin movement in one direction) when compared to the mean value of all the action times measured in that direction (AT_{av}), i.e., DELTA % = 100 $(AT-AT_{av})$/AT_{av}. The values indicated by 0 are for the AT of dorsal fin movements to the left; those indicated by x are for the AT of dorsal fin movements to the right. (Modified after von Holst, 1939.)

First there is a fluctuation of the value of the percent deviations of the action time which depends upon the phase relationship between the two rhythms. When the rhythms of the "phase" plot are in the vicinity of zero or one phase (denoted by open circles), then the action times are long with respect to the mean. When the rhythms are in the vicinity of 0.5 phase, then the action times are short with respect to the mean. Moreover, the curve has a positive slope for phases greater than 0.5 and a negative slope for phases less than 0.5. If the action time of the other half of the cycle, the "counter-phase" portion of the cycle, is measured then the "counter-phase" curve is inverted with respect to the "phase" curve.

Although the precise form of the time table can be altered by measuring the phase or the action time according to a different convention, von Holst observed that there were fixed relationships which could be detected independent of the convention. First, the dependent rhythm is delayed whenever it is ahead of the phase position maintained in absolute coordination, and is advanced whenever it is behind this phase position (ibid., p. 62). Second, there were always two points on the curve in which the action times were equal to the mean action time. One such zero point was termed the stable equilibrium point, since this was the phase observed if the fins were in absolute coordination. Examination of the time table reveals that any deviation from this point evokes a correcting influence which brings the interfin phase back to the stable equilibrium point. This stable equilibrium corresponds to the zero percent point where the time table of the "phase" curve (Figure 2) has a positive slope. The other zero crossing point had a negative slope in the time table and was termed the labile equilibrium point, since any deviation from that point would tend to move the interfin phase further away from this point. No absolute coordination state utilized this phase. These statements can be better understood when the mathematics of coupled oscillator theory are discussed later in this chapter.

This important work demonstrated that phase modulations in the rhythm of one appendage could be correlated with specific movements of another appendage. These modulations, termed the "magnet effect," must be the result of some process within the central nervous system. These experiments do not reveal, however, the specific nature of the coupling signal responsible for the phase modulation (ibid., p. 106). In order to reveal the neural events underlying the regulation of interlimb phase, relative coordination states must be examined while monitoring the discharge of central nervous system cells. Such recording has been accomplished in the last five years from neurons active during walking movements of the cockroach and during swimmeret movements in the crayfish. In both preparations records of neural activity during a relative coordination state revealed interneurons which may be responsible for coupling the rhythms of different limbs. Such cells have been termed coordinating neurons (Stein, 1971).

Insect Legs (Pearson and Iles, 1973)

The behavior of cockroach limbs during walking has been examined (Delcomyn, 1971) over a wide frequency range. For frequencies over 5 Hz. the interlimb phase between neighboring limbs is 0.5. Experimental analysis of the neural output during walking (Pearson, 1972; Pearson and Iles, 1973; Delcomyn and Usherwood, 1973) reveal that in a single limb there is activity in coxal levator motor neurons which alternates with activity in coxal depressor motor neurons.

During forward locomotion, the levator muscles are active during protraction and the activity of levator muscles in one limb alternates with levator activity in a neighboring limb. Such alternation can also be observed in a headless preparation in which the sensory input to the limbs has been removed (Pearson and Iles,

1973). These preparations can display relative coordination (Figure 3 of Pearson and Iles, 1973). During such a state there is a strong tendency for the levator muscles of one limb not to be active when the levator muscles of a neighboring limb are active (Figure 4 of Pearson and Iles, 1973). Pearson and Iles (1973) inferred from these data that there must be some mutual inhibition among the interneurons controlling levation in the neighboring limbs.

In addition, Pearson and Iles (1973) reasoned that if levation in one limb was inhibited during levation of a neighboring limb, then there may be a levation-correlated neural signal in the interganglionic connectives which serve to coordinate interlimb phase. Recordings from fine filaments dissected from the dorso-lateral region of the interganglionic connectives revealed cells whose spike activity was co-active with levation activity in a nearby limb (Figure 7 of Pearson and Iles, 1973). These cells have the appropriate properties to act as coordinators of the limb rhythms. Further tests of whether these cells actually act as coordinating neurons have yet to be reported.

Crustacean Swimmerets (Stein, 1971, 1974)

A second system which has been examined with neural recording techniques is the swimmeret system of crayfish and lobsters. The behavior of the limb movements during rhythmic "beating" movements has been examined (Hughes and Wiersma, 1960; Davis, 1968a). Retraction of a segmental pair of swimmerets occurs in phase with each other. In addition there is a 0.1 to 0.3 phase lag of retraction in a swimmeret when compared to the retraction of the next posteriad ipsilateral swimmeret. This type of activity has been termed a "metachronal" wave since there is the visual impression of a wave of activity which "travels" in the caudal to rostral direction. The neural output during such behavior consists of the discharge of powerstroke motor neurons during retraction of the limb and the discharge of returnstroke motor neurons during protraction of the limb (Hughes and Wiersma, 1960; Davis, 1968b). During normal rhythmic activity there is alternation between powerstroke and returnstroke motor neurons. Activation of powerstroke motor neurons in an anteriad segment occurs with phase lag when compared to the activation of powerstroke motor neurons in a more posteriad segment (Hughes and Wiersma, 1960; Ikeda and Wiersma, 1964; Wiersma and Ikeda, 1964; Stein, 1971; Davis and Kennedy, 1972). This behavior can occur either spontaneously, or in response to tactile stimulation, or in response to stimulation of single axons ("command" fibers) in the intersegmental connectives of the nerve cord (Hughes and Wiersma, 1960; Wiersma and Ikeda, 1964; Davis and Kennedy, 1972).

Many of the command elements are located in the lateral regions of the connective. It is possible to cut these regions and create the "cut command neuron" preparation (Stein, 1971, 1973, 1974; see esp. Figure 3 of Stein, 1974). In such a preparation, stimulation posteriad to the cut can elicit rhythmic swimmeret movements posteriad to the cut; stimulation anteriad to the cut can elicit rhythmic swimmeret movements anteriad to the cut. If stimulation on either side of the cut yields a movement frequency of the respective limbs which is approximately the same in either situation, then stimulation of the command elements on both sides of the cut during the same time period can yield an absolute coordination state of the limbs both anteriad and posteriad to the cut (Figure 4 of Stein, 1971). If, on the other hand, stimulation on one side of the cut can be manipulated to elicit an extremely different limb movement frequency than that obtained by separate stimulation on the other side of the cut, then when stimulation is applied during the same time period to both sides of the cut a relative coordination state can be produced (Figure 5 of Stein, 1971). In this relative coordination state there is a high probability that powerstroke activity in a posteriad limb will be followed by powerstroke activity in the next anteriad limb. Moreover, if powerstroke activity of a posteriad limb begins during the returnstroke portion of activity in the next anteriad limb, then there is a tendency for the period of the anteriad limb to be shorter than the average period. In contrast, if the powerstroke in the posteriad limb begins during the powerstroke of the anteriad limb then there is a tendency for the anteriad limb to display a cycle period which is longer than the average period (Figure 6 of Stein, 1971). This result is similar to the data presented in the von Holst time table.

This result suggests that powerstroke-correlated cells in the intersegmental connectives serve to coordinate the limbs during absolute coordination and are responsible for the magnet effect observed during relative coordination. Such cells are located in the medial portion of the connective (Hughes and Wiersma, 1960; Wiersma and Hughes, 1961; Figure 2 of Stein, 1971). If this tract is cut, then interlimb coordination is severely disturbed (Figure 3 of Stein, 1971). In addition, interlimb phase can still be regulated in the swimmeret system even after the system is totally de-afferented (Ikeda and Wiersma, 1964). This implies that at least some of the cells responsible for interlimb coordination maintain their normal activity after all sensory input is removed. The powerstroke-correlated cells in the medial portion of the connective are co-active with powerstroke activity in a de-afferented preparation (Figure 2 of Stein, 1971). These cells therefore carry a copy of the efferent activity and satisfy the requirement that some cells carry phase information in the deafferented state.

If this tract of cells does act to coordinate the movements of neighboring ipsilateral limbs, then selective activation of this group of cells should modulate the phase of the rhythm of an anteriad swimmeret. Direct extracellular stimulation of the tract does yield such a phase modulation (Stein, Unpublished data) but this technique is not preferred since such extracellular stimulation may also activate other cells, such as the returnstroke-correlated neurons known to lie in the same region of the connective (Table 1 of Stein, 1971). Instead, selective activation of powerstroke-correlated coordinating neurons can be achieved by applying a brief pulse of command neuron stimulation to the posteriad half of a cut command neuron. The central coordinating neuron discharge is activated with the discharge of posteriad powerstroke motor neurons. The stimulus train can be adjusted so that the elicited discharge in both the coordinating and motor neurons is similar to a single burst of naturally-occurring activity. The anteriad swimmeret can be stimulated with constant frequency command neuron pulses delivered to the anteriad half of a cut command neuron (Figure 3; see also Stein, 1974). In this situation the anteriad limb control center is generating the rhythmic movements of the anteriad limb. In this experiment, a phase modulation in the rhythm of the anteriad oscillator is observed in response to a single burst of coordinating neuron activity (Figure 3). This phase modulation is measured as the ratio of the modulated period to the unmodulated period (Figure 4; see also Stein, 1974). The period ratio minus one is equal to the phase shift. This experiment establishes that both phase advances and phase delays can be produced by the same stimulus when delivered at different times of the cycle (Figure 4).

These phase shift measurements are critical to the elucidation of the mechanisms underlying interlimb phase control. These data reveal that central neural activity generated during <u>one</u> movement in <u>one</u> limb can alter the centrally-generated rhythm of <u>another</u> limb. They are consistent with the hypothesis that retraction-correlated activity of a posteriad limb excites retraction of the next anteriad limb.

During natural locomotory behavior, the movement of the posteriad swimmeret is rhythmic. Therefore there will be rhythmic bursts of activity in the ascending powerstroke-correlated cells. The next question is whether these rhythmic phasic discharges ascending from a posteriad swimmeret control center can alter the phase and frequency of an anteriad swimmeret control center so that both limbs are moving with equal frequency and fixed relative phase. Qualitative arguments (Stein, 1974) suggest that the phase shifts can account for the observed behavior but more parametric experiments need be performed in order to provide even stronger support for this hypothesis.

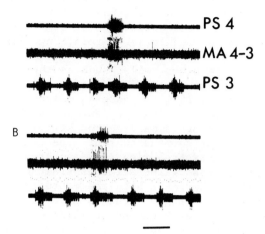

Figure 3. Contingent response of a swimmeret control center to a single burst of coordinating neuron discharge. The "cut command neuron" design (Figure 3 of Stein, 1974) is utilized. In this preparation, the axons of command neurons are cut and the axons of coordinating neurons are left intact. The portion of the command neurons anteriad to the cut are stimulated via the anteriad or STIM 3 electrode; the portion of the command neuron posteriad to the cut are stimulated via the posteriad or STIM 4 electrode. Simultaneous recordings of powerstroke motor neurons from the fourth ganglion (PS 4, upper trace), coordinating neurons from the medial ascending tract of the 4-3 connective (MA 4-3, middle trace), and powerstroke motor neurons from the third ganglion (PS 3, lower trace) are displayed. Constant frequency stimulation (12.5 Hz) was applied via the STIM 3 electrode in order to produce repetitive bursts of PS 3. A short burst (170 msec) of stimulation (45 Hz) was delivered via the STIM 4 electrode in order to produce a single burst of PS 4 discharge and a single burst of MA 4-3 discharge. The STIM 4 electrode was activated at different times of the PS 3 cycle. A, Activation of the STIM 4 electrode during the PS 3-off (returnstroke) portion of the cycle was associated with an advance in the onset of the next powerstroke burst (PS 3). B, Activation of the STIM 4 electrode during the PS 3-on (powerstroke) portion of the cycle results in a more vigorous PS 3 burst, and the onset of the subsequent PS 3 burst was delayed. A systematic plot of these phase shifts is presented in Figure 4. Time mark, 1 sec. (from Stein, 1974).

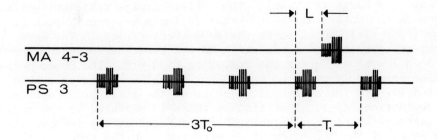

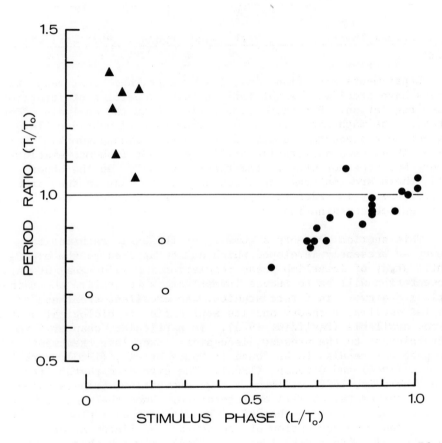

Figure 4. The first transient PRC of the swimmeret system. Plot of the phase modulation of powerstroke motor neuron discharge in the third abdominal ganglion (PS 3) correlated with the discharge of medial ascending coordinating neurons in the 4-3 connective (MA 4-3). The triangles represent points in which the MA 4-3 discharge began and ended (Legend continued on next page.)

in the PS 3 burst (PS 3-on). The filled circles represent points in which the MA 4-3 discharge began during a quiescent epoch of PS 3 (PS 3-off) and ended in the subsequent PS 3-on. The open circles represent points in which MA 4-3 burst began during the PS 3-on and ended during the following PS 3-off. The phase modulation in PS 3 is measured as the period ratio of the perturbed period (T_1) to the arithmetic average of the three prior unperturbed periods (T_0). The stimulus phase is measured as the ratio of latency (L) to T_0. Latency (L) is defined as the interval to the onset of the MA 4-3 burst as measured from the onset of the prior PS 3 burst. The period ratio minus one is equal to the phase shift (from Stein, 1974).

Mathematical Theory (Moore, Perkel and Segundo, 1963; Pavlidis, 1973)

Experiments with fish fins, insect legs, and crustacean swimmerets have provided insight into certain properties of interlimb phase regulation. But there are still gaps in the analysis. For example, the mathematical theory of coupled nonlinear oscillators has not been adequately utilized in the experimental work. Stein (1973, 1974) uses qualitative predictions made from mathematical theory but does not utilize the theory in full. On the other hand, there are improvements which need to be made in the mathematical theory in order to apply the theory rigorously to all data on interlimb coordination.

This section presents a summary of the basic mathematical theory as presently developed which can be applied to the experimental study of interlimb phase regulation. A major goal of this presentation will be to induce investigators to utilize the mathematical framework in future studies. An excellent treatment of coupled oscillator theory and its application to biological systems is now available (Pavlidis, 1973). In particular, chapter 4 is most relevant to the present discussion. Excellent treatments of the problem are also to be found in Moore et al., (1963); Perkel et al., (1964) and Ottoson, (1965). The circadian rhythm literature (Pavlidis, 1973) utilizes a sign convention which is opposite that of the neural oscillator literature. Once the sign convention is accounted for, the mathematical results are equivalent. In this section the sign convention of the neural oscillator literature (Moore et al., 1963; Perkel et al., 1964) in which phase advances are negative and phase delays are positive will be utilized.

The phase shifts in a modulated oscillator which are in response to a single "pulse" from a modulator can be expressed graphically as a function of the relative phase of the "pulse" in the cycle of the modulated oscillator (see Figure 5 for a

INTERLIMB PATTERN GENERATOR

hypothetical example). Such a curve is termed the Phase Response Curve (PRC; Pavlidis, 1973). The present treatment will assume that the entire phase shift is expressed in the first modulated cycle. In the application of this theory to interlimb phase regulation, note that the "pulse" can in fact be the burst of coordinating neuron activity.

It is important to know what conditions are necessary for the existence of a stable phase relationship between the oscillators. It is given that the modulator oscillator will produce a pulse at constant intervals. This pulse will alter the period of the modulated oscillator. If this alteration is such that the new cycle period of the modulated oscillator equals the cycle period of the modulator, then phase stability is possible.

In the present hypothetical example (Figure 6), the modulating oscillator is the "A" oscillator and its period is A seconds. The oscillator which is modulated by "A" is the "B" oscillator. When "A" is not active, the period of the "B" oscillator is B seconds. When a pulse from "A" occurs at latency L in the "B" oscillator

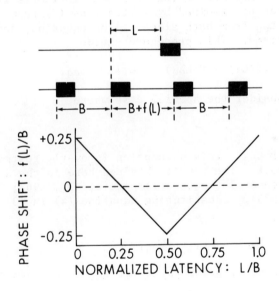

Figure 5. A hypothetical phase response curve (PRC). This curve displays both phase advances and delays in addition to positive and negative slope. A single pulse of the "A" oscillator occurs with latency L when compared to the last pulse of the "B" oscillator. The "B" oscillator displays a change in cycle period, $f(L)$, in response to the pulse from "A." This PRC is a plot of the normalized $f(L)$ as a function of the normalized latency.

cycle, then the modulated cycle period of "B" is $B + f(L)$ seconds. A plot of $f(L)$ as a function of L is the Phase Response Curve (Pavlidis, 1973). For convenience it is useful to normalize the PRC and plot $f(L)/B$ as a function of L/B (Figure 5). Normalization of the plot has the advantage that PRCs measured from different frequencies can be compared. However, there are situations when normalization is not appropriate, e.g., when one portion of the cycle remains fixed while cycle period varies.

If we consider the case where the "A" pulse repeats every A seconds (Figure 6), then it can be seen that

$$A + L_i = B + f(L_i) + L_{i+1} \tag{1}$$

Now if the system is in a state of equilibrium then $L_i = L_{i+1}$. In equilibrium, then it follows from equation (1) that

$$A - B = f(L_{eq}), \tag{2}$$

where L_{eq} is the interoscillator latency at equilibrium. Note that equation (2) provides a necessary condition for an equilibrium state but does not indicate whether the equilibrium is stable or not. Stability can be examined by subtracting (2) from (1), further subtracting L_{eq} from both sides of the equation, and then rearranging the terms. This procedure yields

$$[L_i - L_{eq}] - [f(L_i) - f(L_{eq})] = L_{i+1} - L_{eq}, \tag{3}$$

For simplicity consider the special case where

$$f(L) = mL + b, \tag{4}$$

in the vicinity of L_{eq}. This assumption is overly restrictive, but simplifies the proof. The result obtained here is the same obtained by a more general treatment (Moore et al., 1963; Ottoson, 1965; Pavlidis, 1973). Substituting equation (4) in (3) then yields

$$(1 - m)(L_i - L_{eq}) = L_{i+1} - L_{eq}, \tag{5}$$

If L_{eq} is a stable equilibrium point for absolute coordination then

$$[L_{i+1} - L_{eq}] < [L_i - L_{eq}] \tag{6}$$

must be true. That is, it is required that the absolute difference between the latency of the current cycle and that at equilibrium must diminish with each successive cycle. The inequality (6) will be satisfied if and only if

INTERLIMB PATTERN GENERATOR

$$|1 - m| < 1, \text{ or } 0 < m < 2, \tag{7}$$

Thus it is a necessary condition for stability that the slope (=m) of the PRC at a given L is between zero and two. Note that if m equals one then the equilibrium latency can be reached in the next cycle (see equation 5). If m is between zero and one then the approach to equilibrium is monotonic. On the other hand if m is between one and two then there is a oscillatory approach to equilibrium (see also Moore et al., 1963, Table 1).

Thus, for a <u>stable equilibrium</u> point to exist both relationships 2 and 7 must be satisfied. Consider the example of the hypothetical PRC of Figure 5. For all phases less than 0.5 all possible equilibria are unstable, whereas for all phases greater than 0.5 all possible equilibria are stable. If $A - B < 0$, then stable equilibrium is possible only between phases of 0.5 and 0.75. If $A - B > 0$, then stable equilibrium is possible only between phases 0.75 and 1.0. For the special case of $A = B$, then stable equilibrium can only occur at a phase equal to 0.75. Note that an equilibrium point for $A = B$ also exists at phase equal to 0.25, but that this equilibrium point is unstable.

It is to von Holst's credit that he was able to observe the distinction between stable and unstable equilibria in his time table (von Holst, 1973, p. 62) without having the benefit of a mathematical formalism. Note however that von Holst had a very restricted view of the equilibrium point, that is the zero crossing of the time table. It is only an equilibrium point for the special case where the two oscillators have the same unmodulated frequency (and also where the mean modulated period is equal to the unmodulated period of the dependent rhythm). From a more general point

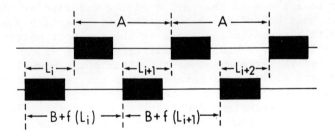

Figure 6. The interrelationship between oscillator period and interoscillator latency. The "A" oscillator modulates the "B" oscillator. The magnitude of the phase shift in the "B" oscillator is defined by the phase response curve (Figure 5). The relationship between A, B, L and f(L) is given by equation 1 of the text.

of view, if the difference between A and B is nonzero then two equilibria may exist (one stable and one unstable) in the hypothetical example in Figure 5. The phase of the system in the stable equilibrium state will vary in a systematic way as the difference between A and B varies (equation 2). The shift in the phase of absolute coordination state as a function of a shift in the difference between the unmodulated periods has been directly measured in the swimmeret system (Stein, 1973).

There are problems with the direct application of this formulation as presently stated. The neural oscillators controlling the swimmerets do not exhibit the entire phase shift in just one cycle in response to a single burst of coordinating neuron discharge. Two or three cycles may be necessary before the period of the modulated oscillator will return to its unmodulated value. In this case then it is necessary to measure a set of PRCs: $f_1(L)$, $f_2(L)$, etc. Thus $f_1(L)$ is the phase shift measured in the first modulated period and can be termed the first transient PRC. Then $f_2(L)$ is the phase shift measured in the next cycle of B and $f_3(L)$ is that of the third cycle. As many $f_i(L)$ should be measured until there is an $f_k(L)$ which equals zero for all L. In this case then $f(L)$ can be defined as the steady state phase shift and is equal to the sum of $f_1(L)$, $f_2(L)$ and so on up to $f_{k-1}(L)$. With these modifications then it may be possible that the equilibrium condition (equation 2) may still apply to the steady state PRC. It is not clear what the appropriate stability condition is under these circumstances. Modifications of the stability equation will be necessary before this formulation can be applied rigorously to interlimb phase stability conditions.

It is important to note, however, that the first transient PRC of the swimmeret system (Figure 4) does show a positive slope less than one for phases greater than 0.5. These phase measurements refer to a phase in the period of the anteriad or modulated oscillator. Most behavioral measurements of interlimb phase have been made with respect to the phase in the cycle of posteriad or modulator oscillator. When these differences are accounted for, it can be observed that the range of interlimb phase normally observed in the swimmeret system does correspond to the region of the first transient PRC with a positive slope less than one. In order to measure the second transient PRC and that of the steady state, averaging techniques need to be applied in the swimmeret system.

Summary

Thus the experimental work with fins, legs and swimmerets does lend itself to the application of coupled oscillator theory. It is clear that future experiments in these and other animals will need protocols that are dictated by the needs of the mathematical equations. When such dicta are followed, then it will be possible to utilize the full power of the mathematics.

But even with measures of phase modulation as different as the time table and the PRC, it is possible to see certain qualitative similarities in the data. These similarities must be noted as correlative statements at the present time, e.g., the correlation between certain slopes of the time table (Figure 2) or slopes of the first transient PRC (Figure 4) and the stable phase points during absolute coordination. With the proper measurements then the PRCs could be utilized in a stronger sense to make predictive statements.

Moreover, it should be possible to utilize the mathematical theory to predict the range of stable entrainment during the absolute coordination state. These predictions could be tested by the appropriate experimental designs. In addition, the application of the mathematical theory will be most useful when intracellular recordings from control center interneurons are made (see Pearson and Fourtner, 1975). In such a preparation the postsynaptic potential change in a control center interneuron could be measured in response to a burst of coordinating neuron discharge. Then it would be interesting to apply a brief polarization to the control center interneuron which is adjusted to match the polarization produced by a single burst of coordinating neuron discharge. Such a direct polarization might yield a PRC which is similar to that produced by the synaptic drive from coordinating neurons. Such an experimental procedure would reveal the properties of the modulated control center which contribute to the form of the PRC.

While this author thinks that the full utilization of the mathematical theory would be an important contribution to the understanding of interlimb phase control, an equally important contribution is the qualitative utilization of the mathematical approach to new experimental situations. One example of such a usage is the identification of the interlimb coupling signal. To date there are many animals whose locomotory behaviors have been examined; there are only a few animals in which the interlimb coupling signal has been identified even tentatively. The PRC in response to a single coupling event has been measured only in the swimmeret system. It is apparent to this author that the conceptual framework of coupled oscillator theory must now be utilized in future work with interlimb phase. In particular there is a strong need to utilize this approach in the limbed vertebrates. For example, work in the author's laboratory has demonstrated that electrical stimulation applied to the dorsolateral columns of the spinal cord can produce coordinated swimming movements in one or several limbs of a turtle with an intact central nervous system (Lennard and Stein, 1974). Moreover, stimulation of dorsolateral columns in a low spinal turtle can elicit swimming movements of a single limb (Lennard, 1975; Lennard and Stein, 1976) and such stimulation delivered in a high spinal turtle can elicit swimming in several limbs (Stein, 1976). In the latter case the interlimb coordination is that seen during swimming behavior in intact turtles. This technique is now being extended to produce the equivalent of the "cut command neuron" preparation in the turtle.

Thus, what is most important in all these studies is the elucidation of a common conceptual framework which can apply both to the arthropods and the vertebrates. This conceptual framework can be viewed as a working hypothesis which needs testing in a number of organisms. The author feels that this approach will reveal deep understanding of the logical features of the neural networks controlling locomotion.

This conceptual framework may be summarized as follows:

(1) During a specific phase of the movement cycle of limb A, coordinating neurons are activated. This phase may correspond to the activation of a specific set of muscles in limb A.

(2) The activity of these coordinating neurons synaptically influences a specific set of interneurons in the control center for limb B. These interneurons may themselves activate a specific set of motor neurons innervating limb B.

(3) The synaptic influence on these interneurons in the control center for limb B by the coordinating neuron discharge can phase modulate the rhythm expressed by the control center of limb B. This phase modulation can be described by a specific phase response curve PRC.

(4) The magnitude of interlimb phase observed during naturally-occurring locomotion exhibiting absolute coordination corresponds to a stable equilibrium position of the PRC. Moreover, the stability properties of interlimb phase observed during absolute coordination are predicted by application of the appropriate equations.

Acknowledgment

The author has been supported by NSF Grant #GB-35534.

REFERENCES

Aschoff, J., (1965) <u>Circadian Clocks</u>. North-Holland Publ., Amsterdam.

Davis, W.J., (1968a) Quantitative analysis of swimmeret beating in the lobster. J. Exp. Biol. 48, 643-662.

Davis, W.J., (1968b) The neuromuscular basis of lobster swimmeret beating. J. Exp. Zool. 168, 363-378.

Davis, W.J. and Kennedy, D., (1972) Command interneurons controlling swimmeret movements in the lobster. I. Type of effects on motoneurons. J. Neurophysiol. 35, 1-12.

Delcomyn, F., (1971) The locomotion of the cockroach <u>Periplaneta americana</u>. J. Exp. Biol. 54, 443-452.

Delcomyn, F. and Usherwood, P.N.R., (1973) Motor activity during walking in the cockroach. I. Free walking. J. Exp. Biol. 59, 629-642.

Gray, J., (1968) <u>Animal Locomotion</u>. Weidenfeld and Nicolson, London.

Gray, J., Lissman, H.W. and Pumphrey, R.J., (1938) The mechanism of locomotion in the leech (<u>Hirudo medicinalis Ray</u>). J. Exp. Biol. 15, 408-430.

Grillner, S., (1975) Locomotion in vertebrates - central mechanisms and reflex interaction. Physiol. Rev. 55, 247-304.

Holst, E. von, (1939) Die relative Koordination als Phänomen und als Methode zentralnervöser Funktionsanalyse. Ergebn. Physiol. 42, 228-306.

Holst, E. von, (1973) <u>The behavioural physiology of animals and man: the collected papers of Erich von Holst</u>. Volume one. Translated by Robert Martin. University of Miami Press, Coral Gables, Printers; Methuen and Co. Ltd., London, Publishers.

Hughes, G.M. and Wiersma, C.A.G., (1960) The coordination of swimmeret movements in the crayfish <u>Procambarus clarkii</u> (Girard). J. Exp. Biol. 37, 657-670.

Ikeda, K. and Wiersma, C.A.G., (1964) Autogenic rhythmicity in the abdominal ganglia of the crayfish: the control of swimmeret movements. Comp. Biochem. Physiol. 12, 107-115.

Kristan, W.B., Jr., Stent, G.S. and Ort, C.A., (1974) Neuronal control of swimming in the medicinal leech. I. Dynamics of the swimming rhythm. J. Comp. Physiol. 94, 97-119.

Kulagin, A.S. and Shik, N.L., (1970) Interaction of symmetrical limbs during controlled locomotion. Biophysics. 15, 171-178 (English translation).

Lennard, P.R., (1975) Neural control of swimming in the turtle. Doctoral thesis. Washington University, St. Louis, Mo.

Lennard, P.R. and Stein, P.S.G., (1974) Control of swimming in the turtle by electrical stimulation of the spinal cord. Abstracts of the Fourth Annual Meeting of the Society for Neuroscience, p. 303.

Lennard, P.R. and Stein, P.S.G., (1976) Swimming movements of the turtle elicited by electrical stimulation of the spinal cord. I. The low spinal and the intact preparation. (Submitted for publication.)

Macmillan, D.L., (1975) A physiological analysis of walking in the American lobster (Homarus americanus). Phil. Trans. R. Soc. London B. 270 (901), 1-59.

Moore, G.P., Segundo, J.P. and Perkel, H., (1963) Stability patterns in interneuronal pacemaker regulation. From Proc. San Diego Symposium for Biomedical Engineering. (Paull, A., ed.), (184-193).

Ottesen, E.A., (1965) Analytical studies on a model for the entrainment of circadian networks, Unpublished B.A. Thesis, Princeton University, Princeton, N.J.; cited in Pavlidis (1973).

Pavlidis, T., (1973) Biological Oscillators: Their Mathematical Analysis. Academic Press, N.Y.

Pearson, K.G., (1972) Central programming and reflex control of walking in the cockroach. J. Exp. Biol. 56, 173-193.

Pearson, K.G. and Fourtner, C.R., (1975) Nonspiking interneurons in walking system of the cockroach. J. Neurophysiol. 38, 33-52.

Pearson, K.G. and Iles, J.F., (1973) Nervous mechanisms underlying intersegmental coordination of leg movements during walking in the cockroach. J. Exp. Biol. 58, 725-744.

Perkel, D., Shulman, J.H., Bullock, T.H., Moore, G.P. and Segundo, J.P., (1964) Pacemaker neurons: effects of regularly spaced synaptic input. Science. 145, 61-63.

Pinsker, H.M., (1976) Aplysia bursting neurons as endogenous oscillators: I. Phase response curves for pulsed inhibitory synaptic input. (Submitted for publication.)

Stein, P.S.G., (1971) Intersegmental coordination of swimmeret motoneuron activity in crayfish. J. Neurophysiol. 34, 310-318.

Stein, P.S.G., (1973) "The relationship of interlimb phase to oscillator activity gradients in crayfish," In Control of Posture and Locomotion. (Stein, R.B., Pearson, K.G., Smith, R.S. and Redford, J.B., eds.), Plenum Press, New York, (621-623).

Stein, P.S.G., (1974) The neural control of interappendage phase during locomotion. Am. Zool. 14, 1003-1016.

Stein, P.S.G., (1976) Swimming movements of the turtle elicited by electrical stimulation of the spinal cord. II. The high spinal preparation. (Submitted for publication.)

Wendler, G., (1966) The co-ordination of walking movements in arthropods. Symp. Soc. Exp. Biol. 20, 229-249.

Wendler, G., (1974) The influence of proprioceptive feedback on locust flight co-ordination. J. Comp. Physiol. 88, 173-200.

Wiersma, C.A.G. and Hughes, G.M., (1961) On the functional anatomy of neuronal units in the abdominal cord of the crayfish, Procambarus clarkii (Girard). J. Comp. Neurol. 116, 209-228.

Wiersma, C.A.G. and Ikeda, K., (1964) Interneurons commanding swimmeret movements in the crayfish, Procambarus clarkii (Girard). Comp. Biochem. Physiol. 12, 509-525.

Willows, A.O.D., Getting, P.A. and Thompson, S., (1973) "Bursting mechanisms in molluscan locomotion," In Control of Posture and Locomotion. (Stein, R.B. et al., eds.), Plenum Press, New York, (457-475).

Wilson, D.M., (1967) "An approach to the problem of control of rhythmic behavior," In Invertebrate Nervous Systems. (Wiersma, C.A.G., ed.), U. Chicago Press, Chicago, (219-229).

BASIC PROGRAMS FOR THE PHASING OF FLEXION AND EXTENSION MOVEMENTS OF THE LIMBS DURING LOCOMOTION

J.M. Halbertsma, S. Miller* and F.G.A. van der Meché

Department of Anatomy, Erasmus University, Rotterdam, Rotterdam, The Netherlands

Observations of movements about the hip joint and of the scapula upon the rib cage in normal and decerebrate cats stepping on a treadmill have expanded and supported earlier conclusions (Miller et al., 1975a, 1975b) that the coordination of the four limbs in alternate locomotion and in phase locomotion results from different combinations of basic programs for the phasing of flexion and extension movements of the homologous limbs (hindlimbs or forelimbs) and homolateral limbs (hind- and forelimb of the same side).

In alternate locomotion, (e.g., pace, trot or swimming), flexion and extension movements of the homologous limbs occur out of phase. The homolateral limbs are coupled in two basic patterns. In the first, seen during pacing, the flexion and extension movements of the hip and scapula occur in phase, with the onset of hip extension occurring in phase with the onset of scapula extension. In the second, seen for example during trotting, the movements of the hip and scapula are out of phase, with the onset of hip extension occurring in phase with that of scapula flexion.

During in phase locomotion the flexion and extension movements of the hindlimbs and, to a lesser extent those

---

*Department of Anatomy, Bristol University Medical School, University Walk, Bristol, England

of the forelimbs, occur in phase. In the transverse, rotatory and half bound gallops and occasionally in jumping, the coupling of the movements of the homolateral limbs is asymmetric. On one side they are coupled out of phase with hip extension occurring together with scapula flexion, a pattern typically seen symmetrically in alternate gaits, e.g., trotting. This is the most consistent homolateral coupling. On the other side the movements of hip and scapula occur with certain restrictions approximately in phase in a pattern which is also observed symmetrically in alternate gaits, e.g., pacing. Although this pattern is well defined, it is more variable than that of the opposite side. In the full bound and usually in the jump, the homolateral limbs on both sides are coupled out of phase. In phase forms of locomotion therefore display the elements of homolateral coupling observed in the alternate forms of locomotion.

Electromyograms from the different limbs have confirmed the patterns of alternate stepping and show that transitions between the different types of coupling occur abruptly. If the stepping cycle is perturbed by internal or external influences resulting temporarily in differential rhythms between the limbs, coincidence of the onsets of flexion and extension appears to be sought where possible.

Observations in spinal cats have shown that these programs of coordination of the four limbs are organized in the first instance within the spinal cord. In decerebrate cats the activity in long and crossed spinal connections may be gated to the flexor or extensor neural mechanisms controlling each limb depending on the phase of the step. It is suggested that these functional spinal connections form the neural substrate for elements of coordinated movement of the limbs which may be further controlled and modulated by segmental afferent input and by supraspinal centres.

Introduction

The structure of locomotor movements within a limb remains remarkably consistent in different situations such as stepping on the ground, jumping, stepping in the air and swimming (dog: Arshavsky et al., 1965; cat: Goslow et al., 1973; Miller and van der Meché, 1975; Miller et al., 1975b). The alternating periods of flexion and extension movements at the different joints form the elements of the locomotor cycle, which may be adapted in force and timing to satisfy the requirements for support of body, balance and direction of progression. These basic elements of movement are produced by the activities of interneurones and motoneurones lying in the respective spinal segments innervating each limb. How far locomotor movements in each limb can be generated independently by the respective half segments is not entirely clear. In the hindlimbs, at least, even quite complex locomotor activity can be produced by lumbosacral segments disconnected from the rest of the nervous system (Grillner and Zangger, 1974; Grillner, 1975.) In his study of narcosis progression in the cat, Brown (1911) reported that rhythmic movements at the ankle remained in one limb when the spinal cord was divided in the upper lumbar region and when the great part of the opposite lateral half of the lumbar spinal cord was removed. Good evidence therefore exists for somewhat independent stepping or locomotor pattern generators for each hindlimb. However, it is not known at present if locomotor movements of the forelimbs in the cat can be produced similarly by cervicothoracic spinal segments, or if with the phylogenetic process of encephalization of motor control forelimb movements are in some way dependent on structures in the brain stem (See Miller and van der Meché, 1975). It should also not be overlooked that in different types of locomotion movements of the head, rib cage and entire vertebral column can occur which are linked in phase with the locomotor patterns of the limbs (Hildebrand, 1959; Stuart et al., 1973). In searching for the basic neural strategies underlying locomotion it is necessary to consider if the concept of four interconnected spinal centres, each controlling one of the limbs (Grillner, 1975; Miller et al., 1975b), is sufficient or if it needs to be enlarged to include movements of the trunk and head.

In previous reports (Miller et al., 1975a, b) it was suggested that all types of locomotion in the cat are achieved by different phasing of flexion and extension movements in the four limbs. Evidence is presented in this chapter that at least some forms of quadrupedal locomotion in the cat can be produced by the spinal cord alone. The previous analysis has also been extended to a comparison of movements of the more proximal parts of the limbs; the hip joint and the scapula upon the rib cage. Movement at these joints differs from that of the more distal and appears to be of particular importance in the elaboration of stepping movements (Sherrington, 1910; Giovanelli Barilari and Kuypers, 1969; Miller et al., 1973; Grillner, 1975; Miller and van der Meché, 1975).

Materials and Methods

The normal, decerebrate and spinal cats described in this chapter formed part of the material analysed in Miller et al.,

(1975a, b), and the surgical and technical procedures are the same. The previous material was obtained from measurements made from 16 mm films and from automatic television analysis (Furnée et al., 1974). For movements of the scapula and hip further measurements have been made on the previous 16 mm films, particularly of cats numbered F6 and 13 in the alternate gaits and cats F6, 7, 8, 12 and 13 in the in-phase gaits. The small number of samples in some of the points in Figures 6-9 and the gaps in the velocities selected are due to a strict selection of sequences in which the stepping pattern remained constant.

In experiments on the phase dependent reversal of interlimb reflexes, 5 high decerebrate cats (method of Miller et al., 1975a) and 4 spinal cats (method of Miller et al., 1973) were used. Electrical stimuli were applied to the skin over the dorsum of the metatarsals or metacarpals at varying times within the step cycle by the use of a delay triggered from one of the EMGs. The electrodes consisted of two 5 mm convex brass discs glued at 20 mm centres onto the skin with celloidin. Electrode paste was injected through a central hole in each disc. 3-5 current pulses at 300 Hz, of 1 msec duration and 1-10 mA were given generally every fourth step; the effects on the EMGs were compared with those of the preceding step.

Simultaneous recordings of movements and EMGs in different types of stepping were made in one decerebrate cat with the automatic television recording technique described by Furnee et al., (1974). In this technique the positions of small white paper discs glued to the skin above the bony landmarks at the different joints are digitized at a sampling rate of 60 Hz. The skin overlying the scapula was stitched tightly to the spine of the scapula, so that the paper discs would give an accurate estimate of the scapula's movement with respect to the horizontal plane. At the same time 4 channels of EMG activity are digitized at 240 Hz. Before digitization the raw EMG (Figure 1a) is high pass filtered at 30 Hz (-3 dB) and treated to full-wave rectification (Figure 1b). The signal is then passed through a low pass filter (Figure 1c) in which frequencies above 100 Hz are suppressed by 40 dB (Halbertsma, 1975). The EMG signals in Figures 3, 4, 5 and 11 have been subjected to further software filtering during computation, as described by Halbertsma (1975).

The following conventions are used: homolateral limbs refer to the hindlimb and the forelimb of the same side of the body, and homologous limbs refer to either the pair of hindlimbs or the pair of forelimbs. Electromyogram is abbreviated to EMG.

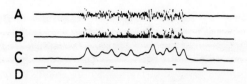

Figure 1. Processing of EMG signals. A. raw EMG with high pass filter at 30 Hz (-3dB); B. after full wave rectification; C. after low pass filtering at 100 Hz (-40dB); D. time scale 100 msec between pulses.

Results
Locomotion of All Four Limbs in the High Spinal Cat

In the high spinal cat coordinated movements of all four limbs resembling stepping sequences have occasionally been observed (Guillebeau and Luchsinger, 1882; Sherrington, 1910; Luttrell et al., 1959; Miller and van der Burg, 1973). From these results Miller et al., (1975a) argued that the basic elements of coordinated locomotor movements in the four limbs could probably be generated by the spinal cord. In four experiments on high spinal cats the earlier observations were repeated that the forelimbs occasionally make stepping movements in register with those of the hindlimbs. The cats in these experiments were treated with the monoamine oxidase inhibitor Nialamid (40 mg/kg i.v.) and after a delay of 30 minutes with a further i.v. injection of a mixture of L-DOPA (50-100 mg/kg) and the peripheral DOPA decarboxylase inhibitor Ro 4-4602/1 (Roche; 25-50 mg/kg) dissolved in 0.9% NaCl. The former two drugs have been used to activate and potentiate the terminals of the noradrenergic reticulospinal pathways which have been associated with the initiation and maintenance of locomotion in the cat (Grillner and Zangger, 1974; Grillner, 1975). When the cat is placed in contact with a motor driven treadmill the hindlimbs begin to step. At first the gait is alternating, but as the velocity increases the hindlimbs begin to gallop. The forelimbs are retracted caudally with a strong extensor tonus at all joints. In 2 out of the 4 spinal cats and occasionally in a third the forelimbs stepped in time with the hindlimbs moving stiffly forward and backward. Unlike the hindlimbs, the forelimbs are not lifted up well during the swing phase and do not support the weight of the body. The electromyograms in Figure 2 illustrate the close register of the homolateral forelimb and hindlimb activity in one of these preparations. In this case the homolateral coupling is in-phase; in other cases it can be out-of-phase and abrupt transitions can occur between the two forms (Miller and van der Meché, In preparation). These observations are important since it can now be concluded that coordinated stepping movements of all four limbs in the cat can be generated by spinal mechanisms.

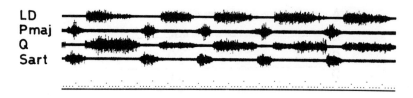

Figure 2. EMGs from homolateral forelimb and hindlimb in a high spinal cat during stepping on treadmill. LD latissimus dorsi, (forelimb extensor), Pmaj pectoralis major, (forelimb flexor), Q quadriceps femoris (hindlimb extensor), Sart sartorius (hindlimb flexor). Time scale: smallest divisions = 100 msec.

Interlimb Coordination in Alternate Forms of Locomotion

The alternate form of locomotion is defined by strict alternation of movements between the homologous pairs of limbs and is seen in all forms of walking and trotting, and during swimming (Miller et al., 1975a, b). The particular type of alternate gait adopted depends on the coupling of the homolateral limbs and may take the form of pacing, in which the extension movements of the homolateral limbs are in-phase, or the form of trotting, in which these movements are out-of-phase. Examples of these types of homolateral limb coupling are illustrated in the computed displays of Figures 3, 4 and 5. These were obtained in a decerebrate cat stepping at various constant velocities on a treadmill. In part A of each figure is a display of the successive steps analysed. In Part B the movements have been superimposed by reference to the onset in each step cycle of the extension phase of the hip. Below the superimposed movement traces, the durations of the EMG activity from the different muscles are given as solid lines. In Figure 3 the cat is stepping at 0.6 m/sec in a pacing gait. Both the flexion and extension phases of the movements at the hip and scapula occur in phase and this is reflected in the EMGs. In Figures 4 and 5 the cat is stepping at 0.75 and 1.5 m/sec, respectively. The scapula and the timing of the respective EMGs are out-of-phase; extension of the hip is correlated with flexion of the scapula. These gaits would be classified on the basis of foot contact patterns as walk and trot (cf. Muybridge, 1957; Roberts, 1967; Gray, 1968).

In previous reports of the coupling of movements of the homolateral limbs (See Miller et al., 1975a) measurements were made of the changeover times of flexion and extension movements at the knee and elbow, since the positions of these joints could be estimated in all four limbs. The measurements have now been extended

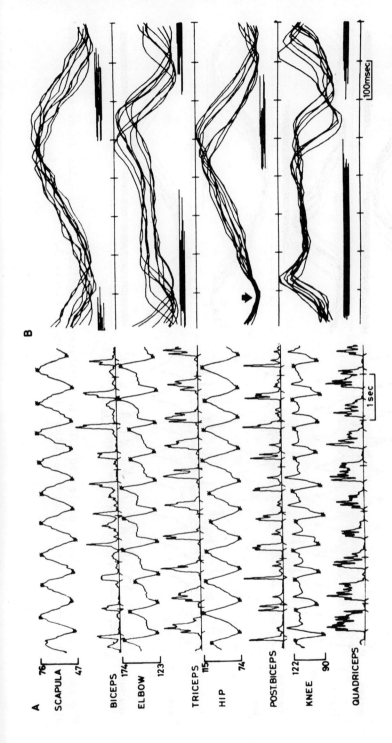

Figure 3. Changes of joint angle and EMGs in homolateral hindlimb and forelimb of decerebrate cat stepping on treadmill at 0.6 m/sec. In the movement traces upward deflection indicates extension, downward flexion. A. Successive steps. B. Superimposition of steps using onset of hip extension as trigger point. The duration of the EMGs is given in the bars under the movement traces (See Halbertsma, 1975). Movements at the knee joint may be slightly distorted in the extension phase by the action of a spring system partly suspending the hindquarters of the cat.

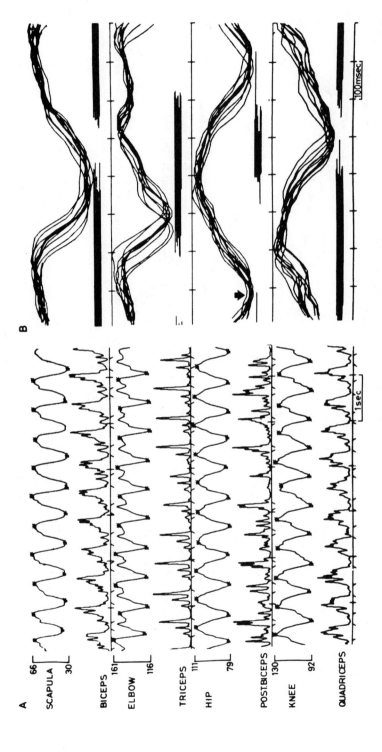

Figure 4. Changes of joint angle and EMGs in homolateral hindlimb and forelimb in decerebrate cat stepping on treadmill at 0.75 m/sec. Other details as in Figure 3.

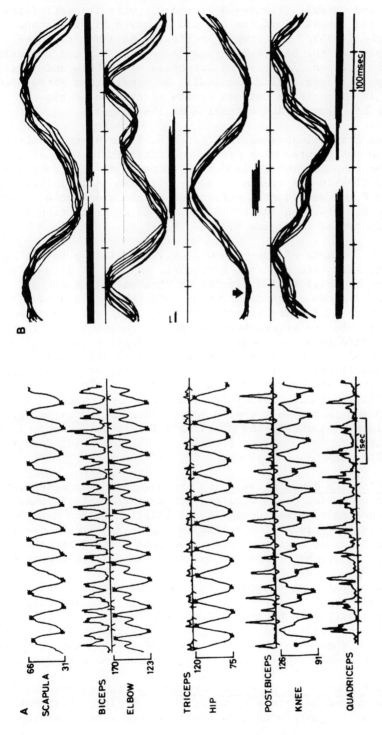

Figure 5. Changes of joint angle and EMGs in homolateral hindlimb and forelimb in decerebrate cat stepping on treadmill at 1.5 m/sec. Other details as in Figure 3.

to the hip and scapula (Figures 6, 7 and 8), with the definition of flexion and extension of the scapula as given by Miller and van der Meché (1975). The graphs in Figures 6A and B show the intervals between the onsets of extension of the hip and extension of the scapula (open circles) and extension of the hip and flexion of the scapula (large filled circles). No significant differences were found in comparisons between onsets of flexion at hip and scapula, or hip flexion and extension at the scapula.

The two cats illustrated in Figures 6, 7 and 8 were analyzed in detail in the figures of a previous article (Miller et al., 1975a). The coupling intervals during stepping at different constant velocities on the treadmill are plotted in real time in Figures 6A and B and scaled as a percentage of the step cycle in Figures 6C and D. Distribution of the intervals as a function of the stepcycle is tested in Figure 8.

Part of the data of knee-elbow couplings relating to the same steps in the same cats reported previously (Miller et al., 1975a) is replotted as small filled circles in Figures 6A and C. As before there are two distinct forms of hindlimb-forelimb coupling below and above a dividing velocity of about 1 m/sec. Note that the dividing velocity refers here to the separation of the in-phase and out-of-phase forms of homolateral limb coupling during alternate locomotion, and not to the separation of trotting and galloping. The homolateral coupling in alternate gaits (e.g., pacing) is rarely seen when the cat is freely moving. In the previous data reported (Miller et al., 1975a) it occurred below 1 m/sec. A pacing coupling is not observed in normal cats during swimming. In one cat (Figures 6A and C) the dividing velocity is 1 m/sec, in the other (Figures 6B and D) it is 1.4 m/sec. The important features are the following.

(1) <u>Below the dividing velocity</u> the onsets of hip extension and scapula extension (open circles) are approximately in-phase (Figure 6: averaged values, range -6 to 13%). This can also be visualized in the graphs of movement in Figure 3. Slight deviations to positive or negative values indicate that either the hindlimbs or the forelimbs, respectively, are leading in the movements.

(2) <u>Above the dividing velocity</u> it is the onsets of hip extension and scapula flexion (filled circles) which are approximately in-phase (Figure 6: averaged values, range -2 to -10%). This can also be visualized in the graphs of movement of Figures 4 and 5. Note that here all the values tend to be negative indicating that the forelimb flexion leads the hindlimb extension.

(3) <u>Distribution of phase relationships.</u> The couplings of movement described in points 1 and 2, above, define sufficiently the two forms of homolateral limb coordination. The remaining

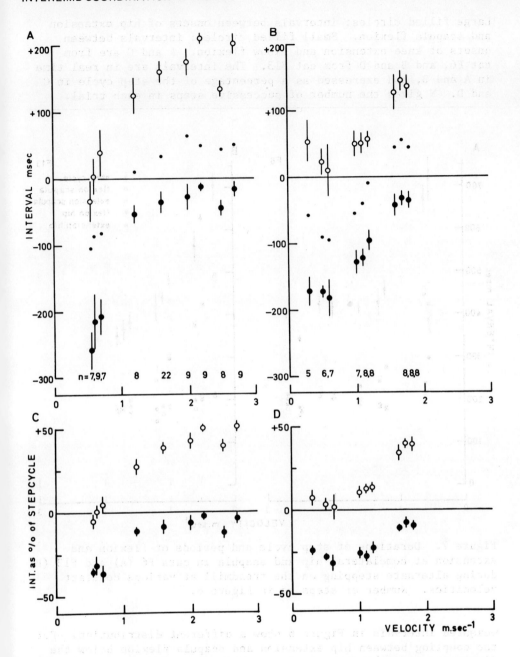

Figure 6. Coupling of movements of homolateral hindlimb and forelimb in normal cats trained to step at different constant velocities on treadmill. Open circles: intervals between onsets of hip extension and scapula extension. (Legend continued on next page.)

Large filled circles: intervals between onsets of hip extension and scapula flexion. Small filled circles: intervals between onsets of knee extension and elbow flexion. A and C are from cat F6, and B and D from cat F13. The intervals are in real time in A and B, and expressed as a percentage of the step cycle in C and D. N gives the number of successive steps in each trial.

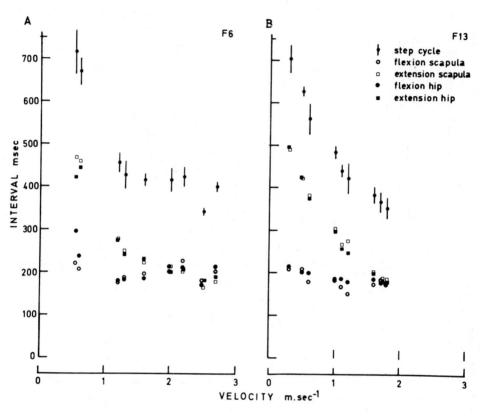

Figure 7. Durations of step cycle and periods of flexion and extension at homolateral hip and scapula in cats F6 (A) and F13 (B) during alternate stepping on the treadmill at various constant velocities. Number of steps as in Figure 6.

measured intervals in Figure 6 show a different distribution. For the coupling between hip extension and scapula flexion below the dividing velocity the range of averaged values, expressed as percentage of the step cycle, is -40% to -23%. For the coupling between hip extension and scapula extension above the dividing velocity the range is 25% to 50%. Since all the measures are referenced to the onset of hip extension these values are

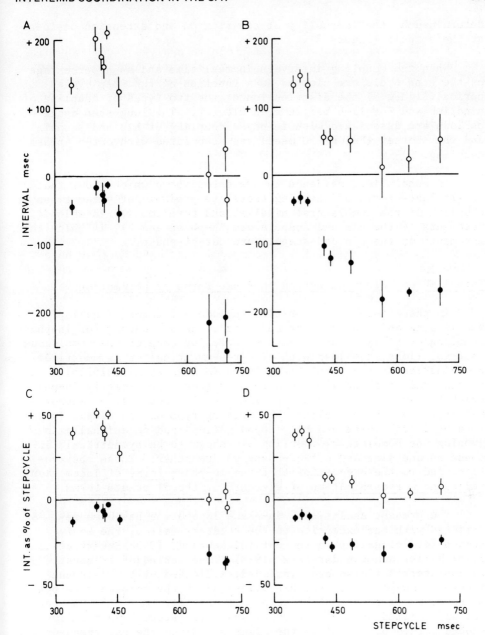

Figure 8. Coupling intervals of homolateral hindlimb and forelimb in real time (A and B) and as percentage of step cycle (C and D) in plots against the step cycle duration. Open circles: intervals between onsets of hip extension and scapula extension. Filled circles: intervals between hip extension and scapula flexion. Same data as in Figure 6.

determined by the inequality of the flexion and extension periods at the scapula (Figure 7).

When the coupling intervals in real time and as a percentage of the step cycle are plotted as a function of the step cycle period (Figure 8) the distribution of the two types of coupling of hindlimb and forelimb become even clearer. The long step cycle periods are associated with in-phase coupling of hip and scapula and the shorter (below 450 msec) are associated with out-of-phase coupling.

In conclusion, evidence is presented above which shows that in alternate forms of stepping there are predominantly two types of coupling of the homolateral hindlimb and forelimb 1), in which movements of the hip and scapula are in-phase and 2), in which the movements of the hip and scapula are out-of-phase.

Interlimb Coordination During In-Phase Forms of Locomotion

In-phase forms of locomotion include all forms of galloping and jumping and can be described broadly as a tendency for in-phase coupling of flexion and extension phases of each of the homologous pairs of limbs, combined with either symmetrical or asymmetrical coordination of the homolateral limbs (Miller et al., 1975b). The homolateral coupling was shown to be either approximately in-phase, which the authors called the pacing type of coupling; or approximately out-of-phase, called the trotting type of coupling. In the rotatory, transverse and half bound gallops and occasionally in jumping the homolateral coupling was shown to be symmetrical; in-phase on one side and out-of-phase on the other. These observations led to the suggestion that the in-phase forms of locomotion represent a recombination of elements of the alternate forms.

The present analysis of in-phase locomotion has been restricted with one exception to the rotatory gallop, the most common form of gallop in the cat (Hildebrand, 1959; Stuart et al., 1973; Miller and van der Burg, 1973). One series of transverse gallop steps has also been included, indicated by a 't' in the figures. Movements of the hip joint and of the scapula have been measured in five cats stepping on a treadmill. One of these cats was also filmed moving overground and a few steps have been included marked by an 'f' in the figures. Since the cat frequently changes the laterality and type of its gallop (Miller and van der Burg, 1973; Stuart et al., 1973), step cycles have only been selected from sequences of galloping in which the laterality and type remained constant for at least five steps. The first and the last steps of the sequences were discarded to avoid the possible influence of changes in the preceding and succeeding steps. For technical reasons movements at the proximal joints could not be filmed simultaneously on both sides of the body. Figure 9 therefore represents the pooled observations of the coupling of the hip and scapula on the left side of the cats during both left and right rotatory gallops and during one right transverse gallop. Comparison with results previously obtained for the couplings between knee and elbow (Miller et al., 1975a, b) suggests this is a reasonable procedure to obtain estimates of the movements on both sides.

In alternate gaits the homolateral limb coupling is bilaterally symmetrical in each of the two forms. During in-phase gaits it is asymmetrical, being on one side out-of-phase and on the other in-phase. The out-of-phase coupling is always the most consistent. In the left rotatory and right transverse gallops the onsets of hip extension and scapula flexion on the left side are in-phase (Figures 9a and c, filled circles, hatched zones, averaged values in range -8% to +5%). The onsets of extension at the hip and scapula are out-of-phase (Figures 9a and c, open circles, hatched zones, averaged values in range 38% to 50%). These interlimb couplings are comparable to those above the dividing velocity in the alternate gaits (Figure 6). The interlimb couplings on the other side of the body are represented by the left limbs in the right rotatory gallop. Here there is a coupling resembling the in-phase homolateral coupling obtained in alternate gaits below the dividing velocity (Figures 9a and c, open zones). The onsets of hip and scapula extension are more in phase (open circles, averaged values in range 8% to 30%). The onsets of hip extension and scapula flexion are more out-of-phase (filled circles, averaged values in range -47% to -27%). The in-phase homolateral couplings in alternate and in-phase gaits are compared in the Discussion section.

Replotting the data as a function of the step cycle period (Figures 9b and d) reveals that the groups of couplings are spread evenly throughout an almost twofold range of stepping frequencies. There is an important difference, however, with the coupling intervals shown in Figures 8A and B. In the alternate gaits it is exclusively at the longer step cycle periods that the hip and shoulder move in-phase, and exclusively at the shorter step cycle periods that the hip and shoulder move out-of-phase. In the gallop the step cycle periods partly overlap the lower end of the range for alternate gaits. Here both types of hip and scapula coupling are found.

In conclusion, observations of movements at the hip and scapula in the rotatory gallop show that the homolateral limb coupling on one side is out-of-phase and resembles the coupling seen in alternate gaits at higher frequencies of stepping. On the other side the coupling is in-phase and resembles the coupling seen

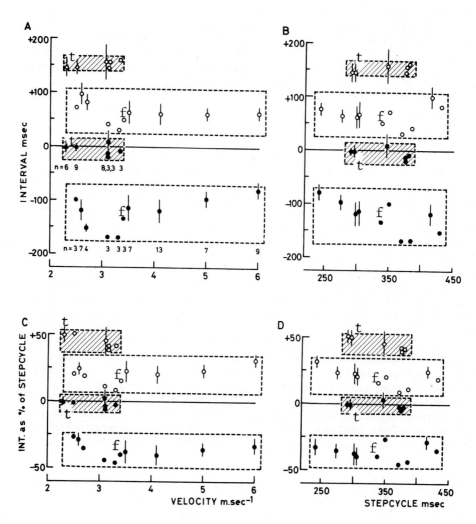

Figure 9. Coupling of movements of homolateral hip and scapula in in-phase locomotion. Open circles: intervals between onsets of hip extension and scapula extension. Filled circles: intervals between hip extension and scapula flexion. Hatched zones: left homolateral limbs in left rotatory and right transverse gallops; open zones: left homolateral limbs in right rotatory gallop. 't' indicates the data referring to the transverse gallop. All the data were obtained in 5 normal cats galloping on treadmill, except in one case, marked 'f' where the cat was freely moving. In A and C the intervals are plotted against the velocity of the treadmill, in B and D they are plotted as a function of the step cycle duration. n = number of trials.

in alternate gaits at lower frequencies of stepping. Of the two types of coupling the out-of-phase type is the more consistent.

Increase of Stepping Frequency of Forelimbs over Hindlimbs

Under natural conditions cats do not step at constant velocity and frequency for more than a few steps. Normal cats and also decerebrate cats may deviate from a regular stepping pattern even while stepping on the treadmill. This occurs particularly at low velocities, those below about 1 m/sec. Sometimes the forelimbs take one step more than the hindlimbs. This is illustrated in Figure 10A for the homolateral limbs in a decerebrate cat. In the first few steps the cat showed an in-phase coupling of hip and scapula, typical of the pace. Suddenly the forelimb step cycle is shortened and that of the hindlimb lengthened (first arrow). This brings the hip and scapula movement out-of-phase. The shift is temporary and by the next step of the hindlimbs the process has been repeated, bringing the hip and scapula back into phase (second arrow). There are two striking features in this whole process: 1) the homolateral coupling springs abruptly between the coincidence of onsets of extension in hip and scapula to those of hip extension and scapula flexion, and then back again; 2) the flexor and extensor EMGs also show the same shift of in-phase to out-of-phase. It would appear, therefore, that this is a situation in which the frequency of stepping of the forelimbs temporarily exceeds that of the hindlimbs. There is a strong tendency for coincidence of the changeover points of flexion and extension in the homolateral limbs in one form or the other, and also correlation of the appropriate flexors and extensors in the limbs.

Asymmetric Homolateral Limb Coupling in an Alternate Gait

In five decerebrate cats stepping on the treadmill and stepping suspended in the air sequences were observed in which the cat appeared to limp. This occurred in an alternate gait with the homologous limb coupling deviation from its value of 50% of the step cycle and with an asymmetric homolateral coupling. In the example of Figure 11 the left homolateral limbs move in-phase and the right are out-of-phase. Once again the out-of-phase homolateral coupling is stricter than the in-phase type: extension of the left knee and flexion of the left elbow are associated more closely than extension of the right knee and extension of the right elbow. The homologous limbs no longer move strictly out-of-phase, as indicated by the arrows which mark the expected 50% point of onset of extension in the right hindlimb and right forelimb. It would therefore appear that the mechanisms coordinating the homolateral limbs displayed not only the two different forms of

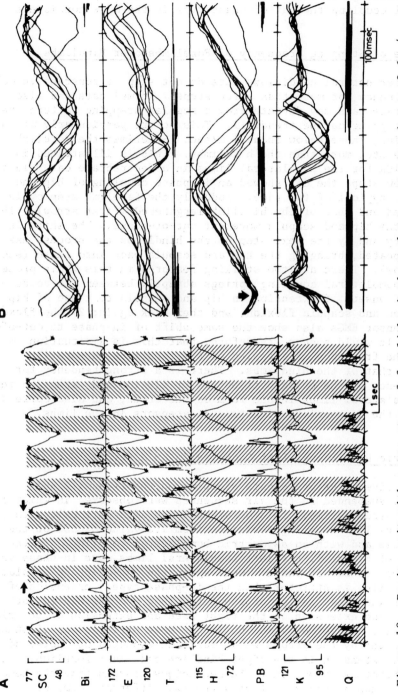

Figure 10. Pacing gait with extra step of forelimbs in decerebrate cat stepping at 0.6 m/sec on treadmill. Arrows indicate occurrence of extra step. SC scapula, E elbow, H hip, K knee, Bi biceps brachii, T triceps brachii, PB biceps femoris posterior, Q quadriceps femoris. For other details see Figure 3.

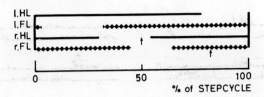

Figure 11. Duration of flexion and extension periods at knee and elbow joints in a decerebrate cat stepping on the treadmill at 1.0 m/sec. Solid and broken lines indicate extension and gaps flexion. Data are averaged from 11 successive steps. l left, r right, HL hindlimb, FL forelimb. Arrows indicate expected onsets of extension in right hindlimb and forelimb: further discussion in text.

coupling, but that they were able to override the tendency for the hindlimbs and especially the forelimbs to step in strict alternation.

Hindlimb-Forelimb Reflex Reversal Dependent on the Phase of the Step Cycle

In the neurophysiological studies of descending long propriospinal pathways in high spinal cats (Jankowska et al., 1974; Miller et al., In preparation) hindlimb motoneurones receive a mixture of excitatory and inhibitory effects directly or indirectly over the long pathways. These data are difficult to interpret functionally unless one assumes that the mutually antagonist effects belong to different functional mechanisms or to different phases of a cyclical process. Miller et al., (In preparation), tested the latter hypothesis further in decerebrate cats stepping on the treadmill by investigating the influence of electrical or mechanical stimulation of the paw of one limb girdle on the EMGs of flexor and extensor muscles of limbs in the other girdle. Some preliminary results of these experiments are shown in Figures 12 and 13. The criteria for excitation were an increase of at least 50% in amplitude of the EMG lasting at least 15 msec and occurring within 30 msec of the stimulus. In Figures 12A and C hindpaw stimulation evokes excitation in biceps and triceps brachii only at certain phases of the step cycle. A similar influence of forepaw stimulation on hindlimb EMGs is shown in Figures 12B and D. In a more recent experiment (Figure 13) the EMGs have been averaged to obtain more precise estimates of the effects. The averaged latencies of these effects could be as low as 13 msec, although they mostly fell in the range 15-22 msec. This investigation is still in progress, but the present results indicate that excitation

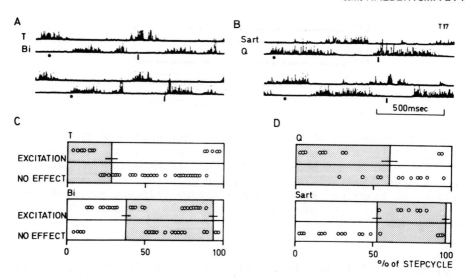

Figure 12. Reversal of long spinal reflexes as a function of step cycle in a decerebrate cat pacing on treadmill at 0.6 m/sec. T triceps brachii, Bi biceps brachii, Q quadriceps femoris, Sart sartorius. A. Effect of electrical stimulation of 3 pulses at 7.5 mA (See Methods) of hindpaw on ipsilateral forelimb EMGs. The EMGs are rectified and filtered with a time constant of 3 msec. Vertical bar indicates onset of stimulus; dot indicates corresponding point in previous step cycle. C. Distribution of excitatory effects during step cycle. Shaded areas indicate average periods of EMG activity; horizontal bars indicate standard deviation. B and D: same as in A and C, but for hindlimb EMGs following ipsilateral forepaw stimulation.

evoked over long ascending and descending paths is gated in a predictable manner to either flexor or extensor motoneurones depending on the phase of the step cycle. Whether the gating is exclusively a function of the accepting limb motor centre, or depends on the spinal segments where the stimulated cutaneous afferents enter is not yet clear. The latencies of the effects below 18 msec would strongly suggest that the effects at least in part are dependent on spinal mechanisms (Miller et al., 1973; Shimamura and Livingston, 1963). However, further attempts to repeat these observations in high spinal stepping preparations will be made.

Discussion

Spinal Generation of Coordinated Locomotor Movements in All Four Limbs. The spinal cord alone can generate coordinated

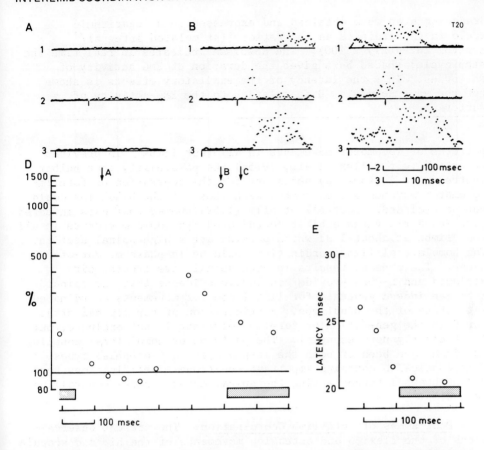

Figure 13. Computer analysis of long spinal reflex reversal dependent on phase of stepping cycle in a decerebrate cat stepping on treadmill at 1.0 m/sec. The EMG's were rectified and filtered with a time constant of 3 msec. Effect of forepaw stimulation (3 shocks at 3 mA: see Methods) on ipsilateral sartorius. A, B and C: averages of 4 samples each of control step cycles (row 1) and stimulated step cycles (rows 2 and 3). The stimulus was given every 4 step cycles and compared with the preceding step cycle. In each of the stimulated step cycles the vertical bar indicates the onset of the stimulus. In rows 1 and 2 the averages last 300 msec and begin 100 msec before the stimulus; in row 3 the averages last 50 msec and start with the stimulus. The vertical gain is arbitrary, but it is the same in A, B and C. For the construction of the graph of D averages were made of 4 samples of EMGs of the period 50 msec after the stimulus or the corresponding point in the previous step. The arrows indicate the point in the step cycle at which the samples in A, B and C were obtained. Integrals of these averages between 9 and 40 msec from the stimulus (or from the corresponding point in the   (Legend continued on next page.)

preceding step were obtained and expressed in a logarithmic scale in the ordinate as the ratio: (stimulated integral/ control integral) x 100). The abscissa indicates in real time the step cycle; shaded bars given the duration of EMG activity of sartorius. In E the latency of the excitatory effects is shown for the same points as D in relation to the EMG activity of sartorius.

---

patterns of locomotor movements in all four limbs. In previous reports (e.g., Miller et al., 1975a) the possibility that bulbar centres were in some way necessary for the generation of forelimb locomotor movements coordinated with those of the hindlimbs could not be excluded. Luttrell et al., (1959) showed that cats infected with Newcastle virus exhibit coordinated myoclonic movements in all four limbs at about 1 Hz which persist after high spinal section. The homolateral limb coordination could be in-phase or out-of-phase. The present results in high spinal cats treated with Nialamid and L-DOPA provide conclusive evidence that the spinal cord can indeed generate forelimb locomotor movements coordinated with those of the hindlimbs. Participation of supraspinal structures in the generation of forelimb movement is not excluded, but it is clearly not essential. The patterns of homolateral coupling obtained have been of both the in-phase and out-of-phase types. This conclusion contains important functional specifications for the pathways interconnecting the spinal motor centres controlling each limb.

Programmes of Interlimb Coordination. The present observations of the flexion and extension movements of the hip and scapula and of flexor and extension EMGs provide further support for the statement that the different forms of alternate locomotion (e.g., walking, trotting and swimming) and in-phase locomotion (galloping and jumping) result from the interaction of basic patterns, or programmes, for coordinating the homologous limbs and for the homolateral limbs (Miller et al., 1975b).

In alternate locomotion flexion and extension movements of the homologous limbs occur strictly out-of-phase (Miller et al., 1975b). The present data of hip and scapula movements has shown more clearly that the homolateral limbs are coupled either approximately out-of-phase, as in trotting and swimming, or in-phase, as in pacing. In-phase locomotion is characterized by a strong tendency for in-phase coupling of movements of the hindlimbs, and to a lesser extent, for those of the forelimbs (Miller et al., 1975b). As presented in this chapter, measurements of the hip and scapula movements during in-phase locomotion have been restricted for technical reasons to one side of the body. The results from the five cats investigated have been very consistent. They have provided further support for earlier observations taken from measurements of knee and elbow movements. In the rotatory and transverse gallops the coupling of movements of the homolateral limbs is asymmetric; on one side coupled out-of-phase and on the other, in-phase.

These programmes of interlimb coupling represent basic patterns of coordination which occur consistently and they are supported further by the abrupt transitions of coupling seen in the electromyograms (Figure 10; and Miller et al., 1975a). If these basic patterns are organized at a spinal level, as suggested by the present observations in high spinal cats a conceptually simple model emerges. Coordinated locomotion of all four limbs is achieved by intrinsic spinal programmes which may be modulated to suit the requirements of the cat by the activity of segmental afferent input and supraspinal descending control (See Grillner, 1975; Miller et al., 1975b).

The programmes do not exclude variants which may be dictated by either external or internal perturbations, by training or by learning. Figure 10 illustrates the situation in which the frequency of forelimb stepping temporarily exceeds that of the hindlimbs. This is in fact a 3:2 change in frequency. It is achieved by a homolateral coupling which shifts by approximately a half cycle from in-phase, to out-of-phase, and back to in-phase.

In experiments of Kulagin and Shik (1970), a 2:1 frequency difference occurred between the two hindlimbs in 7 out of the 15 mesencephalic cats stepping on a longitudinally split treadmill. The threshold for this effect generally exceeded a split belt difference in velocity of more than 2:1. Three features are striking in their results. 1) In their Figure 3, which shows a few steps of the hindlimbs in a 2:1 frequency difference, a tendency exists for coincidence of the onsets of backward movements in the slow hindlimb and of the onsets of every second forward movement in the fast hindlimb. Comparison with the present data can be made if forward and backward movement of the limb is assumed to reflect flexion and extension of the hip joint. 2) Kulagin and Shik (1970) state that a ratio of speed of the left and right belts of the treadmill greater than for a 1:1 rhythm (of the hindlimbs) and less than for a 2:1 rhythm represented an unstable zone. Here, the steps with 1:1 and 2:1 rhythm alternated after 2-3 steps. In other words there were no immediate couplings between the hindlimbs, and the hindlimb coupling at intermediate differential velocities of the belts displayed a jitter between 1:1 and 1:2 ratios. This is reminiscent of the jitter in the coupling between the EMGs of the homolateral limbs at particular intermediate velocities (Figure 6E of Miller et al., 1975a). 3) Only one of the 15 cats reported by Kulagin and Shik (1970) showed a division of the rhythm of the forelimbs and the data of this cat are not published. However, their Figure 2 illustrates a right to left belt velocity of about 2:1. There is no division of stepping frequency, though for about 5 steps the distance covered by the right limbs is twice that of the left. The stepping pattern is of the alternate form of locomotion with the homolateral limbs out of phase. It is remarkable that the homolateral coupling retains, as far as can be judged, this out-of-phase relationship throughout the differential changes of belt velocity.

All of these examples serve to illustrate that deviations from the in-phase or out-of-phase patterns of interlimb coupling may occur in response to internal or external demands. There remains, however, a strong tendency to return to the appropriate patterns of interlimb coupling wherever and whenever possible.

Relative Dominance of Patterns of Interlimb Coordination. The lack of strict in-phase couplings between the different limbs in the gallop may be explained by the following suggestion. The relative dominance of the different programmes for interlimb coordination during a particular locomotor cycle determines the patterns of gait adopted (Miller et al., 1975b). Further support for this concept is demonstrated in Figure 11: the occurrence of strongly linked homolateral coupling, on one side out-of-phase and on the other in-phase, is in conflict with the normal strict out-of-phase homologous limb coupling of alternate gaits. Under the conditions of the experiments reported here the more dominant forms of coupling were the in-phase and out-of-phase couplings of the hindlimbs and the out-of-phase homolateral limb coupling.

Spinal Pathways. Since coordinated locomotor movements of all four limbs can take place in the high spinal preparation it is evident that neural projections between the spinal enlargements are responsible in this preparation for coordination of hindlimb and forelimb movements. Forssberg and Grillner (1973) have shown that lumbosacral segments isolated from the rest of the central nervous system can generate both forms of homologous limb coupling. Similar evidence is not yet available for the forelimbs. But for the hindlimbs, at least, coordination of the two spinal halves must be achieved by short crossed segmental connections, perhaps those described by Holmqvist (1961).

The possible role of long propriospinal pathways in the coordination of the limbs in locomotion has been argued by Miller et al., (1975a). They suggested that the prepotent ipsilateral effects evoked in neurophysiological experiments would fit with the interlimb coupling which favours the association of the hindlimb and forelimb of the same side. The analysis presented above of interlimb coordination during locomotion in the cat provides conceptually relatively simple principles for the underlying neural

organization responsible. It is possible that there are spinal
interneurones concerned with the generation of periods of flexor
and extensor activity in one limb which also compare the phase of
the step cycle in that limb with the corresponding phase in another
limb. Neurones with comparator functions have been described, for
example, in the visual cortex for signalling the optimal orientation of the eyes for binocular vision (Burns and Pritchard, 1968)
and in the inferior olive for comparing cortical and spinal influences (Miller and Oscarsson, 1969). In the context of interlimb
coordination it is particularly interesting that Jankowska et al.,
(1973) found interneurones in lumbar segments receiving excitatory
convergence from long descending propriospinal pathways and from
the contralateral segmental flexor reflex afferents. Long propriospinal pathways also evoke powerful inhibitory effects, at latencies which are sometimes shorter than those of excitatory effects
(Miller et al., In preparation). The possibility should not be
excluded that interlimb coordination is partly achieved by inhibitory signals, as has been shown in insects (Pearson and Iles, 1973).

The association of the coupling of movements of the homolateral
pairs of limbs and the prepotent ipsilateral influences of long
propriospinal pathways is logical. However, it has yet to be
reconciled with the observations from a recent neuroanatomical
study using the horse radish peroxidase technique (Kuypers and
Molenaar, 1975, In preparation) that a large proportion of the
axons ascending or descending between the spinal enlargements have
their cell bodies on the opposite side of the spinal cord. If
these connections are concerned with interlimb coordination it
could mean that each spinal motor centre could receive direct information concerning the other three limbs.

The preliminary results of long spinal reflex reversal dependent on the phase of the step cycle are suggestive that long propriospinal pathways may evoke cyclical effects during locomotion. This
is apparent particularly since the peripheral skin nerves in the
limbs evoke potent effects in these pathways of the immobilized
high spinal preparation (Miller et al., 1973; Miller et al., In
preparation). It still remains to be shown if the phase dependent
reflex reversal utilizes the same spinal pathways as those which
are concerned with interlimb coordination.

If long propriospinal pathways are responsible for homolateral
limb coupling it should be possible to disconnect the forelimbs and
hindlimbs by small lesions in the funiculi at mid or low thoracic
level where the pathways lie (See Miller and van der Burg, 1973).
Afelt (1974) has shown that lesions in the funiculi can uncouple
the rhythm of hind- and forelimbs. The results are difficult to
interpret since the lesions also interrupt descending and ascending
pathways between the spinal cord and brainstem. In two cats bilateral, histologically identified lesions were made in the

ventrolateral funiculus at T10 and T11, which resulted in a disorganization of hindlimb-forelimb coordination lasting about two weeks (Miller et al., 1975, Unpublished observations). For the following 18 months the coordination of their limbs during stepping on the treadmill or on the ground appeared indistinguishable from prelesion controls. The only clue was that one of the cats when trotting or galloping fast over rough ground would fall with its limbs out of sequence and tangled. The lesions would have interrupted the direct ascending, and a portion of the direct descending, long propriospinal projections (See Miller and van der Burg, 1973). Anatomical (Giovanelli Barilari and Kuypers, 1969; Sterling and Kuypers, 1968) and neurophysiological (See Miller and van der Burg, 1973) studies have shown the presence of direct projections between the intumescences and of other projections to and from the intervening segments. Several questions still remain. Are the direct long connections essential for the hindlimb-forelimb coordination, or can indirect pathways take over this function? Do these pathways interconnect the motor centres for the limbs with segments controlling the cyclical movements of the back and neck muscles during locomotion?

Acknowledgment

We would like to acknowledge Hans van der Burg and Jan Ruit for their participation in some experiments; Mrs. Edith Jongbloed for typing the manuscript; Wouter van der Oudenalder and Miss Paula Delfos for photographic assistance; Bob Verhoeven for care of the cats; members of Dr. Sten Grillner's laboratory, Institute of Physiology, University of Goteborg, for discussions, and the European Training Programme in Brain and Behaviour Research for a training grant to our two laboratories; the Dutch Organization for Fundamental Medical Research (FUNGO: Project 13-46-09) for material support and for the position of F.G.A. van der Meché; Gist-Brocades NV for a sample of L-DOPA; and Hoffman-La Roche BV for a sample of Ro 4-4602/1.

REFERENCES

Afelt, Z., (1974) Functional significance of ventral descending tracts of the spinal cord in the cat. Acta Neurobiol. exp. 34, 393-407.

Arshavsky, Y.I., Kots, Y.M., Orlovsky, G.N., Rodionov, I.M. and Shik, M.L., (1965) Investigation of the biomechanics of running by the dog. Biofizika. 10, 665-672 (English transl.)

Brown, T.G., (1911) The intrinsic factors in the act of progression in the mammal. Proc. R. Soc. 84, 308-319.

Burns, B.D. and Pritchard, R., (1968) Cortical conditions for fused binocular vision. J. Physiol. 197, 149-171.

Forssberg, H. and Grillner, S., (1973) The locomotion of the acute spinal cat injected with clonidine i.v. Brain Res. 50, 184-186.

Furnée, E.H., Halbertsma, J.M., Klunder, G., Miller, S., Nieurkerke, K.J., van der Burg, J. and van der Meché, F.G.A., (1974) Automatic analysis of stepping movements in cats by means of a television system and a digital computer. J. Physiol. 240, 3-4P.

Giovanelli Barilari, M. and Kuypers, H.G.J.M., (1969) Propriospinal fibers interconnecting the spinal enlargements in the cat. Brain Res. 14, 321-330.

Goslow, G.E., Jr., Reinking, R.M. and Stuart, D.G., (1973) The cat step cycle: hindlimb joint angles and muscle lengths during unrestrained locomotion. J. Morphol. 141, 1-42.

Gray, J., (1968) Animal Locomotion. Weidenfeld and Nicolson, London.

Grillner, S. and Zangger, P., (1974) Locomotor movements generated by the deafferented spinal cord. Acta Physiol. Scand. 91, 38A-39A.

Grillner, S., (1975) Locomotion in vertebrates: Central mechanisms and reflex interaction. Physiol. Rev. 55, 247-306.

Guillebeau, A. and Luchsinger, B., (1882) Fortgesetzte Studien am Ruckenmarke. Pflügers Arch. 28, 61-69.

Halbertsma, J.M., (1975) Registratie en analyse van bewegingen en spierpotentialen voor het onderzoek naar de neuronale besturing van het lopen bij de kat. Thesis for Degree in Engineering. Department of Applied Physics, University of Technology, Delft, Netherlands.

Hildebrand, M., (1959) Motions of the running cheetah and horse. Journal of Mammalogy. 40, 81-495.

Holmqvist, B., (1961) Crossed spinal reflex actions evoked in somatic afferents. Acta Physiol. Scand. 52, (Suppl. 181), 1-67.

Jankowska, E., Lundberg, A. and Stuart, D., (1973) Propriospinal control of last order interneurones of spinal reflex pathways in the cat. Brain Res. 53, 227-231.

Kulagin, A.S. and Shik, M.L., (1970) Interaction of symmetrical limbs during controlled locomotion. Biofizika. 15, 164-170 (English transl.).

Luttrell, C.N., Bang, F.D. and Luxenberg, K., (1959) Newcastle disease encephalomyelitis in cats. II. Physiological studies on rhythmic myoclonus. A.M.A. Archives of Neurology and Psychiatry. 81, 35/185-41/291.

Miller, S. and Oscarsson, O., (1969) "Termination and functional organization of spino-olivocerebellar paths," In The Cerebellum in Health and Disease. (Fields, W.S. and Willis, W.D. Jr., eds.), Warren H. Green Inc., St. Louis, Missouri, (172-200).

Miller, S., Reitsma, D.J. and van der Meché, F.G.A., (1973) Functional organization of long ascending propriospinal pathways linking lumbosacral and cervical segments in the cat. Brain Res. 62, 169-188.

Miller, S. and van der Burg, J., (1973) "The function of long propriospinal pathways in the coordination of quadrupedal stepping in the cat," In Control of Posture and Locomotion. (Stein, R.B., Pearson, K.G., Smith, R.S. and Redford, J.B., eds.), Plenum Press, New York, (561-578).

Miller, S. and van der Meché, F.G.A., (1975) Movements of the forelimbs of the cat during stepping on a treadmill. Brain Res. 91, 255-269.

Miller, S., van der Burg, J. and van der Meché, F.G.A., (1975a) Coordination of the hindlimb and forelimbs in different forms of locomotion in normal and decerebrate cats, Brain Res. 91, 217-237.

Miller, S., van der Burg, J. and van der Meché, F.G.A., (1975b) Locomotion in the cat: basic programs of movement. Brain Res. 91, 239-253.

Muybridge, E., (1957) Animals in Motion. (Brown, L.S., ed.), Dover Publications, Inc., New York.

Pearson, K.G. and Iles, J.F., (1973) Nervous mechanisms underlying intersegmental co-ordination of leg movements during walking in the cockroach. J. Exp. Biol. 58, 725-744.

Roberts, T.D.M., (1967) Neurophysiology of postural mechanisms. Butterworths, London.

Sherrington, C.S., (1910) Flexion-reflex of the limb, crossed extension reflex, and reflex stepping and standing. J. Physiol. 40, 28-121.

Shimamura, T. and Livingstone, R.B., (1963) Longitudinal conduction systems serving spinal and brain-stem coordination. J. Neurophysiol. 26, 258-272.

Sterling, P. and Kuypers, H.G.J.M., (1968) Anatomical organization of the brachial spinal cord of the cat. III. The propriospinal connections. Brain Res. 7, 419-443.

Stuart, D.G., Withey, T.P., Wetzel, M.C. and Goslow, Jr., G.E., (1973) "Time constraints for inter-limb coordination in the cat during unrestrained locomotion," In Control of Posture and Locomotion. (Stein, R.B. et al., eds.), Plenum Press, New York, (537-560).

FUNCTION OF SEGMENTAL REFLEXES IN THE CONTROL OF STEPPING IN COCKROACHES AND CATS

K.G. Pearson and J. Duysens

Department of Physiology

University of Alberta, Edmonton, Canada

The reflex mechanisms controlling stepping in the cockroach and cat have been compared in an attempt to identify common functional reflexes in the control of terrestrial walking. The main question examined was: what initiates the transition from stance to swing in a single limb? In both animals it has been found that blocking leg extension during the stance phase inhibits the rhythmic movements of that leg but not the rhythm in the other legs. Allowing the blocked leg to extend slowly eventually leads to the initiation of swing. Two mechanisms could explain these observations: 1) the activity in receptors signalling the position of the limb at the transition point causes the switch from stance to swing, and 2) as the leg is extended the load carried by that leg is reduced and decreased activity in receptors detecting the load causes the initiation of swing. In the cat the hip angle at the end of stance remains constant in a variety of behavioural situations, indicating that the signal for swing initiation originates from hip position afferents (mechanism 1). For the cockroach it has been found that activity in cuticular stress receptors (campaniform sensilla) during stance inhibits the system responsible for producing swing. Thus swing is initiated when the inhibition of the swing generating system by load receptors is reduced (mechanism 2).

In an attempt to determine whether unloading of a limb could also be an important factor for initiating swing in the cat, the effect of loading the ankle extensor muscles on the locomotory rhythm was investigated in thalamic and mesencephalic cats walking on a treadmill. Stretching the ankle extensor muscles beyond a certain length in one hindleg during periods of walking caused the abolition of rhythmic contractions of all muscles of that leg (the other three legs continued to step). When the locomotory rhythm was inhibited in this manner the leg extensors contracted tonically; the flexors relaxed. Unloading the tonically contracting ankle extensors restored the locomotory rhythm in both flexors and extensors. Loading the ankle extensors by electrically stimulating a ventral root filament, or the muscles directly, could also prevent the occurrence of rhythmic contractions of ankle flexor and extensor muscles during walking. These observations demonstrate in the cat that loading the ankle extensors inhibits the locomotory rhythm generator, and suggest a necessary condition for swing to be initiated is that the leg extensors be unloaded. Thus for both the cockroach and cat an important reflex functions to prevent swing in a leg when that leg is loaded.

Another functionally important reflex in both the cockroach and the cat is a reinforcing reflex which increases the magnitude of extensor activity if the extension movement is resisted during stance. This reflex presumably functions to help prevent changes in the speed of contraction in extensor muscles under conditions where there can be changes in the resistance to extension movements.

The swing phase is also partially controlled by segmental reflexes. In the cockroach, stick insect and the dog reflexes appear to limit the amplitude of swing so that the leg position at the end of swing remains constant irrespective of the position at the beginning of swing. Why reflex control of swing amplitude is necessary remains uncertain.

Introduction

The importance of reflexes in the control of stepping in mammals was recognized in the early work of Sherrington (1906), Brown (1914) and Philippson (1905). Philippson explained the locomotory rhythm entirely in terms of serial activity in different reflex pathways, whereas Brown (1911) and later Sherrington (1913) considered that the locomotory rhythm was generated centrally and modified by sensory input. There is now considerable evidence for a central rhythm generator (Grillner, 1975 and this volume) and a number of demonstrations that reflex pathways play an important role in the control of walking (this chapter). Plausible functions for some limb receptors during walking have recently been described

(Severin, 1970; Forssberg et al., 1975) but in general the precise functions of most limb sensory receptors (in skin, muscle and joints) are simply not known. Similarly in insects the function during walking of most groups of leg receptors remains unknown despite the fact that the importance of reflexes in controlling leg movements during walking has been recognized for some time (Pringle, 1961; see Wilson, 1966, for references). Quite recently, however, the functions of a number of reflex pathways in insects have been defined (Wendler, 1966; Runion and Usherwood, 1968; Pearson et al., 1973).

The aim of this chapter is to review the data related to the reflex control of walking in insects and mammals. The major emphasis will be on data from the cockroach and the cat since most recent work involves these two animals. The comparison of data obtained from the cockroach, cat and other animals shows a number of common features and readily suggests certain principles for the nervous control of terrestrial walking in all animals. No attempt is made to give functional explanations for all the reflex pathways so far described in these animals nor to assess the significance of the convergence of various reflex pathways within the central nervous system (see Lundberg, 1969; Evoy and Cohen, 1971; McIntyre, 1974). Thus only those intrasegmental reflexes which, at the present time, appear to be functionally important in controlling stepping in a single leg will be discussed in detail.

I. Reflexes Controlling the Rate of Stepping

Stepping movements in a single limb of cats (Creed et al., 1932; Orlovsky and Fel'dman, 1972; Grillner, 1975), toads (Gray and Lissman, 1946), stick insects (Wendler, 1966; Bassler, 1967) and cockroaches (Pringle, 1961; Pearson, Unpublished observations) can be inhibited by preventing the backward movement during stance. The remaining legs continue to step but their coordination may differ from normal. If the held leg is moved backwards relative to the body the swing phase is initiated when the position of the limb is close to that where swing would be initiated during normal walking. This latter observation demonstrates that reflexes arising from leg receptors must provide a signal for initiating the swing phase. Since the duration of the swing phase is relatively constant for different walking speeds it follows that the rate of stepping must depend to some extent upon the sensory input causing the switching from stance to swing. The most obvious demonstration of this is in the mesencephalic or spinal treadmill walking cat where the rate of stepping matches the speed of the treadmill for wide variations in treadmill speed (Shik et al., 1966; Grillner, 1973). There are two possible mechanisms whereby reflex activity could cause the switching from stance to swing: 1) input from receptors signalling position of the leg at the end of stance excites the system of neurons producing swing and inhibits the system producing stance, and/or 2) input from receptors detecting the load carried by the leg is diminished at the end of stance and this diminution in sensory input causes the removal of an inhibitory influence from the system of neurons producing swing.

For the cockroach it has been proposed (Pearson et al., 1973) that the second of these two mechanisms determines the initiation of protraction (note: when referring to the two phases of leg movements in insects the usual terms of protraction and retraction will be used instead of swing and stance). Rhythmic burst activity in motoneurons giving flexion movements of the femur during protraction is inhibited and reset by mechanical deformation of the cuticle of the trochanter (Figure 1). This type of stimulation excites stress receptors (the campaniform sensilla) located within the cuticle. During retraction when the leg is loaded it is probable that these receptors are highly active (Pringle, 1961). Thus the system of neurons producing protraction is strongly inhibited. As retraction progresses this inhibitory influence diminishes due to a reduction in stresses in the trochanter until it is insufficient to prevent a burst of activity being initiated in the flexor motoneurons. The same mechanism probably functions in the stick insect to initiate protraction since continuous deformation of the trochanter of a leg by compression with a small clamp causes that leg to be held continuously retracted while the other legs are stepping normally (Bässler, Personal communication).

Figure 1. Inhibition of rhythmic burst activity in flexor motoneurons of the cockroach metathoracic leg by pressure on the trochanter. The stimulus (indicated by the horizontal line under the record) excites cuticular stress receptors named campaniform sensilla. In a walking animal it has been proposed that input from these receptors when the leg is loaded during stance (leg extension) inhibits the generation of a flexor burst and hence a swing movement. The initiation of swing is presumably due to the removal of this inhibitory effect as a result of unloading of the leg near the end of stance (From Pearson and Iles, 1973).

At present there are no data in arthropods clearly demonstrating that sensory input from leg position receptors is important in initiating protraction. However, removal of some position receptors in the legs of cockroaches and stick insects (the hair plates) can significantly alter the movements of the legs (Wendler, 1966; Wong and Pearson, 1975). Thus it is probable that information about both leg loading and position are used to determine the instant for initiating protraction.

By contrast with the results on insects, the data from studies on cats and dogs suggest that receptors signalling the position of the leg are important in initiating swing. Early work by Sherrington (1910) showed that swing movements in spinal dogs could be initiated by proprioceptors in the proximal parts of the leg. Sudden extension of the hip joint initiated swing whereas extension of the other joints (knee and ankle) did not. More recently, Shik and Orlovsky (1965) found in treadmill walking dogs that if the posterior trunk of the animal was lifted slightly the duration of the stance phase in the hindlegs was prolonged but the hip angle at the beginning of swing did not change. Further evidence that afferents signalling the position of the hip joint are involved in initiating swing comes from recent work in chronic spinal cats (Rossignol et al., 1975; Grillner, 1975). If in a walking animal the hip joint of one hindleg is held stationary, movement of the knee and ankle joints does not initiate leg flexion. If the hip is allowed to extend, flexion is initiated at a hip angle similar to that at the onset of flexion during normal walking. From these studies it may be concluded that reflexes from hip afferents signalling hip position are probably important in initiating swing. The afferents have not yet been identified and could originate in the hip joint, in hip muscles or in both.

A reflex from hip afferents cannot fully explain the initiation of swing in all experimental situations. In some initial observations on mesencephalic cats walking on a treadmill we noticed that rhythmic burst activity in flexor motoneurons could be inhibited by flexion of the ankle joint even when the hip was held at an angle considerably more extended than that occurring at the end of the normal stance phase. Thus this preliminary observation indicated that loading the ankle extensor muscles inhibited the locomotory rhythm and suggested the hypothesis that unloading of the leg extensor muscles at the end of stance is necessary for initiating the swing phase.

To test the hypothesis that loading the ankle extensor muscles can indeed inhibit the locomotory rhythm we have studied the effects of loading these muscles in thalamic cats walking spontaneously on a treadmill. Animals were anaesthetized with halothane and all muscles of one hindleg denervated except the ankle extensors (triceps surae) and the ankle flexor (anterior tibialis). After fixing the animal in an animal frame in which the head, hips and left knee were held rigidly by metal pins, the Achilles tendon of the ankle extensors was severed and connected to a tension transducer mounted on a device for lengthening the muscle. The distal tendons of the ankle flexors were severed. Thus in this preparation the partially denervated left leg was held firmly and arranged so that the triceps surae group of muscles could be stretched. EMG recording electrodes were implanted in the triceps muscles of the left and right legs and the anterior tibialis (ankle flexor) of the left leg. Decerebration was carried out by removing both cerebral hemispheres then by cutting the mesencephalon at an angle of $50°$ to horizontal at a point immediately rostral to the superior colliculi. This gave a preparation which walked spontaneously for varying periods of time (30 minutes to 5 hours) after recovery from the anaesthetic.

During periods of walking of the three intact limbs on the treadmill, the ankle flexor and extensor muscles of the partially denervated left leg contracted rhythmically, reciprocally and $180°$ out of phase with the homologous muscles in the intact right leg. Slowly extending the rhythmically contracting triceps muscles initially led to an increase in the magnitude of the force generated during each burst of extensor activity and to an increase in the magnitude of the extensor EMG (Figures 2 and 3). At a critical length all rhythmic contractions ceased and the extensors remained tonically contracted (Figure 2). Concomitant with the disappearance of burst activity in the extensor muscles all burst activity ceased in the flexor motoneurons and these motoneurons remained silent (Figure 3). The inhibition of the rhythmic burst activity was a sudden 'all-or-none' phenomenon. That is, when the rhythmic bursts disappeared there was no sign of amplitude modulation of extensor EMG, nor was there any sign of the occurrence of flexor bursts. Thus from these observations we concluded that stretch of ankle extensors (triceps surae) inhibits the system producing the periodic inhibition of extensor activity and also the system producing the periodic excitation of the flexor motoneurons. At present it is simplest to postulate that the same interneuronal system periodically inhibits extensors and excites flexors.

The force in the extensor muscles when the rhythm was inhibited was in the range of 1.5 to 4 Kgms, which corresponds closely to the calculated range of force (2 to 5 Kgms) in the ankle extensors during stance in a normal walking animal (Grillner, 1972). The length of the muscle when the rhythm was inhibited was also in the physiological range. This was demonstrated in a number of animals by leaving the Achilles tendon attached to the bone of the foot and observing that holding the ankle joint slightly flexed (an angle of approximately $80°$ relative to the tibia) inhibited the locomotory rhythm. From these observations then we may conclude that the inhibitory reflex onto the rhythm generator with stretch of the

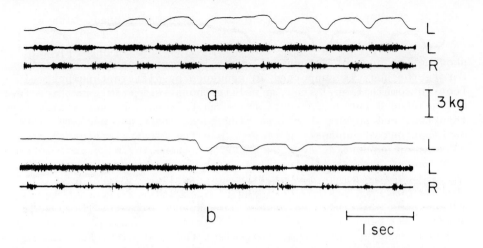

Figure 2. Inhibition of locomotory rhythm by stretching the isolated left (L) hindlimb ankle extensor muscles (triceps surae) in a mesencephalic cat walking on a treadmill. Top traces - tension in left ankle extensor; middle traces - EMG from left ankle extensors; bottom traces - EMG from right ankle extensor. The calibration is for the tension trace. (a) Gradual lengthening of the isolated left ankle extensor muscles initially increases the force of contraction and the EMG amplitude. At a critical length the periodic inhibitory pauses in extensor activity do not occur. Stretching beyond this length (b) results in long periods of maintained contraction but does not inhibit stepping in the intact right leg.

ankle extensors could be functionally important in a normal walking animal. However, these data alone do not show whether the parameter for inhibiting the locomotory rhythm is either muscle force or muscle length. To determine which of these parameters is associated with the inhibition, the ankle extensor muscles were held at a constant length. The load was increased by either stimulating the muscles directly or by stimulating a ventral root filament of either S1 or L7 during periods of rhythmic activity. The locomotory rhythm could be inhibited by briefly stimulating the muscle or filament near the end of a contraction (Figure 4). Since the contractile elements would have shortened in this situation, it is probably an increase in muscle force and not length that inhibits the locomotory rhythm. The receptors in the ankle extensors producing the inhibitory effect on the locomotory rhythm have not yet been identified. Since the muscle force appears to be the important parameter for inhibiting the rhythm, the receptors signalling muscle force, the Golgi tendon organs, are the most logical candidates.

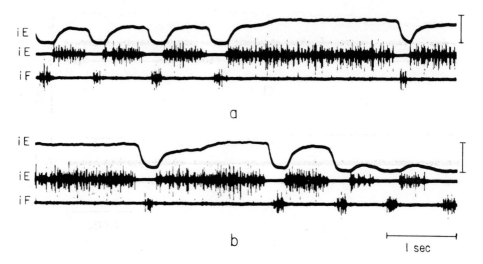

Figure 3. Abolition of burst activity in ankle flexor motoneurons by stretch of the ipsilateral ankle extensor muscles in a mesencephalic cat walking on a treadmill. Top traces - force in ankle extensors (calibration = 2.6 kg); middle traces - EMG recorded from ankle extensor muscles (triceps surae); bottom traces - EMG recorded from ipsilateral ankle flexor muscle (tibialis anterior). (b) is continuous with (a). The ankle extensor muscles were initially stretched so that the inhibitory pauses in extensor activity occasionally did not occur (c.f. Figure 2). Note that the abolition of the inhibitory pause in extensor activity was always associated with the abolition of a flexor burst. Near the end of (b) the extensor muscle length was reduced and there was a corresponding marked decrease in extensor EMG.

II. Reflexes Controlling the Amplitude of Motor Output

Sensory input produced directly or indirectly by the contraction of a given muscle can function to either reinforce or inhibit the activity in the motoneurons producing the contraction in that muscle. These two types of effects will be discussed separately.

A. Reinforcing reflexes. The need for reinforcing reflexes is quite apparent in locomotory systems in which sudden unexpected changes in load can occur, although the need for reflexes for all types of load compensation is not necessary (Grillner, 1972). If the speed of contraction in shortening muscle is to be maintained when presented with an additional load then a greater level of motoneuronal activity is required. This could be produced by the

Figure 4. Inhibition of the locomotory rhythm in a mesencephalic cat walking on a treadmill by stimulation of an S1 ventral root filament. The stimulating frequency was 80/second and the stimulus duration was 0.5 sec. The stimulus artifact can be seen as the dark band on the middle of the third trace. Top trace - force in the ankle extensors ipsilateral to the stimulated ventral root (calibration = 6.5 Kgms); second trace - EMG from ipsilateral ankle extensors; third trace - EMG from ipsilateral ankle flexor (anterior tibialis); bottom trace - EMG from contralateral ankle extensors. Loading the ipsilateral ankle extensor muscles by stimulating the ventral root filament near the end of an isometric contraction prevented the inhibition of extensor activity and abolished the corresponding burst of activity in the ipsilateral flexors. The rhythmic motor activity in the contralateral leg was not inhibited.

recruitment of previously silent motoneurons or by an increase in the discharge rate of the motoneurons already active. Reinforcing reflexes have now been demonstrated in many motor systems (Severin, 1970; Pearson et al., 1973; Kater et al., 1974) and are clearly important in controlling the level of activity in the leg extensor muscles of the walking cockroach and cat.

If the leg extension is resisted in a walking cockroach by letting the animal either walk up an inclined surface or drag a weight, the level of activity in extensor motoneurons is increased (Figure 5). This reflex is mediated via a strong (probably monosynaptic) excitatory reflex pathway from cuticular stress receptors (campaniform sensilla) to the extensor motoneuron producing femur extension movements (Pearson, 1972). These stress receptors are excited during leg retraction and thus facilitate activity in motoneurons producing the retraction movement.

In the mesencephalic cat walking on a treadmill, resistance of leg extension during stance in a single limb gives rise to a very marked increase in the EMG recorded from leg extensor muscles.

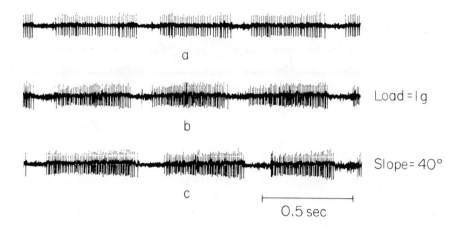

Figure 5. Increased discharge rate of an extensor motoneuron in the metathoracic leg of the cockroach caused by resisting the extension movements of the femur either by allowing the animal to drag a small weight or to walk up a sloped surface. This motoneuron is referred to as motoneuron D_s (Pearson, 1972), and is entirely responsible for giving extension movements of the femur in a slowly walking animal. (a) Walking on a flat surface; (b) walking on a flat surface dragging a weight of 1 gm; (c) walking up a slope inclined at 40° to the horizontal.

Resisting shortening of rhythmically contracting ankle extensor muscles also leads to a marked increase in the extensor EMG (Figure 3). Thus a reinforcing reflex exists from receptors in leg extensor muscles to the extensor motoneurons. The receptors producing this reinforcing effect in extensors have not yet been identified. The muscle spindles in the extensor muscles are the most likely candidates since blockage of conduction in gamma motoneurons significantly diminishes the intensity of extensor activity during stance in a walking cat (Severin, 1970).

Another reinforcing reflex demonstrated by the early work of Sherrington (1910) is the extensor thrust reflex. Here stimulation of foot pad receptors excites extensor motoneurons which results in a strong thrust of the foot downward and backward. The bilateral nature of the effects produced in spinal animals led Sherrington to conclude that it is most probably involved in reinforcing extensor activity in a galloping animal. However, the calculations by Grillner (1972) have shown clearly that time delays in segmented reflex pathways are too great for this reflex to function in a galloping animal. Recent work (Duysens and Pearson, 1975), has shown that weak electrical stimulation of the pad and plantar surface of the foot in a slowly walking thalamic animal results in an increase in the magnitude of the extensor burst when the stimulus is given during the stance phase. Moreover, direct stimulation of cutaneous nerves from the foot with weak electric currents gave a similar result. An increase of extensor EMG has also been reported for stimulation during stance of the dorsum of the foot in chronic spinal cats (Forssberg et al., 1975). Thus reflexes arising from activity in cutaneous receptors in the foot during stance function to reinforce activity in motoneurons discharging during this phase of leg movement.

B. Inhibitory reflexes. In the dog, cockroach and stick insect there are reflex pathways functioning during the swing phase to limit the amplitude of the swing movement. Orlovsky and Shik (1965) showed in the dog that perturbations of movements at the elbow joint had little effect on the magnitude of the swing phase and concluded that flexion of the elbow occurs up to a definite fixed angle irrespective of the initial angle at the beginning of swing. It seems therefore that in the dog certain reflex pathways function to inhibit the motor activity at the end of swing.

In the cockroach and stick insect it has been found that removal of certain groups of hair receptors on the legs leads to exaggerated flexion movements during swing (Wendler, 1966; Wong and Pearson, 1975). The trochanteral hair plate receptors in the cockroach leg are excited during flexion movements of the femur (protraction) and their removal gives the exaggerated femur flexion movements. These hair plate afferents disynaptically inhibit flexor motoneurons and monosynaptically excite extensor motoneurons (Pearson et al., 1975). Thus their removal abolishes an inhibitory input to the flexor motoneurons during flexion and as a result, the flexion movement is increased in amplitude. The importance of the inhibitory reflex from the hair plate afferents onto the flexor motoneurons is uncertain. Perhaps it functions to ensure that any fatigue in flexor muscles does not lead to a decrease in the amplitude of the flexion movement. Another, more probable, function is that it limits the amplitude of protraction so that irrespective of the limb position at the end of retraction there will be no mechanical interference with the anterior leg during protraction. A similar inhibitory reflex has recently been described in the flight system of the locust (Burrows, 1975). Here the inputs from wing stretch receptors excited during wing elevation inhibit the elevator motoneurons and excite the depressor motoneurons. Thus these pathways function to limit the amplitude of wing elevation. Two more systems in which this type of negative feedback has been described are the swimmeret system of the lobster and the masticatory system of the snail (Davis, 1969; Kater and Rowell, 1973). It is perhaps significant that in all these systems (walking, flight, swimmeret and mastication) the inhibitory reflexes function to limit the amplitude of the movement not subject to variations in load, that is, the movement opposite that propelling the animal in the locomotory systems.

III. Interaction of Reflexes with Locomotory Rhythm Generator

We now know that the patterning of motoneuronal activity underlying stepping movements is to a large extent dependent upon the organization and properties of interneurons within the ventral nerve cord of arthropods and within the spinal cord of vertebrates. Central patterning has been discussed by Grillner and Fourtner elsewhere in this volume and consequently it will not be dealt with at length here. Nonetheless, it is important to consider some aspects of the central organization of the stepping system. It is obvious that an understanding of the functioning of reflex pathways depends on a knowledge of the properties and organization of the central elements.

In all locomotory systems studied in detail there is now good evidence demonstrating that rhythmic reciprocal bursts of activity can be generated in motoneurons producing swing (mainly flexors) and stance (extensors) in the absence of sensory input. The cellular mechanisms responsible for producing these centrally generated patterns of activity are unknown. However, three basic models have been discussed in the literature. The first is the 'half center' model initially proposed by Brown (1914) in which the two systems of neurons (half centers) producing swing and stance mutually inhibit each other. The rhythmicity as well as the reciprocity in the motor output is dependent on the coupling between each half center. A second, and somewhat simpler model is the pacemaker model (Grillner, 1975). In this model the rhythmicity is generated by a single set of interneurons (or single interneuron) which excites one group of motoneurons and inhibits the antagonists. The essential difference between the pacemaker and half-center models is that in the latter both sets of motoneurons are directly excited by interneurons belonging to the rhythm generating system whereas in the former only one set of motoneurons is directly excited by the rhythm generating system and the other set becomes active by a release from inhibition. The third model is the 'ring' model (Kling, 1971; Gurfinkel and Shik, 1972) in which the activity is considered to be propagated around a closed loop of interneurons. These neurons in turn excite different motoneurons to give the complex temporal pattern of activity found in a walking animal. The ring model in its simplest form reduces to the half center model.

There is much evidence for the pacemaker model in the walking system of the cockroach (Pearson and Iles, 1970; Pearson and Fourtner, 1975; Fourtner, this volume) while the half center model is favoured for the cat walking system (Lundberg, 1969; Grillner,

1975). In assessing the present data on the central organization of the cat walking system and comparing this data with that from the cockroach we conclude that the pacemaker model can also explain the cat data. Some of the observations leading to this conclusion are as follows.

(1) There is a clear asymmetry in the cyclic nature of leg movements in the cockroach and cat in that the swing phase duration is relatively constant whereas the duration of the stance phase varies directly with walking speed. This asymmetry reflects an asymmetry in the central organization since after deafferentation the flexor burst duration remains relatively constant with variations in cycle time (Pearson and Iles, 1970; Brown, 1911; Grillner, this volume).

(2) In the walking system of the cockroach rhythmic flexor bursts are generated in deafferented preparations in the absence of extensor burst activity but the reverse is never observed (Pearson and Iles, 1970). Similarly in cats under some conditions flexor bursts can be generated without extensor bursts. The clearest example of this phenomenon in cats is during narcosis progression (Brown, 1912). Records of Perret (1973, Figure 7) obtained in deafferented preparations after severing the spinal cord also show flexor burst activity without extensor activity but not the reverse, while the only published records of motor activity in a curarized spinal cat after the injection of DOPA and Nialamide show spontaneously generated flexor bursts but not extensor bursts (Jankowska et al., 1967; Viala et al., 1974). Finally, in mesencephalic cats walking spontaneously on a treadmill, we have observed occasionally that the swing phase is normal but during the stance phase there is no activity in extensor motoneurons and the leg is extended passively by the treadmill. Stimulation of the midbrain locomotory region in these animals causes the extensor motoneurons to become active during stance. Similarly Perret (1973) has observed that curarization of thalamic cats eliminates the extensor part of the nerve activity to the bifunctional sartorius muscle while the rhythmic efferent flexor activity to this muscle remains.

(3) In the cockroach an indication that extensor activity is produced by a release from inhibition is that often extensor motoneurons become active immediately following a flexor burst but the reverse is never seen (Pearson and Iles, 1970). A similar phenomenon has been observed in deafferented hindlimb muscles of the cat (Brown, 1911). Moreover, rebound of extensor activity invariably follows a flexor reflex in a decerebrate animal (Sherrington, 1913).

From these observations we propose that in both the cockroach and the cat there exists a system of interneurons which periodically produces bursts of activity in motoneurons active during swing (mainly flexors) and concomitantly inhibits the motoneurons active during stance (extensors) (Figure 6). This system of interneurons will be referred to as the swing generator. It is assumed that the swing generator is periodically activated by central and reflex inputs and remains active for a relatively constant period of time independent of the cycle time. Between the periods of activity in the swing generator the extensor motoneurons are activated by post-inhibitory rebound, a central command input and reinforcing reflexes. Postulating a swing generator in the walking system of each leg considerably simplifies the explanation for the mechanism of reflex regulation of the motor output.

For swing to be initiated the leg must be unloaded and extended beyond a certain position (section I). Thus sensory input from receptors detecting the load carried by the leg and the leg position at the end of stance must influence the swing generator. In the scheme shown in Figure 6 it is proposed that afferents signalling load inhibit the swing generator while those signalling leg position at the end of stance excite the swing generator. As the leg is extended during stance the inhibitory influence from load sensitive afferents is reduced and the position afferents become more active. Both these effects increase the excitability of the swing generator. When the leg is unloaded and extended the swing generator becomes active and a swing movement is produced.

Reinforcing reflexes are known to facilitate activity in leg extensor motoneurons during stance. These reflexes are represented

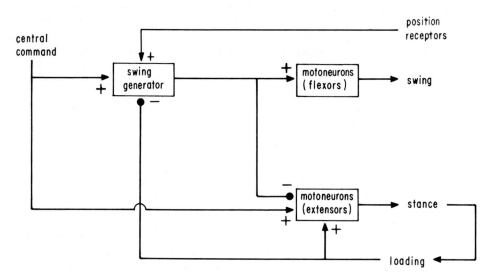

Figure 6. Block diagram summarizing the major features of the central and reflex organization of the system producing stepping movements in a single limb of the cockroach and cat. See section III for discussion of this diagram.

in Figure 6 by an excitatory pathway from load detecting afferents to the extensor motoneurons. These afferents are not necessarily identical to those inhibiting the swing generator. Any increase in resistance to leg extension during stance increases the activity in the afferents exciting the extensor motoneurons and thus increases the extensor activity to help maintain the speed of the stance movement.

Figure 6 does not include those reflexes arising from cutaneous afferents (Forssberg et al., 1975; Duysens and Pearson, 1975), the central and reflex influences from other segments (Miller and van der Burg, 1973; Rossignol et al., 1975), nor nonspecific tonic excitatory influences from peripheral receptors (Orlovsky and Fel'dman, 1972; Pearson, 1972). Nor is any attempt made to explain the complex temporal sequence of activity observed during walking (Grillner and Zangger, 1975). Nonetheless, despite these obvious shortcomings, the proposal of a swing generator for each half segment explains many experimental results to date and may considerably simplify the understanding of stepping movements since the problem reduces to one of understanding how the swing generators are coupled between segments and the required conditions for their activation.

Conclusion

Functionally similar reflexes exist in the legs of the cockroach and the cat for controlling stepping movements during walking. In both animals the transition from stance to swing in a single limb is inhibited when that limb is loaded. Thus a necessary condition for the initiation of swing is an unloading at the end of stance. Resistance to leg extension during stance leads to an increased activity in extensor motoneurons. This reinforcing reflex probably functions to maintain the speed of the extension movement during stance.

REFERENCES

Bässler, U., (1967) Zur Regelung der Stellung des Femur-Tibia-Gelenkes bei der Stabheuschrecke Carausius morosus in der Ruhe und im Lauf. Kybernetik. 4, 18-26.

Brown, T.G., (1911) The intrinsic factors in the act of progression in the mammal. Proc. R. Soc. B 84, 308-319.

Brown, T.G., (1912) The phenomenon of 'narcosis progression' in mammals. Proc. R. Soc. B 86, 140-164.

Brown, T.G., (1914) On the nature of the fundamental activity of the nervous centres: together with an analysis of the conditioning of rhythmic activity in progression, and a theory of the evolution of function in the nervous system. J. Physiol. 48, 18-46.

Burrows, M., (1975) Monosynaptic connections between wing stretch receptors and flight motoneurons of the locust. J. Exp. Biol. 62, 189-219.

Creed, R.S., Denny-Brown, D., Eccles, J.C., Liddell, E.G.T. and Sherrington, C.S., (1932) Reflex Activity of the Spinal Cord. Oxford University Press, London.

Davis, W.J., (1969) Reflex organization in the swimmeret system of the lobster. I. Intrasegmental reflexes. J. Exp. Biol. 51, 547-563.

Duysens, J. and Pearson, K.G., (1975) The role of cutaneous afferents from the distal hindlimb in the regulation of the step cycle of treadmill walking thalamic cats. Exp. Brain Res. (In the press).

Evoy, W.H. and Cohen, M.J., (1971) "Central and peripheral control of arthropod movements," In Advances in Comparative Physiology and Biochemistry. Vol. 4, (Loewenstein, O., ed.), Academic Press, N.Y., (225-266).

Forssberg, H., Grillner, S. and Rossignol, S., (1975) Phase dependent reflex reversal during walking in chronic spinal cats. Brain Res. 85, 103-107.

Gray, J. and Lissman, H.W., (1946) The co-ordination of limb movements in the amphibia. J. Exp. Biol. 23, 133-142.

Grillner, S., (1972) The role of muscle stiffness in meeting the changing postural and locomotor requirements for force development by the ankle extensors. Acta Physiol. Scand. 86, 92-108.

Grillner, S., (1973) "Locomotion in the spinal cat," In Control of Posture and Locomotion. (Stein, R.B., Pearson, K.G., Smith, R.S. and Redford, J.B., eds.), Plenum Press, N.Y., (515-535).

Grillner, S., (1975) Locomotion in vertebrates: central mechanisms and reflex interaction. Physiol. Rev. 55, 247-306.

Grillner, S. and Zangger, P., (1975) How detailed is the central pattern generation for locomotion? Brain Res. 88, 367-371.

Gurfinkel, V.S. and Shik, M.L., (1972) "The control of posture and locomotion," In Motor Control. (Gydikov, A.A., Tankov, N.T. and Kosarov, D.S., eds.), Plenum Press, N.Y., (217-234).

Jankowska, E., Jukes, M.G.M., Lund, S. and Lundberg, A., (1967) The effect of DOPA on the spinal cord. 5. Reciprocal organization of pathways transmitting excitatory action to alpha motoneurons of flexors and extensors. Acta Physiol. Scand. 70, 369-388.

Kater, S.B., Heyer, C. and Kaneko, C.R.S., (1974) "Identifiable neurons and invertebrate behavior," In Neurophysiology. (Hunt, C.C., ed.), (Physiology series/MTP Int. Rev. Sci.), University Park Press, Baltimore, (1-52).

Kater, S.B. and Rowell, C.H.F., (1973) Integration of sensory and centrally programmed components in the generation of cyclical feeding activity of Helisoma trivolis. J. Neurophysiol. 34, 142-155.

Kling, U., (1971) Simulation neuronaler Impulsrhythmen. Zur Theorie der Netzwerke mit cyclischen Hemmverbindungen. Kybernetik. 9, 123-139.

Lundberg, A., (1969) Reflex control of stepping. The Nansen Memorial Lecture, Universitetsforlaget, Oslo, (1-42).

McIntyre, A.K., (1974) "Central actions of impulses in muscle afferent fibres," In Handbook of Sensory Physiology III(2). (Hunt, C.C., ed.), Springer-Verlag, Berlin, (235-288).

Miller, S. and van der Burg, J., (1973) "The function of long propriospinal pathways in the co-ordination of quadrupedal stepping in the cat," In Control of Posture and Locomotion. (Stein, R.B. et al., eds.), Plenum Press, N.Y., (561-577).

Orlovsky, G.N. and Shik, M.L., (1965) Standard elements of cyclic movement. Biophysics. 10, 935-944.

Orlovsky, G.N. and Fel'dman, A.G., (1972) Role of afferent activity in the generation of stepping movements. Neurophysiology. 4, 304-310.

Pearson, K.G. and Iles, J.F., (1970) Discharge patterns of coxal levator and depressor motoneurons of the cockroach, Periplaneta americana. J. Exp. Biol. 52, 139-165.

Pearson, K.G., (1972) Central programming and reflex control of walking in the cockroach. J. Exp. Biol. 56, 173-193.

Pearson, K.G., Fourtner, C.R. and Wong, R.K., (1973) "Nervous control of walking in the cockroach," In Control of Posture and Locomotion. (Stein, R.B. et al., eds.), Plenum Press, N.Y., (495-514).

Pearson, K.G. and Iles, J.F., (1973) Nervous mechanisms underlying intersegmental co-ordination of leg movements during walking in the cockroach. J. Exp. Biol. 58, 725-744.

Pearson, K.G. and Fourtner, C.R., (1975) Nonspiking interneurons in the walking system of the cockroach. J. Neurophysiol. 38, 33-52.

Pearson, K.G., Wong, R.K. and Fourtner, C.R., (1975) Connections between hair plate afferents and motoneurons in the cockroach leg. J. Exp. Biol. (In press).

Perret, C., (1973) Analyse des mécanismes d'une activité de type locomoteur cher le chat. Doctoral Thesis, University of Paris.

Philippson, M., (1905) L'autonomie et la centralisation dans le système nerveux de animaux. Trav. Lab. Physiol. Inst. Solvay (Bruxelles) 7, 1-208.

Pringle, J.W.S., (1961) "Proprioceptors in arthropods," In The Cell and the Organism. (Ramsay, J.A. and Wigglesworth, V.B., eds.), Cambridge University Press, Cambridge, (256-282).

Rossignol, S., Grillner, S. and Forssberg, H., (1975) Factors of importance for initiation of flexion during walking. Proc. Neuroscience Abstracts. 5, 181.

Runion, H.I. and Usherwood, P.N.R., (1968) Tarsal receptors and leg reflexes in the locust. J. Exp. Biol. 49, 421-436.

Severin, F.V., (1970) On the role of γ-motor system for extensor α-motoneuron activation during controlled locomotion. Biophysics. 15, 1138-1145.

Sherrington, C.S., (1906) The integrative Action of the Nervous System. Yale University Press, New Haven.

Sherrington, C.S., (1910) Flexion-reflex of the limb, crossed extension-reflex, and reflex stepping and standing. J. Physiol. 40, 28-121.

Sherrington, C.S., (1913) Reflex inhibition as a factor in the coordination of movements and postures. Quart. J. Exp. Physiol. 6, 251-310.

Shik, M.L. and Orlovsky, G.N., (1965) Coordination of the limbs during running in the dog. Biophysics. 10, 1148-1159.

Shik, M.L., Severin, F.V. and Orlovsky, G.N., (1966) Control of walking and running by means of electrical stimulation of the mid-brain. Biophysics. 11, 756-765.

Viala, D., Valin, A. and Buser, P., (1974) Relationship between the "late reflex discharge" and locomotor movements in acute spinal cats and rabbits treated with DOPA. Arch. Ital. Biol. 112, 299-306.

Wendler, G., (1966) "The coordination of walking movements in arthropods," In Symp. Soc. Exp. Biol., No. 20, Nervous and Humoral Mechanisms of Integration. (229-250).

Wilson, D.M., (1966) Insect walking. Ann. Rev. Entomol. 11, 103-122.

Wong, R.K. and Pearson, K.G., (1975) Properties of the trochanteral hair plate and its function in the control of walking in the cockroach. J. Exp. Biol. (In press).

THE ROLE OF VESTIBULAR AND NECK RECEPTORS IN LOCOMOTION

T.D.M. Roberts

Institute of Physiology, University of Glasgow

Glasgow G12 8QQ, Scotland

Locomotion involves complex sequences of motor command. To find the role of particular receptors in such sequences we need to consider what information those receptors might convey to the central nervous system and what is the nature of each of the various subtasks which might be influenced.

Motor activities of all types may be assembled from a repertoire containing only six basic categories of subtask: set, hold, drive, punch, catch and throw; the last three of which consist of sequences of elements from the first three. Prop and paddle actions of the limbs are distinguishable according as forces are exerted along or at right angles to the line of the limb from proximal joint to ground contact.

The required commands to individual muscles involve not just the control of tension or of length but rather the continuous adjustment of the dynamic stiffness with which each muscle interacts with the load imposed upon it. To develop adequate forces at specified lengths some muscles need to be activated before they come under load so as to take advantage of the additional force available during the stretching of active muscle.

In locomotion the support phases in individual limbs are reduced in duration when the gait gets faster. This calls for corresponding increases in the peak upthrusts to maintain the mechanical impulse needed to counterbalance the effects of gravity. Because the body consists of a sprung lattice structure with distributed masses, the intermittent leg thrusts produce propagated disturbances as the acquired momentum is transmitted and distributed among the parts.

The moment of inertia of the skull, about the junction of neck to thorax, may be used as a store from which momentum may be borrowed, to be repaid later. Receptors in the labyrinth report changes in both angular and linear momentum of the skull. The neck receptors shift the reference frame to the trunk rather than the head, which is thus freed from reflex restraint.

Reflex mechanisms necessarily involve time delays, whereas successful locomotion calls for anticipation. A study of the generation of reflex and voluntary hopping in man leads to the suggestion that locomotion involves sequences of learned <u>anticipatory pre-emptive actions</u>, triggered off from early stages in patterns of sensory signals before the thresholds have been reached for the initiation of regular reflexes.

Introduction

Locomotion is defined here as the translocation of the body of an animal over the earth's surface by the manipulation of direct contact forces exerted against the ground. In this chapter it is only the locomotion of the terrestrial vertebrates that will be considered.

In many previous studies the emphasis has been on description of the motion of animals as recorded by an external observer or by a camera supported on the ground. The results have been presented as sequences of footfalls or of the time-courses of the changes in the angles at various joints (for references see Gambaryan, 1974). In this chapter emphasis will be shifted from displacements to forces. Since the action of gravity accelerates the body toward the ground, the body needs to exert appropriate contact forces to prevent this and to avoid collisions. If the magnitude, direction and timing of these contact forces pushing against the ground are not adjusted correctly, the animal's head will strike the ground. In consequence, any strategy for translocation must include provision for avoiding such punishing collisions of the head with the planet.

The machinery available for putting into effect any chosen strategy consists of muscles and bones arranged in triangulated lattices (Roberts, 1967). The mass of the body is distributed

throughout the sprung structure so that the application of an
impulse to any one part leads to a succession of relative movements
of all the parts, in propagated oscillation, as the acquired
momentum becomes partitioned among the masses. If the compliant
linkages were purely elastic, the oscillations would continue
indefinitely. In practice, the compliances of the individual muscles are adjustable under nervous control, and there is an important element of hysteresis (Roberts, 1963) which contributes
greatly to the suppression of oscillation (Figure 1).

It is tempting to consider, as a starting point, a rest position in which the forces supporting the body are in equilibrium
with the action of gravity. However, the forces generated by
individual skeletal muscle fibres each vary with the time since
their most recent activation by a nerve impulse, and in waking life
the parts of the body are set in continual relative motion by
respiratory and cardiac activity. There is thus no true rest
position of the body. Again, because the points of contact with
the ground are usually below the centre of gravity, there is an
inherent tendency to instability that periodically calls for
corrective action to avoid toppling.

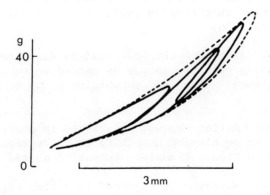

Figure 1. Tension/length diagram for a reflexly active extensor
muscle (soleus) in a decerebrate cat. The tension is varied
sinusoidally over various ranges with a period of about 3 sec.
The hysteresis loops are traced in a clockwise direction. Note
that the loops for the smaller ranges fit neatly between the upper
and lower parts of the largest loop (broken line). The widths of
the loops are not velocity-dependent within the range of physiologically important velocities. The effect resembles that of distributed simple friction. (From Roberts, 1963.)

Successful standing implies an appropriate distribution of upthrusts from the available points of support. If the adjustment in one of the upthrusts is either exaggerated or deferred, the centre of gravity of the trunk will execute a correspondingly exaggerated sway. It is by a succession of deliberately exaggerated sways that an animal initiates and performs locomotory progressions.

Motor Control

Each adjustment in limb thrust is achieved by a sequence of motor commands issued to alpha- or to gamma-motoneurones or to both. Because muscle is a compliant tissue it is necessary not only to adjust the working point for each muscle in its tension/length diagram, but also to fix the dynamic stiffness with which the muscle will interact with an imposed load. The requirements will be different according to the task that the muscle is to perform and the traditional choices - direct alpha drive, fusimotor drive with follow-up length servo, alpha-gamma co-activation - seem to offer too restricted a description of the necessary repertoire.

Consider the following different modes of motor control each of which calls for a different strategy of command.

(1) SET. A part of the body is moved freely from one position to another without encountering resistance. Example: head rotation about a vertical axis.

(2) HOLD. A part of the body resists displacement by external forces, developing opposing forces as needed to conform with a selected compliance. Example: supporting a cup to be filled from a jug.

(3) DRIVE. A limb segment is shifted in position, in spite of opposition, by developing enough force to overcome that presented by obstacles to motion. Example: thrusting a hand into a glove.

(4) PUNCH. Momentum is imparted to a limb segment by forces greater than those just needed for setting from one position to another. The momentum is delivered up by impact with an obstacle. Example: hitting a punch bag, kicking a football.

(5) CATCH. Momentum is absorbed by muscles performing negative work after a preliminary stage of setting to a starting position which is to be held with a chosen compliance. Example: ball games, catching a ball.

(6) THROW. A part of the body is driven in a chosen direction with excess force so as to impart momentum by positive work, this momentum being later absorbed by catching at the end of the stroke. Example: In throwing a ball, momentum is imparted to the arm as well as to the ball. The arm momentum is absorbed after the ball has been released.

It is hard to predict what proportion of alpha or gamma activation would be appropriate for each of these tasks. DRIVE might be achieved by pure alpha command. HOLD looks more like servo action. SET might involve either or both. PUNCH, CATCH and THROW may be considered to be assembled from sequences of elements each of which is an example of one of the first three. Clearly there is no simple choice between coactivation and servo action.

Each of these six modes of motor control differs from the others, yet between them and in combination they seem to satisfy all demands. In locomotion there are examples of SET, HOLD and DRIVE in the grazing walk: SET during the swing phase, HOLD in stance, and DRIVE in moving the centre of gravity forward over the feet. In gaits faster than the walk the body is THROWN up at take-off and is CAUGHT again on landing. The PUNCH mode does not appear to be represented in locomotor activity as, even in fast galloping, some shock-absorbing phase is needed.

Limb tasks

In developing forces between the body and the ground the limbs can act in two different ways, separately or in combination. They can act as PROPS to exert compressive thrust in the line between the point of foot contact and the point of attachment to the trunk at the proximal joint (hip or shoulder), or they can act as PADDLES to apply torque across the proximal joint. These two separate functions are achieved by separate groups of muscles. Compressive thrust in the line of a limb is met by extensor muscles each spanning a single joint. Paddle action in the most commonly required "retract" direction is provided by two-joint muscles such as biceps femoris and gastrocnemius, which may contribute little or nothing to the resistance of the limb to compression (Figure 2). Two-joint muscles spanning adjacent joints in opposite senses, such as sartorius and tensor fasciae latae, serve to fold up the limb during the swing phase of locomotion. The prop and paddle functions seem to be separately controlled from different parts of the nervous system. The limbs of a mid-collicular decerebrate cat are strong props that rigidly resist compression, while their paddle action is so deficient that the preparation cannot stand without lateral support.

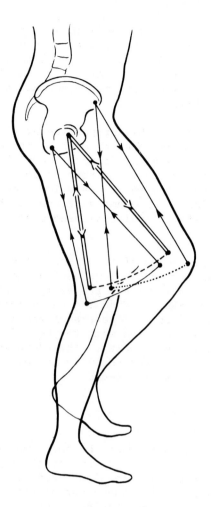

Figure 2. Sketch to indicate that some two-joint muscles are not well-placed to resist compressive thrusts on the legs. The muscular links indicated by single lines do not suffer any change in length when the leg length is reduced by raising the foot. (From Roberts, 1971.)

If one leg of a pair, say one hind leg, is to be lifted, it is essential that torque be applied to the trunk from the opposite leg, because the upthrust in that leg, applied at the proximal joint (hip), (Figure 3) is offset from the line connecting the point of support with the centre of gravity. In the slow walk, the weight is taken on three legs while the fourth is lifted. The torque applied to the trunk by each of the supporting legs varies as the body is moved forward over the ground in a succession of surges.

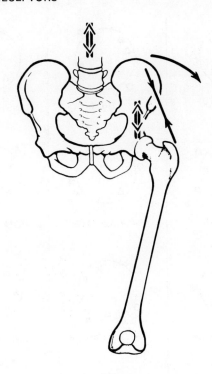

Figure 3. Torque must be applied across the hip joint when the pelvis is supported on a single leg because the upthrust is applied at a point, the acetabulum, which is displaced from the midline. (From Roberts, 1969.)

As the speed of progression increases through the various gaits, the total duration of the locomotor cycle decreases at the expense of the durations of the support phases for the individual limbs (Figures 4 and 5), the durations of the swing phase remaining relatively unchanged (Goslow et al., 1973). For progression on level ground the sum of the impulses in the upthrust direction over the period of each complete cycle must be equal to the gravitational impulse over the same time. ("Impulse" is taken as the integral of force with respect to time.) Accordingly, as the duration of each support phase decreases, the magnitude of the upthrust during that phase must increase proportionally. Corresponding increases must consequently occur in the torsional stresses within the trunk.

Since the trunk consists of a sprung lattice with distributed mass, each limb upthrust must be expected to produce some distortion of the trunk. For example in the trot, the thrust of the left foreleg tends to rotate the forequarters toward right side down, to pitch the trunk toward nose up and to pitch the head toward

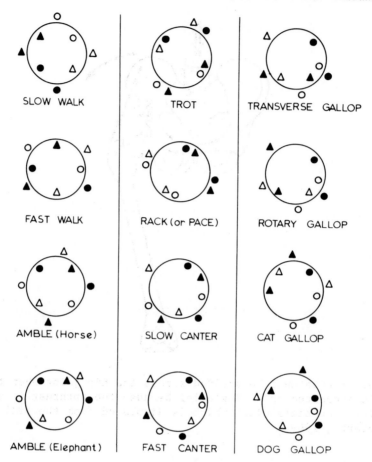

Figure 4. Sequences of footfalls and liftings in various gaits. Each circle is to be read as a clockface traversed clockwise by a pointer indicating the current moment in a repetitive sequence. A symbol within the circle indicates that the corresponding foot touches down at this point in the cycle. The moments of lifting are indicated by the symbols outside the circle. O = left hind, Δ = left fore, ● = right hind, ▲ = right fore. (From Roberts, 1967.)

nose down. At the same time the thrust of the right hind leg tends to rotate the hindquarters toward left side down and to pitch the trunk toward nose down. However, if one studies the trunk of a trotting horse it may be seen that the motion does not correspond exactly with this simple expectation. Toward the end of the unsupported phase, when the animal is about to land, say on the left diagonal, the trunk is actively twisted to throw the left

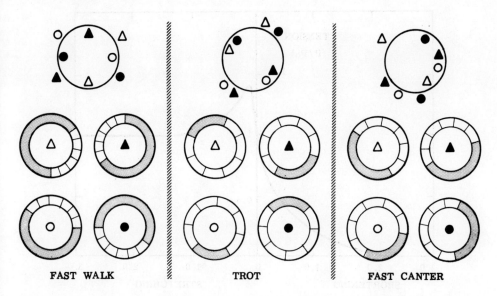

Figure 5. Locomotor cycles for three gaits from Figure 4, expanded to indicate the relative durations of support phase (stippled) and swing phase for each limb. Note that the absolute duration of the swing phase is approximately the same in each case. Correspondingly, as the cycle duration shortens with increase of speed, the durations of the support phases decrease both relatively and absolutely.

shoulder and right hip downward. The abdominal muscles are tightened to resist the trunk flexion and the dorsal muscles of the neck throw the head upward relative to the trunk to produce the appearance that the forequarters seem to fall while the head does not.

Thus, at the moment of impact, hip and shoulder are already moving toward the ground faster than the rest of the trunk. They move to meet the impending impact, just as the limb extensor muscles become active before the foot actually touches down (Engberg and Lundberg, 1962; Melvill Jones and Watt, 1971). The combined effect of these changes occurring during the free-fall, unsupported, phase is to ensure that when the impact does occur, the muscles are set ready to act as very stiff springs. If a muscle is stretched just after activation, it can develop more tension even than the maximum tetanic isometric tension (Aubert, 1956; Curtin and Davies, 1972), (Figure 6). Note that the natural period of oscillation of a mass-spring system is proportional to the square root of the ratio of mass to stiffness: the larger the mass, the longer the period. The period of vertical oscillation of the trunk of a horse at the trot is shorter than can be comfortably followed by the

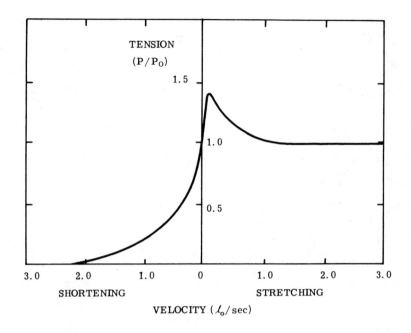

Figure 6. Force/velocity curve for tetanised muscle. Note that the muscle is able to develop a force greater than the isometric tetanic tension if it is forcibly extended at a moderate speed. (Redrawn after Curtin and Davies, 1972.)

rider even though his mass is much less than that of the horse. The implication is that the horse's leg and trunk muscles must act as exceedingly stiff springs.

The head as an inertia paddle

The action of the neck muscles on the head is important at other stages of locomotion. The mass of the head at the end of the neck presents substantial moment of inertia to pitching motion about the junction of neck with thorax. Thus, if the head is thrown upwards by the neck muscles, the trunk will be subjected to pitching torque in the nose down direction. To resist this, the upthrust in the forelegs needs to be greater than that in the hindlegs. Indeed it may become possible to dispense altogether with the support of one hindleg, permitting it to be lifted and moved forward. The sequence can be seen clearly in the head movements of cattle walking briskly. The head is thrown up, and is also pitched toward nose-up, each time a hindleg is due to be lifted. Once the head has acquired upward momentum, the neck muscles can be used later to transfer this momentum to the forequarters, to facilitate the lifting of a foreleg.

The head may also be used as an inertia paddle in the preparation for take-off over a jump. The head is thrown upward while the forefeet are on the ground. The animal then pulls down on the head to raise the forelegs and to incline the trunk into a suitable attitude for launch by the extra thrust of the hindlegs. In manoeuvres such as this it is important for the animal to be able to assess, at each moment, how much momentum is available to be borrowed from the head. Here it is the momentum relative to the trunk that is relevant rather than the momentum relative to the ground attributable to the action of gravity.

Assessment of skull momentum

For the assessment of angular momentum the receptors in the semi-circular canals appear well suited. Because of the large amount of damping present in the canals, the signal from the canal receptors corresponds closely in time-course to the angular velocity of the head, at any rate for movements having a duration within the normal range. This is in spite of the fact that the auxiliary structures operate as angular accelerometers (Roberts, 1967; Melvill Jones and Milsum, 1970) (Figure 7).

The otolith organs have a structure suited to operation as differential-density linear accelerometers, but it has been traditional to regard them as concerned with assessing the direction of the gravitational vertical rather than with head movements. When the skull is at rest, as it can be when fastened to the experimenter's stereotaxic frame, there is a simple relationship between the action of gravity and the effect of contact forces. The earth's gravitational field attracts all the particles of which the skull is composed in a direction toward the mass centre of the earth, and contact forces prevent the particles from falling.

Effects attributable to gravity

If one considers two equal volumes having different densities, the gravitational forces acting on these volumes will be different and proportional to their masses. If the two volumes of material are both free to move, they will accelerate toward the earth's centre under the action of the gravitational forces. But for a given force, the acceleration of a body is inversely proportional to its mass. In consequence, as Galileo pointed out, bodies of different masses undergo identical accelerations in a particular gravitational field. From this one deduces that our notional two

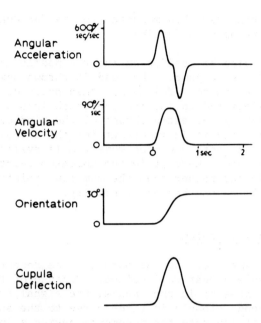

Figure 7. Time-courses of angular acceleration, angular velocity and orientation of the head during a naturally occurring head movement, together with the cupula deflection predicted by analogue computation from the equation of motion of the cupula. Note that the cupula deflection, and therefore the significant content of the neural signal from the canal, corresponds closely in time-course to that of the angular velocity of the skull. (From Roberts, 1967.)

equal volumes, even though they are of different density, will not develop any relative displacement as a result of the action of gravity alone.

It also follows that within a region where the gravitational field can be regarded as uniform, no mechanism can be constructed to detect directly either the intensity or the direction of the gravitational field. To do this one must have access to another region where the gravitational field is different. We may, for example, use a frame of reference based on stellar observations, or we may compare the motion of a small object near the earth with that of the earth itself. The motion of the small object in free fall toward the earth is dominated by the earth's gravitational field, while the motion of the earth itself, although it equally depends on the field generated by the totality of all the celestial bodies, is influenced much less by objects as small as a man.

Effects attributable to contact forces

Turning now to the supporting contact forces, if our two notional volumes of material form part of an animal's head which is being held in a stereotaxic frame, they will need to be supported by pressure gradients in a direction opposite to that of the gravitational pull. Equal volumes of different density need different support forces and therefore different pressure gradients, (Figure 8). Consequently, within a cavity in the skull, such as the labyrinth, regions of different density tend to arrange themselves, as in a centrifuge tube, in such a way that the density gradient lines up with the direction of the contact forces.

With the skull supported at rest in a stereotaxic frame, the resultant of the contact forces is in line with, but opposite in direction to, the gravitational vector and we have a simple sedimentation situation. However, in the case of the otolith organs, the jelly mass loaded with otoconia is not entirely free to move, but is restrained within the cavity of the otolith organ by fine filaments attached to the wall. These filaments are compliant, so that small amounts of reversible temporary slippage can occur, proportional in amount to the magnitude of the shearing force. It is these shearing deformations that excite the haircells and generate the neural signal.

If the skull is not at rest, (and in the normal life of the animal the skull is seldom truly at rest) the contact forces acting on the skull are continually fluctuating both in magnitude and in direction. There will be corresponding fluctuations in the shearing forces acting on the otoliths and in the intensity of the resulting neural signal. Meanwhile, the gravitational field continues unchanged. It is unaffected by movements of the head. Indeed the ascent to a spacecraft orbiting 200 miles up involves only a ten percent diminution in the magnitude of the gravitational field with no change in its direction, whereas a small hop can involve increased upthrusts of nearly twice body weight at take-off and landing, separated by a diminution to zero upthrust during the actual hop itself. Variations in the direction of the contact forces on the skull also occur continually as the animal moves its head either to change the momentum of the skull or to change its orientation with respect to the gravitational vertical.

Because of the way that the contact force is transformed into an effective stimulus at the receptor, it is clear that the same change in haircell deformation can be produced either by a change in the magnitude or by a change in the direction of the resultant contact force applied to the skull. Some form of recognition process, which involves the comparison of neural messages from many receptors and which takes account of the way each signal is

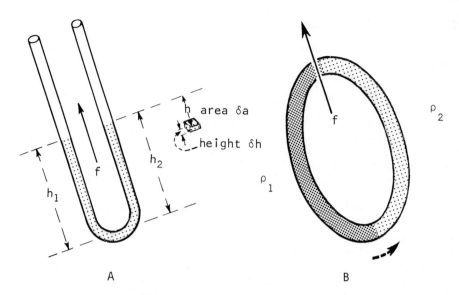

Figure 8. Scheme to explain the action of the otolith apparatus. a) A U-tube containing fluid of density ρ is subjected to acceleration f in the direction of the arrow. A small volume of fluid of height δh, measured in the direction of the arrow, area δa at right angles to this, and placed at a depth h below the surface is supported by a pressure difference $(\delta a.\delta h.\rho.f)/\delta a$ so that the pressure at the depth h is $\rho.f.h$. There is no movement round the bend if $h_1 = h_2$. b) The closed tube is filled with two fluids of different densities. A linear acceleration, f, applied in the direction of the arrow, produces a tendency for the fluids to move round anticlockwise if ρ_1 is greater than ρ_2. In the otolith organs, an inequality of density is provided by the otoconia imbedded in jelly. Movement of the jelly is constrained by compliant filaments. The relative displacement of the jelly, which is monitored by the haircells, depends on the magnitude as well as on the direction of the acceleration. These, in turn, depend on the resultant of the contact forces applied to the skull. (From Roberts, 1975b.)

changing with time, becomes necessary before the animal can "draw any conclusions," so to speak, about what is happening to the skull. However, such complex recognition processes are needed also in other sensory modalities and we should not be put off by the need for such a process in this context. We may suppose that, from the information received from the sensory units in different parts of the otolith apparatus, the animal is able to gauge the current values of the magnitude and direction of the resultant of the contact forces acting on the skull. As with other sensory modalities also, the estimated values of the relevant parameters will be coloured by the way those parameters are changing with time.

Reflex consequences of head movements

A traditional method for determining the significance to an animal of a particular sensory signal is to study the reflex effects produced by changing the stimulus situation. Thus we may look for changes in the limbs associated with tilting the head in various directions. A complication arises from the fact that if the head is tilted while the body remains still, the neck will be bent, and, as Magnus and de Kleijn showed (Magnus and de Kleijn, 1912), bending the neck itself produces reflex changes in the limbs. The two sets of reflexes may be separated by cutting the nerves to the first two intervertebral joints and then manipulating skull and axis vertebra independently, while keeping the body still (Roberts, 1963). For pitching and rolling movements of the skull alone, the downhill limbs extend and the uphill limbs relax, whereas corresponding movements of the axis vertebra produce opposite effects. Side down tilting leads to ipsilateral relaxation from the neck reflexes but ipsilateral extension from the labyrinth reflexes (Roberts, 1970; Lindsay et al., 1972; Rosenberg and Lindsay, 1973). The combined effect of these reflex patterns is to adjust the limbs in such a way as to stabilize the trunk in relation to the direction of the resultant upthrust transmitted to the skull, regardless of the orientation of the skull itself (Roberts, 1973).

When these labyrinth reflexes upon the limbs are studied in the decerebrate cat, fairly large tilting movements are usually required, say 10° to 15°. The latencies are not particularly short, ranging from 40 msec to about 4 sec. This means that although the effects are in the right direction for stabilization, the timing is not appropriate for the reflex effects alone to be entirely adequate to preserve the animal's balance. One must also remember that the efferent nerve supply to the labyrinth causes a reduction in the intensity of those signals associated with voluntary movement (Klinke, 1970). Thus any "dead reckoning" system of computation of the direction of the vertical from the afferent signals from the labyrinth is liable to considerable error. Short-term estimations of changes in momentum are all that can be expected. This leaves still unanswered the question how animals stand and move about without continually falling and hitting their heads on the ground.

An alternative to reflex action

Some hint of a possible solution may be provided from observations on human subjects standing on one leg and caused to overbalance by the application of external horizontal forces. The subject stands on a force-plate incorporating an array of strain gauges from which three independent parameters are obtained: the magnitude of the upthrust, the displacement of its point of application from a reference line, and the magnitude of the lateral force at the foot. Horizontal displacements of the skull and of the hips are obtained from ultrasonic rangefinders which avoid the possibility of introducing proprioceptive cues by way of a mechanical measuring link. The measured parameters are processed by online computer and displayed as a matchstick figure (mannikin) on an oscilloscope screen. Data may be stored and replayed at different speeds. The time-courses of the changes in the parameters may also be displayed in conventional fashion. This provides an opportunity for linking specific moments in the time-course display with the corresponding snapshot appearance of the oscilloscope mannikin (Figure 10).

The subject may be induced to hop by one of a number of manoeuvres. A belt round the hips may be attached to a horizontal rope which is slowly drawn in by a winch (Figure 9); the rope may be attached to the subject's shoulders instead of to his hips; alternatively the rope, attached either to hips or to shoulders, may be secured to the wall while a trolley supporting the force-plate is moved slowly away from the wall.

The induced hops are of two distinct kinds, characterized by the way the upthrust changes. The moment of take-off is in each case preceded by a substantial increase in the upthrust. This increase does not occur in a step, where the weight is transferred smoothly from one leg to another. In some recordings, the increase in upthrust is preceded by a dip: the leg is flexed momentarily just before the extra thrust at take-off. Momentarily reducing the support thrust has the effect of allowing the body to fall so that the muscles that are to generate the lifting force are being extended passively at the moment of activation. This ensures that they are well placed to develop maximum tension for the thrust at take-off. The subjects usually report that they felt the need to hop and that they hopped deliberately. Indeed very similar traces are obtained when a subject hops voluntarily on request. This type of hop will be referred to as "voluntary".

In the other type of recording there is no dip in the upthrust trace. To produce such a trace the subject needs to be relaxed or distracted in some way so as not to be preparing for the pull of the rope. It proves very difficult to reproduce traces of this type by voluntary hopping, particularly from the straight-knee starting position. For purposes of discussion the hops in which the extra upthrust appears without a preparatory dip will be termed "reflex". The performance of such a hop is not accompanied by the feeling that the hop was deliberate. Some subjects produce only voluntary hops.

It is still not certain what is the adequate stimulus for the reflex hop, or even whether the receptors are in the foot, in the hip, or in the labyrinth. A possible candidate is the inclination of the support vector as displayed in our computer-generated mannikin (Figure 10). This vector represents the support force at

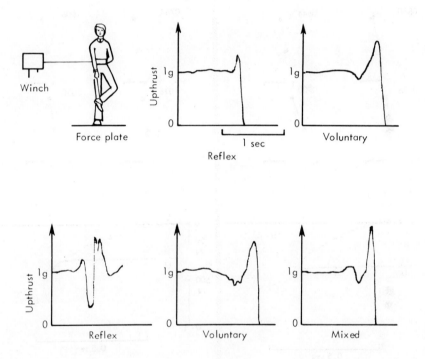

Figure 9. Time-courses of representative changes in upthrust in the hop. The subject stands on one leg on a force-plate (top left). A rope round his waist is slowly drawn in by a winch until he is constrained to hop. In some cases (top right) the upthrust decreases momentarily before the hop, in which case the subject is aware of the need to hop. In other cases (top centre) the upthrust rises without a preparatory dip. Bottom left: a small hop after which the subject lands on the force-plate. This indicates the time available for repositioning the foot. Bottom centre: the unsteady upthrust before the hop is associated with adjustments of balance. Bottom right: a voluntary hop takes over after the start of a reflex hop. (From Roberts, 1975a.)

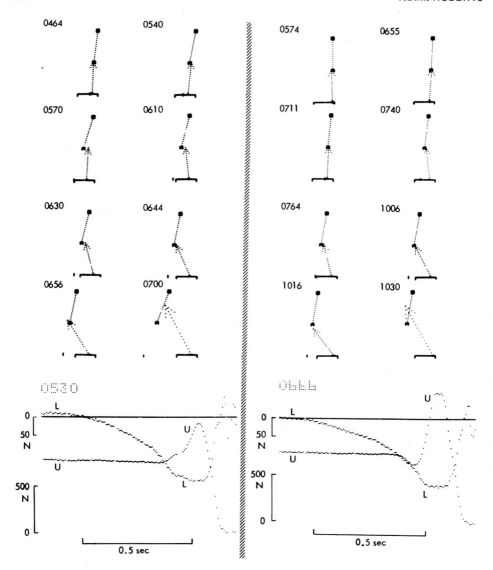

Figure 10. Examples of computer-generated displays for a reflex hop (left) and a voluntary hop (right). Above: 8 stages in the movement of the mannikin, for the data points indicated (in Octal). Horizontal scales are expanded x 4, compared with vertical scales. Below: time courses of upthrust (U) and lateral force (L). The traces have been shifted vertically to indicate, by their point of crossover, the moment taken as threshold. When the lateral force is above its baseline this means that the subject is resisting the pull of the rope. When L is below baseline, the rope is slack and the subject is unbalanced.   (Legend continued on next page.)

VESTIBULAR AND NECK RECEPTORS

the foot, but in the conditions prevailing before take-off in the hop there are no significant movements of the head relative to the trunk, and consequently the displayed vector must be similar in direction and related in magnitude to the vector representing the support force on the skull. It serves then as a direct indicator of the stimulus to the otolith organs.

We may take as a marker for the threshold conditions that moment of time at which the upthrust trace first visibly deviates from the rest position, either upward or downward. It is interesting to notice that the inclination of the support vector at this threshold moment is always less for the voluntary hops than it is for the reflex hops. Up to this point the change in inclination of the vector is brought about entirely by an increase in the horizontal component of force, with no change in the vertical component. In one series, 44 voluntary hops (3 subjects) gave a mean lateral force at threshold of 55 ± 2.66 (S.E.) Newtons as against 124 ± 11.3 N for 18 reflex hops ($\overline{3}$ subjects).

Anticipatory pre-emptive action

The implication is that the nervous system is able to recognise when conditions are changing in a direction which might ultimately lead to development of a reflex response, and that the nervous system then interposes a voluntary corrective action which takes effect before the stimulus has actually reached the threshold for initiating the reflex. Motor control is pre-empted by voluntary processes and does not pass to the reflex mechanisms.

For such anticipatory pre-emptive actions to be developed it is necessary that the messages signalling the stimulus conditions should occupy extended periods of time with a gradual build-up toward the threshold for reflex action. This happens in many of the subtasks of standing and of locomotion. Furthermore, the inevitable delays involved in the neural mechanisms of reflex action are too long for smooth balancing in the conditions encountered in locomotion. The intervention of voluntary anticipatory pre-emptive actions speeds up the corrective movements and allows scope for moment-to-moment modifications of the command-patterns to take account of obstacles, unevennesses of terrain and changes in course and speed.

Scale for L is expanded x 4 compared with scale for U. The run starts at the data point indicated (in Octal). Time scale: 6.8 msec per data point, or 146 (= 222 Octal) data points/sec.

The process of learning strategies of anticipatory pre-emptive actions presumably starts very early in life. In the herbivores of the open plains, the young have to stand and run within a few minutes of birth. These abilities are not inborn. One may observe trials and failures in the initial struggles of a newborn foal. Skill in maintaining balance is attained progressively. The various component tasks of locomotion are rehearsed so often that the successful strategies soon become habitual so that they can be performed without conscious supervision. A well-known example of the development of locomotor habits later in life is that of learning to ride a bicycle. After a while, casual cycling requires little or no attention. Certain manoeuvres, such as the avoidance of obstacles, may call for close supervision and the execution of such tasks may rank as a skill rather than a habit.

Conclusion

It is suggested that the complex patterns of motor command by which the various gaits of locomotion are executed are assembled from learned habits of anticipatory pre-emptive actions triggered off by recognized sequential patterns in the neural signals reporting changes in the relative momentum of the parts of the body, and particularly of the head.

Acknowledgment

It is a pleasure to acknowledge the collaboration of Dr. George Stenhouse in the computer programming and in the conduct of the hopping experiment.

REFERENCES

Aubert, X., (1956) Le couplage énérgetique de la contraction musculaire. Thèse d'aggrégation, Université Catholique de Louvain. Editions Arscia, Brussels.

Curtin, Nancy A. and Davies, R.E., (1972) Chemical and mechanical changes during stretching of activated frog skeletal muscle. Cold Spring Harbor Symp. Quant. Biol. 37, 619-626.

Engberg, I. and Lundberg, A., (1962) An electromyographic analysis of stepping in the cat. Experientia. 18, 174.

Gambaryan, P.P., (1974) How Mammals Run. John Wiley, New York.

Goslow, G.E., Reinking, R.M. and Stuart, D.G., (1973) The cat step cycle: hind limb joint angles and muscle lengths during unrestrained locomotion. J. Morphology. 141 (1), 1-42.

Klinke, R., (1970) Efferent influence on the vestibular organ during active movements of the body. Pflügers Arch. ges. Physiol. 318, 325-332.

Lindsay, K.W., Roberts, T.D.M. and Rosenberg, J., (1972) Asymmetric tonic labyrinthine reflexes. 2nd Internat. Symposium on Motor Control, Varna, Bulgaria.

Magnus, R. and de Kleijn, A., (1912) Die Abhängigkeit des Tonus der Extremitatenmuskeln von der Kopfstellung. Pflügers Arch. ges. Physiol. 145, 455-548.

Melvill Jones, G. and Milsum, H.J., (1970) Characteristics of neural transmission from the semicircular canal to the vestibular nuclei of cats. J. Physiol. 209, 295-316.

Melvill Jones, G. and Watt, D.G.D., (1971) Observations on the control of stepping and hopping movements in man. J. Physiol. 219, 709-727.

Roberts, T.D.M., (1963) Rhythmic excitation of a stretch reflex, revealing (a) hysteresis and (b) a difference between the responses to pulling and to stretching. Quart. J. Exp. Physiol. 48, 328-345.

Roberts, T.D.M., (1967) Neurophysiology of Postural Mechanisms. Butterworths, London.

Roberts, T.D.M., (1969) The mechanics of the upright posture. Physiotherapy. 55 (10), 398-404.

Roberts, T.D.M., (1970) Changes in stretch reflexes in limb extensor muscles during positional reflexes from the labyrinth. J. Physiol. 211, 5-6P.

Roberts, T.D.M., (1971) Standing with a bent knee. Nature (Lond.). 230 (5295), 499-501.

Roberts, T.D.M., (1973) Reflex balance. Nature (Lond.). 244 (5412), 156-158.

Roberts, T.D.M., (1975a) Reflex contributions to the assessment of the vertical. Acta Astronautica. 2, 59-67.

Roberts, T.D.M., (1975b) "Vestibular Physiology", In Scientific Foundations of Otolaryngology. (Harrison, D. and Hinchcliffe, R., eds.), Heinemann, London.

Rosenberg, J.R. and Lindsay, K.W., (1973) Asymmetric tonic labyrinthine reflexes. Brain. Research. $\underline{63}$, 347-350.

NON-RHYTHMIC SENSORY INPUTS: INFLUENCE ON LOCOMOTORY OUTPUTS IN ARTHROPODS

Jeffrey M. Camhi

Section of Neurobiology and Behavior

Cornell University, Ithaca, New York

Locomotory rhythms have several variable parameters. Of these, variations in frequency and in the amplitude of some or all of the component movements are consistent with known interactions between command fibers and central oscillators. Variation in phase relations among rhythmic motor neurons often cannot be explained by such interactions. Phase changes suggest complexities of central oscillators which have not yet been demonstrated empirically. In one model system for studying the control of locomotion, the cercal giant fibers of the cockroach evoke depolarizing psp's, which may be monosynaptic epsp's, in at least one depressor motor neuron of the leg. Different giant fibers encode different information about wind direction, and perhaps wind velocity.

Introduction

In order to understand the mechanisms by which tonic sensory inputs influence rhythmic motor outputs, one would need an understanding of the cellular composition and physiology of at least five neural subsystems:

(1) the central oscillator producing the rhythmic output in the motor neurons,

(2) the neuromuscular apparatus which transduces this rhythm into movement,

(3) rhythmic feedback influencing the motor output,

(4) the system of command interneurons which turns on and modulates the oscillator, and

(5) the connections by which tonic afferents exert a biasing influence on each of these levels.

Although small portions of this picture can be sketched in, most aspects are still unclear. This chapter reviews briefly the current conception of the central oscillators for locomotion (number 1 above). Next to be considered are the three different kinds of sensory biasing which can influence rhythmic locomotory outputs. Suggestions as to the cause of this sensory biasing will be offered. These include the influence of sensory inputs on command fibers and of command fibers on central oscillators. Finally, new data on sensory and interneuronal control of an escape system, namely interactions between cercal giant interneurons and leg motor neurons in the cockroach, will be presented.

LOCOMOTORY OSCILLATORS

Locomotory oscillators only recently have been studied intracellularly. In particular it is known in locusts that rhythmic sensory feedback, delayed excitation among motor neurons, and unknown interneuronal oscillators contribute to the pattern generation in flight (Burrows, 1973, 1975). In cockroaches, non-spiking interneurons appear to comprise part of the central oscillator for walking (Pearson and Fourtner, 1975), while rhythmic sensory feedback also may contribute to this pattern (Wilson, 1965). Each of the non-spiking interneurons in the cockroach influences primarily only levator, or only depressor motor neurons, rather than both antagonistic sets, as in a crustacean respiratory oscillator (Mendelson, 1971). Mutual inhibition among antagonistic sets of neurons, predicted from studies of motor outputs (Wilson, 1967) has not yet been found in these two systems. This inhibition is present, however, in the oscillator for swimming in the mollusc Tritonia (Willows et al., 1973) and in some non-locomotory oscillators in arthropods (Bentley, 1969; Wyse, 1972). In summary, a multiplicity of synaptic mechanics operating at several synaptic sites may cooperate to generate a rhythmic output. A complete description has not been provided for any one behavior.

Control of Frequency

Locomotory oscillators differ greatly in their capacity to operate at different frequencies. The flight of mature locusts operates between about 17 and 25 Hz., a 1.5 fold range of frequency (Kutch, 1974); lobster swimmerets between 0.8 and 3.3 Hz., a 4 fold range (Davis, 1968a); and cockroach walking from presumably below 0 up to 25 Hz., a greater than 25 fold range (Delcomyn, 1971). Since frequency appears to be continuously variable over its entire range, one presumes that a single neural system, however complex, operates at all frequencies, rather than a separate oscillator operating for different frequencies. Since flight requires at least a moderate wingbeat rate in order to generate positive lift, a relatively small frequency range is not surprising in flight behavior. This is in contrast to walking and perhaps to swimming.

Many sensory cues can control the speed of an animal's locomotion. Speed is often controlled partly by changing the frequency of limb movements, a parameter which therefore is often under strong sensory influence. Locomotory speed and frequency are indeed positively correlated in cockroach walking (Delcomyn, 1971). In locust flight, however, the connection between speed and frequency is much less clear for several reasons. Most of the locomotory force is delivered as lift, not thrust. Lift and thrust are interdependent, both being functions of the forewing angle of attack (Wilson and Weis-Fogh, 1962). A further complication results from the fact that at high wingbeat frequencies, the movement of a wing lags behind the motor output driving it. Thus, for example, a burst of spikes in the depressor motor neurons occurs before the wings fully are elevated. Consequently, contraction of the depressor muscles serves to decrease wingbeat amplitude, resulting in less sensory feedback from receptors in the wing hinge which fire when the wing is maximally elevated. Since one role of this feedback is to increase wingbeat frequency, the entire system operates as a homeostatic control loop stabilizing frequency (Wilson and Weis-Fogh, 1962). Yet another factor is that the average wingbeat frequency, recorded from deafferented or intact locusts, increases with age. In spite of all these factors, a locust's flight speed and its wingbeat frequency do co-vary during prolonged flights, so that there may be some causal relationship between the two (Gewicke, 1974).

In several animals, feedback loops influence locomotory speed as well as the frequency at which locomotory appendages move. Stimulation of a locust's cephalic sensory hairs by the flight wind results in a higher maintained flight speed (and thus possibly wingbeat frequency) than when the hairs are covered, a positive feedback relationship (Gewicke, 1974). Similarly, stepping frequency increases in lobsters and in crayfish if stripes are moved posteriorly past the eyes (Davis and Ayers, 1972). On the other hand, flight wind stimulation of receptors at the bases of the antennae results in lower flying speeds than if these appendages are immobilized, a negative feedback situation (Gewicke, 1974).

Command interneurons can influence the frequency of rhythmic motor outputs in several animals (Wiersma and Ikeda, 1964; Davis and Kennedy, 1972a, b; Bowerman and Larimer, 1974; Bentley and Hoy, 1974; Wyse, 1975). The most quantitative studies, on the lobster swimmeret system, show that stimulation of putatively single command fibers evokes normal rhythmic outputs in motor neurons, even under deafferented conditions (Davis and Kennedy, 1972a). Different command fibers are specialized in evoking different ranges of frequency, and pairs of stimulated command fibers evoke higher output frequencies than do individual fibers (Davis and Kennedy, 1972a, b).

One would anticipate that in the swimmeret system, sensory inputs could express themselves on oscillatory frequency by controlling the frequency of spikes in command fibers, as well as by recruiting different sets of command fibers. This idea is difficult to test in lobsters, because the swimmeret frequency is not manipulated easily by sensory inputs. Nevertheless, some swimmeret command fibers do appear to be excited by tactile stimulation of the ventral surface of the body. Unfortunately, command fibers for locust flight and cockroach running, two other systems in which this scheme could be tested, have not been discovered.

A persistent difficulty in extending these studies may be one of definition. There are probably many interneurons which are physiologically very similar to command fibers, but are not so called because they do not evoke pronounced behaviors under the testing conditions (Burrows and Rowell, 1973; and this chapter). Intracellular recording of synaptic potentials could reveal interactions that are too weak under experimental conditions to evoke spikes, but which nevertheless may be employed in the intact animal. Unfortunately, the current size limitation of cells for intracellular penetration renders most command fibers inaccessible to this approach.

Variations in Amplitude

As mentioned above, the frequency and amplitude of a locust's wingbeat vary inversely within its limited frequency range (Wilson and Weis-Fogh, 1962). But frequency varies directly with motor neuronal burst strength (loosely defined as the number of spikes per burst in a motor neuron, and the number of motor neurons recruited). Burst strength can be taken as a rough indication of the amplitude of the underlying neuronal oscillation. The strange conclusion is that the amplitude of the neural oscillation which causes the wingbeat, and that of the wingbeat itself, vary inversely.

In the lobster swimmeret system, both stroke amplitude and burst strength of the motor neurons vary directly with frequency (Davis, 1968a, 1969). At the highest frequencies, the swimmeret's motion lags behind the motor neuronal bursts. Therefore, as in locusts, each burst probably serves to decelerate and then accelerate the appendage, although no decrease in amplitude at high frequency was reported (Davis, 1968a).

Among the tonic sensory cues which control an appendage's amplitude is increased air speed, which decreases the wingbeat amplitude in flying locusts and other insects (Gewicke et al., 1974). Wind receptors at the antennal bases are among the inputs for this reflex (Gewicke, 1970, 1974). Other examples probably can be found in most terrestrial arthropods which walk faster, perhaps employing a combination of higher frequency and greater amplitude, when disturbed.

Many locomotory steering movements are achieved in part by changes in the burst strength (amplitude) of a subset of motor neurons. If one imposes an abnormal roll orientation on a lobster, the direction of the swimmeret beat changes as a result of alterations in the burst strength of several motor neurons. The response is controlled through the statocysts (Davis, 1968b). In locusts, sensory stimuli for flight direction evoke changes in burst strength in the motor neurons controlling wing twisting. Early reports stated that these changes developed slowly over tens of wingbeats. The stimuli included a rotated dorsal light or horizon line (Waldron, 1967), and imposed changes in the body's angle of pitch (Wilson and Weis-Fogh, 1962; Gettrup, 1966). Also, interference with afferent input from receptors near the wing hinge slowly decreased the wingbeat frequency (Wilson and Gettrup, 1963).

More recently, several forms of sensory stimuli have been shown to produce cycle-to-cycle changes in the wingbeat, and thus presumably effect rapid changes in the flight path. These stimuli include manipulation of the wind speed, which is detected by the cephalic hairs and evokes changes in the angle of attack of the forewings on the downstroke (Camhi, 1969b); and manipulations of either wind angle or velocity, detected by the same sensory hairs, which evoke steering movements of the abdomen by altering the strength of abdominal flight bursts (Camhi, 1970a, b; Camhi and Hinkle, 1972, 1974). Also, forced rhythmic movements of a single wing at about the flight frequency can entrain the other wings within a few cycles (Wendler, 1974). Reafferent excitation of the stretch receptor at each wing hinge can increase phasically the burst strength of depressor motor neurons and decrease that of the elevators through probably monosynaptic excitation and inhibition of flight motor neurons (Burrows, 1975).

Rhythmic outputs evoked experimentally by stimulating command fibers show all the types of variation in burst strength described here. The type of covariation of burst strength and cycle frequency observed in locust flight and lobster swimmeret beating, is seen upon stimulation of command fibers in the swimmeret system. Increasing the stimulus frequency; or stimulating two command fibers together; or stimulating a command fiber having a higher range of output frequencies evoke an increase in both burst strength and cycle frequency (Davis and Kennedy, 1972a, b).

Also, different command fibers evoke the rhythm in different constellations of swimmerets, exciting primarily either the swimmerets on one side or those on a few neighboring segments. Some command fibers also appear to bypass the oscillators and to project more directly toward specific sets of motor neurons (Davis and Kennedy, 1972a). Either of these mechanisms could explain the steering behaviors in locust flight described above, where some motor neurons increase their burst strength while others are unaffected. Such an increase of burst strength could result from interneurons exciting either a subset of oscillators among a pool of coupled oscillators, or a subset of outputs from one oscillator. A slight phase advance with increasing burst strength, predicted by either model, has been observed (Waldron, 1967). Moreover, both these rapid and the slow changes in the locust flight bursts as described here could be accounted for by command fibers like those found in lobsters. The rapid changes may derive from activity in command fibers tightly coupled to the oscillator. This activity produces an almost immediate effect. The slow changes may derive from active loosely coupled command fibers. These effects on the output develop over a period of seconds (Davis and Kennedy, 1972a). Although command fibers for locust flight have not been reported, experiments with whole cord stimulation (Wilson and Wyman, 1965) and recordings from wind-responsive interneurons (Camhi, 1969a) indicate they probably exist.

Variation in Phase

In many rhythmic behaviors, the phase[1] of one motor neuron with respect to another can change drastically, usually producing some change in behavior. Extreme phase changes suggest some alteration in the central oscillator itself, rather than merely a biasing of the motor output from the oscillator. This is especially true if the burst structure is approximately the same at all

---

[1]Phase is defined as the fraction of the interval between the onset of two consecutive bursts in one motor neuron which has elapsed at the time a second motor neuron begins its burst.

phases. The question of phase variation has recently been reviewed (Camhi, 1974; Kammer, 1975), so the following presentation will be restricted to some of our recent results on the walking system of cockroaches (Sherman et al., 1975).

In the cockroach Gromphadorhina portentosa walking is presumed to be controlled partly by a central oscillator, by analogy with another cockroach, Periplaneta americana (Pearson and Iles, 1970; Pearson and Fourtner, 1975) and with most other invertebrates and several vertebrates (this volume). Electromyograms taken from a mesothoracic leg of Gromphadorhina during walking record the main coxal remotor muscle ("CR")[2] and the main femoral extensor muscle ("FE")[2] as nearly synchronous in their bursts of spikes. Since the muscle spikes reflect the motor neuronal spikes one-for-one, the CR and FE motor neurons are active almost synchronously. If the insect is turned on its back, it uses the legs rhythmically as in walking, to right itself. In both behaviors, the same motor neurons appear to be employed, as evidenced both by the constancy of the heights of muscle spikes and of motor neuronal spikes recorded within the muscles. As further verification of this, only two excitatory axons innervate the muscle homologous to FE in Periplaneta, and only two distinctly different spike heights are recorded in this same preparation, being the same during either behavior. The smaller amplitude spike is active in bursts which are almost identical in both behaviors, and closely resembles motor neuron Ds which innervates the homologous muscle 135e of Periplaneta. The larger amplitude spike is active only occasionally during high frequency walking or righting, and thus closely resembles motor neuron Df to muscle 135e of Periplaneta (Pearson and Iles, 1970). Similar evidence for the use of the same motor neurons in both behaviors is available for muscle CR as well. Moreover, the structure of the bursts in both muscles, and the curves relating burst length to cycle frequency, are very similar for the two behaviors. Also, the same tripod pattern of interleg coordination is employed, although only the legs on one side of the body engage in the rhythm during righting. These factors suggest a basic similarity or identity in the central oscillatory network underlying walking and righting.

The clearest difference between the two behaviors is that the legs move forward to back during walking, but medially to laterally during righting. Underlying this change in direction is a change in the phase of CR with respect to FE (Figures 1B, 2). The phases for walking and for righting are stable under a variety of sensory

---

[2]CR is probably homologous to muscle 130 and FE to 135e of Periplaneta.

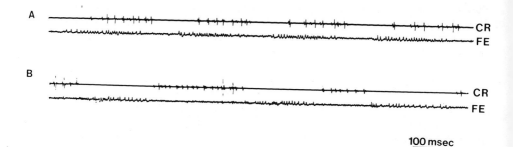

Figure 1. Electromyograms from mesothoracic leg muscles of the cockroach Gromphadorhina portentosa. A. During walking. B. During righting. Both records from the same tethered animal. CR -- main remotor of the coxa. FE -- main extensor of the femur. Gain of CR channel is reduced by half in B. All the spikes in FE are presumed to result from motor neuron Ds. Df was not active during this recording.

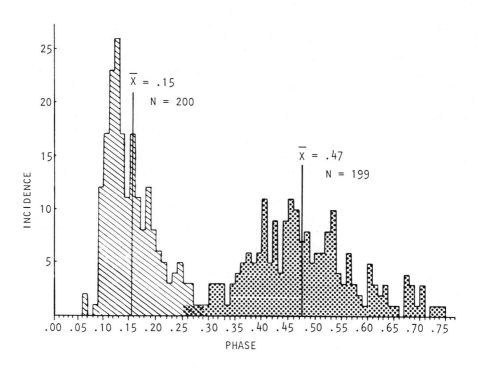

Figure 2. Histogram of phases for bursts of CR in FE during walking (cross hatched) and righting (dotted). The data of this figure are from the same animal as Figure 1.

conditions, including walking in an inverted position or up and down an incline, and righting while tethered with the body in any orientation with respect to gravity.[3] In this animal then, a change in the direction of leg movements is initiated primarily by a change in the phase relations among bursting motor neurons, rather than by variation in the strengths of different motor neuronal bursts, as in the lobster swimmeret (Davis, 1968b).

Though other examples of phase shifting have been reported, the mechanism is unknown in all cases. Theoretical work would suggest two central oscillators, in this case one for each muscle CR and FE; the oscillators would interact with each other by more than one set of couplings with only one set working at any one time (Wilson, 1967). It is known that oscillators can be coordinated by interneuronal couplings (Stein, 1971) employing, in different animals, either excitatory or inhibitory interactions among antagonistic sets of neurons (Willows et al., 1973; Burrows, 1974). Individual oscillatory neurons of Periplaneta have a strong effect on either the levator or the depressor motor neurons, but not both (Pearson and Fourtner, 1975). This means that changes in cross-couplings among these cells, if such couplings exist, could alter the phase relations of the outputs. In Periplaneta, as in Gromphadorhina, marked phase differences between levator and depressor motor neurons do occur in righting as compared with walking (Reingold, 1975).

Changes in phase also occur when Periplaneta walks with different stepping frequencies (Delcomyn, 1971). As the frequency decreases, the time occupied by leg remotion (the backward power stroke) increases much more than the time for promotion (the return stroke). The same is true of Gromphadorhina, where in addition appropriately placed tactile stimuli can evoke brief bouts of stepping in a backwards or sideways direction. (Sideways stepping is employed during locomotory turns.) Thus the power stroke can be directed posteriorly (normal walking), anteriorly (backward walking), laterally (contralateral turning), or medially (ipsilateral turning). For each direction of stepping, at low stepping frequencies a leg spends most of the time of each cycle in the power stroke, although this stroke must employ different muscles for each direction of walking (Camhi, Unpublished data). Since the phase shifting which accompanies forward walking at different frequencies occurs in deafferented preparations (Pearson and Iles, 1970) the complex phase changes implied by the above observations suggest an enormous flexibility in the central oscillatory networks.

---

[3]Righting is evoked by loss of leg-to-ground contact, and not by gravitational cues (Camhi, 1975).

The concept of a fixed oscillator which can be treated as a simple black box is not consistent with these observations.

The only way to study these properties directly is by intracellular recordings from identified central neurons. The cockroach Periplaneta offers a favorable opportunity for such studies, because earlier investigations have identified individual leg motor neurons (Cohen and Jacklet, 1967; Pearson and Fourtner, 1973), central oscillatory cells (Pearson and Fourtner, 1975) and giant interneurons which have been implicated (Roeder, 1948), though with some question (Dagan and Parnas, 1970; Iles, 1972), in the generation of walking. Preliminary studies directed at clarifying the interactions among some of these neurons are in progress. Although this work does not address specifically the questions of cycle frequency, amplitude or phase, it offers the future possibility of uncovering the cellular basis of these and other properties of a whole locomotory system.

ESCAPE BEHAVIOR AND GIANT INTERNEURONS IN THE COCKROACH, PERIPLANETA

The cercal interneurons of the cockroach Periplaneta comprise a classic example of a system of giant fibers evoking escape behavior (Roeder, 1948, 1967). By analogy with crayfish, one might expect that the giant fibers would excite directly motor neurons driving escape muscles (leg depressors, in the cockroach) and that giant or other interneurons also would excite a central locomotory oscillator (Schramek, 1970; Wine and Krasne, 1972; Zucker, 1972; Atwood and Pomeranz, 1974). However, recently reported experiments disagree with this view. These experiments involved: stimulation of the whole cord with single shocks at low current strength, or stimulation of a penetrated giant fiber at frequencies up to 200 Hz. for one second, while recording from leg motor axons in the periphery or while observing legs visually (Dagan and Parnas, 1970); and stimulation of the whole cord with pulses of low current strength at a repetition rate of 5 Hz. while recording synaptic potentials in the neuropile from motor neuron Df, the fast depressor of the coxa (Iles, 1972). These authors state that the giant fibers do not excite any of the numerous leg motor neurons from which recordings were made. However, preliminary experiments by Ritzmann and Camhi (Unpublished data) on this system point to a different conclusion, as I shall now describe.

Periplaneta has about seven bilateral pairs of giant axons (maximum diameter 50 μm) which run from the last abdominal to the metathoracic ganglion or beyond (Parnas and Dagan, 1971). Of these, four giant axons can be found in a ventral cluster on each side (Figure 3A, C). The axons are sometimes dispersed in an irregular pattern in the connectives (Figure 3B), but return to a regular

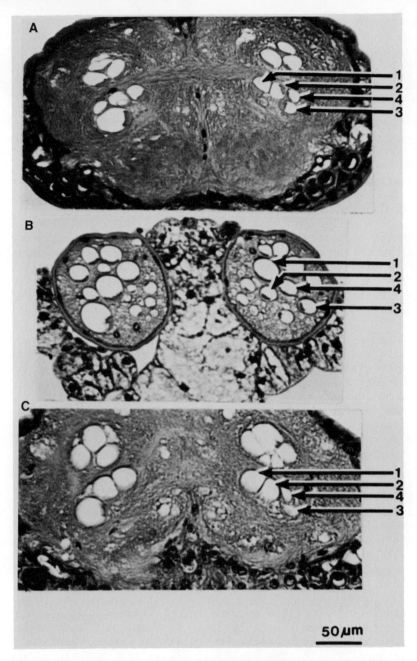

Figure 3. Giant interneurons of <u>Periplaneta americana</u>, cross section. A. Fifth abdominal ganglion. B. Connective from 4th to 5th abdominal ganglia. C. Fourth abdominal ganglia. Each axon in the two ventral clusters (numbered on right) was followed in 10 mm serial sections through   (Legend continued on next page.)

these two ganglia and connectives. The characteristic array, seen in each ganglion, became disrupted in the right connective but not in the left. (Numbering after Harris and Smyth, 1971.) The dorsal cluster contains three large axons, tentatively identified individually by numbers 5 (medial), 6 (central) and 7 (lateral).

array at each ganglion (Figure 3A, C). This statement is based upon serial sections through eight neighboring pairs of ganglia and their connectives, in three animals. The remainder of the giant fibers, contained in two bilateral dorsal clusters, are almost always arranged in a row within the ganglia (Figure 3). We tentatively identify these axons individually by the numbers 5 (medial), 6 (central), and 7 (lateral).

In the metathoracic ganglion, most of the motor neuronal somata are on the ventral surface. Whole ganglia stained with toluidine blue reveal an unusually large bilateral pair of cells (Figure 4, arrows), each previously identified as motor neuron Df, the fast depressor of the coxa (Pearson and Fourtner, 1974); also termed cell 28 (Cohen and Jacklet, 1967).

After impaling with a procion-filled microelectrode a putative giant fiber, we record spikes in the fiber in response to stimulation of the whole cord through hook electrodes. Then we drive the fiber by intracellular current pulses through the microelectrode, and record the ascending spikes from the single giant fiber with the hooks (Figure 5). Simultaneously, we record from the soma of motor neuron Df. (Fiber optic illumination usually reveals this neuron.) The latter electrode is a cobalt-filled, bevelled micropipette, to which a backing current of 1-3 na is applied.

Stimulating through the hook electrodes with a brief suprathreshold shock produces a spike in the fiber and evokes a depolarizing psp in the motor neuron. In the experiment shown (Figure 6) only those stimuli which evoke a spike also evoke a psp. Thus the impaled fiber, and/or other axons having the same threshold, evoke the psp. In many other experiments, the threshold for the giant fiber has been either slightly above, or slightly below or identical with that for the psp.

Stimulating the fiber intracellularly evokes a single large spike recordable from the whole cord, followed by a small depolarizing psp in the motor neuron. Subthreshold stimuli evoke no psp's (Figure 7). The low psp amplitude is attributable to the great electrical distance from the dendrites to the recording site, and to the loading effect of the large soma. Computer averaging of 32 motor neuronal recordings just above, and 32 just below threshold for the spike reveal this difference more clearly (Figure

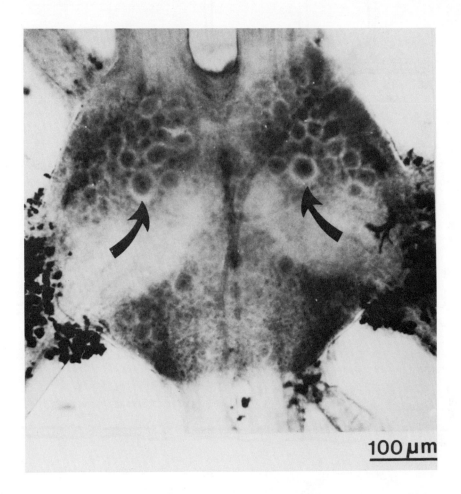

Figure 4. Metathoracic ganglion of Periplaneta; toluidine blue stain, whole mount, ventral view. Arrows point to the bilateral pair of somata of motor neurons Df. This identification is based on cobalt back-filling from nerve 5 Br 1 (Pearson and Fourtner, 1973) or the whole nerve 5 (Ritzmann and Camhi, Unpublished).

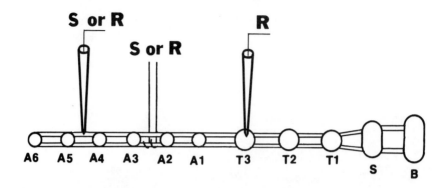

Figure 5. Schematic diagram of central nervous system of <u>Periplaneta</u>, showing electrode positions. S -- stimulate; R -- record; B -- brain; S -- suboesophageal ganglion; T_{1-3} -- thoracic ganglia; A_{1-6} -- abdominal ganglia. Wax platforms support connective A_{4-5} and ganglion T_3. These two regions exposed to 1% pronase for 5 to 10 min. Perfused with oxygenated saline (Callec and Satelle, 1973).

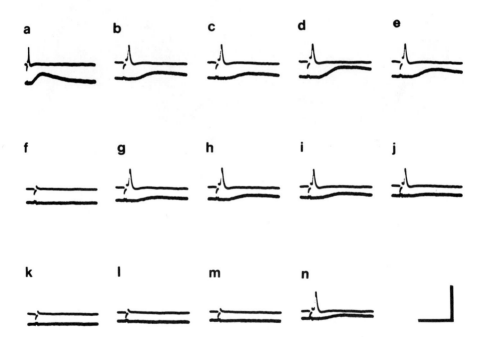

Figure 6. Intracellular recordings from an axon in connective A_{4-5} (top traces) and the soma of a metathoracic motor neuron (bottom traces) while stimulating both A_{2-3} connectives. Cells subsequently defined (Legend continued on next page.)

NON-RHYTHMIC INPUTS IN ARTHROPODS 575

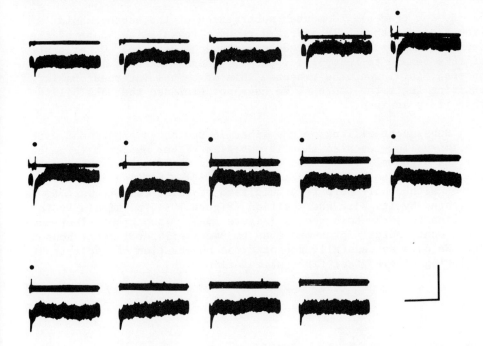

Figure 7. Intracellular recordings from soma of same motor neuron as in previous figure (bottom traces). Impaled axon in connective A_{4-5} was stimulated and its spike recorded from whole cord at A_{2-3} (top traces). Stimulus voltage increased and decreased throughout the series beginning at top left and proceeding by horizontal rows. Stimuli were suprathreshold for the impaled axon in seven traces where the extracellularly recorded spike is indicated with a dot. Psp's are discernible in most or all of these records. Records are consecutive at 1 sec. intervals. Calibration: 5 mv, 50 msec.

as right giant fiber number 1 and right motor neuron Df. Stimulus was adjusted to just suprathreshold for the axonal spikes, and then gradually decreased, so that in records f, k, l and m the stimulus failed to evoke a spike. In only these four records is there no visible psp. Records are consecutive, at 1 sec. intervals. Calibration -- vertical; top traces 40 mv, bottom traces 20 mv. horizontal; record a, 50 msec., all others, 25 msec.

8A, B, top traces). Computer subtraction of the subthreshold from the suprathreshold averaged records reveals the shape of the depolarizing psp, which suggests a chemical synapse. Other tests of this remain to be applied. Also, though the event is presumed an epsp, in view of the expected role of cercal interneurons in exciting leg motor neurons, we have not excluded the possibility that it is an ipsp.

Subsequent filling of the fiber in the above experiment with procion yellow revealed that it was giant fiber number 1 (Figure 9A) unambiguously identified by following it in serial section to the next more anterior ganglion (Figure 9B). Filling the motor neuron with cobalt sulfate precipitate revealed that it was Df, as indicated by its size and position (Figure 10A), particularly in relation to surrounding cells (Figure 10B). Thus giant fiber number 1 contributes a presumed epsp to the ipsilateral motor neuron Df. We have successfully repeated the observation of psp's in Df following single giant fiber stimulation.

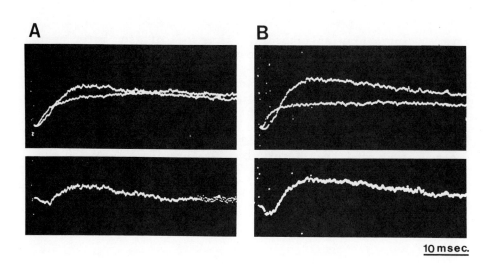

Figure 8. Computer average of motor neuronal psp's like those of Figure 7, from the same experiment. A and B -- averages of two separate groups of 32 responses. For each group, top superimposed pair of traces is the average of motor neuronal responses to stimuli which were just suprathreshold (higher curve of the pair) and just subthreshold (lower curve of the pair) for an axonic spike. Bottom single trace -- computer subtraction of the above pair.

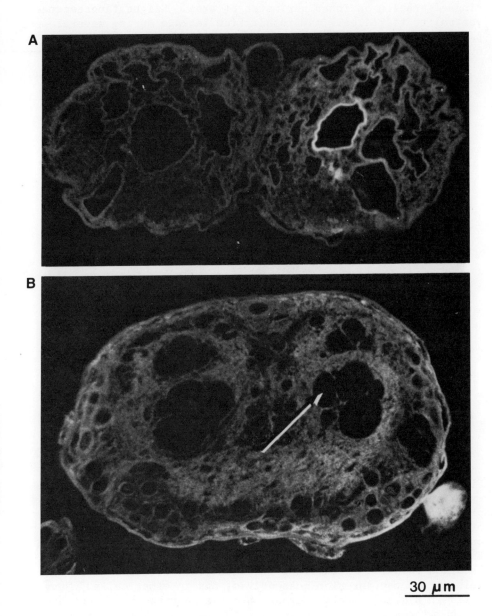

Figure 9. Cross sections of nerve cord from same experiment as Figures 6-8, viewed in fluorescence. A. Connective A_{4-5}, showing a procion filled giant axon. B. Ganglion A_4, to which the filled axon was followed in 10 μm serial sections, to identify it as giant fiber number 1 (arrow). Giant axons largely devoid of cytoplasm, resulting in a   (Legend continued on next page.)

ring of procion on inner perimeter of axon. Slight fluorescence ventral to filled cell may result from leakage during several passes of microelectrode from the ventral direction.

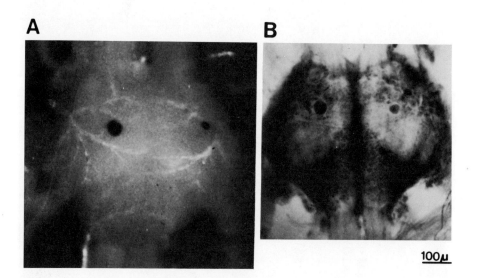

Figure 10. Whole metathoracic ganglion, ventral view, from same experiment as Figures 6-9, after developing for cobalt sulfate precipitate. A -- before and B -- after toluidine blue staining of whole ganglion. Large cell on left (animal's right side, seen on left in ventral view) is site of motor neuronal recordings in Figures 6-8. It is identified as Df based on its size and its location relative to the large trachea just posterior to it and relative to the anterior connectives, and to other cells. Many cells in B didn't stain, perhaps because many had been punctured by the microelectrode on several prior passes. The electrical records in Figures 6-8 could not have been taken from the smaller cobalt-filled cell which had been filled earlier, since this cell is on the opposite side of the ganglion. Shrinkage after fixation, between A and B, is about 15%.

The latency from a giant fiber spike in the cord to the start of the psp is approximately 2.5 msec., based upon measurements from single and averaged records (Figures 7, 8). Allowing 1 msec. for conduction over the 6 mm. distance from the hook electrodes to the ganglion, and about .45 msec. for conduction two-thirds of the way through ganglion T3 (Spira et al., 1969) leaves about 1 msec. for synaptic transmission.[4] This is consistent with, but not proof of, a monosynaptic interaction.

Thus far we have not recorded spikes in the soma or in the peripheral axon of motor neuron Df in response to single giant fiber spikes, consistent with earlier reports (Dagan and Parnas, 1970; Iles, 1972). It is possible that motor neuron Df requires convergent input from several modalities, or the setting of some sort of attentiveness, before it will respond with a spike to input from giant fibers, a situation similar to that of motor neuron FETi of locusts. This neuron, largely responsible for the powerful jump of the hindlegs, receives monosynaptic input from large, identified visual interneurons. Nevertheless, during no experiment can these neurons be made to evoke a motor neuron spike, apparently requiring convergent input from several modalities (Burrows and Rowell, 1973). Other explanations for the absence of spikes in cell Df are the generally depressed property of the nervous system and behavior during the experiment, the rapid defacilitation of psp's and the fact that each giant fiber responds to wind puffs with several spikes, whereas our experiments so far have involved single spikes in one axon.

In order to understand the integration in this system, one would need to clarify how the seven or more giant fibers of each side differ in terms of both their inputs and their outputs. The responsiveness of most of these neurons to controlled wind puffs of different directions, has been determined by employing intracellular recording with procion-filled electrodes (Langberg et al., In Preparation). Both giant fibers 1 and 2 are relatively non-directional, but differ in the number of spikes evoked by a standard wind puff (Figure 11A, B). Cell 5 responds selectively to wind from the ipsilateral, posterior quadrant (Figure 11C). Cell 6

---

[4]Among the difficulties with calculating transmission time are the inaccuracy in measuring distances in a distorted and slightly stretched cord, the unknown location of the synapse within the ganglion and disagreement over conduction velocity in the abdominal region of the cord (Roeder, 1948; Spira et al., 1969).

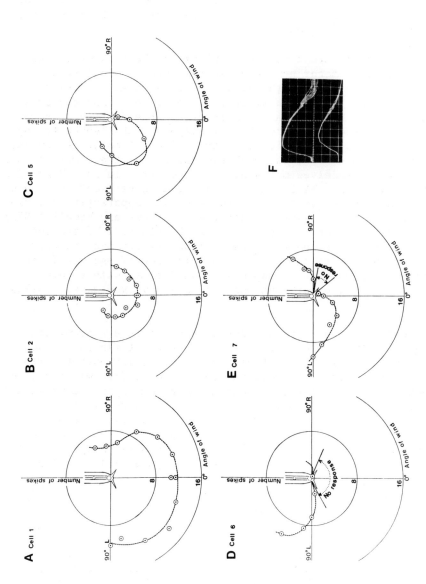

Figure 11. Responses of identified giant fibers to standard wind puffs from different directions. Puffs delivered by movable, precision wind tunnel. In the polar plots A–E, the cerci of the cockroach are represented at the zero coordinates. (Legend continued on next page.)

responds to a wide range of angles, but not to wind from behind. Cell 7 responds to a wide range of angles, but not to those angles which would excite contralateral cell 5. No data are yet available from cells 3 or 4. Thus each giant fiber appears to abstract different information about the wind stimulus. This information is probably employed in executing an appropriate escape behavior. Our preliminary behavioral studies show that a cockroach turns away from an oncoming puff of wind.

It is known that non-giant interneurons, also excited by wind on the cerci, can initiate locomotion (Dagan and Parnas, 1970). If it can be demonstrated that the giant fibers do evoke an evasive movement under natural conditions, then the cockroach escape system would come to resemble in some respects that of crayfish. In crayfish, giant interneurons directly excite motor neurons for rapid abdominal flexion (Zucker, 1972; Atwood and Pomeranz, 1974) and most subsequent swimming cycles are driven by non-giant interneurons (Schramek, 1970; Wine and Krasne, 1972). In both animals, one of the unsolved problems would be the basis for proper timing of subsequent locomotory movements following the initial escape motions evoked by the giant fibers. In cockroaches, the question is complicated by the alternation in the stepping motions of opposite legs, and by the possibility of directional biasing by the dorsal giant fibers.

All graphs are plotted as though the recording was from the left connective (i.e., right connective recordings are redrawn as mirror images). Each point represents the mean number of spikes in response to five identical puffs from the particular direction. Standard deviations, all less than "1," are omitted for clarity. Each curve has been obtained from the same identified cell at least four times. The cells whose curves are shown are representative of all those recorded. Owing to steric interference, puffs could be delivered only from a range of about $200°$. The wind source allowed inward or outward puffs to be generated at the tunnel's open end, about 1 cm. from the cerci. All data shown are for outward puffs. Cells 1, 2 and 5 generally do not respond to inward puffs when the wind tunnel is located near $0°$ (i.e., posteriorly flowing wind). Cell 6 does so respond, with a maximum number of spikes at $0°$ (i.e., head on wind), so the graph of cell 6 should probably be closed in the front. There is no comparable information yet for cell 7. F - Standard wind puffs, as recorded just inside the open end of the wind tunnel (top record) and at the site of the cerci (bottom record). Each record consists of 20 superimposed traces. Each square division equals 10 msec. horizontally. The peak of the bottom curve represents a wind velocity of 1.9 m/sec. The anemometer generating the upper curve is not calibrated.

Conclusion

   This chapter has presented a brief outline of several of the major types of control which non-rhythmic sensory inputs exert on locomotory behavior. Suggestions have been made on the basis of studies on a few model systems presented here, of the possible form of neural control involved in these behaviors. Two types of sensory influence on locomotion can be explained in principle in terms of known properties of command fibers. These include frequency and amplitude modulation of all or only some of the components of the output. No mechanism is known for a third type of sensory modulation; that which evokes a profound phase shift among the outputs. Finally, the demonstration that giant fibers in cockroaches depolarize leg motor neurons, partially resolves an old controversy. Moreover, this method offers another approach to the study of interactions among several central neurons involved in driving the stepping motions of the legs.

Acknowledgment

   I thank Drs. Ron Hoy and Roy Ritzmann for critically reading the manuscript, and Carol Monroe for technical assistance. The work described was supported by Grant No. NS 09083-05 from National Institutes of Health.

REFERENCES

Atwood, H. and Pomeranz, B., (1974) Crustacean motor neuron connections traced by backfilling for electron microscopy. J. Cell Biol. 63, 329-334.

Bentley, D., (1969) Intracellular activity in cricket neurons during generation of song patterns. Z. vergl. Physiol. 62, 267-283.

Bentley, D. and Hoy, R., (1974) The neurobiology of cricket song. Sci. Amer. 231, 34.

Bowerman, R. and Larimer, S., (1974) Command fibers in the circumoesophageal connective of crayfish. II. Phasic fibers. J. Exp. Biol. 60, 119-134.

Burrows, M., (1973) The role of delayed excitation in the coordination of some metathoracic flight motoneurons of a locust. J. Comp. Physiol. 83, 135-164.

Burrows, M. and Rowell, C.H.F., (1973) Connections between descending visual interneurons and metathoracic motoneurons in the locust. J. Comp. Physiol. 85, 221-234.

Burrows, M., (1974) Monosynaptic connexions between wing stretch receptors and flight motoneurons of the locust. J. Exp. Biol. 62, 189-219.

Callec, J. and Sattelle, D., (1973) A simple technique for monitoring the synaptic actions of pharmacological agents. J. Exp. Biol. 59, 725-738.

Camhi, J.M., (1969a) Locust wind receptors. II. Interneurons in the cervical connective. J. Exp. Biol. 50, 349-362.

Camhi, J.M., (1969b) Locust wind receptors. III. Contribution to flight initiation and lift control. J. Exp. Biol. 50, 363-374.

Camhi, J.M., (1970a) Yaw-correcting postural changes in locusts. J. Exp. Biol. 52, 533-537.

Camhi, J.M. and Hinkle, M., (1972) Attentiveness to sensory stimuli: central control in locusts. Science. 175, 550-553.

Camhi, J.M., (1974) "Neural mechanisms of response modification in insects," In Experimental Analysis of Insect Behavior (Barton Browne, L., ed.), Springer-Verlag, Berlin, (60-86).

Camhi, J.M. and Hinkle, M., (1974) Response modification by the central flight oscillator of locusts. J. Exp. Biol. 60, 477-492.

Camhi, J.M., (1976) Behavioral switching in cockroaches. I. Tactile reflexes during standing and righting behavior. J. Comp. Physiol. (In press).

Cohen, M. and Jacklet, J., (1967) The functional organization of motor neurons in an insect ganglion. Phil. Trans. B. 252, 561-572.

Dagan, D. and Parnas, I., (1970) Giant fibre and small fibre pathways involved in the evasive response of the cockroach Periplaneta americana. J. Exp. Biol. 52, 313-324.

Davis, W., (1968a) Quantitative analysis of swimmeret beating in the lobster. J. Exp. Biol. 48, 643-662.

Davis, W., (1968b) Lobster righting responses and their neural control. Proc. R. Soc. Lond. B. 170, 435-456.

Davis, W., (1969) The neural control of swimmeret beating in the lobster. J. Exp. Biol. 50, 99-117.

Davis, W. and Ayers, J., (1972) Locomotion: Control by positive feedback optokinetic responses. Science. 177, 183-185.

Davis, W. and Kennedy, D., (1972a) Command interneurons controlling swimmeret movements in the lobster. I. Types of effects on motoneurons. J. Neurophysiol. 35, 1-12.

Delcomyn, E., (1971) The locomotion of the cockroach, Periplaneta americana. J. Exp. Biol. 54, 443-452.

Gettrup, E., (1966) Sensory regulation of wing twisting in locusts. J. Exp. Biol. 44, 1-16.

Gewicke, M., (1970) Antennae: another wind-sensitive receptor in locusts. Nature, Lond. 225, 1263-1264.

Gewicke, M., (1974) "The antennae of insects as air-current sense organs and their relationship to the control of flight," In Experimental Analysis of Insect Behavior. (Barton Browne, L., ed.), Springer-Verlag, Berlin, (100-113).

Gewicke, M., Heinzel, H.G. and Philippen, J., (1974) Role of antennae of the dragonfly Orthetrum cancellatum in flight control. Nature. 249, 584-585.

Harris, L. and Smyth, T., (1971) Structural details of cockroach giant axons revealed by injected dye. Comp. Biochem. Physiol. 40, 295-303.

Iles, J., (1972) Structure and synaptic activation of the fast coxal depressor motoneurone of the cockroach Periplaneta americana. J. Exp. Biol. 56, 647-656.

Kammer, A., (1975) "Generation of rhythmic motor outputs in insects," In Neural Control of Respiration. (In press).

Kutch, W., (1974) "The development of the flight pattern in locusts," In Experimental Analysis of Insect Behavior. (Barton Browne, L., ed.), Springer-Verlag, Berlin, (149-158).

Langberg, J., Westin, J. and Camhi, J.M., (1976) Response of individual giant fibers of the cockroach to varying wind direction. (In Preparation).

Mendelson, M., (1971) Oscillator neurons in crustacean ganglia. Science. 171, 1170-1173.

Parnas, I. and Dagan, D., (1971) "Functional organizations of giant axons in the central nervous system of insects: New aspects," In Advances in Insect Physiology. (Beaumont, J.W.L., Treherne, J. and Wigglesworth, V.B., eds.), Academic Press, New York and London, (95-143).

Pearson, K. and Iles, J., (1970) Discharge patterns of the coxal levator and depressor motoneurons of the cockroach, Periplaneta americana. J. Exp. Biol. 52, 139-165.

Pearson, K. and Fourtner, C., (1973) Identification of the somata of common inhibitory motoneurons in the metathoracic ganglion of the cockroach. J. Canad. Zool. 51, 859-866.

Pearson, K. and Fourtner, C., (1975) Nonspiking interneurons in walking system of the cockroach. J. Neurophysiol. 38, 33-52.

Reingold, S., (1975) Rhythmic leg movements during different behaviors in the cockroach: Implications for central oscillators. Ph.D. Thesis, Cornell University.

Roeder, K., (1948) Organization of the ascending giant fiber system in the cockroach, Periplaneta americana. J. Exp. Zool. 108, 243-261.

Roeder, K., (1967) Nerve Cells and Insect Behavior. Harvard University Press, Cambridge, Mass.

Schramek, J., (1970) Crayfish swimming: alternating motor output and giant fiber activity. Science. 169, 698-700.

Sherman, E., Novotny, M. and Camhi, J.M., (1976) Behavioral switching in cockroaches. II. Leg rhythms during walking and righting. J. Comp. Physiol. (In press).

Spira, M., Parnas, I. and Bergmann, F., (1969) Organization of the giant axon of the cockroach, Periplaneta americana. J. Exp. Biol. 50, 615-627.

Stein, P., (1971) Intersegmental coordination of swimmeret motoneuron activity in crayfish. J. Neurophysiol. 34, 310-318.

Waldron, J., (1967) Neural mechanism by which controlling inputs influence motor output in the flying locust. J. Exp. Biol. 47, 213-228.

Wendler, G., (1974) The influence of proprioceptive feedback on locust flight co-ordination. J. Comp. Physiol. 88, 173-200.

Wiersma, C.A.G. and Ikeda, K., (1964) Interneurons commanding swimmeret movements in the crayfish, Procambarus clarkii (Girard). Comp. Biochem. Physiol. 12, 509-525.

Willows, A., Dorsett, D. and Hoyle, G., (1973) The neural basis of behavior in Tritonia. III. Neuronal mechanism of a fixed action pattern. J. Neurobiol. 4, 255-285.

Wilson, D. and Weis-Fogh, T., (1962) Patterned activity of co-ordinated motor units, studied in flying locusts. J. Exp. Biol. 39, 643-667.

Wilson, D. and Gettrup, E., (1963) A stretch reflex controlling wingbeat frequency in grasshopper. J. Exp. Biol. 40, 171-185.

Wilson, D. and Wyman, R., (1965) Motor output patterns during random and rhythmic stimulation of locust thoracic ganglia. Biophys. J. 5, 121-143.

Wilson, D., (1965) Proprioceptive leg reflexes in cockroaches. J. Exp. Biol. 43, 397-409.

Wilson, D., (1967) Central mechanisms for the generation of rhythmic behavior in arthropods. Symp. Soc. Exp. Biol. 20, 199-228.

Wine, J. and Krasne, E., (1972) The organization of escape behavior in the crayfish. J. Exp. Biol. 56, 1-18.

Wyse, G., (1972) Intracellular and extracellular motor neuron activity underlying rhythmic respiration in Limulus. J. Comp. Physiol. 81, 259-266.

Wyse, G., (1975) "Central and proprioceptive control of gill ventilation in Limulus," In Neural Control of Respiration. (In press).

Zucker, R., (1972) Crayfish escape behavior and central synapses. I. Neural circuit exciting lateral giant fiber. J. Neurophysiol. 35, 599-620.

NEURAL CONTROL OF LOCOMOTION IN THE DECORTICATE CAT

C. Perret

Laboratoire de Neurophysiologie comparée

Université Paris VI, Paris, France

The acute decorticate (thalamic) cat exhibits spontaneous locomotor hindlimb movements which can be demonstrated by observing patterns of rhythmic EMG bursts. Corresponding but simpler efferent nerve discharges can be found after suppression of all phasic afferent inputs by curarization, showing that there exists a central locomotor program. Complete deafferentation of the hindlimb, suppressing tonic afferent inflow, further simplifies the locomotor pattern: in the "knee flexors," activity during the flexion phase disappears while activity during the extension phase remains in relation to the hip extensor function of these biarticular muscles. It might be concluded that the central program acts both directly on motoneurones and, indirectly, on flexor reflex pathways through modulation. The fusimotor activity, with its partial independence on reflex actions and its low threshold appears as a sensitive expression of the generator output. It confirms that the locomotor cycle is rather simple, with an immediate activation of flexor motoneurones and a more progressive one of extensor motoneurones, and without the strict alternation generally seen in alpha motoneurones.

The spontaneous occurrence of locomotion is dependent on the presence of caudal hypothalamic regions. However, activities similar to the ones of the decorticate cat can be induced for a brief period by mechanical stimulation of the spinal cord in the acute spinal cat. This favoured the hypothesis of a spinal rhythm generator and

of its command by a tonic descending action. Recording
extrapyramidal discharges in the curarized decorticate
cat shows that they effectively display variations
corresponding to the locomotor periods. Thus, their
participation in the command cannot be excluded from
other monoaminergic pathways. These discharges also
display a modulation related to the locomotor rhythm
itself; in rubrospinal cells, for instance, a burst
occurs after the beginning of each contralateral hindlimb flexor discharge, with no evidence of its command
in a simple way. Conversely, a central spinal origin of
the supraspinal modulation was likely since a rhythmic
ascending activity is found in ventral as well as in
ventrolateral fasciculi of the spinal cord. Spinoretic-
ular as well as spinocerebellar pathways can thus participate in spino-supraspinal central feed-back loops,
modulated rubro- and reticulospinal cells remaining
after cerebellar ablation. The relative participation
of these loops in the control of locomotion is not
known, especially when sensory information from the
periphery is present.

Introduction

It was known that spontaneous locomotor movements may appear after partial or total removal of telencephalic structures (decorticate or thalamic cats) in acute (Mella, 1923; Laughton, 1924; Hinsey et al., 1930) as well as in chronic preparations (Dusser de Barenne, 1920; Magoun and Ranson, 1938; Langworthy and Richter, 1939; Kennard, 1945). On the other hand, hindlimb locomotor-like movements can be induced in the spinal cat either in acute (Roaf and Sherrington, 1910; Sherrington, 1910; Brown, 1911, 1913; Miller, 1923; Maling and Acheson, 1946) or in chronic conditions (Sherrington, 1910; Ranson and Hinsey, 1930; Shurrager and Dykman, 1951). These experiments did not give clear conclusions concerning the respective afferent and central participation in locomotion because adequate methods had not yet been formulated. The aim of this chapter is to survey the results obtained during the last ten years concerning the neurophysiology of the stereotyped spontaneous rhythmic locomotor-like activity which usually develops in the acute decorticate cat. Two main questions will be considered here:

(1) What is the relative importance of afferent information from moving limbs and of central patterning in the efferent organization?

(2) Which structures are necessary for their "spontaneous" onset and continuation and through which mechanisms do such structures act?

Efferent Activities of the Decorticate Cat

The study was restricted to the hindlimbs, mainly because they display stepping movements in the spinal cat and thus allow comparisons with locomotor movements of the decorticate cat. However, some features of the movements of the other parts of the body also will be given.

(1) <u>General characteristics of the locomotor activity</u>. When the effects of anaesthesia were no longer evident, regular high movements appeared spontaneously either without interruption for minutes or hours, or more often as periodic sequences of movements separated by periods of immobility (Figure 2C, Figure 4A). Generally, hindlimb and forelimb movements were present with correspondence between forelimb flexion and ipsilateral hindlimb extension however the forelimbs may show faster rhythms.

In some cats, locomotor movements were accompanied by other activities: lashing of the tail (see Figure 1D), extrusion of the claws, erection of the tail hairs, pupillary dilatation, tachycardia, hypertension, tachypnea, sweating from the toe pads and even urination, all signs of sham-rage behaviour (Cannon and Britton, 1925; Bard, 1928). Sometimes the cat remained motionless. In this case, various stimulations such as the displacement of one limb or a light pinch of the skin generally induced appearance of one or several motor sequences quite similar to the spontaneous ones. On the other hand, locomotor activities could be blocked in a reversible way (Figure 1D) by pressing the surface of the back (see Viala and Buser, 1965, in the rabbit).

No clear difference was found which could be linked to the extent of the ablation (which could include the striatum, the thalamus and the rostral part of the hypothalamus) as long as the caudal hypothalamic region was spared. Effectively, similar features have been described in the thalamic preparation by Orlovsky (1969). More caudal lesions were followed by disappearance of the spontaneous activity. For instance, in the mesencephalic cat, locomotor movements only appear during electrical stimulation of a "locomotor region" of the brain (Shik et al., 1966a).

(2) <u>Activities of the intact limbs</u>. It was first necessary to verify if the locomotor movements elicited appeared as normal. When the cat was fixed by rachis and pelvis above a treadmill, its successive movements of flexion and extension alternating on each side made the belt move. They looked very similar to the locomotor movements in the intact animal. However, some weakness and anomalies could be seen in the extension phase: wrong contacts of the foot upon the belt were not corrected due to lack of placing reactions; extension was not powerful enough to support the body

when the axial fixation was removed; postural reactions did not recover during the few hours following the ablation. When the speed of the belt was changed by means of a motor it was generally possible to make the frequency and amplitude of the movements adapt. Moreover, at high speed, alternate activity could be converted into synchronous behaviour. Without the treadmill, when the cat was lying on a table allowing only sliding of the limbs, movements nevertheless appeared. They were generally alternating in both limbs (Figure 1A) with a frequency around 2 Hz, but abrupt changes of "gait" could be seen (Figure 1B, C). If the plantar surface of one foot made contact with an obstacle during the limb extension, this extension was maintained. When the obstacle was removed, the movements of the limb reappeared.

Electromyographic (EMG) recording of various hindlimb muscles showed a rather pure alternation between short bursts in flexor muscles and longer bursts in extensors. However, in biarticular muscles like the sartorius lateralis (Figure 2A) and rectus femoris which are hip-flexor, knee-extensor muscles, both flexor and extensor bursts could occur during each locomotor cycle, as described in the intact cat (Engberg and Lundberg, 1969) and in the mesencephalic preparation (Orlovsky, 1972a).

(3) <u>Effect of suppression of phasic afferences</u>. Sherrington (1910) thought that spinal reflexes participated in the generation of stepping. However, Brown (1913) obtained spinal rhythmic hindlimb activities after suppression of the reflexes through complete deafferentation. If the rhythm generation is a chain of local reflexes induced by the afferents activated during the movement itself, it would be suppressed by reduction of this movement through fixation of the limbs, by suppression of the kinetic afferent inputs through curarization, and by deafferentation. Such experimental situations in which one or the other category of afferent inflow is eliminated have been used here in the decorticate cat and in the spinal cat (see below). Preliminary indications were obtained when the limb was fixed by drills inserted in the bones. EMG bursts did appear in flexor and extensor muscles, with the alternation described above. But, in double-joint muscles (Figure 2B) the EMG was more simple than before fixation, with only one burst for each locomotor cycle. Curarization (with Flaxédil at supramaximal doses blocking extra- and intrafusal motor innervation) allowed complete suppression of phasic afferences from all parts of the body. Under these conditions, efferent activity could only be recorded from the central cut end of muscle nerves. Rhythmic bursts alternated in nerves to flexors and nerves to extensors on one side, and in symmetrical homonymous nerves (Figure 3A, B) in a way similar to the muscle activations of the non-paralysed cat, but with lower frequency (.5-1 Hz). The burst amplitude could be

modified by changing the limb position (Figure 3D) thereby altering the tonic afferent inputs. The main conclusion from these observations is a locomotor command of central origin remains after Flaxédil treatment.

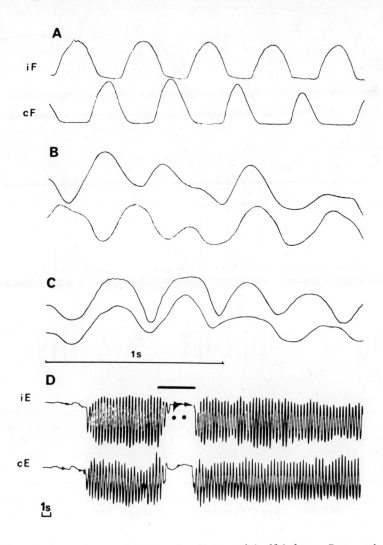

Figure 1. Locomotor movements in intact hindlimbs. Decorticate cat lying on a table. Right (iF, iE) and left (cF, cE) movements. A-C: flexion upward, D, extension upward. Notice in A alternate movements, in B and C, beginning and end of phase of symmetrical movements, in D beginning of a spontaneous sequence; pressure upon the back (line) transitorily suppressed the movements but did not block tail waving which produced mechanical artifacts (dots).

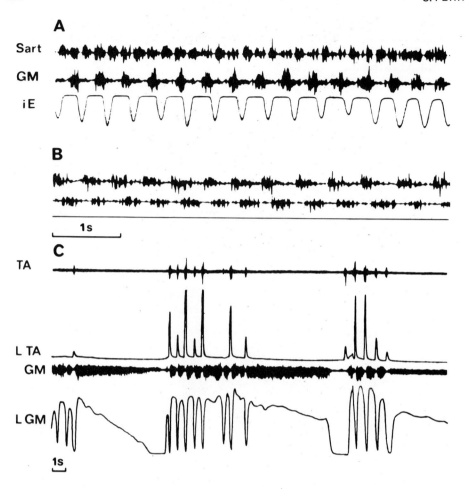

Figure 2. Effect of reduction or suppression of the hindlimb afferent inputs. Decorticate cat. A-B: EMG of sartorius lateralis (Sart) and gastrocnemius medialis (GM) muscles before (A) and after (B) fixation of the limb. iE: limb movements (extension upward). Notice simpler EMG when movement is blocked. C: EMG and length (L, contractions upward) of tibialis anterior (TA) and GM in the deafferented limb. Notice successive sequences and alternation between short flexor (TA) bursts and longer extensor (GM) bursts.

Flaxédil paralysis seemed little favourable to development of rhythmic activities. In about half of the cats, such activity was absent at least in the recorded nerves.

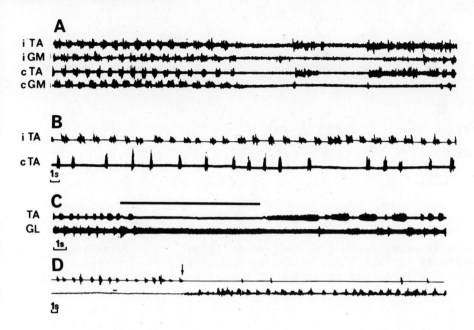

Figure 3. Efferent nerve discharges in the curarized decorticate cat. A-B: discharges in nerves to right (i) and left (c) tibialis anterior (TA) and gastrocnemius medialis (GM). C-D: activity in nerves to TA and gastrocnemius lateralis (GL); blocking effect of pressure upon the back (line) in C; effect of increase of flexion of the limb (arrow) in D: notice flexor nerve bursts in extended limb, extensor nerve bursts in the flexed limb.

(4) <u>Effect of suppression of tonic and phasic afferences from the limb</u>. Deafferentation of one limb by section of all corresponding dorsal roots (from L3 to S4) did not affect the activities on the contralateral side but reduced the participation of the muscles in the ipsilateral intact limb (Figure 4). Alternate bursts were still found in flexor and extensor muscles (Figure 2C) or nerves, however with a frequent overlap between the end of the flexor burst and the beginning of the extensor burst. This was also noticed during locomotion induced in the mesencephalic cat (Grillner and Zangger, 1975).

The case of biarticular muscles has been studied recently in some detail by Perret and Cabelguen (1975, and in preparation). They showed that the "knee flexor" muscles (posterior biceps, semitendinosus, gracilis and tenuissimus), which are in fact hip extensors as well, can receive a command as extensors in the deafferented limb. When tonic afferent inputs are present (e.g.,

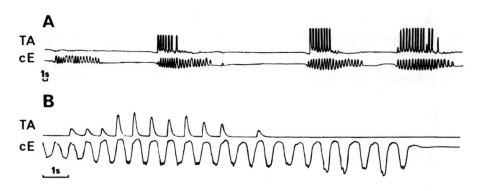

Figure 4. Effect of deafferentation on the spontaneous activities of the hindlimbs. Decorticate cat. TA: contractions (upward) of tibialis anterior in the deafferented limb. cE: contralateral movements (intact limb, extension upward). Notice successive sequences in A, and reduction of activities on the deafferented side.

intact limbs fixed or paralysed, or a deafferented limb with sustained dorsal root stimulation) their extensor burst is reduced while the flexor burst is prolonged. This burst is likely to result from interaction of the central program and reflex pathways, probably those activated by flexor reflex afferents which are excitatory to these muscles. Thus, it appears that the central locomotor program can act upon motoneurones both directly and by control of reflex pathways (see also Forssberg et al., 1975). As a consequence, afferents are able to influence not only the amplitude (Grillner and Zangger, 1975) but also the timing of muscle activities, by the selection of one expression of the command or by the other. In the mesencephalic cat, Gambarian et al., (1971) found a pure extensor burst in "knee flexors" in the intact limb during fast locomotion. Grillner and Zangger (1975) found after deafferentation either a pure extensor burst or mainly a short burst at the transition between extension and flexion. In our preparation, such a burst was found only in some cases in the tenuissimus muscle and corresponded to a short late extensor burst occurring before the next flexion phase, especially when the locomotor rhythm was slow.

What appears clearly in the deafferented decorticate cat is a different behaviour of flexors and extensors. A rhythmic sequence is made of successive cycles with a flexor and an overlapping extensor burst, separated by a variable pause between end of extension phase and following flexion phase. These observations suggest a theory concerning the organisation of the generator for locomotion which is: a rhythmic command would be sent toward motoneurones, especially those of flexor muscles; the state of excitability of the intermediate pathways (which probably include interneurones of reflex pathways) and of the motoneurones themselves would lead to a sequential activation instead of a synchronous one. Afferent actions could modify this pattern and allow adaptation to different situations (biomechanical properties of the limb, external conditions). This hypothesis is intermediate between two main other ones: Brown's (1911) theory considers a simple rhythm generation by interaction of two antagonistic centres; the other supposes a detailed central program and is supported by Grillner and Zangger's results (1975).

(5) <u>Characteristics of the fusimotor activity</u>. Gamma activity is known to be more independent of reflex actions than alpha activity (see Matthews, 1972) and hence might give further details on the central command during locomotion. Severin et al., (1967) and Severin (1970) gave examples of fusimotor control during locomotion induced in the mesencephalic cat. Here, a precise analysis was possible on the fixed deafferented limb where a locomotor activity persists (see above). Moreover, the alpha activity being often reduced in these conditions let the fusimotor actions appear more clearly.

Different complementary methods (Figure 5) were used for this study (Perret, 1971). (1) Direct recording from a filament of an otherwise intact muscle nerve (Figure 6A) showed besides large spikes corresponding to a contraction of the muscle (alpha spikes), rhythmic discharges of small spikes even when the muscle did not contract. Furthermore, they corresponded to spindle afferent discharges, and so were probably gamma spikes. From these preliminary experiments, it was inferred that the fusimotor activity underwent variations related to locomotor movements. (2) Recording potentials on the surface of the tenuissimus muscle showed besides large spikes linked to a contraction, small local potentials (Figure 6B) similar to the intrafusal potentials described by Bessou and Laporte (1965) and identified in some cases as such through stimulation of fusimotor fibres in ventral root filaments. Increases in the rhythmic frequency (from 0-20 up to 30-50 ips) were related to similar changes in the spindle afferent ones. (3) Recording from spindle afferents showed frequency variations which could not be explained solely by passive responses to length variations and therefore were attributed to fusimotor actions. In spindle afferents from the tibialis anterior muscle, this was clearly the case in most experiments. Instead of a pause (see Figure 7A), an increase of the discharge took place during shortening of the muscle (Figure 7B) or even when contractions were small (Figure 7B) or absent. Not only primary but also secondary afferents displayed this behaviour, showing that there was at least a static (Appelberg

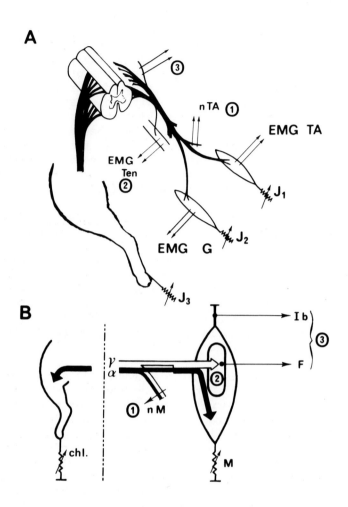

Figure 5. Methods used for study of the fusimotor activity. A: experimental set up. 1: recording of efferent discharges in muscle nerve filament (nTA). Muscle contraction of the muscle recorded through EMG or strain gauge (J_1). 2: recording of intrafusal potentials on the surface of the tenuissmus. 3: recording of muscle afferents in dorsal root filaments. Further indications are given by recording contraction of other muscles (G, etc.) and contralateral movements (chl) through strain gauges (J_2, J_3) or EMG. B = interpretation. 1: comparison between nerve activity nM (alpha+gamma) and muscle contractions (alpha). 2: intrafusal potentials are a direct manifestation of fusimotor activity. 3: comparison between passive responses of muscle afferents (e.g., Ib: tendon afferent) and fusorial (F) discharges under gamma control.

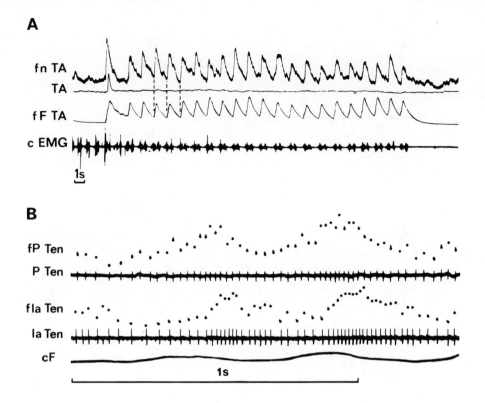

Figure 6. Fusimotor activity in the decorticate cat. A: mean frequency (increase upward) of efferent discharge (large and small spikes: see text and Figure 5-1) in a filament of nerve to tibialis anterior (fnTA) and of spindle afferent discharge from the same muscle (fF TA): contractions of the muscle (upward). c EMG: activity of a muscle in the contralateral limb. Notice frequency increases in efferent activity and spindle afferent discharge in spite of absence of contraction, at the rhythm of contralateral locomotor activity. B: instantaneous frequency (increase upward, f P Ten, f Ia Ten) of a local propagated potential (P Ten) of tenuissimus (see Figure 5-2) and of a spindle primary afferent (Ia Ten) probably belonging to the same spindle. c F: contralateral movements (flexion upward).

et al., 1966) rhythmic fusimotor action (see Perret and Buser, 1972). Conversely, tendon afferents never showed any other change of activity than the one related to contraction of their own muscle. In gastrocnemius spindle afferents (Figure 7D), the discharge frequency progressively increased during the contraction of flexors and before onset of gastrocnemius contraction. There was

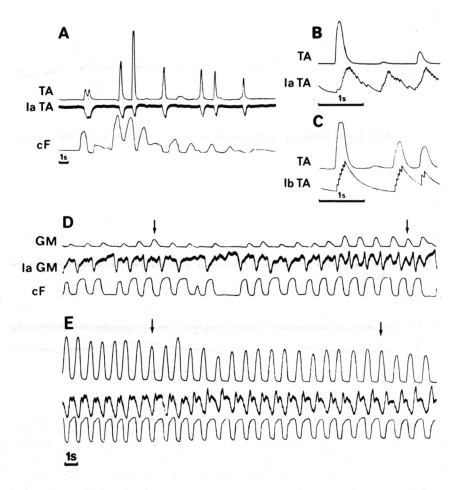

Figure 7. Activity of flexor and extensor afferents. Decorticate cat. TA, GM: contractions of respectively tibialis anterior and gastrocnemius medialis (upward). Mean frequency (increase upward) of spindle (Ia, TA, Ia GM) and tendon (Ib TA) primary afferents. cF: contralateral movements (flexion upward). Notice in A: pure passive behaviour of a spindle afferent deprived of fusimotor control (exceptional case), in B: spindle afferent with fusimotor control; notice discharge instead of pause during contraction, even when weak, in C: Ib discharges only during large contractions. In D, two contractions of similar amplitude (arrows) did not give the same frequency decrease which indicates a variation in static fusimotor action. In E, similar contractions (arrows) did not give the same increase during lengthening (indicating a variation in dynamic fusimotor action).

a more or less clear decrease which occurred during muscle shortening. This corresponded to a precession of the fusimotor activity and to a weak static gamma action. Indications of dynamic actions were also obtained (see Figure 7E). Perret and Berthoz (1973) studying the responses to stretching of muscles in spindle afferents during locomotion found that a rhythmic dynamic action existed in the gastrocnemius muscle but could not be seen, if present, in the tibialis anterior. In the knee flexor spindle afferents, Perret and Cabelguen (1975 and in preparation) found rhythmic discharges with a time course similar to the one in the gastrocnemius spindle afferent discharge. This confirmed that knee flexor muscles, similar to the extensors, receive a direct central command of their alpha and gamma motoneurones as extensors.

Finally a rhythmic fusimotor control accompanied the contractions (alpha-gamma linkage) but might be present alone, indicating a higher sensitivity to the central command. There was (Perret, 1974) a real alpha-gamma coactivation in the tibialis anterior at the onset of the flexion phase and soon after a progressive gamma activation in gastrocnemius followed by an alpha one (gamma precession). Conversely, acceleration in the flexor spindle discharge did not begin before the end of the extensor activity (Figure 8). The reciprocal innervation of flexor and extensor muscles suggested by the alternating EMG bursts in the intact limb does not seem to be strictly programmed by the generator for locomotion.

The gamma activations are not the cause of the alpha ones, the latter being still present after deafferentation and so both originate from central pace-makers. But, a functional significance of the differences in timing and intensity of the fusimotor drive to flexor (tibialis anterior, extensor digitorum longus, peroneus longus) and extensor muscles (gastrocnemius, soleus, etc.) can be

Figure 8. Relations between fusimotor actions in a flexor and an extensor muscle. Decorticate cat. Ia afferent discharge from tibialis anterior (Ia TA) and gastrocnemius medialis (Ia GM). Simultaneous recording of length variations showed that the muscles did not contract. Notice that extensor spindle discharge overlaps with end or preceding flexor one, but not with beginning of next one.

proposed. The flexor phase of the locomotor cycle is known to be more centrally programmed as indicated by its constant duration (Shik et al., 1966b; Goslow et al., 1973): during this phase a powerful command would reach flexor alpha and static gamma motoneurones, leading to a servo-assisted movement (see Stein, 1974). Regulatory reflexes act mainly during the extension phase (see Arshavsky et al., 1965; Orlovsky and Shik, 1965): they could act earlier thanks to the gamma precession and stronger because of the increased sensitivity of the primary spindle afferents under dynamic fusimotor control.

Efferent Activities of the Spinal Cat

In the decorticate preparation, supraspinal influences could modify or even replace completely spinal locomotor mechanisms. So, we tried to verify if locomotor characteristics similar to those present in the decorticate cat could be obtained in the spinal cat. Unfortunately, since no spontaneous activity remains after spinalization, it was necessary to use stimulation. We chose Brown's (1911) technique of mechanical stimulation of the spinal cord which allowed immediate observations at the end of experiments with decorticate cats. Sometimes we used Forssberg and Grillner's (1973) method of injection of Clonidine (500 microg/kg). Rhythmic activities could thus be induced in the acute spinal cat. All their characteristics (Perret et al., 1972b) looked very similar to those of the decorticate cat with alternate or symmetrical movements, similar frequencies, corresponding alternate contractions in flexor and extensor muscles (Figure 9A), efferent nerve discharges after curarization (Figure 9C), similar pattern of EMG activations in the deafferented limb, effects of pressure on the back (Figure 9B), rhythmic fusimotor control with alpha-gamma linkage and cases of gamma precession (Figure 10). Grillner and Zangger (1974) and Sjöström and Zangger (1975) did not find evidence of an extensor behaviour of the semitendinosus and of the tenuissimus in the deafferented spinal cat. This might indicate that Clonidine, Nialamide - DOPA or dorsal root stimulation used in these experiments induce motor effects slightly different than those in the deafferented decorticate, mesencephalic or "Brown's spinal" preparation. Let us mention that any method of stimulation of the spinal cord probably induces side effects.

Structures responsible for the main features of the locomotor activity, namely central pace-makers for alpha and gamma discharges are thus present at the spinal lumbo-sacral level. The only difference is that this activity is not spontaneous and one could imagine that supraspinal structures only give a tonic descending command (see Sherrington, 1910) to put the spinal rhythm generator in action: cord stimulation in the spinal cat would activate more

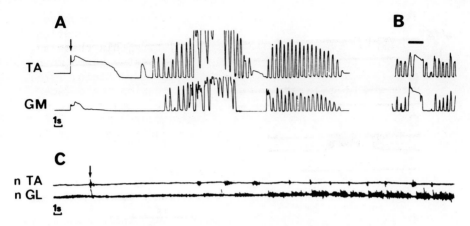

Figure 9. Efferent activities in the spinal cat, induced by incision (arrows) of the spinal cord at lumbar (L1) level. A-B: contractions (upward) of tibialis anterior (TA) and gastrocnemius medialis (GM). Blocking effect of pressure upon the back (line) in B. C: efferent discharges in nerves to TA and gastrocnemius lateralis (GL) in the curarized cat.

or less specifically the corresponding descending fibres. An attempt to verify this possibility was made by recording from descending pathways in the decorticate cat.

CENTRAL CHARACTERISTICS

A spinal origin of locomotor rhythm does not rule out the role of supraspinal pathways either eliciting the spontaneous appearance of this rhythm or even modifying or replacing the spinal generator. It was interesting to know if, in the absence of phasic afferent inputs, a central rhythmic activity related to the efferent activity was also present in supraspinal structures, especially those from which descending pathways originate.

In order to eliminate the possibility of recurrent actions from the periphery via sensory pathways, experiments were performed on paralysed cats (Figure 11). Special attention was given to those cells in which variations of the discharge were detected. Proportional changes could not be determined due to variations which occurred among the cats throughout each experiment; and due to the limited number of units recorded in each preparation.

(1) <u>Descending activities (Perret, 1968, 1973)</u>. The first striking finding was that a rhythmic activity linked to the

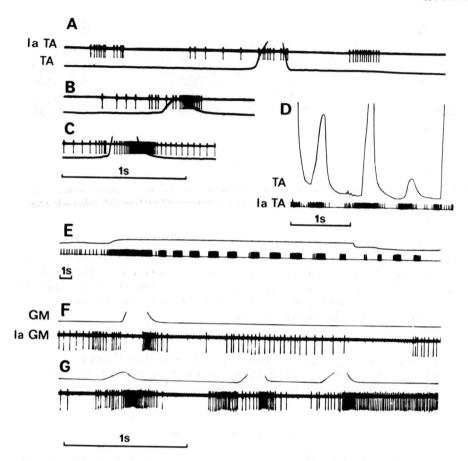

Figure 10. Spindle afferent discharges in the spinal cat. A–C and F–G after mechanical stimulation of the spinal cord. D–E: under Clonidine, with pinch of the foot in E. Ia discharge from tibialis anterior (Ia TA) and gastrocnemius medialis (Ia GM). TA, GM: muscle contractions (upward). Notice in E that stimulation gave a mechanical artifact on length recording and no contraction but induced rhythmic spindle afferent discharges.

efferent discharges was present in supraspinal structures. No attempt was made to explore all of them. Modulation was not found at the diencephalic level but in most of the structures below that level, including the reticular formation, cerebellum (see also Viala et al., 1970, in the rabbit), lateral vestibular nucleus and red nucleus. In the medial bulbar reticular formation, especially in the case of reticulospinal cells whose axons reach the lumbar

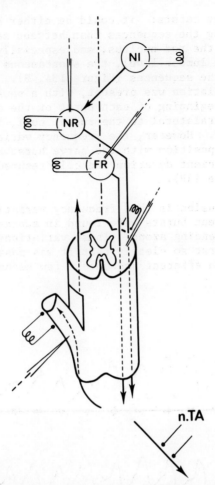

Figure 11. Electrophysiological methods used for study of the descending and ascending activities. Decorticate curarized cat. Stimulation of nucleus interpositus (NI) for identification of rubral units (Tsukahara et al., 1967), of the dorsolateral fasciculus for identification of rubrospinal cells recorded in the red nucleus (NR), of NR and medial bulbar reticular formation (FR) for identification of rubro and reticulospinal fibres in the spinal cord of the dissected fasciculi for identification of ascending fibres. n TA: recording of nerve to tibialis anterior.

level (see Figure 12), neighbouring units could have opposite behaviours with frequency increasing either synchronously or alternately with the recorded flexor nerve bursts. The general activity of the cells was also often related to the presence or absence of efferent bursts: it could be either higher or lower (Figure 12C) during the sequences than between sequences of rhythmic nerve bursts. In the red nucleus, and especially in rubrospinal cells reaching the lumbar level, the spontaneous activity generally increased during the sequences (Figure 13A, B). In all these units (Figure 13) a modulation was present, with a pause of the discharge before or at the beginning of each burst of the corresponding flexor nerve (contralateral to the rubral cell, ipsilateral to the rubrospinal fibre). However, the frequency variations were not really in phase opposition with the nerve burst because when the latter had a sufficient duration the cell resumed activity before its end (see Figure 13B).

A first conclusion is that frequency variations related to the sequences of efferent bursts were found in suprasplnal units and corresponding descending axons. These variations sometimes preceded the first burst so clearly that it was possible to predict the beginning of an efferent sequence a few seconds in advance.

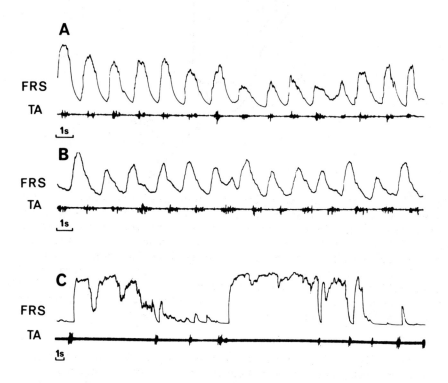

Figure 12. Activity of reticulospinal fibres of the lateral fasciculus. Decorticate curarized cat. FRS: mean frequency (increase upward) of unit discharge. TA: nerve to tibialis anterior.

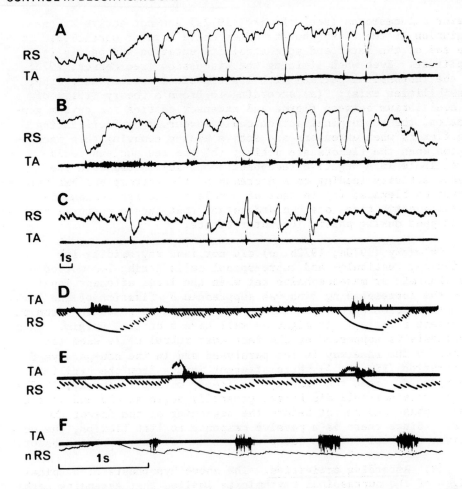

Figure 13. Activity of rubrospinal fibres. Decorticate curarized cat. RS: mean frequency (increase upward) of unit discharge; TA: nerve to ipsilateral tibialis anterior. A, B, D correspond to the same unit, C and E to another one. F: non rubrospinal fibre of dorsolateral fasciculus.

It might indicate that supraspinal centres really send a tonic descending command to the spinal rhythm generator. Since they also exhibited a modulation related to the rhythm itself, the hypothesis that they command this rhythm could be considered. An opposing hypothesis, namely that the supraspinal modulation is not the cause but the consequence of the spinal activity, is supported by the fact that rubrospinal cells which are known to give excitation to flexor motoneurones (see Orlovsky, 1972a) are not active at the beginning of the flexor burst. They could however participate in the end of the burst and so control its duration and partly its amplitude. Even when limiting the discussion (see Perret, 1973) to the rubrospinal pathway with its classical action, many other possibilities exist: (a) according to Brown's theory (1911) of an equilibrium between flexor and extensor centres for rhythm generation, the rubrospinal action could favour it by facilitating the flexors when extensors are active and by removing this facilitation when the flexors are active; (b) the red nucleus could participate in a negative feed-back loop, any increase in the flexor activity leading to a decrease of the rubrospinal excitatory action on flexors; (c) in case of a role of the red nucleus in postural activities, one can understand that such a postural action is stopped during phasic locomotor bursts.

Orlovsky (1970a, 1972b, c) did not find rhythmicity in reticulo-, vestibulo- and rubro-spinal cells during locomotion of the thalamic or mesencephalic cat when the local afferent input from the corresponding limb was suppressed by fixation of the limb. If a modulation persisted, it was interpreted as resulting from an imperfect fixation. It might as well have a central origin. This hypothesis is supported by the fact that rubral cells seem to behave in the same way in the paralysed and in the non-paralysed preparation (Figure 14) where afferent inputs from the hindlimb are reduced. When the hindlimb is freely moving (Orlovsky, 1972c) the rubrospinal cell discharges generally begin at the end of the stance phase and so not before the beginning of the flexor EMG burst. Since there is a passive response to limb flexion, the cell activity probably occurs later when this afferent input is absent.

(2) <u>Ascending activities</u>. The above hypothesis of a spinal origin of the supraspinal rhythmicity implied that ascending pathways carry information of the central spinal events. Actually, during explorations in the spinal cord, fibres were observed to have precise responses to afferent stimulation and a direct response to stimulation of the reticular formation. Therefore, they were probably ascending fibres and their discharge was modulated in relation with the efferent rhythm. Their responses to natural stimulation of the hindlimbs could not be studied accurately since reverberated responses through supraspinal loops were also present. Especially responses to stimulation of the rostral parts of the body indicated a descending action upon these ascending units. Another method was then used, recording from fasciculi (Perret et al., 1972a, b) dissected in order to suppress the descending activities (Figure 12). No modulation was found in the dorsal and dorsolateral fasciculi, i.e., mainly the lemniscal and dorsal spinocerebellar pathways. This was confirmed in the detailed study of Arshavsky et al., (1972a). Conversely, a clear modulation (Figure

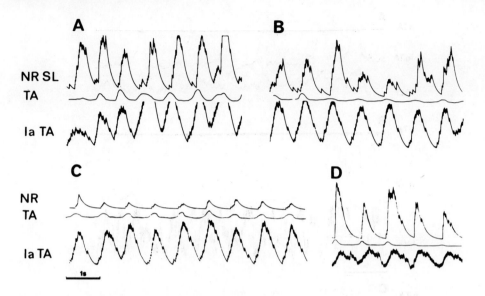

Figure 14. Activity of rubral cells. Decorticate cat. A-B: mean frequency (increase upward) of discharge in rubrospinal cells reaching the lumbar level (NRSL). C-D: id. in rubral cells which did not reach the lumbar level (NR). Notice that frequency increases begin after the contraction of the contralateral tibialis anterior (TA) and after the frequency increase (upward) of Ia afferent discharge from this muscle (Ia TA).

15A, C, D) was present in the ventrolateral fasciculus. The corresponding pathway could be the ventral spinocerebellar tract as later shown by Arshavsky et al., (1972b). But other pathways were known which carried information on both peripheral and central events (see Oscarsson, 1967). In fact, a rhythmic ascending activity was also found in the ventral fasciculus (Figure 15B) which contains no direct spinocerebellar fibres. It could explain that after cerebellar ablation (Figure 16) rhythmic descending discharges were still displayed (Orlovsky, 1970b, 1972c; see also Cabelguen et al., 1973) with the same characteristics as before cerebellectomy. Orlovsky somewhat neglected those cells because of their apparently small proportion. But the proportion of cells receiving a central modulation through the ventral spinocerebellar pathway is unknown and might not be larger. Moreover, the proportion of modulated cells is probably variable in different experiments. In a few of our cats, which displayed quite typical rhythmic efferent discharges, no corresponding modulation could be found in supraspinal structures. It confirms that this modulation is not necessary for generation of the rhythm.

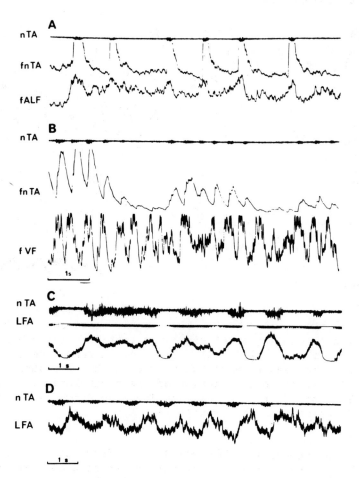

Figure 15. Ascending activities. Decorticate curarized cat.
n TA: efferent activity in tibialis anterior nerve and (fnTA) its integration. Integration of mass discharge of ventrolateral (fALF) and ventral (fVF) fasciculi. LFA: discharge of an ascending fibre in lateral fasciculus and its mean frequency.

Conclusion

A simple interpretation of the results is that tonic descending influences from meso-diencephalic structures allow the spinal generator to function. Unmyelinated monoaminergic pathways have been suggested for this action (see Grillner, 1975) but fast conducting descending pathways have been shown to carry information of this command and their participation cannot be excluded.

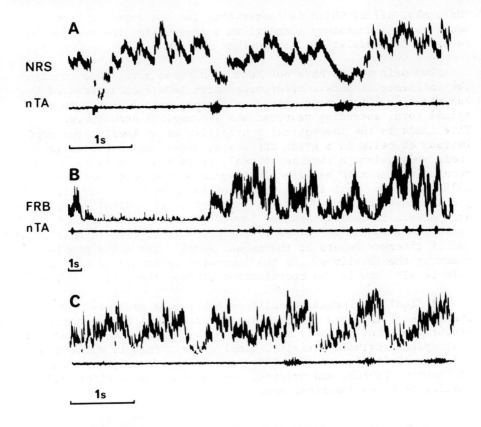

Figure 16. Activity of supraspinal cells. Decorticate cerebellectomized cat. Mean frequency of a rubrospinal cell reaching the lumbar level (NRS) and of a cell (FRB) in the bulbar reticular formation (increases upward). n TA: nerve to contralateral tibialis anterior.

Moreover, in locomotion of the intact conscious animal, other pathways activated from the cortex may be important.

Several spino-supraspinal feed-back loops (involving the cerebellum or not) may control the spinal functions, either in a positive or in a negative way. No clear effect of this control appeared here, since a few cats (without cerebellum) in which no supraspinal rhythmic activity could be found, displayed typical efferent discharges. An extensive study of the time relations between activities in different centres will probably give more information on the function of the loops. For instance, the delay between mean activities in nucleus interpositus and red nucleus

(Orlovsky, 1972d) which is longer than the one expected from monosynaptic excitatory connections suggests that the red nucleus receives its modulation from another source.

Two main points have not been considered here. One concerns the influence of phasic afferences which have been suppressed by curarization. They may interact with the central rhythm in the spinal cord, ascending neurones and supraspinal structures.
This leads to the theoretical possibility of at least three populations of cells in a given structure, according to the origin of their modulation: either peripheral, or central, or both. Such populations seem to be effectively present (see e.g., Orlovsky, 1972b, c). So, what is the result of the convergence of the multiple ascending pathways carrying information of central and peripheral origin at the supraspinal level? What is the result of the convergence on the spinal generator activity of descending actions and of afferent inputs at the spinal level? The other problem concerns the forelimbs. Is the locomotor mechanism identical at this level? How is the coordination between the girdles achieved?

A final interpretation will probably come not only from a synthesis of the data obtained during locomotion in different preparations but probably also from a comparison with another central spinal rhythmic activity, namely scratching, in which many efferent (Deliagina et al., 1975) and central characteristics (Arshavsky, Pavlova and Orlovsky, Personal communication) look similar to those described here.

Acknowledgment

We wish to thank Dr. Sten Grillner for his numerous suggestions to improve the presentation of this paper.

REFERENCES

Appelberg, B., Bessou, P. and Laporte, Y., (1966) Action of static and dynamic fusimotor fibres on secondary endings of cat's spindles. J. Physiol., London. $\underline{185}$, 160-171.

Arshavsky, Yu.I., Kots, Y.M., Orlovsky, G.N., Rodionov, I.M. and Shik, M.L., (1965) Investigation of the biomechanics of running by the dog. Biophysics. $\underline{10}$, 737-746 (Translated from Russian).

Arshavsky, Yu.I., Berkinblit, M.B., Fukson, O.I., Gelfand, I.M. and Orlovsky, G.N., (1972a) Recordings of neurones of the dorsal spinocerebellar tract during evoked locomotion. Brain Research. $\underline{43}$, 272-275.

Arshavsky, Yu.I., Berkinblit, M.B., Fukson, O.I., Gelfand, I.M. and Orlovsky, G.N., (1972b) Origin of modulation in neurones of the ventral spinocerebellar tract during locomotion. Brain Research. 43, 276-279.

Bard, P., (1928) A diencephalic mechanism for the expression of rage with special reference to the sympathetic nervous system. Amer. J. Physiol. 84, 490-515.

Bessou, P. and Laporte, Y., (1965) Potentiels fusoriaux provoqués par la stimulation de fibres fusimotrices chez le chat. C.R. Acad. Sci. (Paris) 260, 4827-4830.

Brown, T.G., (1911) The intrinsic factors in the act of progression in the mammal. Proc. R. Soc. Lond. s. B. 84, 308-319.

Brown, T.G., (1913) The phenomenon of "narcosis progression" in mammals. Proc. R. Soc. Lond. s. B. 86, 140-164.

Cabelguen, J.M., Millanvoye, M. and Perret, C., (1973) Caractéristiques centrales et efférentes de l'activité rhythmique de type locomoteur chez le chat decortiqué après ablation du cervelet. J. Physiol. (Paris) 67, 253A.

Cannon, W.B. and Britton, S.W., (1925) Studies on the conditions of activity in endocrine glands. XI-Pseudoaffective medulliadrenal secretion. Amer. J. Physiol. 72, 283-294.

Deliagina, T.G., Feldman, A.G., Gelfand, J.M. and Orlovsky, G.N., (1975) On the role of central program and afferent inflow in the control of scratching movements in the cat. Brain Research. 100, 297-313.

Dusser de Barenne, J.G., (1920) Recherches expérimentales sur les fonctions du système nerveux central faites en particulier sur deux chats dont le néopallium avait été enlevé. Arch. néerl. Physiol. 4, 31-123.

Engberg, I. and Lundberg, A., (1969) An electromyographic analysis of muscular activity in the hindlimb of the cat during unrestrained locomotion. Acta Physiol. Scand. 75, 614-630.

Forssberg, H. and Grillner, S., (1973) The locomotion of the acute spinal cat injected with clonidine I.V. Brain Research. 50, 184-186.

Forssberg, H.S., Grillner, S. and Rossignol, S., (1975) Phase dependent reflex reversal during walking in chronic spinal cats. Brain Research. 85, 103-107.

Gambarian, P.P., Orlovsky, G.N., Protopopova, T.J., Severin, F.V. and Shik, M.L., (1971) The activity of muscles during different gaits and adaptative changes of moving organs in family Felidae. Proc. Inst. Zool. Acad. Sci. USSR. 48, 220-239 (In Russian).

Goslow, G.E., Jr., Reinking, R.M. and Stuart, D.G., (1973) The cat step cycle: hind limb joint angles and muscle lengths during unrestrained locomotion. J. Morphol. 141, 1-41.

Grillner, S. and Zangger, P., (1974) Locomotor movements generated by the deafferented spinal cord. Acta Physiol. Scand. 91, 38A-39A.

Grillner, S., (1975) Locomotion in Vertebrates: central mechanisms and reflex interaction. Physiol. Rev. 55, 247-304.

Grillner, S. and Zangger, P., (1975) How detailed is the central pattern generation for locomotion? Brain Research. 88, 367-371.

Hinsey, J.C., Ranson, S.W. and Mc. Nattin, R.F., (1930) The role of the hypothalamus and mesencephalon in locomotion. Arch. Neurol. Psychiat. 23, 1-43.

Kennard, M.A., (1945) Focal autonomic representation in the cortex and its relation to sham rage. J. Neuropath. exp. Neurol. 4, 295-304.

Langworthy, O.R. and Richter, C.P., (1939) Increased spontaneous activity produced by frontal lobe lesions in cats. Amer. J. Physiol. 126, 158-161.

Laughton, N.B., (1924) Studies on the nervous regulation of progression in mammals. Amer. J. Physiol. 70, 358-384.

Magoun, H.W. and Ranson, S.W., (1938) The behaviour of cats following bilateral removal of the rostral portion of the cerebral hemispheres. J. Neurophysiol. 1, 39-44.

Maling, H.M. and Acheson, G.H., (1946) Righting and other postural activity in low-decerebrate and in spinal cats after d-amphetamine. J. Neurophysiol. 9, 379-386.

Matthews, P.B.C., (1972) Mammalian muscle receptors and their central actions. Monographs of the Physiological Society (E. Arnold, London), Volume 23.

Mella, H., (1923) The diencephalic centers controlling associated locomotor movements. Arch. Neurol. Psychiat. 10, 141-153.

Miller, F.R., (1923) V. Studies in mammalian reflexes. Trans. R. Soc. Canada. 17, 29-32.

Orlovsky, G.N. and Shik, M.L., (1965) Standard elements of cyclic movements. Biophysics. 10, 935-944 (Translated from Russian).

Orlovsky, G.N., (1969) Spontaneous and induced locomotion of the thalamic cat. Biophysics. 14, 1154-1162 (Translated from Russian).

Orlovsky, G.N., (1970a) Work of the reticulo-spinal neurones during locomotion. Biophysics. 15, 761-771 (Translated from Russian).

Orlovsky, G.N., (1970b) Influence of the cerebellum on the reticulo-spinal neurones during locomotion. Biophysics. 15, 928-936 (Translated from Russian).

Orlovsky, G.N., (1972a) The effect of different descending systems on flexor and extensor activity during locomotion. Brain Research. 40, 359-371.

Orlovsky, G.N., (1972b) Activity of vestibulospinal neurons during locomotion. Brain Research. 46, 85-98.

Orlovsky, G.N., (1972c) Activity of rubrospinal neurons during locomotion. Brain Research. 46, 99-112.

Orlovsky, G.N., (1972d) Work of the neurones of the cerebellar nuclei during locomotion. Biophysics. 17, 1177-1185 (Translated from Russian).

Oscarsson, O., (1967) "Functional significance of information channels from the spinal cord to the cerebellum," In Neurophysiological Basis of Normal and Abnormal Motor Activities. (Yahr and Purpura), Raven Press, N.Y., (93-113).

Perret, C., (1968) Relations entre activités efférentes spontanées de nerfs moteurs de la patte postérieure et activités de neurones du tronc cérébral chez le chat décortiqué. J. Physiol., Paris. 60, 511-512.

Perret, C., (1971) Activités des fibres fusimotrices statiques et dynamiques au cours de mouvements locomoteurs chez le chat. J. Physiol., Paris. 63, 139A.

Perret, C. and Buser, P., (1972) Static and dynamic fusimotor activity during locomotor movements in the cat. Brain Research. 40, 165-169.

Perret, C., Millanvoye, M. and Cabelguen, J.M., (1972a) Messages spinaux ascendants pendant une locomotion fictive chez le chat curarise. J. Physiol., Paris. 65, 153A.

Perret, C., Cabelguen, J.M. and Millanvoye, M., (1972b) Caractéristiques d'un rythme de type locomoteur chez le chat spinal aigu. J. Physiol., Paris. 65, 472A.

Perret, C., (1973) Analyse des mécanismes d'une activité de type locomoteur chez le chat. Thèse Doct. Sci. Paris. CNRS AO 8342.

Perret, C. and Berthoz, A., (1973) Evidence of static and dynamic fusimotor actions on the spindle response to sinusoidal stretch during locomotor activities in the cat. Exp. Brain Research. 18, 178-188.

Perret, C., (1974) Activites efferentes et generateur de rythme locomoteur chez le chat. J. Physiol., Paris. 69, 284A.

Perret, C. and Cabelguen, J.M., (1975) A new classification of flexor and extensor muscles revealed by study of the central locomotor program in the deafferented cat. Exp. Brain Research. Suppl. 23, 160.

Ranson, S.W. and Hinsey, J.C., (1930) Reflexes in the hindlimbs of cats after transection of the spinal cord at various levels. Amer. J. Physiol. 94, 471-495.

Roaf, H.E. and Sherrington, C.S., (1910) Further remarks on the mammalian spinal preparation. Quart. J. exp. Physiol. 111, 209-211.

Severin, F.V., Orlovsky, G.N. and Shik, M.L., (1967) Work of the muscle receptors during controlled locomotion. Biophysics. 12, 575-586 (Translated from Russian).

Severin, F.V., (1970) The role of the gamma motor system in the activation of the extensor alpha motor neurones during controlled locomotion. Biophysics. 15, 1138-1145 (Translated from Russian).

Sherrington, C.S., (1910) Flexion-reflex of the limb, crossed extension-reflex, and reflex stepping and standing. J. Physiol., London. 34, 1-50.

Shik, M.L., Severin, F.V. and Orlovsky, G.N., (1966a) Control of walking and running by means of electrical stimulation of the mid-brain. Biophysics. 11, 756-765 (Translated from Russian).

Shik, M.L., Orlovsky, G.N. and Severin, F.V., (1966b) Organization of locomotor synergism. Biophysics. 11, 1011-1019.

Shurrager, P.S. and Dykman, R.A., (1951) Walking spinal carnivores. J. comp. physiol. psychol. 44, 252-262.

Sjöström, A. and Zangger, P., (1975) α-γ linkage in the spinal generator for locomotion in the cat. Acta Physiol. Scand. 94, 130-132.

Stein, R.B., (1974) Peripheral control of movement. Physiol. Rev. 54, 215-243.

Viala, G. and Buser, P., (1965) Décharges efferentes rythmiques dans les pattes postérieures chez le lapin et leur mécanisme. J. Physiol., Paris. 57, 287-288.

Viala, G., Coston, A. and Buser, P., (1970) Participation de cellules du cortex cérébelleux aux rythmes "locomoteurs" chez le lapin curarisé en absence d'informations somatiques liees au mouvement. C.R. Acad. Sc. Paris. 271, 688-691.

MODULATION OF PROPRIOCEPTIVE INFORMATION IN CRUSTACEA

William H. Evoy

Laboratory for Quantitative Biology, University of

Miami, Coral Gables, Florida

Several studies of locomotion and postural regulation in different Crustacea indicate that utilization of the sensory information in apparently simple proprioceptive reflexes is subject to considerable variability during both centrally and peripherally commanded changes in motor output. Proprioception in Crustacea is mediated primarily by sensory cells whose dendritic processes insert into connective tissue associated with joints between limb or body segments. These systems generally signal movement or position. In a few instances, the sensory structures are mechanically linked to specialized receptor muscles or to the muscles involved in locomotion and can thus serve to monitor muscle length or tension, as in vertebrates. Other mechanoreceptors signal deformation of the somewhat elastic exoskeleton or of the nerve cord sheath. Activation of these receptors may evoke reflex feedback to the segment of origin or may influence motor outputs in other segments. Consideration will be given to the roles that these receptors play in naturally occurring motor response and to interactions between centrally initiated motor activity and proprioceptive reflexes.

A majority of studies on proprioceptive reflexes of Crustacea have been technically restricted to responses in totally or partially restrained preparations. The quantitative information from these studies is often difficult to interpret in terms of naturally occurring movements. Although a number of motor scores for crustacean movements can be evoked by selective stimulation of interneurons in the CNS, neuronal connections of these units with final motor pathways are largely unknown. Little more is known about central synaptic connections of even the simplest reflexes. Both muscle and joint receptors appear to interact with central programs for control of position, movement and adjustments to varying load conditions, fulfilling many of the functions ascribed to vertebrate proprioceptive reflexes.

An evaluation will be made of the means by which afferent signals are modified by central motor commands or by inputs from other receptors in movements involving crayfish abdominal muscle receptor organs and receptors of walking legs in several Crustacea. Both central and peripheral modifications of proprioceptive signals that alter the gain of a reflex will be considered.

Introduction

Crustaceans and other arthropods possess, as do most higher animals, proprioceptors that signal various aspects of movement and position. Some of these arthropod receptors lend themselves to electrical recording or ablation. They compare in complexity to vertebrate systems by their association with muscle, connective tissue and joints. Many of them are susceptible to modification by several different efferent control mechanisms. Although data are not yet available directly correlating arthropod motor control with vertebrate cortical and cerebellar systems for modification and control of movement, it is possible now to demonstrate a number of instances in which central pathways influence the effectiveness of proprioceptive inputs. Some of the means by which modification may occur are by inhibitory central gating of the sensory input, by efferent modulation of the receptor cells, or by efferent control of receptor muscles (small muscles whose contraction can alter the discharge of an associated proprioceptor).

It was once assumed that invertebrates were simple reflex animals, operating by a sequence of peripherally inhibited responses akin to the mammalian knee jerk. More recent studies of motor outputs of invertebrates suggest that these animals possess hard-wired central motor scores triggered by inputs to appropriate central neurons resulting in stereotyped sequences of muscle contraction (Wilson, 1961; Wiersma and Ikeda, 1964; Kennedy et al., 1966b; Willows, 1967). In spite of this apparently machine-like basic program for movement, evidence for considerable plasticity in responses of invertebrates has accumulated. These include such phenomena as habituation, learning and variability of response to sensory inputs. Proprioceptors that monitor the various aspects of movement and changes in muscle tension appear to modulate the central scores and possibly the basic pattern of movement. Also they can compensate for variability encountered during the movement, as in encountering a load. Although invertebrates have provided some simple systems that serve as models for neuronal organization (e.g., the cardiac and stomatogastric ganglia of Crustacea and the segmental ganglion of the leech), emphasis here is placed on aspects of control of movement and posture that provide a basis for comparison with vertebrate systems.

Invertebrate systems for control of movement must be subjected to the same sort of conceptual treatment as those of higher animals and they offer some very real advantages for study of the neural and sensory factors in locomotion. The greatest advantage is probably the reduced number of neurons that serve an individual function, so that morphologically and physiologically identifiable cells may be examined repeatedly in successive preparations. In several instances now, an investigator familiar with a particular system can accurately predict not only which neurons will be active but also their patterns of activity for a number of given moments. In several arthropod preparations, it is also possible to estimate the sensory information that accompanies the movement. Although a number of reflex responses may be described in partially or completely restrained preparations, there is generally less precise information about the role of this sensory information in modifying movements.

In arthropods, the most easily described reflex is the recruitment of motoneurons that innervate a muscle antagonistic to an applied movement, often called the resistance reflex (Eckert, 1959; Bush, 1962). In an alert, but stationary animal, applied movement of a leg or body segment will generally evoke distinct contractions of antagonistic muscles, although they are rarely as strong as those encountered during ongoing movements. But, what function does such a response serve in normal control of posture and movement? One suggestion comes from an hypothesis for compensation to loads that are encountered during centrally commanded movement or during change to a new angle of the segment. If a sense organ is present that can signal resistance to such a movement, the motor output that is responsible for achieving the movement can be boosted to overcome the obstacle via motor excitation. This idea has been discussed in some detail by Kennedy (1969) and tested in some of the preparations described here. In addition to serving as a detector for encountered loads during a movement, gain of some proprioceptors can be controlled independently via central control of their excitability. Several crustacean proprioceptors are coupled to independent contractile elements so that they can also serve as peripheral references for determination of a stopping point, or set position. When shortening of the contractile elements of a muscle receptor organ (MRO) is driven simultaneously with contraction of a parallel working muscle, the sensory signal is effectively nulled out unless a resistive load is encountered during a movement.

During some types of movements, it may be disadvantageous to the animal to respond to sensory factors that influence the motor output. One way to achieve insensitivity to modulating influences would be to reduce the sensory input. In other cases, it may be desirable to enhance the gain of a reflex by increasing activity in a sensory pathway. In an attempt to assess the various situations in which proprioceptive information is modified, a discussion follows on the modulation of proprioceptive information during a variety of movements in two systems - the abdomen and the walking legs of crustaceans.

Sensory and Motor Control of Abdominal Movements

Movement of the abdomen of lobsters and crayfish is effected by contractions of antagonistic flexor and extensor muscles that span one or more abdominal segments. Each of the antagonistic muscle groups can be divided into two types - slow and fast - on both physiological and morphological grounds and they are innervated by separate motoneuron populations. The fast flexors and extensors are responsible for the tail flip and make up the bulk of the abdominal musculature. These muscles develop rapid twitches in a more-or-less all-or-none manner.

During an escape response, backward swimming is achieved by either a single fast flexor response or by alternating contractions of the antagonistic muscles of the fast system and involves recruitment of activity in at least part of the population of motoneurons to the muscles of each half-segment.* Central nervous activation of fast flexors occurs both by way of the giant interneurons in escape responses, and by other interneuronal pathways in swimming (Schrameck, 1970). The former responses are subject to considerable habituation (Wine and Krasne, 1972). Much smaller antagonistic slow abdominal flexors and extensors are responsible for most of the fine postural adjustments of the abdominal segments. These

---

*The fast flexors in each abdominal segment are innervated by 9 to 11 excitatory neurons and a peripheral inhibitor. The fast extensors each receive a total of about 7 excitors with neuromuscular properties similar to the fast flexors, as well as one or two peripheral inhibitors (Parnas and Atwood, 1966).

muscles develop tension by summation of nonpropagated post synaptic responses from five excitatory motoneurons and a single inhibitory motoneuron to the muscles of each side of an abdominal segment. Suggested pathways for control over the postural system are intersegmental (command) interneurons and segmental reflexes, as summarized by Kennedy et al., (1966a) and Evoy and Cohen (1971).

A wide variety of segmental and extrasegmental sensory influences impinge upon these systems. The focus of the first part of this discussion will be on two types of receptors intimately associated with modulation of motor activity: the flexion-sensitive muscle receptor organs (dorsal MRO's) and the extension-sensitive cord stretch receptors (CSR's), found in the connective tissue of the ventral nerve cord. These two sensory systems evoke significant reflexes, and in conjunction with the central programs seem to provide much of the control for movements of the abdomen and for development of tension in the muscles under a variety of conditions.

Each abdominal segment possesses two pairs of MRO's, located on either side of the animal's dorsal midline. Each MRO spans a segment and has a single multipolar sensory neuron with its distortion sensitive dendrites inserted into the mid-region of the muscle bundle. The phasic MRO appears to be associated primarily with rapid abdominal movements; its sensory discharge adapts rapidly and has a higher threshold to stretch than that of the tonic MRO. The latter provides a reliable signal of either receptor muscle length or tension due to its slowly adapting discharge. Both MRO's are arranged mechanically so that they would tend to be unloaded mechanically by extension movements and to be activated by contraction of their own receptor muscle or by flexion movements (Figure 1).

Reflex connections of the MRO with the parallel extensors form part of a complex control system that appears to be significant in both movement and posture. One important reflex effect of tonic MRO activity is excitation of at least motoneuron #2 that innervates most of the fibers of the slow abdominal extensors of the same segment (Fields, 1966). Stretch of the phasic MRO has no effect on motor discharge to the slow abdominal extensors (Fields and Kennedy, 1965). MRO stimulation also does not appear to have any effect on motor output to the fast system in otherwise unstimulated preparations (Zucker, 1972a). However, the possibility remains that the phasic MRO input may summate with central excitation of the fast musculature during rapid movements.

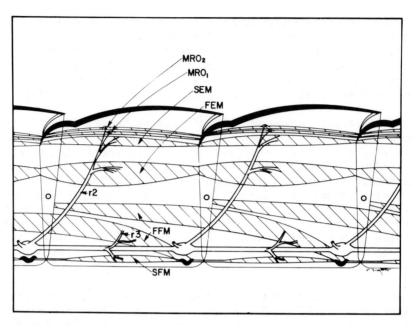

Figure 1. Approximate location of major muscle groups, nerves and the dorsal muscle receptor organs in abdominal segments of lobster and crayfish, diagrammed as a mid-sagittal section. Contraction of the extensor muscles straightens the abdomen unloading the receptor muscles; whereas contraction of the flexors produces a ventral curvature stretching the receptor muscles. The small circles represent the approximate location of the hinge points between the abdominal segments. FEM, fast extensor muscles; FFM, fast flexor muscles; MRO_1 and MRO_2, tonic and phasic muscle receptor organs; r_3, 3rd nerve root to flexors; r_2, 2nd nerve root to extensors and MRO's; SEM, slow extensor muscle; SEM, slow flexor muscles. Mechanical aspects of this system are discussed in Fields et al., (1967) and Rayner and Wiersma (1967).

Central Control of MRO Activity

Efferent modulation of extensor muscle contractions by motoneuron to the receptor muscle. Motor innervation of the receptor muscles provides one of the functional pathways for central nervous control over the MRO's (Kuffler, 1954). The receptor muscle of the tonic MRO in the crayfish, Procambarus clarkii receives excitatory innervation from at least one motoneuron (axon #4 of Fields et al., 1967) that also innervates some of the parallel slow extensors, although it produces relatively little tension in these fibers by comparison with other members of the motoneuron population.

Interneurons in the abdominal cord activate motoneuron #4 so that slack in the receptor muscle is taken up during extension movements. Thus, efferent control of the receptor muscle is likely to establish a set point via reafferent excitation of the extensors and is thought to serve to maintain the gain in load-detecting systems (Kennedy, 1969). The same pathway probably accounts for observations that tonic MRO discharge is often maintained during a centrally evoked extension movement and sometimes increases preceding an extension (Fields, 1966).

The receptor muscle of the phasic MRO shares motor innervation with both fast and slow extensor muscles (Fields and Kennedy, 1965). Although the functional significance of this pathway in normal movements has not been worked out, there is some reason to propose that excitability is maintained in the phasic MRO during relatively rapid extension of the abdominal segments. Fields and Kennedy (1965) describe an increase in the discharge of the phasic MRO along with the slow receptor when tactile stimuli that evoke extension are applied in a restrained preparation. If, as suggested earlier, there is summation of the phasic MRO inputs onto the fast extensor system, it is likely to be integrated over many cycles of the repeated tail flip in a manner similar to sensory maintenance of excitability in locust flight (Wilson and Gettrup, 1963). During the repeated tail flips of the escape response, the antagonistic flexors and extensors usually show single, closely phased bursts of electrical activity during each stroke of the cycle (Schrameck, 1970). Thus, it seems unlikely that the phasic MRO could play any role in response to encountered loads or determination of motoneuron excitability within a single cycle, but it may exert a modulating influence on subsequent cycles. Answers to these questions must await development of techniques for recording sensory and motor activity in peripheral abdominal nerves during rapid movements of a portion of the animal's body that includes a majority of the somatic musculature.

There is also an indication of control over MRO excitability by peripheral inhibitory endings on the receptor muscles described anatomically by Florey and Florey (1955) and Alexandrowicz (1951, 1967). These peripheral connections have not as yet been worked out physiologically due to difficulties in obtaining intracellular recordings from the receptor muscle.

Peripheral inhibition of the MRO's by efferent neurons that synapse onto the sensory neuron. The more well-studied pathway of inhibitory control is the direct inhibitory connection to the MRO sensory neurons, described by Kuffler and Eyzaguirre (1955). There appear to be species - specific differences in functional inhibitory control of the MRO. In the European freshwater crayfish, Astacus fluviatilis, there is evidence that the tonic MRO receptor cell receives inhibitory connections from three efferent fibers (Jansen et al., 1970a). However, in several species of the freshwater crayfish Procambarus endemic to regions of the southern United States, only a single accessory nerve (AN) has been demonstrated as functional. The situation is compounded further by morphological evidence (Nakajima et al., 1973) of axo-axonal connections onto peripheral inhibitory terminals of the tonic MRO of Procambarus sp. in a conformation that suggests presynaptic inhibition of one inhibitory synapse by another. If these terminals do represent endings of separate inhibitory fibers, the situation in American crayfish may be more similar to that in the European species.

Alexandrowicz (1967) has summarized the morphological and physiological evidence regarding innervation of MRO's of several crustacea in the most complete review of the functional connections of this system to date. Physiological and morphological studies have not yet settled the extent to which differences in peripheral innervation of sensory neurons and receptor muscles exist among species, but there is evidence from both approaches that the sensory neurons and the slow receptor muscle may share these inhibitory neurons (Alexandrowicz, 1967).

Central control of AN. Jansen et al., (1970b) demonstrate that two of the inhibitory efferent fibers are activated by MRO stretch; these are thought to correspond to the thick and thin accessory nerves of the lobster, Homarus vulgarus described by Alexandrowicz (1967). The third inhibitory fiber, activated by contralateral MRO's, is thought to be fiber "x" of Alexandrowicz that innervates the slow extensor muscles and RM, as well as both sensory neurons. Most of this section deals with the thick accessory nerve in the so-called accessory nerve reflex in which inhibition of receptor discharge is initiated by tonic MRO activity in the same or adjacent abdominal segments. However, the circuitry involved in this receptor system, at least in some of the animals used in the investigations reported here, is clearly more complex than the present treatment might indicate. Some of the better established aspects of the central and peripheral circuitry are diagrammed in Figure 2.

The most readily observed central nervous connection between tonic MRO activity and the accessory nerve is activation of AN upon stretch of the tonic receptor muscle, with a resulting suppression of activity in the primary sensory neuron (Eckert, 1961a). In some cases, the fast MRO also can contribute to activity in AN (Jansen et al., 1970c). Various proposals have been presented for this apparent self-inhibition, depending on the design of the experiments. Critical factors in interpretation seem to be the extent to which the abdomen is free to move and the conditions of resistance to movement that are encountered.

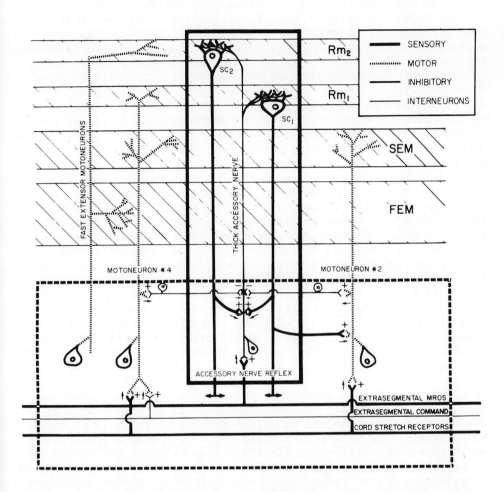

Figure 2. Summary of some of the established and inferred neural connections involved in reflex and central nervous control of movements of the crayfish abdomen, with emphasis on connections involving the dorsal MRO's discussed in the text. Excitatory synapses are represented by +; inhibitory synapses by − and arrows represent the direction of transmission. The portion within the heavy solid line represents both central and peripheral connections involved in peripheral inhibition of the MRO sensory neurons and the portion within the broken line represents connections within the abdominal ganglia. FEM, fast extensor muscle; Rm_1, Rm_2, receptor muscles of tonic and phasic MRO's; SC_1 and SC_2, sensory cells of tonic and phasic MRO's; SEM, slow extensor muscle.

A further complication in approaching the MRO-abdominal muscle control system of crayfish is that reflexes are not restricted to a single segment. Activation of the thick accessory nerve upon activation of MRO's in adjacent and particularly posterior segments has been reported by several investigators (Eckert, 1961a; Fields et al., 1967; Jansen et al., 1970c; Jansen et al., 1971; Page and Sokolove, 1972; Sokolove, 1973). Variations in experimental techniques and perhaps in species used result in some disagreement as to the rostrally directed nature of intersegmental spread of the MRO-AN reflex. Recordings from nerves to slow extensor muscles and MRO's when the abdomen moves freely show that suppression of sensory discharge during flexion movements is at least in part a result of intersegmental reflex driving of the thick accessory fiber by MRO activity from posterior segments (Page and Sokolove, 1972). During slow flexions of the abdomen, usually initiating in posterior segments, the MRO response to flexion movements is effectively blocked before it is stimulated. This suppression of the sensory discharge would prevent reflex excitation of the antagonistic extensors. However, during extension of the abdomen, the accessory fiber is not active and the MRO is active because its receptor muscle contracts in parallel with the slow extensors. The accessory fiber is not active during excitation of extensor motoneurons of the same ganglion due to a central inhibitory control over its activity (Sokolove, 1973). Antidromic stimulation of the smaller extensor excitatory motoneurons (Tatton and Sokolove, 1975) reveals that the accessory fiber is inhibited actively by extensor motoneuron discharge in the same segment. Although there is no direct evidence concerning the nature of the inhibitory cross-connection, an interneuron activated by motoneuron collaterals is suggested (Figure 2). During centrally commanded extension movements, the gain of the MRO-extensor loop will be maximized as it is released from peripheral inhibitory suppression via the accessory fiber reflex and can serve as a set device for determination of position and in the load compensating system proposed by Kennedy (1969). On the other hand, the accessory fiber functionally is synergistic with central activation of flexion movements (Sokolove and Tatton, 1975), so that any ability of the MRO to detect flexion probably will be suppressed during centrally commanded flexion movements. Thus, a situation exists in which central nervous connections convert an apparently simple mechanical system for detection of flexion movements into one that is functional primarily during extension, particularly when the movement is impeded.

High frequency bursts of slow receptor sensory discharge also produce a silent period in thick accessory nerve activity. Ilyinsky et al., (1974) suggests this results from a monosynaptic inhibitory reflex connection from the MRO. However, the presence of an inhibitory cross connection between one or more of the intermediate sized slow extensor excitatory motoneurons and the thick accessory nerve suggests a more complex route for suppression of AN. If the extensor motoneuron that is driven by MRO stretch (Fields et al., 1967) is responsible either partially or wholly for thick accessory nerve suppression, this could account for the apparent inhibition of accessory nerve activity during strong stretch of the MRO.

Other central influences on MRO excitability. Part of the apparently central setting of MRO excitability during extension movements may be due to positive feedback from extension-sensitive receptors in the nerve cord (the CSR's) (Hughes and Wiersma, 1960; Grobstein, 1973a, b). Activation of CSR's by stretch of the connective tissue sheath of the ventral nerve cord in posterior segments excites slow abdominal extensors and inhibits the antagonistic slow flexors. At the same time, the tonic MRO is excited via the excitatory motoneuron (Axon 4) to the slow receptor muscle. If it is functional during a centrally commanded extension movement, input from these receptors would maintain tension in the dorsal receptor muscles to insure their parallel contraction with the extensor muscles, thereby increasing the gain of the reflex during the movement. Although extension movements do not behave as though controlled by pure positive feedback that would result in regenerative increases in extensor tension, the CSR's are likely candidates for control of gain in the motor system.

During the rapid contractions of the fast flexors that produce the tail flip and backward swimming, both slow and fast extensor and receptor muscles are stretched. In its relaxed state, crustacean muscle is extremely plastic. In its active state while being stretched, muscle tissue and its attachments would be susceptible to damage. Likely adaptations to prevent stress on the dorsal musculature would be: 1) to provide central reciprocity between fast extensors and flexors, 2) to provide peripheral inhibition of extensor and receptor muscle contraction during rapid flexions, and 3) to suppress extensor excitation from the MRO's via the accessory nerve reflex. Eckert (1961b) gives evidence that the first and third mechanisms function during swimming in two crayfish species. Accessory nerve discharge is maintained until just before full flexion is reached during the tail flip. Direct evidence is lacking that the MRO is effectively silenced during the tail flip and the violent stretch that occurs may still override the inhibitory influence of AN. However, records from nerves to the abdominal muscles show a discharge of the slow extensor peripheral inhibitor preceding the tail flip, followed by a discharge in AN accompanying the rapid flexion (P.G. Sokolove, Personal communication). At present, it is not known whether AN activation originates centrally or is due to an intersegmental reflex. During repeated applied stretches of the tonic MRO, maintained bursts of AN occur, suggesting that any stretch of the receptor during fast flexion would result in self inhibition of that segment (Ilyinsky et al., 1974). At least part of the apparent suppression of motor output to the extensor system that occurs during the rapid flexion phase of the tail flip and swimming may be due to unloading of the cord stretch receptors so that any phasic or tonic excitation of the extensors is inactive (Fields, 1966). Grobstein (1973a) found ipsilateral inhibitory connections to the CSR's in the abdominal cord. Inputs that might evoke activity in this inhibitory pathway have not been found, but there is a suggestion that MRO and extensor excitability is affected indirectly at some phase of centrally command abdominal movements by central suppression óf the CSR's.

Central Modulation of Sensory Input During Rapid Abdominal Movements

In addition to modification of the load-and position-related sensitivity of the MRO and CSR, the effect of sensory input during rapid movements of the abdomen of crayfish can be varied at synapses onto interneurons that mediate inputs to the motor systems. Slow, graded movements associated with postural adjustments are better understood in terms of interactions with the peripheral reference system, while the phasic movements of the escape response (tail flip) have been studied primarily in terms of neuronal connections within the CNS. In the escape response, the large, intersegmental lateral and medial giant interneurons are pathways for excitation of the motoneurons that innervate the fast flexor muscles of the abdominal segments. Because these interneurons are large and are identified easily visually, they serve as excellent reference points for intracellular recording.

One pathway for eliciting the tail flip can be activated experimentally by stimulation of sensory hairs, referred to as tactile or touch receptors in most of the literature, that synapse chemically onto uni- and intersegmental first order interneurons. These interneurons in turn have excitatory, apparently electrical synapses onto the lateral giant interneurons (Zucker et al., 1971; Zucker, 1972a; Selverston and Remler, 1972). One of the motoneurons to the fast flexor muscles, the motor giant cell, is excited by an electrotonic junction with the lateral giant interneuron (Furshpan and Potter, 1959). The other eight motoneurons to the same muscle receive excitatory electrical synapses from the lateral giants as well as excitation from other interneurons that form an additional pathway for control of escape and swimming (Kennedy and Takeda, 1965; Kennedy et al., 1969; Schrameck, 1970; Larimer et al., 1971; Wine and Krasne, 1972). A parallel pathway of sensory inputs directly excites the lateral giant interneurons by excitatory electrical synapses that faithfully follow stimulation frequencies of up to 200 Hz with negligible delay (Zucker, 1972a).

In spite of the apparently all-or-none nature of the tail-flip response to certain sensory stimuli, the central pathway that mediates this reflex shows considerable habituation. This behavioral habituation is characterized by waning of the response or an apparent increase in its threshold upon repeated stimulation and the recovery of sensitivity following a rest period of 5 minutes or more (Krasne, 1969; Krasne and Woodsmall, 1969). Sensory inputs to the population of interneurons that form the less direct excitatory pathway to the lateral giant evoke antifacilitating, chemically mediated EPSP's in the first order interneurons (Kennedy, 1971; Zucker, 1972a). In a detailed study of transmission at the various synapses in this circuit, Zucker (1972b) provides direct evidence that the antifacilitation responsible for habituation of the tail flip is due primarily to a process in the presynaptic terminals of the tactile afferents. Upon repeated stimulation of afferent roots, the quantal content of the EPSP's recorded in the first order interneurons declines steadily, although there is some variation in the rate of antifacilitation at different synapses onto the three tactile interneurons. This central waning of efficacy of synaptic transmission closely matches the frequency-dependent behavioral habitation mentioned above. Thus there is a usage- and time-dependent mechanism built into the escape response central pathway in crayfish so that they will not respond to continuing sources of stimulation, but will initiate escape when there is sudden disturbance. Some of these connections are summarized in Figure 3.

The presence of depression of central excitability caused by repetitive stimulation of the sensory inputs is of adaptive value in the response when environmental disturbances continue. However, during the repeated tail-flips of backward swimming, which often bypasses the lateral giant interneurons, (Schrameck, 1970), or during several repeated threats that initiate the tail-flip, it would be disadvantageous for the animal to undergo severe depression of the central pathway due to activation of the tactile afferents by violent abdominal movements in water. A recurrent inhibitory pathway from the tail-flip motor system appears to operate by presynaptic inhibition at the tactile afferent-interneuron synapses, as demonstrated by Krasne and Bryan (1973). Inhibition of the lateral giant fiber response to tactile stimuli was maximally effective at 30-50 msec. following electrical stimulation of that interneuron (Wine et al., 1975). This inhibitory feedback has been called protection. It appears to block the synapses of the tactile afferents onto the 1st order interneurons so that antifacilitation is reduced during the tail flip or swimming. The result is that subsequent stimuli to the sensory system in a quiescent animal can initiate escape responses. Direct evidence for a presynaptic inhibitory mechanism has been obtained by intracellular recording of a depolarization in the afferent fibers that corresponds to

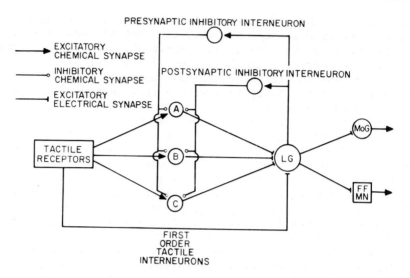

Figure 3. Central synaptic connections involved in the tail-flip response of crayfish. Only the connections involved in modulation of the sensory inputs are included, and no attempt is made to account for the temporal factors involved. The important feature is that inputs from the tactile receptors, or sensory hairs, can be inhibited pre- and post-synaptically by activity in the lateral giant interneuron (LG) at points of convergence on tactile interneurons A, B and C. FFMN and MoG are the non-giant and giant motoneurons to the fast flexor muscles, the final pathways of this network. (Adapted from Zucker, 1972a and Krasne and Bryan, 1973).

inhibition of the interneuronal response in a manner similar to primary afferent depolarization in the dorsal root fibers of the vertebrate spinal cord (Kennedy et al., 1974). During the course of the primary afferent depolarization, there is a marked shunting of conducted impulses in the tactile afferents. In this case at least, presynaptic inhibition appears to operate by a conductance increase in the presynaptic terminals of the sensory fibers near the region of their synapses onto the first order interneurons. Presynaptic inhibition is evoked by activity in the lateral giant interneurons, and is apparently mediated by inhibitory internuncials.

In addition to the presynaptic protection pathways there is also a short-term inhibitory feedback from the tail-flip motor system (pathway 2 of Krasne and Bryan, 1973), that is due to postsynaptic inhibition of the tactile interneurons. Postsynaptic inhibitory feedback might serve as a phasic cut-off for the fast flexor motor system, but does not appear to enter into control of behavioral habitation which may last as long as two hours.

Physiological and morphological probing of this system has revealed at least one example of modulation of sensory effects by inhibitory feedback from central elements during a behaviorally significant although highly specialized movement. The presence of identifiable cells favorable for intracellular recording helped detect these pathways. Analogous mechanisms probably exist in other pathways involved in posture and locomotion of crustaceans. Habitation also occurs in sensory pathways onto the fast extensors which are involved in the tail flip and in the swimming. There is no evidence regarding possible modifying influences at these synapses (Treistman and Remler, 1974).

MODIFICATION OF REFLEXES IN WALKING LEGS

Locomotion and Postural Changes

Proprioception in the five pairs of limbs of a decapod crustacean is a function of several types of receptor. These include joint receptors that consist of elastic tissue and sensory cells that respond to joint movements and changes in position; several types of muscle receptor and sensory cells associated with major muscle tendons that respond to evoked tension and cuticular receptors that respond to shearing stresses. The majority of investigations dealing with central nervous effects of activation of these proprioceptors have dealt primarily with motor outputs initiated by mechanical stimulation of the receptors. In most of these studies, suggestions have been presented about the roles of these reflexes in control or modification of movement. Many of these ideas have been treated in considerably more detail than in the present discussion (Evoy and Cohen, 1971). Information regarding several major aspects of integration of these reflexes is lacking. One problem is that little is known about the central nervous connections of the sensory input with the motor systems. The sensory input is composite and consists of individual sensory cells that respond to different aspects of rate and direction of movements and changes in position of the leg joints. The motoneurons are diffuse in morphology with many fine fiber branches that extend over the surface of the neuropil in the ganglion (Wilson and Sherman, 1975; Hofmann and Evoy, Unpublished observations). Therefore, they are not as amenable to intracellular recording as the larger motor and interneurons discussed in previous sections. The clean resistance reflexes observed upon moving a joint or receptor of a restrained animal do not serve an obvious behavioral function. Although there is a tendency for these responses to return the joints to a resting angle if they are disturbed in some way, it is also possible that they are utilized in maintenance of normal walking and posture in the presence of factors that impede movements. The majority of the proprioceptors are linked mechanically with joint movements, although one recently described type, the tendon organs (MacMillan and Dando, 1972) are associated with the major muscle apodemes and respond to changes in active muscle tension. Most joint receptors do not seem to have specific efferent connections that could alter their excitability during locomotion and therefore must provide a reliable signal of joint movements during walking or postural adjustments. However, a receptor that in some animals appears to respond reliably to joint movements is affected by contraction of a specialized muscle in other species (Clarac and Vedel, 1971).

In most cases, any ongoing movement would seem to be opposed by the resistance reflex if the central reflex connections remain functional during centrally initiated movements. One possible function of these reflexes is to counteract forces that oppose a centrally commanded movement. Mechanical opposition to an ongoing movement at distal walking leg joints of a crab, Cardisoma guanhumi, resulted in increased excitatory activity (Barnes et al., 1972). Upon consideration of these factors, it can be proposed that the resistance reflexes are inactive during centrally initiated movements that do not meet any appreciable resistance. In the studies above on walking, central inhibition of proprioceptive inputs to the walking leg motor system was favored. The possibility does exist that the relatively weak resistance reflexes are overridden by the central program. A model for central inhibition of resistance reflexes during activity of a hypothesized central oscillator that drives the motoneurons to the leg has been proposed by Barnes (1975). However, in studies on resistance reflexes of locust walking legs, Burrows and Horridge (1974) argue that the low gain of the reflex easily could be masked by more intense excitatory inputs to the motor pathway. Resistance reflexes also have been studied in relation to complex movements that occur during rhythmic cleaning movements initiated by stimulation of chemoreceptors in a hermit crab (Field, 1974a, b). In this case, the resistance reflex does not seem to be active during the movements. It is precisely opposite to the responses recorded in the movements and again inhibition of the reflex pathway is proposed. In the studies mentioned, there is a suggestion that the reflexes are involved in some sort of modulatory control of the motor program to adjust to conditions of load, tension and position.

A second major aspect of motor and sensory control of limb movements is the nature of the central nervous pathways that initiate and maintain coordinated muscle tension at the several joints. Considerable attention has been paid to the role of intersegmental interneurons, which when stimulated electrically, evoke precise tonic or rhythmic movements of the limbs (Wiersma, 1952; Atwood and Wiersma, 1967; Bowerman and Larimer, 1974a, b). Although excitation of these pathways always evokes changes in motor output, normally phased muscle contractions occur only when the proprioceptive signal is appropriate, such as when the legs can contact the ground. This observation further corroborates the suggestion that there is a complex interaction between central or extrinsically initiated programs for movements and proprioceptive feedback from the movements.

An important aspect of the interaction between reflex and centrally or extrasegmentally initiated motor output is the nature of the neuronal connections in the resistance reflex. On the basis of input-output relations between imposed movements and both the responses in the motoneurons and the close coupling of activity in several motoneurons to distal joints of crab walking legs, Spirito et al., (1972) suggested that at least one pathway of connection between joint receptors and motoneurons is via interneurons that summate these inputs. Estimates of central delays for the resistance reflexes mostly are based on measures of total reflex latency between sensory discharge and peripheral motor response recorded in efferent fibers or muscle. The composite input from the joint receptors makes it difficult to obtain a discrete starting point for the sensory input. The few successful intracellular recordings from limb motoneurons obtained to date show complex postsynaptic potentials and small, electronically conducted spike components (Field, 1974a). These recordings are probably from penetrations of fiber processes in neuropil, although they are not dissimilar from soma recordings in insect ganglia (Hoyle and Burrows, 1973). Allowing for 8-10 msec of EPSP summation before spike initiation, synaptic latencies of 4-6 msec were calculated in the resistance reflex of the fiddler crab cheliped. The several sources of inaccuracy in these measurements allow for anywhere from one to several synapses between sensory terminals and motoneurons but the measurements do not provide an adequate test of models based on earlier studies.

Although sensory signals from complex movement- and position-sensitive joint receptors of crustacean walking legs are probably the most reliable signal of ongoing movements, there seems to be considerable lability in the response of the final motor pathways to proprioceptive inputs. Centrally commanded muscular contractions evoke movements that are opposed directly by reflexes initiated by the joint receptors. The movements do persist in a regular and coordinated manner unless the movement is opposed mechanically and compensatory changes are seen. It is tempting to suggest the presence of a highly sophisticated comparator system, in which the sensory signal is neutral when it matches a centrally determined template and compensates for differences between the command and the resulting movement (Hoyle, 1964; Barnes et al., 1972). The best analogy to a voluntary movement in a crayfish, lobster or crab is initiation of walking or swimming or an observable change in posture initiated by visual, tactile or chemosensory stimulation, possibly mediated by command interneurons. Visual inputs initiate the defense posture: chelae raised, walking legs depressed and abdomen extended; this response is mimicked by stimulation of selected fibers in the circumoesophageal connectives (Wiersma, 1952; Bowerman and Larimer, 1974a). Olfactory inputs initiate the cleaning reflex in hermit crabs (Field, 1974a), equilibrium and visual receptors modify motor output to the walking legs of several animals (Roye, 1972; Davis and Ayers, 1972). Although patterns of connectivity between these several sources of initiation of motor response are unknown, the complex inputs to interneurons with connections to selected motor outputs (Kennedy, 1971) suggest the presence of wired in pathways for discrete responses. Commands that coordinate discrete movements could also interact with proprioceptive feedback in several ways. If the joint receptors connect monosynaptically with limb motoneurons, sensory inputs could be inhibited presynaptically as in the case of the tactile afferents to the fast flexion system. Interposition of one or more interneurons between proprioceptive inputs and the final motor output provides additional sites for inhibitory override of the reflex. Available physiological and morphological evidence does not allow a choice between these possibilities.

CENTRAL INFLUENCE OVER OTHER PROPRIOCEPTIVE PATHWAYS IN WALKING LEGS

Muscle Receptor Organs

Two proprioceptive organs in crustacean walking legs consist of movement- or distortion-sensitive sensory cells that are linked mechanically with small contractile elements separate from the major musculature of the limb segment. In a number of studies on morphology, sensory responses, reflex connections and alterations of sensory discharge by receptor muscle contraction, similarities with the dorsal MRO as well as with other muscle receptors, including vertebrate spindles have been considered. Minor modifications in walking or posture have been detected in several instances following removal or disturbance of these muscle receptors, but no clear function has been attributed to them. Details of established or suggested function of these muscle receptors will not be discussed here. Rather, only possible functions of central control over receptor excitability mediated by motor control of the receptor muscles relevant to the present discussion will be considered.

The myochordotonal organs of the meropodite of the walking legs and chelipeds consist of several populations of sensory cells attached by connective tissue to the receptor muscle (accessory flexor) which is arranged mechanically in parallel with the main flexor muscle of the same segment. Components of the sensory population respond to flexion and extension of the next distal joint, vibration and to contraction of the receptor muscle (Cohen, 1963, 1965; Clarac and Vedel, 1971; Horch, 1971). Reflex effects of myochordotonal organ stimulation appear in the motor outflow to the main flexor and extensor muscles and also in feedback to the receptor muscle (Evoy and Cohen, 1969).

The mechanical linkage of the receptor muscle is complex; there are two portions of the muscle, the proximal and the larger distal head. These muscular components are linked by a small tendon that inserts onto the apodeme of the main flexor muscle. Contraction of the distal head pulls the proximal structures distally with sensory and reflex effects similar to joint extension. The role of the proximal head of the receptor muscle remains obscure; it receives the same innervation as the distal head and apparently contracts in series with it. Excitation of the proximal head may simply serve to assure the linkage of distal head contraction to the receptor cells by a concomitant increase in muscle stiffness. The receptor muscle is innervated by an excitatory and an inhibitory axon; the excitatory axon is independent of the innervation to the mechanically parallel M-C flexor muscle while the inhibitory axons to the receptor muscle and antagonistic extensor are shared (Wiersma, 1961; Cohen, 1963; Dorai Raj et al., 1964; Evoy and Cohen, 1969; Angaut-Petit et al., 1974). Certain tactile stimuli evoke discharge in the shared extensor-receptor muscle inhibitor of the crayfish cheliped, often accompanied by driving of main flexor motoneurons (Angaut-Petit, Personal communication). This finding suggests that there are central pathways that suppress myochordotonal organ sensitivity during certain flexion movements, assuming that the peripheral inhibitory branch to the receptor muscle is effective in decreasing the contractile response of this muscle.

Contraction of the receptor muscle in crab walking legs mimics the effects of joint extension in the resistance reflex. This would appear to indicate a role in a length-detection servo similar to the load-compensating system suggested for the dorsal MRO. During ongoing walking movements, myograms of muscle activity suggest that the myochordotonal organ receptor muscle contracts simultaneously with the main flexor (Evoy and Cohen, 1971). The excitatory pathways to these two muscles are excited in near synchrony during resistance reflexes and tactile reflexes, which suggests common coupling in the central nervous system (Angaut-Petit et al., 1974). Proprioceptive inputs from a more proximal joint in the walking leg also evoke excitation of the flexor and receptor muscle, although the excitatory motoneuron to the receptor muscle is more sensitive to these extrasegmental influences (Bush and Clarac, 1975).

While it has been suggested in a number of cases that the myochordotonal organ could be the detector element in a load compensating system, observations to date have failed to support this hypothesis. In one animal (Cancer magister) removal or manipulation of the myochordotonal system resulted only in changes in resting limb angles (Evoy and Cohen, 1971); in Cardisoma guanhumi similar manipulations resulted in changes in the amplitude of joint movements during walking but did not provide clear evidence of a role in response to load (Fourtner and Evoy, 1973). This system, which has the appropriate reflex and efferent connections to serve a very specific control function, has not yet been shown to fit any of the currently available models. An exploration of central pathways that could influence selectively the sensory signal in the myochordotonal organ may reveal significant functions. Studies to date of reflex and ongoing activity have failed to provide direct evidence for central control over this muscle receptor during limb movements.

A second system that has functional connections for efferent control over reflex excitability is the coxal MRO, a muscle receptor located proximally in legs of decapod crustaceans in parallel with muscles that produce the forward-directed stroke (promotion), (Alexandrowicz and Whitear, 1957; Ripley et al., 1968). The resistance reflex initiated by stretch of the receptor muscle of this MRO is mediated by a graded, electronically conducted generator potential in two large sensory fibers and evokes increased motor discharge to the promotor musculature (Bush and Roberts, 1968; Roberts and Bush, 1971). In crayfish, the receptor muscle receives motor innervation that apparently is not shared with the innervation of other muscles; discharge in the motor fiber to the receptor muscle is inhibited by stretch of the MRO (Moody, 1970, 1972). A definitive role for this muscle receptor system has not been suggested. There is no direct evidence that the central nervous system can alter the sensitivity of the receptors via the efferent pathway to the receptor muscle, although activity in the receptor muscle nerve is affected slightly by sensory inputs from distal joints.

Tendon Receptor Organs

Sensory cells associated with the tendons of major muscles have been found in a number of crustacean walking leg segments (MacMillan and Dando, 1972). Although they do not appear to possess independent neural connections that would allow the central nervous system to vary their excitability, they are arranged so that increases in muscle tension transmitted via the apodeme bring about increases in sensory discharge. Also, joint movements that must oppose a load resisting the centrally evoked movement result in increased sensory discharge from these receptors proportional to the total tension in the tendon (Hartman, Personal communication). Lobsters and crayfish show changes in motor output to walking leg muscles that are related to the gravitational load they encounter during stepping and that are related to dragged weights that impede forward walking (MacMillan, 1975 and Grote, In preparation). The manner in which the tendon receptors integrate centrally initiated increases in muscle tension with resistance to the movement suggests that these receptors are situated ideally to evoke compensatory motor discharge that will overcome loads encountered during leg movements. However, the reflexes evoked by the tendon organs are complex and appear to reflect several components of the sensory population that may serve very different control functions. Some components of the tendon organ afferent input evoke a central inhibitory effect on motor output to the muscle that is contracting, whereas other components have excitatory feedback (Clarac and Dando, 1973). Integration of tendon receptor inputs with other controls of walking leg motor output remains to be worked out.

Cuticular Stress Detectors

Another receptor system that is linked mechanically to muscular contraction as well as to sources of load consists of sensory cells associated with soft cuticle in the proximally located basi-ischiopodite of several decapod crustacean walking legs. This system was first described by Wales et al., (1971). These stress detectors feed back positively to levator muscles that participate in posture, walking and in the autotomy response (limb loss due to specific sequences of muscular contraction) (Clarac et al., 1971; McVean, 1974; Moffett, 1975). The same receptors also bring about excitation of more distal muscles of the same walking leg (Angaut-Petit et al., 1974) including the receptor muscle of the myochordotonal organ discussed above. Thus, similar to the tendon receptors, these receptors integrate functions of the centrally commanded movement and external factors that affect the movement. They may play a role in sensing contact with the ground during walking.

Conclusions

Crustacean movements involve interactions between central commands for coordinated motor output and the proprioceptive and other sensory signals that provide inputs for initiation or modification of the movement. In only two of the several cases examined here is there sufficient evidence regarding the central nervous connectivity to propose summary models of interactions between central and sensory factors. In the connections for control of the dorsal muscle receptor of the abdomen, there is clear-cut evidence for central regulation of the reflexes by way of peripheral inhibitory and excitatory connections to the receptor. In the case of initiation of activity in the fast flexors of the abdomen during escape and swimming by afferent signals, the well defined inhibitory gating of the central pathway suggests functions for control of excitability in this pathway relevant to the observed behavior of the animals.

In some of the other systems under study there are indications of pathways for modulation of receptor influences, but there are insufficient physiological and morphological data to assemble complete summaries of connectivity and function. In none of the crustacean examples is there sufficient correlation of the function of these control systems with ongoing behavior in unrestrained animals to provide definitive conclusions about the roles of the systems in normal and varied locomotor situations. All of the systems have proven more complex than originally anticipated, but at least identification of most of the major components involved has begun. Basic features of these systems that are shared with the more intensively studied vertebrate locomotor and postural systems will continue to be examined. The goal is to establish general principles of the mechanisms that have evolved for control and modulation of locomotion.

Acknowledgment

I thank Drs. J.D. Marrelli, P.G. Sokolove and Donald Kennedy for suggestions and a critical reading of the manuscript. The figures were prepared with the assistance of Christy Duquette. Original research has been supported by grants GB 8847 and GB 30605 from the National Science Foundation.

REFERENCES

Alexandrowicz, J.S., (1951) Muscle receptor organs in the abdomen of Homarus vulgaris and Palinurus vulgaris. Quart. J. Micr. Sci. 92, 163-199.

Alexandrowicz, J.S. and Whitear, M., (1957) Receptor elements in the coxal region of decapod Crustacea. J. Mar. Biol. Assoc. U.K. 36, 603-628.

Alexandrowicz, J.S., (1967) Receptor organs in thoracic and abdominal muscles of Crustacea. Biol. Rev. 42, 288-326.

Angaut-Petit, D., Clarac, F. and Vedel, F.P., (1974) Excitatory and inhibitory innervation of a crustacean muscle associated with a sensory organ. Brain Res. 70, 148-152.

Atwood, H.L. and Wiersma, C.A.G., (1967) Command interneurons in the crayfish central nervous system. J. Exp. Biol. 46, 249-261.

Barnes, W.J.P., Spirito, C.P. and Evoy, W.H., (1972) Nervous control of walking in the crab Cardisoma guanhumi. II. Role of resistance reflexes in walking. Z. vergl. Physiol. 76, 16-31.

Barnes, W.J.P., (1975) "Nervous control of locomotion in Crustacea," In Simple Nervous Systems. (Usherwood, P.N.R. and Newth, D.R., eds.), Arnolds, London, (415-441).

Bowerman, R.F. and Larimer, J.L., (1974a) Command fibers in the circumoesophageal connectives of crayfish. I. Tonic fibers. J. Exp. Biol. 60, 95-117.

Bowerman, R.F. and Larimer, J.L., (1974b) Command fibers in the circumoesophageal connectives of crayfish. II. Phasic fibers. J. Exp. Biol. 60, 119-134.

Burrows, M. and Horridge, G.A., (1974) The organization of inputs to motoneurons of the locust metathoracic leg. Phil. Trans. R. Soc. London. B. 269, 49-94.

Bush, B.M.H., (1962) Proprioceptive reflexes in the legs of Carcinus maenas (L.). J. Exp. Biol. 39, 89-106.

Bush, B.M.H. and Roberts, A., (1968) Resistance reflexes from a crab muscle receptor. Nature. 218, 1171-1173.

Bush, B.M.H. and Clarac, F., (1975) Intersegmental reflex excitation of leg muscles and myochordotonal efferents in decapod Crustacea. J. Physiol. 246, 58P-60P.

Clarac, F., Wales, W. and Laverack, M.S., (1971) Stress detection at the autotomy plane in the decapod Crustacea. II. The function of the receptors associated with the cuticle of the basi-ischiopodite. Z. vergl. Physiol. 73, 383-407.

Clarac, F. and Vedel, J.P., (1971) Etude des relations fonctionelles entre le muscle flechisseur accessoire et les organes sensorielles chordotonaux des appendices locomoteurs de la langouste Palinurus vulgaris. Z. vergl. Physiol. 72, 386-410.

Clarac, F. and Dando, M.R., (1973) Tension receptor reflexes in the walking legs of the crab Cancer pagurus. Nature. 243, 94-95.

Cohen, M.J., (1963) The crustacean myochordotonal organ as a proprioceptive system. Comp. Biochem. Physiol. 8, 223-243.

Cohen, M.J., (1965) The dual role of sensory systems: Detection and setting central excitability. Cold Spring Hbr. Symp. Quant. Biol. 30, 587-599.

Dando, M.R. and MacMillan, D.L., (1973) Tendon organs and tendon organ reflexes in decapod Crustacea. J. Physiol. 234, 52-53P.

Davis, W.J. and Ayers, J.L. Jr., (1972) Locomotion: Control by positive-feedback optokinetic responses. Science. 177, 183-185.

Dorai Raj, B.S. and Cohen, M.J., (1964) Structural and functional correlations in crab muscle fibers. Naturwiss. 9, 224-225.

Eckert, B., (1959) Uber das Zusammenwirken des erregenden und des hemmenden Neurons des M. abductor der Krebsschere beim Ablauf von Reflexen des myotätischen Typus. Z. vergl. Physiol. 41, 500-526.

Eckert, R.O., (1961a) Reflex relationships of the abdominal stretch receptors of a crayfish. I. Feedback inhibition of the receptors. J. Cell. Comp. Physiol. 57, 149-162.

Eckert, R.O., (1961b) Reflex relationships of the abdominal stretch receptors of a crayfish. II. Stretch receptor involvement during the swimming reflex. J. Cell. Comp. Physiol. 57, 163-174.

Evoy, W.H. and Cohen, M.J., (1969) Sensory and motor interaction in the locomotor reflexes of crabs. J. Exp. Biol. 51, 151-169.

Evoy, W.H. and Cohen, M.J., (1971) "Central and peripheral control of arthropod movements," In Advances in Comparative Physiology and Biochemistry. Vol. 4, (Lowenstein, O., ed.), Academic Press, N.Y., (225-266).

Field, L.H., (1974a) Sensory and reflex physiology underlying cheliped flexion behavior in hermit crabs. J. Comp. Physiol. 92, 397-414.

Field, L.H., (1974b) Neuromuscular correlates of rhythmical cheliped flexion behavior in hermit crabs. J. Comp. Physiol. 92, 415-441.

Fields, H.L. and Kennedy, D., (1965) Functional role of muscle receptor organs in crayfish. Nature. 206, 1235-1237.

Fields, H.L., (1966) Proprioceptive control of posture in the crayfish abdomen. J. Exp. Biol. 44, 455-458.

Fields, H.L., Evoy, W.H. and Kennedy, D., (1967) Reflex role played by efferent control of an invertebrate stretch receptor. J. Neurophysiol. 30, 859-874.

Florey, E. and Florey, E., (1955) Microanatomy of the abdominal stretch receptors of the crayfish (Astacus fluviatilis L.). J. Gen. Physiol. 39, 69-85.

Fourtner, C.R. and Evoy, W.H., (1973) Nervous control of walking in the crab, Cardisoma guanhumi. IV. Effects of myochordotonal organ ablation. J. Comp. Physiol. 83, 319-329.

Furshpan, E.J. and Potter, D.D., (1959) Slow post-synaptic potentials recorded from the giant motor fibre of the crayfish. J. Physiol. 145, 326-335.

Grobstein, P., (1973a) Extension-sensitivity in the crayfish abdomen. I. Neurons monitoring nerve cord length. J. Comp. Physiol. 86, 331-348.

Grobstein, P., (1973b) Extension-sensitivity in the crayfish abdomen. II. The tonic cord stretch reflex. J. Comp. Physiol. 86, 349-358.

Horch, K., (1971) An organ for hearing and vibration sense in the ghost crab, Ocypode. Z. vergl. Physiol. 73, 1-21.

Hoyle, G., (1964) "Exploration of neuronal mechanisms underlying behavior in insects," In Neural Theory and Modelling. (Reiss, R.F., ed.), Stanford University Press, California, (346-376).

Hoyle, G. and Burrows, M., (1973) Neural mechanisms underlying behaviour in the locust Schistocerca gregaria 1. Physiology of identified motoneurons in the metathoracic ganglion. J. Neurobiol. 4, 3-41.

Hughes, G.M. and Wiersma, C.A.G., (1960) Neuronal pathways and synaptic connections in the abdominal cord of the crayfish. J. Exp. Biol. 37, 291-307.

Ilyinsky, O.B., Spivachenko, D.L. and Shtirbu, E.I., (1974) Efferent regulation of the abdominal stretch receptors of the crayfish. J. Exp. Biol. 61, 781-798.

Jansen, J.K.S., Njå, A., Ormstad, K. and Walløe, L., (1970a) IPSPs in the slowly adapting stretch receptor of the crayfish. Acta Physiol. Scand. 79, 14A-15A.

Jansen, J.K.S., Njå, A. and Walløe, L., (1970b) Inhibitory control of the abdominal stretch receptors of the crayfish. I. The existence of a double inhibitory feedback. Acta Physiol. Scand. 80, 420-425.

Jansen, J.K.S., Njå, A. and Walløe, L., (1970c) Inhibitory control of the abdominal stretch receptors of the crayfish. II. Reflex input segmental distribution and output relations. Acta Physiol. Scand. 80, 443-449.

Jansen, J.K.S., Njå, A., Ormstad, K. and Walløe, L., (1971) Inhibitory control of the abdominal stretch receptors of the crayfish. III. The accessory reflex as a recurrent inhibitory feedback. Acta Physiol. Scand. 81, 472-483.

Kennedy, D. and Takeda, K., (1965) Reflex control of abdominal muscles in the crayfish. I. The twitch system. J. Exp. Biol. 43, 211-227.

Kennedy, D., Evoy, W.H. and Fields, H.L., (1966a) The unit basis of some crustacean reflexes. Symp. Soc. Exp. Biol. 20, 75-109.

Kennedy, D., Evoy, W.H. and Hanawalt, J.T., (1966b) The release of coordinated behavior in crayfish by single central neurons. Science. 154, 917-919.

Kennedy, D., (1969) "The control of output by central neurons," In The Interneuron. (Brazier, M.A.B., ed.), UCLA Forum in Medical Sciences No. 11. University of California Press, California, (21-36).

Kennedy, D., Selverston, A.I. and Remler, M.P., (1969) An analysis of restricted neural networks. Science. 164, 1488-1496.

Kennedy, D., (1971) Crayfish Interneurons. Physiologist. 14, 5-30.

Kennedy, D., Calabrese, R.L. and Wine, J.J., (1974) Presynaptic inhibition: Primary afferent depolarization in crayfish neurons. Science. 186, 451-454.

Krasne, F.B., (1969) Excitation and habituation of the crayfish escape reflex: The depolarization response in lateral giant fibers of the isolated abdomen. J. Exp. Biol. 50, 29-46.

Krasne, F.B. and Woodsmall, K.S., (1969) Waning of the crayfish escape response as a result of repeated stimulation. Anim. Behav. 17, 416-424.

Krasne, F.B. and Bryan, J.S., (1973) Habituation: Regulation through presynaptic inhibition. Science. 182, 590-592.

Kuffler, S., (1954) Mechanisms of activation and motor control of stretch receptors in lobster and crayfish. J. Neurophysiol. 17, 558-574.

Kuffler, S. and Eyzaguirre, C., (1955) Synaptic inhibition in an isolated nerve cell. J. Gen. Physiol. 39, 155-184.

Larimer, J.L., Eggleston, A.C., Masukawa, L.M. and Kennedy, D., (1971) The different connections and motor outputs of lateral and medial giant fibers in the crayfish. J. Exp. Biol. 54, 391-402.

MacMillan, D.L. and Dando, M.R., (1972) Tension receptors on the apodemes of muscles in the walking legs of the crab, Cancer magister. Mar. Behav. Physiol. 1, 185-208.

MacMillan, D.L., (1975) A physiological analysis of walking in the American lobster (Homarus americanus). Phil. Trans. R. Soc. B. 270, 1-59.

McVean, A., (1974) The nervous control of autotomy in Carcinus maenas. M. Exp. Biol. 60, 423-436.

Moffett, S., (1975) Motor patterns and structural interactions of basi-ischiopodite levator muscles in routine limb elevation and production of autotomy in the land crab, Cardisoma guanhumi. J. Comp. Physiol. 96A, 285-305.

Moody, C.J., (1970) A proximally directed intersegmental reflex in a walking leg of the crayfish. Amer. Zool. 10, 501.

Moody, C.J., (1972) Some aspects of the reflex organization of a crustacean limb. Ph.D. Dissertation, University of Miami.

Nakajima, Y., Tisdale, A.D. and Henkart, M.P., (1973) Presynaptic inhibition at inhibitory nerve terminals. A new synapse in the crayfish stretch receptor. Proc. Nat. Acad. Sci. 70, 2462-2466.

Page, C.H. and Sokolove, P.G., (1972) Crayfish muscle receptor organ: role in regulation of postural flexion. Science. 175, 647-650.

Parnas, I. and Atwood, H.L., (1966) Phasic and tonic neuromuscular systems in the abdominal extensor muscles of the crayfish and rock lobster. Comp. Biochem. Physiol. 18, 701-723.

Rayner, M.D. and Wiersma, C.A.G., (1967) Mechanisms of the crayfish tail flick. Nature. 213, 1231-1233.

Ripley, S.H., Bush, B.M.H. and Roberts, A., (1968) Crab muscle receptor which responds without impulses. Nature. 218, 1170-1171.

Roberts, A. and Bush, B.M.H., (1971) Coxal muscle receptors in the crab: The receptor current and some properties of the receptor nerve fibers. J. Exp. Biol. 54, 515-524.

Roye, D.B., (1972) Evoked activity in the nervous system of Callinectes sapidus following phasic excitation of the statocysts. Experientia. 28, 1307-1309.

Schrameck, J.E., (1970) Crayfish swimming: alternating motor output and giant fiber activity. Science. 169, 698-700.

Selverston, A.I. and Remler, M.P., (1972) Neural geometry and activation of crayfish fast flexor motoneurons. J. Neurophysiol. 35, 797-814.

Sokolove, P.G., (1973) Crayfish stretch receptor and motor unit behavior during abdominal extensions. J. Comp. Physiol. 84, 251-266.

Sokolove, P.G. and Tatton, W.G., (1975) Analysis of postural motoneuron activity in crayfish abdomen. I. Coordination by premotor connections. J. Neurophysiol. 38, 313-331.

Spirito, C.P., Evoy, W.H. and Barnes, W.J.P., (1972) Nervous control of walking in the crab, Cardisoma guanhumi. I. Characteristics of resistance reflexes. Z. vergl. Physiol. 76, 1-15.

Tatton, W.G. and Sokolove, P.G., (1975) Analysis of postural motoneuron activity in crayfish abdomen. II. Coordination by excitatory and inhibitory connections between motoneurons. J. Neurophysiol. 38, 332-346.

Treistman, S.N. and Remler, M.P., (1974) Antifacilitating and simple following responses in a single motoneuron. J. Neurobiol. 5, 581-584.

Wales, W., Clarac, F. and Laverack, M.S., (1971) Stress detection at the autotomy plane in decapod Crustacea. I. Comparative anatomy of the receptors of the basi-ischiopodite region. Z. vergl. Physiol. 73, 357-382.

Wiersma, C.A.G., (1952) Neurons of arthropods. Cold Spring Hbr. Symp. Quant. Biol. 17, 155-163.

Wiersma, C.A.G., (1961) "The neuromuscular system," In The Physiology of Crustacea, Vol. II, (Waterman, T.H., ed.), Academic Press, New York, (191-240).

Wiersma, C.A.G. and Ikeda, K., (1964) Interneurons commanding swimmeret movements in the crayfish, Procambarus clarkii (Girard). Comp. Biochem. Physiol. 12, 509-525.

Willows, A.O.D., (1967) Behavioral acts elicited by stimulation of single, identifiable brain cells. Science. 157, 570-574.

Wilson, A.H. and Sherman, R.G., (1975) Mapping of neuron somata in the thoracic nerve cord of the lobster using cobalt chloride. Comp. Biochem. Physiol. 50A, 47-50.

Wilson, D.M., (1961) The central nervous control of flight in a locust. J. Exp. Biol. 38, 471-490.

Wilson, D.M. and Gettrup, E., (1963) A stretch reflex controlling wingbeat frequency in grasshoppers. J. Exp. Biol. 40, 171-185.

Wine, J.J. and Krasne, F.B., (1972) The organization of escape behavior in the crayfish. J. Exp. Biol. 56, 1-18.

Wine, J.J., Krasne, F.B. and Chen, L., (1975) Habituation and inhibition of the crayfish giant fiber escape response. J. Exp. Biol. 62, 771-782.

Zucker, R.S., Kennedy, D. and Selverston, A.I., (1971) Neuronal circuit mediating escape responses in crayfish. Science. 173, 645-650.

Zucker, R.S., (1972a) Crayfish escape behavior and central synapses. I. Neural circuit exciting lateral giant fiber. J. Neurophysiol. 35, 599-620.

Zucker, R.S., (1972b) Crayfish escape behavior and central synapses. II. Physiological mechanisms underlying behavioral habituation. J. Neurophysiol. 35, 621-637.

PHASIC CONTROL OF REFLEXES DURING LOCOMOTION IN VERTEBRATES

H. Forssberg, S. Grillner, S. Rossignol* and P. Wallén

Department of Physiology, University of Göteborg

Göteborg, Sweden[o]

The neural control of locomotion in animals should incorporate mechanisms for coping efficiently with obstacles or painful stimuli occurring at any moment of the locomotor cycle. These mechanisms would generate patterns of responses best adapted to the current phase of locomotion. Evidence that such mechanisms exist is provided by the following two observations: 1) in chronic spinal cats walking with the hindlegs on a treadmill, touching the dorsum of a foot at swing phase results in a short latency activation of flexor muscles for all joints of that limb. The same stimulus applied at support phase induces, on the contrary, a marked extensor activity. This switch between flexor and extensor responses in synergy with the current locomotor phase was also repeatedly observed with electrical stimulation of the skin; 2) similarly, in the spinal dogfish, the same electrical stimulation applied to the tail fin at the midline yields opposite responses according to the phase of swimming. In the right half cycle, the stimulation produces a powerful synergistic contraction on the right side and the reverse in the left half cycle. Discussion follows concerning different neuronal mechanisms possibly responsible for these phase dependent reflex reversals.

---

[*]Present address: Département de physiologie, Faculté de médecine, Universite de Montréal, Case postale 6208, Succursale A, Montréal, Québec, H3C 3T8.

[o]Names published in alphabetical order. (Presenter: S. Rossignol.)

Introduction

The study of the interaction between locomotor activity and supraspinal or peripheral influences is furthered since the spinal cord has been shown capable of generating on its own the basic pattern of locomotion. Indeed locomotion can be achieved without supraspinal influences (Shurrager and Dykman, 1951; Grillner, 1973, 1974, 1975; Forssberg and Grillner, 1973) or peripheral inputs (Grillner and Zangger, 1974, 1975). This relative independence is necessary to accomplish the main task of progression and implies that there should be some control exerted on the flow of incoming sensory signals and even a selection of compensatory limb responses well adapted to the continuously changing mechanical conditions of the walking animal.

One type of such controls during locomotion consists in tonic changes of transmission in certain pathways. At the brain stem level, the vestibulospinal system response to tilt is reduced (Orlovsky and Pavlova, 1972) while at the spinal cord level the transmission in short latency inhibitory pathways (Grillner and Shik, 1973) and recurrent inhibition are also reduced (Severin et al., 1968). The transmission in other neuronal circuits is enhanced as exemplified by the appearance of long latency and long duration discharges in preparations capable of walking (Grillner and Shik, 1973; Forssberg and Grillner, 1973).

Brain stem centers have also been shown to exert an efferent control of both primary and secondary neurons in the lateral-line system (Russell, 1974) of some fishes during swimming (Roberts, 1972; Roberts and Russell, 1972; Russell and Roberts, 1972, 1974).

Another type of control can be seen in that stimulation of various descending pathways is effective only in certain phases of locomotion. This effect appears quantitative rather than qualitative since it does not disturb the orderly sequence of EMG activation but only the output amplitude (Orlovsky, 1972).

It thus appears that the neuronal transmission is tonically altered throughout locomotion in some pathways while in other pathways it is effective only during specific phases. These controls can allow the animal to weigh the importance of incoming signals and help to maintain locomotory stability.

However, how does a moving animal react to an external stimulus identical in all respects except that it occurs at any phase of locomotion and requires an immediate compensatory response? It is obvious for simple mechanical reasons, that some response patterns would be appropriate during certain phases but not in others. The present report describes such control of an entirely phasic

CHRONIC SPINAL CAT

Contact Reactions During Locomotion

Cats spinalized at low thoracic level, 1-2 weeks after birth, can walk on a treadmill with adequate weight support and at a speed determined by that of the treadmill belt (Grillner, 1973). In addition, they present tactile placing reactions. When the dorsum of the foot is brought into contact with the edge of a table the limb flexes and then extends to place the foot on the surface of the table (Forssberg et al., 1974).

Investigations were conducted into the pattern of response to tactile stimulation during locomotion of two similar preparations and during all phases of the step cycle (49 experiments). At swing phase, when the dorsum of the foot is touched lightly with a pencil, the whole limb performs a brisk supplementary flexion bringing the foot well above and in front of the pencil (cf. Figure 1). Similarly various obstacles placed on the treadmill belt such as wooden blocks cause this reaction upon contact. The animal effectively overcomes the obstacle and continues to walk undisturbed. Figure 1 is an illustration of part of this reaction, redrawn from selected frames of a 16 mm film synchronized to EMG recordings. The sequence starts with the right leg initiating its swing phase (frame 1). When the foot touches the obstacle (frame 2) represented by the black square (see also the contact trace) the limb flexion is markedly enhanced (frames 3 and 4) and the foot is lifted well above and in front of the obstacle. The EMG traces show a short latency burst of activity both in the ipsilateral knee flexor Semitendinosus (St) and the contralateral knee extensor Quadriceps (Q). Thus during this supplementary ipsilateral flexion the contralateral leg develops more force for support and possibly for propulsion of the body (frame 5).

To determine if the tactile stimulus or the activation of muscle or joint receptors caused this response the effect of subcutaneous infiltration of a local anesthetic (Xylocain) at the dorsum of the foot was investigated. In Figure 1C and D are represented two sets of records of the foot being touched before and after anesthesia during walking. In C, just after the second St locomotor burst, the contact elicits a sharp flexion best seen in the rectified and filtered version of the EMG (int.st). After local anesthesia (D), no such reaction is evoked and the foot remains in contact with the obstacle for a much longer time until

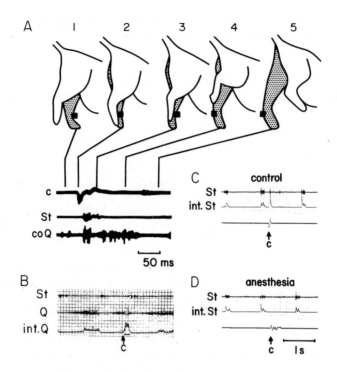

Figure 1. Contact reactions during different phases of locomotion.
A: The reaction occurring during swing is displayed by five frames
(1 to 5) selected from a 16 mm film (64 fr/s) synchronized to EMG
activity photographed on running oscilloscope film. The contact
(black square) is made by a strain gauge held by the investigator.
The contact trace (c) is shown in the records below together with
the activity in the ipsilateral knee flexor semitendinosus (St)
and the contralateral knee extensor quadriceps (Q) recorded by
means of implanted copper wires insulated except at the tip
(Forssberg and Grillner, 1973). B: Extensor response recorded at
low speed on an ink writer with a straight frequency response up to
1200 Hz. During the second Q burst, a similar contact (c), made
with a metal plate forming a microswitch, gives rise to a large
activity in the ipsilateral Q, best seen in the rectified and filtered (Gottlieb and Agarwal, 1970) EMG trace (int. Q). No short
latency response occurs in St. However, the increased extensor
activity is followed by a strong flexion. Time calibration is in
D. C: Flexion response obtained as a control before anesthesia.
The short contact (c) occurring right after the second St burst is
followed by a response in St also represented in int.St trace.
D: After subcutaneous infiltration with a local anesthetic (Xylocain 2%) of a 2 x 3 cm area on the dorsum of the foot, the same
contact does not elicit a flexion response. Instead the foot
remains in contact with the   (Legend continued on next page.)

the leg resumes its locomotion. Hence one may conclude that the response depends upon tactile stimuli.

The same contact applied to the foot during the stance phase gave no flexor response. On the contrary, a marked extensor activation appeared in the stimulated limb. This is shown in Figure 1B; contact occurs during the second Q burst. Note that during contact, a large activity is elicited in Q (see especially int. Q trace) which shortens the overall period of Q activity and is followed by an enhanced flexion.

In conclusion, a light cutaneous stimulation applied to the dorsum of the foot during the swing or the stance phase yields two functionally opposed responses which however occur in synergy with the current locomotor phase. These responses are highly reproducible.

Phasic Responses to Electrical Stimulation of the Skin

A mechanical stimulation applied to a moving limb during walking can hardly be made identical for all phases of locomotion. For further investigation of these responses a weak electrical stimulation (generally 1 to 5 mA, 5 ms pulse) was applied through two silver plates taped on the shaved dorsum of the foot. Such a stimulation at swing phase results in a flexion response in all respects identical to that described for adequate stimulation of the same area. It is noteworthy that stimulation of intensities barely evoking responses at rest can be very effective at the swing phase of locomotion. In records of antagonist knee muscles such as in Figure 2A, an obvious response occurs in St while Q remains silent. A few step cycles later, the same stimulation delivered during stance yields a large response in Q (Figure 2B) and no concomitant flexor activity. Figure 2C and D shows on an expanded time scale the short latency of these responses.

To systematically explore the effect of the same stimulation applied at different points in the whole step cycle a variable delay circuit triggered by one EMG signal was used. In Figure 3 the mean amplitude of the integrated (rectified and filtered) responses observed in St, Q and contralateral Q was plotted against the phase of the normalized step cycles as determined by the onset of St EMG. During the first half cycle corresponding to the swing phase, responses occur in St and contralateral Q (cf. Figure 1A).

---

strain gauge (see prolonged contact trace) for some time before the leg resumes locomotion. Time calibration applies to B, C, and D. Treadmill speed 0.25 m/s.

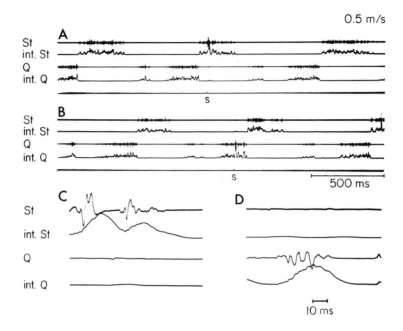

Figure 2. EMG responses to electrical stimulation of the dorsum of the foot at different phases of locomotion. The electrical stimulation (2mA, single pulse 5 ms) is delivered through two 1 cm² silver plates taped on the shaved dorsum of the foot. A: The stimulus (s) occurs during the second St locomotor burst and evokes a discharge in St but not in Q. B: 4 steps later, the same stimulus occurs during stance and a response is seen in Q. Time base for both A and B. C and D: St and Q responses of A and B at higher time resolution.

At <u>around midcycle</u> or beginning of the stance phase these responses decline sharply; instead ipsilateral Q responses appear and these then can be elicited throughout the second half of the cycle. The corresponding average duration of the locomotor EMGs are represented above the response plot. It is clear that responses still occur in St after termination of its main locomotor burst. However when St presents a second burst of activity in early extension (Engberg, 1964), no response occurs in that muscle. In Figure 3 there is also a considerable overlap between the two Q, mainly due to activity in the left Q during the later part of the swing. Despite the presence of this activity, no extensor response appears during swing. Thus, the presence or absence of EMG activity in a given muscle is not the decisive factor for the appearance of responses in that muscle.

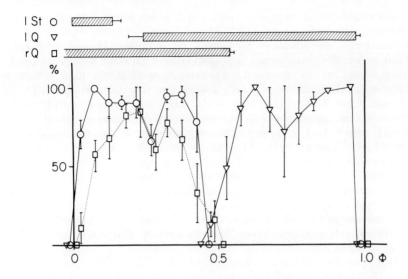

Figure 3. Amplitude and distribution of responses in knee muscles to electrical stimulation at different points of the step cycle. The abscissa represents the normalized duration of 116 step cycles (mean 1070 ms) during which stimuli to the dorsum of the left foot were delivered at different points. The peak integrated EMG responses in the left St, left and right Q were expressed as a percentage of the maximal response recorded in the respective muscle. The responses were grouped in 0.05 phase and the mean and standard error plotted. From 0 to 0.45-0.5 phase (swing), responses are seen in the left St and right Q. From 0.5 to 0.95-1 (stance) the responses are localized in left Q. The EMG duration (mean + S.D.) of the various muscles plotted above were taken from 15 normal cycles preceding the test series. This schematic representation overemphasizes the overlap between the two quadriceps. In fact LQ starts with a small activity which usually peters out (as in Figure 2) or remains very weak before the larger activity is brought in at around 0.5 phase.

The Reversal Points

In Figure 3, the reversal from flexor to extensor responses in the stimulated limb coincides roughly with the midcycle point 0.5 for this particular speed of walking. At this point of transition, either no responses are seen in the antagonist groups or very small responses of equal amplitude appear in both. The reversal point back to flexor responses is close to phase 0 or its equivalent phase 1 which arbitrarily is set here by the onset of EMG activity in St.

To understand the mechanism of this reversal a study was pursued of the time of occurrence of the reversal point and its relationship to the relative duration of stance and swing (Figure 4). The reversal point from flexor to extensor response was examined at different speeds of locomotion. It is known that in the intact dog (Arshavsky et al., 1965) and cat (Goslow et al., 1973) as well as in the spinal cat (Grillner, 1973) the swing phase has a largely constant duration at different speeds while the stance phase shortens with increasing speed. That is, the swing phase occupies a greater proportion of the step cycle at higher speeds. The reversal point was determined at 6 different speeds ranging from 0.25 m/s to 1.5 m/s. The phase of the reversal points are plotted against the speed in Figure 4. This linear increase in the phase of the reversal point with increasing speed corresponds well with what could be expected if reversal from flexor to extensor responses followed the proportional increase of the swing phase with increasing velocity.

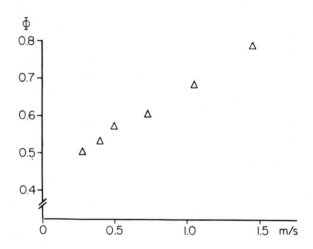

Figure 4. The change in the phase of reflex reversal to electrical stimulation with increasing speed of walking. The phase (ordinate) at which reversal from flexor response (St) to extensor response (Q) occurs (as in Figure 3) has been plotted for different walking speeds with steady gait patterns ranging from 0.25 meter/sec to 1.5 meter sec. Stable gait patterns are necessary to evaluate the step cycle adequately. Therefore higher speeds were not systematically investigated in that fashion since the animal often alternates between walk and gallop. The phase of reversal was defined as the point in the cycle where there is either no response or small responses in both groups of muscles.

Some recordings were performed during gallop. Stimulation during swing yields responses in ipsilateral flexors only, without contralateral extensor activation so that the in-phase gait is not disturbed. During stance, the extensor EMG activity of both limbs is short and because one limb is usually leading the other, the two EMG bursts overlap only for a short period. It was possible to find responses to stimulation in the ipsilateral extensors and sometimes apparently in the contralateral limb with or without ipsilateral responses.

Responses with adequate and electrical stimuli were also investigated during stepping in the air. The cycle during stepping is subdivided in a rather prolonged flexion occupying more than half the cycle and a short extension. Responses were similarly reversed from flexion to extension.

According to the above results, the reversal of the response pattern from flexors to extensors follows the changes occurring in the individual limb step cycle at different speeds and at different gait patterns. It is related then to the actual movements of the limb.

Organization and Latency of the Responses in Different Muscles

Other antagonist pairs of muscles at different joints were studied in a similar manner. At the ankle, responses are switched from Tibialis Anterior (TA) to Gastrocnemius (G) in accordance with the phase. Recordings of Adductor Magnus (AM), an extensor of the hip and of Iliopsoas (IP), a hip flexor, show the same pattern. Responses in Extensor Digitorum Brevis (EDB), a toe flexor active during the stance phase (Engberg, 1964) occur only during the stance phase. From these analyses, it should be convincing that the pattern of response of muscles acting at all joints of the limb to a stimulation of the dorsum of the paw reverses according to the phase of locomotion.

Measuring the latency of responses during ongoing EMG activity presents some difficulties. However, the responses showed an amplitude well above the base line level as judged from the integrated EMG and fell within a certain time range as determined on oscilloscope photographs with fast sweep.

Although in some muscles the latency could change in different parts of the cycle, when mean latencies of responses in different muscles are plotted in increasing order a clear pattern of organization is revealed. As seen in Figure 5 the earliest responses always occur in the knee followed by the ankle and foot and then the hip. This organization appears meaningful since during swing, the early knee flexion withdraws the foot from any obstacle placed in front; the ankle flexion at some ms later, elevates the foot above the obstacle; and the hip carries the limb forward.

It is also clear from Figure 5 that extensor responses at all joints occur later than flexor responses. Another remarkable feature is that Quadriceps is activated earlier during a crossed extension response than during an ipsilateral extensor response. That might indicate the importance of a fast crossed extensor reflex for support of the added weight during such perturbations.

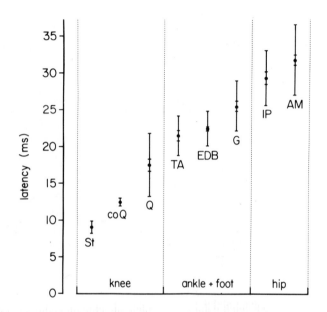

Figure 5. Latency of responses to electrical stimulation in different muscles of the leg. The latency of responses were measured from oscilloscope films and the means with standard deviation and standard error plotted in increasing order. Flexors at any one joint are activated prior to extensors. Knee muscles semitendinosus (St; N = 29), Quadriceps (Q; N = 16) and contralateral Q (coQ; N = 11) respond earlier than ankle muscles tibialis anterior (TA; N = 17) and gastrocnemius (G; N = 23) and foot muscle extensor digitorum brevis (EDB; N = 44). Hip muscles iliopsoas (IP; N = 24) and adductor magnus (AM; N = 42) are activated later. The statistical significance of differences in means was calculated, with the Student test as $p < 0.005$ for St-TA; TA-IP; St-coQ; Q-G; G-AM; Q-EDB; G-EDB; St-Q; TA-G; Q-coQ; and $p < 0.025$ for IP-AM.

Movement Analysis

What does the entire limb actually do during these responses? A detailed movement analysis of a typical flexor and an extensor response to the same stimulation is represented in Figures 6 and 7.

In Figure 6 are shown the angular movements of the hip, knee and ankle for 2 cycles beginning with the swing phase in the cycle prior to stimulation. The swing phase is also represented in the stick diagram in the upper right which illustrates 40 consecutive film frames of a 16 mm film. Note the smooth flexion at all joints in both diagrams. During the second swing phase, a stimulation occurs in midflexion (dotted line). This is followed by a brisk flexion especially marked at the knee and ankle although some flexion also is present at the hip. Note that peak flexion is later at the ankle than at the knee. The stick diagram in the lower right depicts the actual movement of the limb. The beginning of the swing phase is identical to that in the upper diagram. At the frame indicated by the arrow, hyperflexion occurs with an overall lifting of the cat corresponding to the hyperextension occurring on the contralateral side. In the angle plot, the swing phase has the same duration as in the previous cycle although the amplitude is larger. The schematic EMG representation does not display the increased EMG amplitude occurring during the response but does show that the EMG time relationship remains rather undisturbed. This corresponds also to the stable pattern of foot contact as seen below the angle plot.

A similar movement analysis is presented in Figure 7 for an extensor response. The analysis starts here with a normal stance phase. In the upper right stick diagram, the data are arranged to represent the stance phase in a conventional manner; the foot stabilized with respect to the ground, the inverse of the treadmill situation. Upon stimulation, a clear extension of the ankle and to a lesser extent of the knee occurs. This corresponds in the diagram (lower right) to an increased upward and forward movement of the cat and an upward deflection of the respective angular traces. Note in the ankle a clear triangular zone corresponding to an increase of ankle angular velocity. Responses during stance produce less dramatic kinematic readjustments than during the swing phase since the extensor musculature is primarily engaged in supporting the weight. However, the reaction is undoubtedly purposeful in that it stabilizes the stimulated limb and propels the animal forward at a faster speed. In this case again, the EMG cycle and the foot contact pattern are stable (see above and below the angle plot).

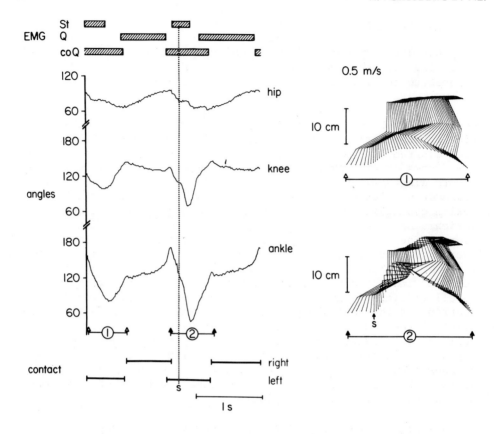

Figure 6. Movement analysis of a flexion response to electrical stimulation during the swing phase. Films of the cats were taken in synchrony with EMG recordings by means of pulses registered on paper and stroboscopic flashes on the film at every 20 frames. The camera speed was calculated for each sequence and varied from 64 to 97 frames/sec. Spots of light reflecting material were glued on the skin of the cat at the iliac spine, the femoral head, the knee joint, the external molleolus and the metatarso-phalangeal joint. The x and y coordinates of each point of a given sequence (the start and end of which are determined by EMG landmarks) were digitized by projecting the frames with suitable enlargement on a digitizer table. The angles of the hip, knee and ankle were calculated from these original coordinates stored in a matrix and could be plotted together with the synchronized EMG. Stick diagrams were also plotted with lines connecting the original coordinates to display the whole limb movements in a selected part of the cycle as represented in the two diagrams of the right hand side. The frames of the sequence (floating average of 3) are displaced from left to right. In this figure, the angular motion of each joint (Legend continued on next page.)

REFLEX CONTROL IN VERTEBRATES

Influences on the Step Cycle and the Gait Pattern

The two detailed movement analyses shown above were chosen because the EMG and foot contact remained relatively stable although quantitative differences existed in EMG and angular amplitudes. A more detailed evaluation of changes induced by the stimulation on the duration of the step cycle was made by comparing each stimulated cycle with the immediately preceding normal cycle. These normal cycles varied from 1 to 1.2 sec. When stimuli were delivered in F, the cycles were prolonged slightly by about 100 ms but when delivered in E2 and early E3 the cycles remained unchanged. Stimuli in late E3 shortened the cycles significantly by approximately 200 ms. During F, E2 and E3 there was no change in the contralateral limb step cycle; the effects being confined to the stimulated limb.

However, when stimuli were given in E1, the cycle and interlimb coordination changed drastically to an in-phase gait or gallop. The sequence of movements resembles the action of a man jumping over a small brook. The flexion response in E1 prolongs the current swing phase while the body is propulsed by a strong extension of the contralateral leg (the actual jump). Upon ground contact of the stimulated leg, the contralateral leg performs a rapid flexion, briefly contacts in a double stance, and thereafter the alternate gait is resumed.

ACUTE SPINAL DOGFISH

A study of the interaction between reflex responses and locomotion in the more primitive vertebrate dogfish was undertaken.

---

is plotted for 2 cycles starting with the flexion phase, (determined by the St onset) of the normal step cycle preceding the step cycle where a stimulation occurs. The swing phase i.e., flexion and first extension E 1 of the right leg is displayed in the upper right stick diagram. Line 1 under the angle plot corresponds exactly to line 1 in the stick diagram. During the swing phase of the second cycle (see foot contact pattern of the two legs under the angular plot) a stimulation occurs in midflexion (dotted line). An increased flexion particularly marked at the ankle and the knee follows the stimulation. The whole swing phase of the stimulated right limb is also displayed in the lower right stick diagram (line 2). The arrow indicates the frame corresponding to the stimulation. Note the ample flexion of the limb compared to the upper right stick diagram of a normal swing. The vertical rise of the hip is due to the contralateral leg extension. Note that the EMG coupling as well as the footfall pattern is not grossly disturbed by this reaction.

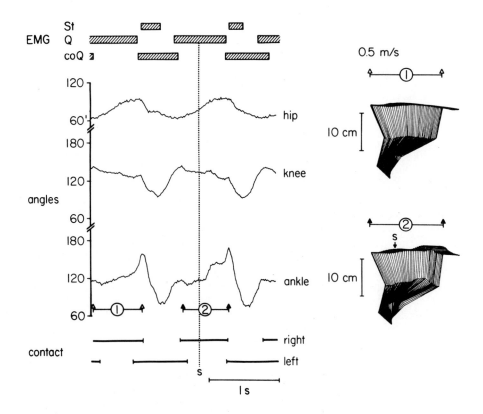

Figure 7. Movement analysis of an extensor response to electrical stimulation during the stance phase. The movement analysis was performed as described in Figure 6. The step cycle preceding stimulation is analyzed starting from the stance phase of the right leg. This first stance is also shown in the stick diagram of the upper right. Manipulations of the coordinates of the points allowed to fix the foot relatively to the treadmill belt as to display the stance phase in the usual manner when the cat is walking on solid ground. Note that after stimulation (dotted line) a clear extension occurs particularly at the ankle (compare with preceding stance phase). In the lower right stick diagram the hip is seen to be accelerated upward and forward. Note at the ankle joint a small triangular clear zone representing an increased angular velocity. Here again the EMG coupling and foot contact pattern is undisturbed.

It has the remarkable ability to swim continuously after transection of the spinal cord 4-8 segments below the foramen magnum (Gray and Sand, 1936; Grillner, 1974). Touching the tail fin during swimming produces a brisk body movement. The direction of this movement depends on the instantaneous position of the fish. When the body is completely curved to the left and starting to move to the right, the pinch induces a powerful movement to the right and vice versa.

This phenomenon was systematically investigated in 12 dogfishes by delivering electrical stimulation through two thin copper wires (100u) inserted at the base of the tail fin on both sides, while recording the EMG of segmental red muscles (see Figure 9 for approximate levels) with synchronized filming of fish movements.

Figure 8 represents an undisturbed swimming movement (thin dotted lines) at specific phases of one cycle together with movements (solid line) at the corresponding phases of the subsequent cycle after stimulation in the right and the left half cycle respectively. In the upper graph the analysis of the movement starts with a complete curvature to the left and in the lower graph with a curvature to the right. The response of the whole body to the stimulation bends the fish more to the right or left depending on the current direction of the movement.

Phase Dependence of EMG Responses

Figure 9 shows EMG recordings obtained during swimming. Each cycle is divided into a right and a left half cycle corresponding to the strict alternation between the two sides. The anterior segments on both sides are activated prior to the more caudal segments (Grillner, 1974; Grillner and Kashin, this volume). The stimulation (Figure 9A) occurs during the EMG burst of the right rostral segment, inducing a large amplitude burst on this side, apparent at both levels of recordings. The left side is silent. The large response shortens the right half cycle and thereby results in a powerful bending to the right. Note that at the onset of EMG in the right rostral segment the fish is bent almost completely to the left (see also top graph of Figure 8) so that the response accelerates the bending to the right.

Correspondingly, the same stimulation occurring during the left half cycle, when the fish starts bending to the left from its right curvature (see also bottom graph of Figure 8), results in a powerful response in the left side musculature (Figure 9B). The same reversal is shown in Figure 9C and D together with the integrated version of the EMG. Thus, as in the cat, the same stimulation can yield opposite responses in synergy with the current locomotor phase.

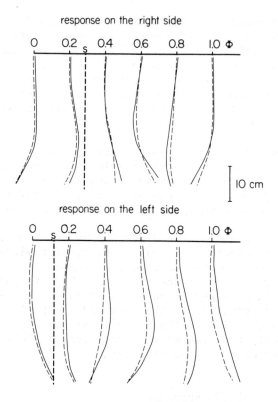

Figure 8. Swimming movements in response to stimuli applied in different phases. Movement analysis of the fishes were performed essentially as indicated in Figure 6 using spots of reflecting material sewn on the midline of the fish (see the fish drawing in Figure 9). The swimming cycle was normalized from the EMG cycle determined in one EMG segment. In both graphs, the thin dotted lines represent the position of the fish in 6 characteristic points of the swimming cycle and the solid lines the corresponding positions after stimulation at the base of the tail fin. The heavy dotted lines represent the stimulation in both graphs. In the upper graph, the analysis starts with the fish bent to the left (i.e., the point where the EMG activity starts in the rostral right segments). After stimulation, there is a strong bending to the right if one compares, for instance, the dotted and solid lines in phase 0.6. In the lower graph, the analysis starts with the fish bent to the right and initiating the left curvature. After stimulation (see phase 0.4 and 0.6) the bending to the left is more accentuated. Thus the same stimulation increases the ongoing movement whether to the right or left. Note that in this case the fish is fixed by the snout and unable to propulse forward, but in ordinary conditions a larger amplitude of movement would result in an increased forward speed.

REFLEX CONTROL IN VERTEBRATES

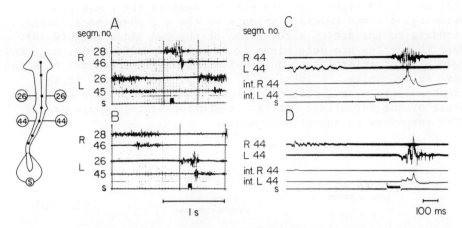

Figure 9. Reversal of EMG responses on the two sides of the body in the swimming dogfish. The EMG activity during swimming was recorded bipolarly with thin (100u) copper wires, inserted into the red lateral musculature at different levels on both sides of the body. In the left drawing of the fish the numbers indicate the approximate level of recording. Electrical stimulation was delivered through 100u copper wires with the insulation removed for 3 mm at the tip and inserted superficially at the base of the tail fin, one electrode on each side. A and B show EMG recordings at the indicated segments. It can be seen that the two sides are alternating and that there is a delay from rostral to caudal segments. The dotted lines divide the recordings in right and left half cycles. In both sets the stimulation (s) is a train of 60 ms (18 pulses of 1 ms) at 2.5 mA. Note time calibration below (B). When a stimulus is delivered within the right half cycle (A), a large amplitude response occurs at both levels on the right side only. The duration of the EMG burst is shorter and so is the corresponding half cycle. When the same stimulation is given a few seconds after the previous one but in the left half cycle (B), the pattern of response is reversed to the left side. C and D show at higher resolution (see time calibration below D) oscilloscope traces of the raw EMGs and their rectified and filtered versions from segment 44 in another preparation. The stimulation (s) is in both sets a train of 100 ms (16 pulses of 5 ms) at 0.8 mA. When the stimulation occurs just before the onset of the right locomotor burst (C), and well within the right half cycle (cf. A) there is a powerful response on the right side, especially apparent in the integrated EMG trace. No activity is seen on the left side. Again, when giving the same stimulation during the left half cycle (D; cf. B) the response pattern is reversed.

When the integrated response amplitudes recorded either on the left or on the right side are plotted against the phase of swimming, a clear reversal appears on the 0.5 phase mark corresponding to the strict alternation of the two sides (Figure 10). Here the cycles are determined from tail EMG (segment 44) and amplitude plotted in arbitrary units with the maximal response on either side as 100 units. A total of 97 stimuli are included in this graph and the mean amplitude of response is plotted for tenths of phase of the normalized cycles. A similar picture would emerge with responses in anterior segments.

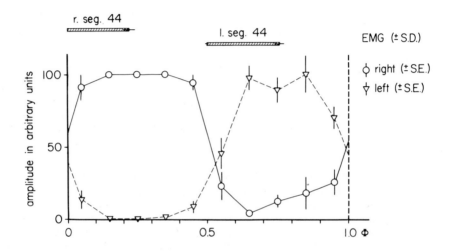

Figure 10. Reversal of EMG responses to electrical stimulation of the tail fin in opposite muscular segments during swimming in the dogfish. Stimuli (3.5 mA, train 50 ms, 1 ms pulse, 3.3 ms interval) were delivered in different phases of 97 swimming cycles and the amplitude of responses on either side was plotted against the phase of the normalized swimming cycle. The phase value for each stimulation was calculated by dividing the interval between the onset of EMG activity on the right side and the stimulus by the duration of the preceding cycle. The amplitude of the responses was measured from the rectified and filtered EMG and normalized with respect to the response of maximal amplitude on either side set to 100 arbitrary units. The mean (± S.E.) amplitude of responses to the right (circles) and the left (triangles) was calculated for all stimuli occurring within tenths of the cycle. Since the responses to the right from 0.1 to 0.3 phase were maximal, the S.E. could not be calculated. The mean (± S.D.) duration of EMG activity on the two sides is plotted above. The period of responsiveness of either side exceeds the period of locomotor EMG activity.

Two types of responses occur either on the left or on the right. In the examples of Figure 9A and B the responses appear first in the rostral segments and then in caudal segments. In another type of response, the EMG burst appears first in more caudal segments and progresses rapidly to the rostral segments. It still has to be determined whether these two patterns are phase dependent and if so, for which mechanical reason.

Discussion

The main outcome of this study is the demonstration of the remarkable dynamic property of the nervous system to reverse the pattern of response to an identical stimulation in synergy with the current phase of locomotion. This phasic reversal is essentially the same during swimming in spinal dogfishes or walking in chronic spinal cats and might thus represent a dynamic control that has evolved as part of the neural control of locomotion throughout the vertebrate and presumably invertebrate series.

It is meaningful that a slight touch to the dorsum of the foot during swing would give rise to an increased flexion in order to overcome an impeding obstacle. On the other hand, it would be most inadequate if a cutaneous stimulation applied during the stance phase would invariably elicit a flexion since with the other limb in flexion the cat would collapse. Instead, an extensor response with shortening of the stance is more appropriate for the stabilization of the weight supporting limb and an increased forward velocity. During swimming, it is perhaps more evident that the fastest means of escaping a nociceptive stimulation of the tail fin is to reinforce the ongoing movement thus increasing the forward speed away from an attacking animal. Which spinal pathways might be subserving these compensatory mechanisms?

I. SPINAL PATHWAYS

A. Ipsilateral Flexor and Contralateral Extensor Responses

The flexion response elicited during swing resembles the tactile placing reaction which indeed is present in the chronic spinal cat (Forssberg et al., 1974). In static conditions, tactile placing is induced also by touching the dorsum of the foot and consists in a flexion of the limb followed by an extension placing the foot on top of the contacting edge. The same pattern is apparent during locomotion although the sequence of events is much accelerated. Indeed latencies of 25-30 ms in St were reported by Lundberg (1972). Similar or longer latencies are found in the hindlimb of the chronic spinal cats (Forssberg et al., 1974) and in the forelimbs of intact cats (Amassian et al., 1972; Smith and Massion, 1972). The mean latency of St responses recorded in the present study is 9.1 ms ± SD. 0.8. If indeed these responses are mediated through the same spinal pathways responsible for tactile placing it must be concluded that the transmission is more effective in these pathways during locomotion. This is also supported by the high reproducibility of the flexion response during locomotion compared to the variability of the static tactile placing in chronic spinal cats (Forssberg et al., 1974) and intact cats (Bogen and Campbell, 1962; Smith and Massion, 1972).

The latency of the flexion response recorded during locomotion is inseparable from that of the flexion reflex (Creed et al., 1932). In one of the chronic spinal cats used in this study it was found that stimulation of the dorsum of the foot at the strength used to elicit flexion responses during locomotion yielded a field potential in the cord at a latency of 4 ms (Edgerton, Grillner, Rossignol and Sjöström, Unpublished data). With an efferent conduction time of 2-3 ms (Creed et al., 1932), there would be about 3-4 ms left for the central delay, which is comparable to that of a flexion reflex.

Similarly the crossed extension found here during the ipsilateral flexion is reminiscent of a crossed extensor reflex. The short latency excludes that this crossed extensor EMG is secondary to a weight transfer to the other limb and that it represents a load compensation response. More likely, the reflex pattern is preprogrammed to support a greater load at the time of contralateral flexion. It should be recalled that this crossed extensor pathway seems to be closed during the swing phase of gallop.

It is certainly of interest to see that known patterns of short latency spinal reflexes can be fully integrated as purposeful adaptive mechanisms during locomotion.

B. Ipsilateral Extensor Responses

From the present results in the cat, it should be clear however that the above reflex pattern of ipsilateral flexion and contralateral extension appears only during the swing phase of locomotion. During the stance phase, the same cutaneous stimulation evokes a pattern of ipsilateral extensor response at all joints of the limb.

Although it is customary to associate flexor responses to cutaneous stimulation, the appearance of ipsilateral extensor responses has often been described. Brown and Sherrington (1912) observed extensor responses to cutaneous nerve stimulation and considered these 'exceptions' as the expression of reflexes usually concealed by the predominant flexion reflex. Verzár (1920, 1923) interpreted such reversed reflexes as due to a central fatigue mechanism. It was shown later that certain hindlimb skin areas are excitatory to extensor muscles but that the nerve supply to these areas also contained fibres innervating surrounding areas inhibiting the same extensor motoneurones (Hagbarth, 1952) thus explaining the concealed effects on extensors obtained with whole nerve stimulation. Similar observations in man showed that electrical stimuli applied to the dorsum of the foot sometimes yielded extensor responses and that cutaneous receptive fields for flexor and extensor reflexes are very close if not overlapping (Kugelberg et al., 1960). Finally it was shown by Wilson (1963) with intracellular studies that stimulation at 1.2 to 2.2 times the threshold of the largest afferents in the suralis nerve frequently gave EPSPs in ipsilateral extensor motoneurones with a central latency ranging from 2.5 to 5 msec. Excitatory action by superficial peroneal nerve also was encountered.

Hence, there are short latency cutaneous pathways from the foot dorsum leading to excitation of ipsilateral extensor motoneurones and it is likely that these are activated during the stance phase of locomotion during which responses with a total latency as low as 12-15 msec during midstance have been found. The stance phase of locomotion is thus a specific condition during which cutaneous excitatory effects on ipsilateral extensors can be revealed while in the swing phase, ipsilateral flexion and contralateral extension responses are the rule.

II. ALTERNATE SPINAL PATHWAYS: REFLEX REVERSAL

The above phenomenon of response appearing in antagonistic muscle groups depending on the current phase of locomotion has been coined in the expression "phase dependent reflex reversal" (Forssberg et al., 1975). The expression 'reflex reversal' implies, historically, that although a given stimulation usually evokes a stereotyped response pattern, sometimes the response is 'exceptionally' the opposite: a flexor response becoming an extensor or an inhibitory response becoming an excitatory response. Thus 'reversal' often implies that a certain pattern is the reverse of the 'normal' pattern. In the present context, the expression reflex reversal is not used in that sense but merely to describe that responses are the reverse of one another in the two different phases of locomotion and that both patterns are equally 'normal'.

As suggested above, the concept of reflex reversal is not new and different aspects of it have been discussed in connection with many experimental findings. Reflex reversals could be classified in two groups, one in which excitatory and inhibitory effects are interchanged in the same muscle group and the other in which the pattern of response appears in antagonistic muscles.

A. Reflex Reversal to the Same Muscle

Cutaneous stimulation can not only excite functionally opposed muscles but it can also inhibit or excite the same muscle group in various conditions. For instance it was shown that strychnine and chloroform induce opposite responses in Quadriceps to cutaneous nerve stimulation (Sherrington and Sowton, 1911). Similarly, excitatory or inhibitory responses of Quadriceps appear with head rotation in opposite directions (Sassa, 1921; Pi-Suñer and Fulton, 1929). Holmqvist (1961) and Holmqvist and Lundberg (1961) have demonstrated conclusively the existence of alternate inhibitory or excitatory pathways to flexor motoneurones from ipsilateral and contralateral cutaneous and high threshold muscular afferents. These pathways are controlled independently at different brain stem levels.

Although this type of reversal is not related directly to the present findings, the concept of alternate spinal pathways is certainly of great relevance.

B. Reflex Reversal to Antagonistic Muscles

Here again, types of reflex reversal should be distinguished according to the static or dynamic conditions eliciting the reversal.

Static or passive conditions. Sherrington (1900) has described in spinal dogs and frogs that the initial limb posture may determine the response pattern of the limb to cutaneous stimulation. If the limb is initially flexed, the response pattern is extension and if it is extended the response is flexion. Magnus (1924) showed the same phenomenon by using a knee tap on the contralateral leg which undoubtedly resulted also in a cutaneous stimulation and recording in the ipsilateral leg whose initial posture, especially at the hip and knee, was changed passively (Umkehr). In acute spinal cats capable of walking after injection of Clonidine (Forssberg and Grillner, 1973) stimulation of ipsilateral (Grillner, 1973) and contralateral (Grillner and Rossignol, Unpublished data) nerve stimulation gives long lasting EMG responses with long central latency in flexor or extensor muscles depending on whether the hip is initially extended or flexed. It is not known if such reversal is due to stretch of hip muscles or to joint afferents. Of

relevance perhaps is the observation by v. Uexkull (1904) that in
the isolated arm of the Brittle-Star (Ophyoglypha) the same stimulation flips the appendage in a direction opposite to its initial
position. The same applies for the tail of chronic spinal cats
(Magnus, 1924).

A common feature of reflex reversal determined by the initial
position is that the responses tend to bring the limb or the appendage in an opposite position. It should be recalled that the
reflex reversal described here during locomotion is the contrary
in that the current locomotor movement is enhanced and not opposed
as during static conditions. However such responses which tend to
bring the limb in another position might be functional during
locomotion at turning points where the limb changes from extension
to flexion or vice versa.

<u>Dynamic or active conditions</u>. Other types of reversal are
determined by the state of activity of the organism. It was shown
in the intact guinea-pig (Brown, 1911), contrary to the observations
on the influence of passive limb positions, that if the animal
maintains actively a flexion or extension of the limbs the response
to cutaneous stimulation is an increased flexion or extension respectively reinforcing the active posture.

Another situation which involves the passage from a static to
a dynamic condition has been described by Lisin et al., (1973). In
hemiplegic patients and decorticate cats a painful and prolonged
stimulation of the skin area innervated by the suralis nerve gives
a flexion response when standing but a long lasting extensor response during locomotion with, however, an arrest of locomotion.

Finally, our present results in both the spinal cat and dogfish should be classified as a dynamic reflex reversal of phasic
nature.

III. SELECTION OF ALTERNATE SPINAL PATHWAYS

From the above discussion it should appear that there are
alternate routes of transmission from cutaneous inputs to both
flexor and extensor motoneurones (I). Different conditions can
select these different routes (II). Which mechanism(s) can likely
account for the selection during locomotion?

<u>Firstly</u>, it might be suggested that cutaneous afferents
always reach both flexor and extensor motoneurones through polysynaptic pathways and, depending on the excitability level of one
or the other group of motoneurones the response will appear in
flexors or extensors during their respective period of high

excitability. This mechanism is unlikely since the period during which responses appear in certain muscles does not correspond precisely with their period of locomotor activity. Furthermore, St being a double joint muscle (mainly knee flexor but also hip extensor) often presents a second burst of activity during the extension phase. Apparently the excitability of St motoneurones is still high during that period, but no flexion response occurs during the second burst of St activity in the stance. This is also seen clearly in the fish in which responses in the tail EMG can appear before, during or after the main locomotor burst.

Secondly, the selection of pathways can be achieved at the interneuronal level by concomitant peripheral inputs such as from the joints (hip), muscle or tendon or other skin afferents which might, by their central pre- or postsynaptic actions, block or facilitate the transmission in one or the other pathway. Such selection could be aided in normal conditions by supraspinal inputs to these interneurons (Lundberg, 1972). Considering all the possible interactions, this selection mechanism can certainly not be discarded. However, certain interactions can be ruled out at present such as the presence or absence of simultaneous foot pad stimulation since the observed reflex reversal can be obtained as well during stepping in the air.

Thirdly, the central generator for locomotion could phasically change the transmission in these alternate pathways while at the same time inducing rhythmic activation of antagonist muscle groups. Changes of transmission in spinal pathways have already been discussed here. It is thus highly probable that a system capable of generating locomotor patterns in detail (Grillner and Zangger, 1975) can also determine the appropriate response to a peripheral cutaneous stimulus at each phase and not depend only on other peripheral inputs (second hypothesis) which might in themselves require a response or be conflicting. Recently evidence for changes in transmission in cutaneous pathways has been acquired during spontaneous rhythmic activity in one acutely operated and paralyzed chronic spinal cat of this series. Intracellular recordings of motoneurones showed phase dependent variations in the postsynaptic potentials in response to stimulation of nerves and the dorsum of the foot (Edgerton, Grillner, Rossignol and Sjöström, Unpublished data).

The advantage of a selection in transmission of pathways by the central generator of locomotion would be to insure the best possible match between reflex responses and the current phase of the locomotion and thus a great stability of the locomotor process would follow. Indeed it was seen that in the greatest part of the step cycle, the reflex response would change only quantitatively the ongoing movement. In some cases such as in E1 (see also

Orlovsky and Shik, 1965) one or two gallop steps follow the stimulation. In such cases, although the interlimb coordination is changed, another preferential mode of locomotion is temporarily adopted so that, in all cases, the progression of the animal is assured despite the great variety of obstacles encountered.

Acknowledgment

This work was supported by the Swedish Medical Research Council project No. 3026 while S.R. received a postdoctoral fellowship from the Canadian Medical Research Council. The indispensable role of Mrs. Margreth Svanberg in the care of the chronic spinal cats and during the experiments is gratefully acknowledged.

REFERERENCES

Amassian, V.E., Weiner, H. and Rosenblum, M., (1972) Neural systems subserving the tactile placing reaction: A model for the study of higher level control of movement. Brain Res. 40, 171-178.

Arshavsky, Yu.I., Kots, Ya.M., Orlovsky, G.N., Rodionov, I.M. and Shik, M.L., (1965) Investigation of the biomechanics of running by the dog. Biophysics. 10, 737-746.

Bogen, J.E. and Campbell, B., (1962) Recovery of foreleg placing after ipsilateral frontal lobectomy in the hemicerebrectomized cat. Science. 135, 309-310.

Brown, T.G., (1911) Studies in the physiology of the nervous system. VIII. Neural balance and reflex reversal with a note on progression in the decerebrate guinea-pig. Quart. J. Exp. Physiol. 4, 273-288.

Brown, T.G. and Sherrington, C.S., (1912) The role of reflex response in the limb reflexes of the mammal and its exceptions. J. Physiol. (Lond.). 44, 125-130.

Creed, R.S., Denny-Brown, D., Eccles, J.C., Liddell, E.G.T. and Sherrington, C.S., (1932) Reflex Activity of the Spinal Cord. Oxford Univ. Press, London.

Engberg, I., (1964) Reflexes to foot muscles in the cat. Acta Physiol. Scand. 62, (Suppl. 235).

Forssberg, H. and Grillner, S., (1973) The locomotion of the acute spinal cat injected with Clonidine i.v. Brain Res. 50, 184-186.

Forssberg, H., Grillner, S. and Sjöström, A., (1974) Tactile placing reactions in chronic spinal kittens. Acta Physiol. Scand. 92, 114-120.

Forssberg, H., Grillner, S. and Rossignol, S., (1975) Phase dependent reflex reversal during walking in chronic spinal cats. Brain Res. 85, 103-107.

Goslow, G.E., Jr., Reinking, R.M. and Stuart, D.G., (1973) The cat step cycle: hind limb joint angles and muscle lengths during unrestrained locomotion. J. Morphol. 141, 1-41.

Gottlieb, G.L. and Agarwal, G.C., (1970) Filtering of electromyographic signals. Amer. J. Phys. Med. 49, 142-146.

Gray. J. and Sand, A., (1936) The locomotory rhythm of the dogfish (Scyllium canicula). J. Exp. Biol. 13, 200-209.

Grillner, S., (1973) "Locomotion in the spinal cat," In Control of Posture and Locomotion. (Stein, R.B., Pearson, K.G., Smith, R.S. and Redford, J.B., ed.), Plenum Press, New York, (515-535).

Grillner, S. and Shik, M.L., (1973) On the descending control of the lumbosacral spinal cord from the "mesencephalic locomotor region." Acta Physiol. Scand. 87, 320-333.

Grillner, S., (1974) On the generation of locomotion in the spinal dogfish. Exp. Brain Res. 20, 459-470.

Grillner, S. and Zangger, P., (1974) Locomotor movements generated by the deafferented spinal cord. Acta Physiol. Scand. 91, 38-39A.

Grillner, S., (1975) Locomotion in vertebrates - central mechanisms and reflex interaction. Physiol. Rev. 55, 247-304.

Grillner, S., Rossignol, S. and Wallén, P., (1975) Phase dependent reflex reversal during swimming in the spinal dogfish. Acta Physiol. Scand. (In press).

Grillner, S. and Zangger, P., (1975) How detailed is the central pattern generation for locomotion? Brain Res. 88, 367-371.

Hagbarth, K.E., (1952) Excitatory and inhibitory skin areas for flexor and extensor motoneurones. Acta Physiol. Scand. 26, (Suppl. 94).

Holmqvist, B., (1961) Crossed spinal reflex actions evoked by volleys in somatic afferents. Acta Physiol. Scand. 52, (Suppl. 181).

Holmqvist, B. and Lundberg, A., (1961) Differential supraspinal control of synaptic actions evoked by volleys in the flexion reflex afferents in alpha motoneurones. Acta Physiol. Scand. 54, (Suppl. 186).

Kugelberg, E., Eklund, K. and Grimby, Z., (1960) An electromyographic study of the nociceptive reflexes of the lower limb. Mechanism of the plantar responses. Brain. 83, 394-409.

Lisin, V.V., Frankstein, S.I. and Rechtmann, M.B., (1973) The influence of locomotion on flexor reflex of the hind limb in cat and man. Exp. Neurol. 38, 180-183.

Lundberg, A., (1972) The significance of segmental spinal mechanisms in motor control. Proceedings of symposial papers 4th International Biophysics Congress, Moscow.

Magnus, R., (1924) In Korperstellung. Verlag von Julius Springer, Berlin, (24-49).

Orlovsky, G.N. and Shik, M.L., (1965) Standard elements of cyclic movement. Biofizika. 10, 935-944, (Engl. transl.).

Orlovsky, G.N., (1972) The effect of different descending systems on flexor and extensor activity during locomotion. Brain Res. 40, 359-371.

Orlovsky, G.N. and Pavlova, G.A., (1972) Response of Deiters' neurones to tilt during locomotion. Brain Res. 42, 212-214.

Pi-Suñer, J. and Fulton, J.F., (1929) The influence of the proprioceptive system upon the crossed extensor reflex. Amer. J. Physiol. 88, 453-467.

Roberts, B.L., (1972) Activity of lateral-line sense organs in swimming dogfish. J. Exp. Biol. 56, 105-118.

Roberts, B.L. and Russell, I.J., (1972) The activity of lateral-line efferent neurones in stationary and swimming dogfish. J. Exp. Biol. 57, 435-448.

Russell, I.J. and Roberts, B.L., (1972) Inhibition of spontaneous lateral-line activity by efferent nerve stimulation. J. Exp. Biol. 57, 77-82.

Russell, I.J., (1974) Central and peripheral inhibition of lateral-line input during the startle response in goldfish. Brain Res. 80, 517-522.

Russell, I.J. and Roberts, B.L., (1974) Active reduction of lateral-line sensitivity in swimming dogfish. J. comp. Physiol. 94, 7-15.

Sassa, K., (1921) On the effects of constant galvanic currents upon the mammalian nerve, muscle and reflex preparations. Proc. R. Soc. B. 92, 341-355.

Severin, F.V., Orlovsky, G.N. and Shik, M.L., (1968) Recurrent inhibitory effects on single motoneurones during an electrically evoked locomotion. Bull. Exp. Biol. Med. 66, 3-9, (In Russian).

Sherrington, C.S., (1900) On the innervation of antagonistic muscles. Sixth note. Proc. R. Soc. B. 66, 66-67.

Sherrington, C.S. and Sowton, S.C., (1911) Chloroform and reversal of reflex effect. J. Physiol. (Lond.). 42, 383-388.

Shurrager, P.S. and Dykman, R.A., (1951) Walking spinal carnivores. J. comp. Physiol. Psychol. 44, 252-262.

Smith, A.M. and Massion, J., (1972) Ajustement postural et movement au cours de la reaction de placement. J. Physiol. (Paris). 65, 306A.

Uexkull, J.v., (1904) Die ersten Ursachen des Rhythmus in der Thierreihe. Ergebn. Physiol. 3, 1-11.

Verzár, F., (1920) Reflexumkehr (paradoxe Reflexe) durch Ermudung und Shock. Pflügers Arch. ges. Physiol. 182-183, 210-234.

Verzár, F., (1923) Reflexumkehr (paradoxe Reflexe) durch zentrale Ermüdung beim Warmblüter. Pflügers Arch. ges. Physiol. 199, 109-124.

Wilson, V.J., (1963) Ipsilateral excitation of extensor motoneurones. Nature. 198, 290-291.

MOTOR BEHAVIOR FOLLOWING DEAFFERENTATION IN THE DEVELOPING AND
MOTORICALLY MATURE MONKEY

Edward Taub

Institute for Behavioral Research

Silver Spring, Maryland

Following deafferentation of both forelimbs by dorsal rhizotomy, motorically mature rhesus monkeys are able to make extensive use of the affected extremities, both for patterned activity (e.g., ambulation, climbing) and for independent use of the deafferented members. Limb deafferentation does, however, result in a clearly apparent motor deficit, particularly in the fine control and timing of movement.

In contrast to the effects of bilateral forelimb deafferentation, unilateral forelimb deafferentation results in an effectively useless extremity in the free situation. However, use of a single deafferented limb can be induced by increasing motivation and employing training techniques. Data suggest that nonuse of a single deafferented limb in the free situation is a learned phenomenon.

When bilateral forelimb deafferentation was carried out on the day of birth, infants spontaneously developed use of the affected members for support of body weight, ambulation, clasping, and reaching for objects. Thumb-forefinger prehension did not emerge spontaneously but could be trained. In further experiments, monkey fetuses were exteriorized two-thirds of the way through gestation, given forelimb deafferentation, and then replaced in utero for the remainder of gestation. Infants that survived through Caesarian delivery and whose spinal cords were protected by a prosthetic device substituting for the dorsal portions of vertebrae removed during surgery displayed motor capacity similar to that of infants deafferented at birth.

Thus, results indicate that the central nervous system of the motorically mature monkey is capable of autonomously generating movements of almost all types in the absence of guidance either from the environment or from sensory cues originating in the organism's body. It would also appear that neither spinal reflexes nor local somatosensory feedback is necessary for the ontogenetic development of many types of movement performed by the forelimb musculature in monkeys.

Introduction

There are a number of different experimental strategies for studying the role of somatic sensation and spinal reflexes in the control of movement: (1) the sensory motor correlation experiment, (2) the sensory recombination experiment, and (3) the deafferentation experiment.

The sensory motor correlation experiment. In the animal with dorsal roots intact, one can correlate such consequences of stimulation of peripheral receptors as spinal reflex activity and input to the central nervous system (CNS) with any of a variety of measures of nonreflexive movement. This can be accomplished either in the fully intact organism or, when appropriate to the specific nature of a given problem, in a "reduced" preparation, part of whose nervous system has been extirpated, or whose neuraxis has been interrupted at some level.

The sensory recombination experiment. These procedures alter normal central-peripheral relations by effecting some constant geometrical transformation of normal patterns of sensory input to the CNS. This can be accomplished surgically by rearranging the anatomical distribution of peripheral innervation, as in tendon-crossing, nerve transposition, or, in amphibia, inversion of the eyeball or reversal of a limb in its socket. When the distance receptors are involved, sensory rearrangement can be achieved by interposing devices, such as wedge prisms, inverting or reversing lens systems, or pseudophones, between the sense organs and the environment. For effective movement to be carried out after sensory recombination, the animal must change the directionality of its movements in space to adjust for the geometrically altered sensory consequences of movement. This type of experiment thus indicates which of an organism's movement patterns are fixed with respect to a given type of sensory input and which are modifiable (Weiss, 1941; Sperry, 1945; Taub, 1968; Taub and Goldberg, 1974).

The deafferentation experiment. By serially sectioning all of the dorsal roots innervating a given portion of the body, one can eliminate all somatosensory input from that region without seriously affecting the motor innervation. (A discussion of the evidence that dorsal rhizotomy completely abolishes spinal reflexes and position-sense input is presented in Taub, 1976, Section V-A.) For movements which the animal continues to perform after deafferentation, spinal reflexes and somatosensory input are obviously not necessary. They may be necessary for movements that are lost, although there are other consequences of dorsal rhizotomy that could be responsible for the loss, such as tonic imbalance, loss of CNS drive, and certain secondary effects (Taub, 1976, Section I-A).

None of these strategies is sufficient in itself to provide a complete picture of the role of somatosensory input in movement. They inform one another and are complementary. Indeed, too heavy a reliance on one type of data can give a distorted view. For example, the fact that a given type of sensory activity normally precedes a given mode of performance does not necessarily signify that the CNS is actually making use of that input in the mediation of the movement; the input could be epiphenomenal or associated with some other, highly correlated activity. Conversely, the fact that an organism can carry out a given movement following deafferentation is not necessarily evidence that the animal does not employ somatic sensation for the performance of that function when it is available. Deafferentation research does, however, provide one type of baseline or null-point information relevant to ascertaining the role of somatosensory input in normal activities. Moreover, if a given type of movement survives deafferentation, the possibility is raised that in the intact animal somatic sensation, though present, is not essential to its performance. This possibility must then be evaluated experimentally. For example, deafferentation research first indicated that somatic sensation is not necessary for ambulation; later work with afferented organisms (Forssberg and Grillner, 1973; Grillner, 1973a, b, 1975; Goslow et al., 1973; Herman et al., 1974; Stuart et al., 1974) suggested that somatosensory feedback and spinal reflexes are indeed less important, and central programming more important, than had generally been thought.

DEAFFERENTATION AFTER MOTOR PATTERNS ARE MATURE

Unilateral Deafferentation

It has long been known that deafferentation of a single limb in monkeys results in an essentially useless extremity in the free situation (Mott and Sherrington, 1895; Lassek, 1953; Twitchell, 1954). While the limbs are not paralyzed, the motor activity is seemingly random, and was once thought to be <u>associated movement</u> elaborated as reflex effects based on movements in other, intact portions of the body. Some rudimentary, purposive use of the proximal musculature was observed, but it was confined to incomplete or largely ineffective movements made for postural support at rest, and in defense against threatening objects. Early in the twentieth century, Munk (1909) had shown that a seated, hungry monkey could be induced to use a unilaterally deafferented forelimb to grasp food and bring it accurately to its mouth if the intact forelimb was restrained. The status of locomotor movements was not described. Munk's report appears to have received little attention, and was effectively lost in the literature.

<u>Initial Conditioned Response Experiment.</u> In 1957 this author, in association with two co-workers (Knapp et al., 1958, 1963) began an attempt to employ conditioned response (CR) procedures to evaluate the generally accepted nonuse by monkeys of a single deafferented limb. It was felt that these techniques involved several features that might force the expression of a latent capacity for purposive movement that was masked in the free situation. These features were as follows: (1) motivation to use the limb could be increased, maintained at a high level over a long period of time, and controlled in a precise fashion, (2) motor requirements could be kept simple, and (3) repeated trials with the same performance requirements would afford time and opportunity for learning to aid in achieving the behavioral objective.

These three features were embodied in the design of the first set of experiments. Monkeys seated in restraining chairs were required to move the deafferented limb within 3.5 sec after the onset of an auditory stimulus in order to avoid electric shock. The shock was an intense (3.5 ma, 60 Hz) stimulus that remained on for 3.5 sec unless avoided or terminated by the appropriate response. The behavioral requirement was a movement of the right forelimb between a lamp and a photoelectric receiver located 5 inches above a horizontal, waist-level board. This was usually accomplished by flexion at the shoulder and elbow, but a protraction of sufficient excursion could also be effective, and the topography of a successful response could vary considerably. Each session consisted of 20 trials, and the training series was to be of long duration, if necessary.

An extension of the neck restraint board prevented view of the body. Movement of the left arm was restricted by securing it to a portion of the apparatus. The conditioned stimulus (CS) was either a response-terminated buzzer (Knapp et al., 1958, 1963) or a click whose duration was too brief for response termination (Taub and

Berman, 1963; Taub et al., 1965). Half of the animals had received preoperative training; the other half had no experience in the testing situation prior to operation. Dorsal rhizotomy was carried out intradurally from C_3 through T_2 inclusive. Testing began two weeks after surgery. All animals were able to establish a pattern of successful avoidance responding. Each of the preoperatively trained monkeys was initially unable to perform the response, but was able to regain this ability within the next 45 testing days (9 weeks). The initial impairment in avoidance responding could represent a deficit in retention, but is more probably related to a simple inability to make the motor response so soon after surgery. The gradual nature of the recovery after deafferentation was not discovered until later in the experimental sequence (Taub and Berman, 1968). It seems quite possible that, had postoperative testing been deferred until some later time, avoidance responding would have been as effective as after preoperative acquisition.

Postdeafferentation learning of the response by animals with no preoperative training required a longer time than was required by normal animals. Later data (e.g., Taub et al., 1966c) indicates that, even when sufficient time is allowed for maximum recovery of function, deafferented animals require more time to learn new tasks than do intact animals.

Straitjacket Restraint of the Intact Limb. After CR testing was complete, six of the unilaterally deafferented animals were placed in straitjackets which restrained the intact limb while leaving the deafferented extremity free. In conformity with the observations of Munk (1909), five of the six animals were observed to use the affected forelimb to reach outside the cage bars to secure food. Two of the animals batted the food back into their cages with cupped palm, but the other three were able to grasp the food, though crudely and without participation of the thumb. In addition, two animals were observed to use the deafferented limb for postural support, both at rest and in moving around the limited confines of their cages. However, the normal ambulatory posture was not assumed, due--as later work showed (Taub et al., 1972a)--to mechanical constraints imposed by the straitjacket. Shortly after purposive movement had been displayed, the straitjackets were removed, and in each case the animal reverted to nonuse of the deafferented limb in the free situation. Similar results for short-term immobilization of the intact limb have been observed by Stein and Carpenter (1965).

It was found more recently, however, that if the straitjacket was left on for three days, then animals continued to use the deafferented limb extensively in the free situation (Taub et al., 1972a; Taub, 1976, Section II-A). It was also found that the monkeys exhibited the maximum function that they would achieve in that condition shortly after being placed in the straitjacket—usually within an hour or two. Even this brief delay in the appearance of maximal function appeared to be due as much to the emotionalizing effect of the novel restraint as to any other factor. This observation is of considerable importance for determining the nature of the mechanisms responsible for recovery of function following deafferentation. Its significance will be discussed in the section below devoted to this subject.

Learned Nonuse of a Single Deafferented Limb. The monkey's nonuse of a single deafferented limb in the free situation was in contrast to its ability to use it when the intact limb was restrained by a straitjacket, or in the conditioning situation (where the intact limb was also restrained). The nonuse was originally explained by the operation of an interlimb inhibition mechanism (Taub and Berman, 1968). It was assumed that this mechanism normally is held in check by the ipsilateral afferent input to the spinal cord, but that when this influence is abolished, as by dorsal rhizotomy, the interlimb inhibition is released so that movements of the intact limb would then have the effect of preventing movements of the deafferented limb. This hypothesis was proven inaccurate. (1) The monkeys demonstrated the ability to make a simultaneous flexion response of the intact and deafferented forelimbs in order to avoid electric shock (Taub et al., 1972a). (2) With sufficiently long straitjacket restraint of the intact limb, the animals could make extensive use of the deafferented limb in the free situation after removal from the restraining device.

The following formulation accounts for all the presently available data. Immediately after operation, monkeys cannot use a deafferented limb; recovery of function requires considerable time, as data from animals with bilateral forelimb deafferentation have shown (see below). An animal with one deafferented limb tries to use that extremity in the immediate postoperative situation, but finds that it cannot. Moreover, it gets along quite well in the laboratory environment on three limbs, while continued attempts to use the deafferented limb often lead to aversive consequences. The monkeys, therefore, learn not to try to use the deafferented limb. This habit persists, and consequently they never learn that, several months after operation, the limb has become potentially useful. This may be viewed as a form of learned helplessness (Seligman and Maier, 1967; Seligman et al., 1971). In contrast, animals with both limbs deafferented (and incapacitated) do not get along very well in the laboratory environment. Motivation to use the deafferented extremities remains very high (in contrast to the unilateral case), and as soon as utility begins to return, the limbs are used to the extent possible.

When the intact limb is immobilized several months after unilateral deafferentation, motivation to use the deafferented limb increases sharply, thereby overcoming the learned nonuse of that limb. The animal then uses the deafferented limb. However, if the straitjacket is removed a short while after the initial display of purposive movement (as was the case in the initial straitjacket experiment), the newly learned use of the deafferented limb acquires little strength and is, therefore, quickly overwhelmed by the well-learned habit of nonuse. If the straitjacket is left on for several days, however, use of the deafferented limb acquires strength and is then able to compete successfully with the learned nonuse of that limb in the free situation.

Bilateral Deafferentation

The situation is considerably less complex following deafferentation of both forelimbs (Knapp et al., 1963; Taub and Berman, 1968). In the immediate postoperative period, the limbs are useless. They exhibit little movement and typically hang limply, almost without tone. Subsequently, there is a gradual recovery of function. After the restitution process has gone to completion, animals are capable of using the limbs effectively for a wide variety of purposes, including ambulation, climbing, reaching toward objects, and grasping. These movements are frequently clumsy, particularly if they are rapid or if they involve the fingers. They often lack the fluency of control and precise timing characteristic of fully afferented rhesus monkeys. However, there is no single category of movement normally exhibited in the free situation that is completely abolished by somatosensory deafferentation. Moreover, almost all of these movements continue to remain possible if the view of body and limbs is obstructed by blindfolding or other means. An exception is those activities whose performance requires either intact vision or somatic sensation for the perception of external events (for example, avoidance of obstacles which the deafferented limb encounters but which the monkey cannot feel or see). In general, bilaterally deafferented animals display the maximum motor ability of which they are capable whenever it is appropriate. Neither restraint of portions of the body nor other special motivating circumstances are required for its evocation, as in the case of monkeys with unilateral deafferentation. However, deafferented limbs frequently sustain severe injuries, due to either self-mutilation or trauma incurred because incipient tissue damage cannot be perceived. Broken bones, ankylosed joints, lacerations, and ulcerations are common. Edema of the hands is a particular problem, often making effective movement of the digits impossible. Consequently, in order to observe maximum recovery of function, it is necessary to keep the deafferented limbs free of injury or to promote expeditious healing. This often involves intensive nursing care, frequent bandaging, and the use of a variety of mechanical restraints. When this is accomplished, recovery to the maximum function possible requires 2-10 months for most types of behavior, with a mean of 6 months. However, improvement in manual dexterity can continue to occur well into the second postoperative year, especially if there is no swelling and all joints retain free movement.

Interest in the present volume is focused primarily on the movements involved in locomotion. Ambulation in intact rhesus monkeys is almost always quadripedal, with the forequarters held somewhat higher off the ground than the hindquarters. However, monkeys commonly engage in a number of other types of locomotion appropriate to an arboreal environment, such as climbing, leaping from branch to branch, and, to a lesser extent, brachiation. Consequently, any discussion of the status of locomotion following deafferentation must include consideration of such categories of movement as ambulation, climbing, reaching toward objects, and prehension.

Ambulation. Immediately after bilateral deafferentation, when the forelimbs are useless, the rhesus monkey typically progresses by pushing itself along entirely by the use of the hindlimbs, the face and ventrum in contact with the floor. The forelimbs are frequently maintained in extension backwards along the torso. In the next, or intermediate, stage of recovery, the monkey employs one of several different means of progression. These can coexist temporally, or one of them can be characteristic of a given monkey for a period of time.

(1) The animal continues to push itself along with hindlimbs as in the first stage, but the hands are positioned at the level of the face or in front of the head with greater and greater frequency. A rudimentary alternating rhythm begins to develop, but no weight is borne, and the forelimbs cannot be said to be participating usefully in progression. With time the forelimbs do begin to bear weight, and the forequarters are lifted off the ground, minimally at first, and then higher and higher as recovery progresses. Alternating rhythm of the forelimbs and coordination with the hindlimbs are initially poor and highly variable, but gradually improve. This sequence seems intuitively to be a simple and direct method for progressively approximating normal ambulatory movements, but in addition the monkeys employ three other modes of locomotion during the recovery period that are rarely or never seen in intact animals.

(2) Duck-walk progression. Locomotion takes place while the monkey is in a sitting position. The hindlimbs are kept extended in front of the body, with the heels in contact with the floor.

The torso is then slid forward along the floor on the seatpads. The focus of traction is at the heels, with flexion taking place at the knees. The legs are then returned to an extended position in front of the animal, usually with an alternating rhythm, and the sequence is then repeated. The forelimbs are used first for balance and for defense when falling; with further recovery, they are employed for continuous postural support, frequently in alternating rhythm. Initially they bear little weight and are placed little, if any, forward of the torso. As support increases, the buttocks are raised slightly from the floor, permitting foot placement along the entire plantar surface. At this point the monkey's progression resembles that of a duck walking. Subsequently, more and more weight is borne on the forelimbs, they are placed further forward, and the body's center of gravity shifts rostrad. The back then becomes more nearly parallel to the floor, and the feet are placed to an increasing extent beneath the torso until they assume the correct position beneath the pelvic girdle.

(3) Rolling-over progression. The monkey sits up and then rolls forward or to one side across its shoulders, attempting to achieve a sitting position at the end of the roll. As the forelimbs begin to gain utility, they are used to help in pushing the forequarters up to achieve the final sitting posture. Later the animal propels itself forward with a simultaneous thrust of both hindlimbs and uses the forelimbs to prevent its face from hitting the floor. Further development is similar to that in mode (1).

(4) Bipedal progression. On rare occasions, this can be observed in normal monkeys. Two types are quite common in the early stages of recovery and become progressively less frequent as forelimb function returns. (a) The torso is held erect, which is difficult because of the pelvic structure. The animal therefore moves slowly, maintaining a precarious balance, frequently on the metatarsal pads with heels off the floor. The forelimbs are not used. (b) The animal keeps the torso inclined at an angle intermediate between the upright and the normal position during quadripedal ambulation, but the forelimbs are not employed for postural support. In this mode, the animal either falls frequently or moves for only short distances before stopping.

In the three modes of ambulation in which the forelimbs are employed during the intermediate recovery period, they buckle frequently at first, and with the failure of forelimb support the face and forequarters often hit the floor. Initially the hands are placed almost exclusively on the wrist dorsum and later the hand dorsum. Eventually, palmar placement appears, becomes predominant, and finally supplants dorsal placement on almost every step. The base of support is not particularly broad during the intermediate stages of recovery, as might be expected, but it is highly variable.

The balance between flexor and extensor muscles appears to remain normal (in contrast to the postdeafferentation condition of carnivores and rodents), and the affected extremities have no greater tendency to lag behind than to anticipate their placement in the normal temporal sequence of alternating progression. Again, the spatial and temporal relations are highly variable. As recovery proceeds, this variability decreases greatly so that, when maximal function is recovered, it is often difficult to distinguish an intact monkey from a bilaterally deafferented monkey in slow ambulation. The movements are smooth and well controlled, and cannot be characterized as ataxic. However, when the deafferented animal attempts to move more rapidly, the coordination tends to become degraded. At the most rapid speeds, movement sometimes becomes completely disorganized. The hands are misplaced beneath the body and the animal slips and falls. A motor deficit is also apparent when a deafferented monkey is required to walk on a 1-inch pipe suspended above the floor. The hands frequently slip off the pipe; nevertheless, monkeys usually are able to cross a 20-foot length of pipe without falling off.

Climbing. Climbing is a safe response for rhesus monkeys and is a typical reaction in a threat situation. When released in a colony room, a rhesus monkey normally climbs the nearest vertical structure--for instance, the outside mesh of cages. Following bilateral deafferentation, of course, monkeys cannot climb until they develop loose grasp capable of bearing some weight. In initial attempts at climbing cage mesh, the animal frequently helps steady itself by gripping a strand of mesh between its teeth. One hand is slowly moved above the head and the fingers are thrust through the mesh, often on the basis of visual guidance. The other arm is then moved above the head in sequence. As recovery progresses, the animal develops the ability to insert its fingers through the mesh rapidly and without visual guidance. Grasp becomes stronger and the fingers become more dexterous. Finally, the animal can climb up a bank of cages with almost the same speed as a normal animal. Lateral climbing on cage mesh is more difficult for rhesus monkeys than vertical climbing. Here, there is a clear difference between intact and deafferented animals, with the latter moving much more slowly and with frequent glances at the hands and fingers, even after maximum recovery has taken place.

Reaching Toward Objects. Rhesus monkeys usually begin attempts at reaching toward visual targets, such as food objects, within the first 2 weeks after deafferentation. At first these movements are phasic in nature. The limb jerks back to its starting position after only a short excursion in the intended direction, much as if it were held in place by a rubber band. The desired food can be reached only if it is nearby, and even then the limb is not under sufficiently good control to bat the food back toward the mouth. The curious oscillatory movement of the limb usually resolves itself into a simple ataxia within 2 weeks of its first appearance. Movements remain phasic, but they are now ballistic and tend to overshoot their mark. Subsequently, some months after operation, all tendency toward ataxia or tremor departs, and movements appear to be smooth and reasonably well controlled throughout their entire trajectory.

Reaching or pointing following full recovery is surprisingly accurate, though not normally as good as in intact animals. The difference has been quantified in an experiment (Taub et al., 1975b) in which intact and fully recovered deafferented monkeys were required to point at one of three randomly presented visual targets. The apparatus permitted different amounts of visual feedback from the responding limb, and in one condition view of the limb was prevented entirely. It was more difficult to train the deafferented animals than the intact animals to make the appropriate movements; the deafferented animals required approximately three times as long to reach their maximal level of performance. When it was attained, they did not perform with as great precision as the intact animals, either when vision was unimpeded or when view of the responding limb was obstructed, but there was overlap in the two distributions of accuracy. The training did not seem to effect any improvement in movement in the free situation.

Other investigators also have found that accurate directional pointing is possible in deafferented monkeys, both when view of their responding limbs is permitted (Gilman, 1970) and when it is prevented (Bossom and Ommaya, 1968; Liu and Chambers, 1971; Vierck, 1976).

Prehension. Prehension typically is carried out in intact rhesus monkeys by means of a movement that has been termed primitive pincer grasp (Jensen, 1961), in which the thumb is brought against the radial side of the index finger at the junction of the middle and distal phalanges. True opposition of thumb and forefinger (neat pincer grasp) usually occurs only when a task demands the finest dexterity of which the monkey is capable. In intact animals it can be evoked by placing small food objects such as raisins in deep wells (20 mm in diameter by 8 mm deep) in a dexterity board (Cole, 1952). A dexterity board with shallower wells (4 mm deep) promotes the performance of primitive pincer grasp.

When in the course of recovery an animal first employs its deafferented arm to reach for visual targets such as food objects, it typically sweeps them back along the ground within its loosely cupped hand. It does not usually develop grasp until well after effective reaching can be accomplished. Initially, grasp is crude. All four fingers are used in unison without participation of the thumb. When the thumb is first employed, it does not move independently but serves primarily as a support function. Primitive pincer grasp develops in all monkeys whose fingers remain relatively free of injury. Initially, it is performed entirely under visual guidance. When the goal object is small and is blocked from view by the hand, the animal frequently rotates the hand in order to see whether the grasp has been successful before bringing the object to its mouth. As manual dexterity increases and unsuccessful attempts at prehension become less frequent, this inspection procedure tends to drop out, and visual guidance in general decreases.

When primitive pincer grasp first develops, the hand is usually brought in from the side to permit support of the wrist by the flat surface upon which the target object rests. Subsequently, the animals develop the ability to adopt a normal posture of the wrist during reaching, and the hand is brought toward its objective from above.

In the author's laboratory, primitive pincer grasp has been observed as early as the second month after surgery, but it sometimes does not appear until the end of the first postoperative year. Neat pincer grasp eventually emerges in some animals, but by no means all of them, even when the fingers have sustained no apparent major damage. Moreover, it is never nearly so precise as in intact animals, is not evinced on every occasion when it would be appropriate, and is not always successful.

Complete Spinal Deafferentation

Deafferentation of the entire spinal cord has been accomplished successfully in three rhesus monkeys (Taub and Berman, 1968). Surgery was carried out in three stages, each involving approximately a third of the cord. The first stage was always performed on the most rostral portion of the cord and resulted in forelimb deafferentation. The three procedures were separated by sufficient time to allow maximal recovery of function to take place. Deafferentation of the middle third of the spinal cord produced no additional deficit in the forelimbs. While there was some impairment in the animals' general locomotor ability, it was compensated rapidly. Deafferentation of the final third of the spinal cord also produced no apparent effect on the ability to use the forelimbs. Although dorsal rhizotomy had abolished spinal reflex activity throughout the entire cord, the motor capacity of the forelimbs remained as extensive as that observed following recovery from deafferentation of the forelimbs alone. The monkeys were unable to use the hindlimbs effectively, but none of them survived more than 17 days after the final procedure, which is too brief a time for substantial recovery to take place.

Movements with Vision Obstructed, and Conditioned Response Experiments

Blindfolding in no way impaired the ability of fully recovered monkeys with bilateral forelimb deafferentation to ambulate when leashed to prevent running and placed on an unobstructed flat surface. The blindfold invariably emotionalized the rhesus monkeys, but after habituation they could walk for hundreds of feet without inappropriate stepping or falling. Their performance in no way resembled that associated with locomotor ataxia in man, where view of the legs is necessary as an aid in walking. On the basis of this observation, it was suggested (Taub and Berman, 1968) that the latter disability may be due less to loss of position sense as such than to a tonic imbalance or to hyperactivity of some reflex or supersegmental influence.

Blindfolded monkeys with bilateral forelimb deafferentation can also climb the mesh of an 8-foot bank of cages. However, in contrast to ambulation, there is a marked impairment initially, and the activity requires considerable practice for effective performance.

In conditioned response and learning experiments carried out in this and other laboratories, monkeys have demonstrated the ability to execute a variety of movements with deafferented limbs. Phenomena studied include grasp (Taub et al., 1963; Levine and Ommaya, Personal communication, 1971); regulation of force emission during forelimb flexion (Taub et al., 1967); maintenance of tonic or sustained limb flexion movement for extended durations (Taub et al., 1967); learning of a no/no-go auditory discrimination followed by discrimination reversal (Taub et al., 1966c); compensation for lateral displacement of vision produced by wedge prisms (Taub et al., 1966b; Bossom and Ommaya, 1968; Taub, 1968; Taub and Goldberg, 1974); performance of limb flexion during a very brief, prescribed temporal window (Wylie and Tyner, Personal communication, 1974); and performance of lateral arm movements across the mid-sagittal plane (Lamarre, Personal communication, 1974). It is of particular significance that in many of these studies the required motor activity was performed while the monkey was prevented from seeing the responding limb.

In conformity with the results for monkeys, it has been found that deafferented rats, cats, and dogs are capable of learning a variety of instrumental or operant conditioned responses, both when able and when unable to see the responding extremities (Gorska and Jankowska, 1959, 1961; Jankowska, 1959; Yamamoto, 1960;

Gorska et al., 1961; Konorski, 1962; Kozlovskaya in Skipin et al., 1969; Goldberger and Murray, Personal communication, 1974).*

Species Differences in Extent of Recovery of Function

Liu and Chambers (1971) have reported greater recovery of function in the stumptailed macaque monkey (Macaca speciosa) for the distal portion of deafferented forelimbs than for the proximal portion. For example, the precision of prehension was relatively good, while fixation of the limb for reaching and ambulatory movements was comparatively poor. However, in one rhesus monkey (Macaca mulatta) that was studied as part of the same experiment, the proximo-distal distinction was not so clear-cut (Personal communication, 1971). In this author's laboratory, Macaca mulatta have not been noted to recover more fully in the proximal than in the distal portions of deafferented limbs.

Thus, there may well be a true difference in the nature of the recovery of function following limb deafferentation in two primate species of the same genus. However, it is small compared with the differences that occur between more widely separated taxonomic categories within the vertebrate series (Taub, 1975). These large differences in the effect of deafferentation make it unwise to compare the results of experiments with different species in terms of one another, without an explicit awareness of the problem that may be involved. It should be emphasized that the descriptions of movement following deafferentation presented herein are quite specific to rhesus monkeys. (For additional descriptions of movements in deafferented rhesus monkeys, see Bossom, 1972, 1974, and Denny-Brown, 1966.)

DEAFFERENTATION AT BIRTH AND PRENATALLY

Deafferentation research with very young rhesus monkeys was undertaken to determine whether somatosensory feedback and spinal reflexes might not be more important for the development of species-characteristic postures and movements than for their performance after motor patterns are mature, or for their recombination in learning situations. Experiments of two types have been performed. In one series, monkeys were removed from their mothers on the day of birth and were given bilateral forelimb deafferentation. In addition, two of these deafferented animals were blinded

---

*Results with respect to the ability to accomplish classical or Pavlovian conditioning of movements in deafferented portions of the body are at present unclear (Taub, 1976, Section III-G).

by sewing the eyelids closed. In a second series, monkey fetuses were deafferented two-thirds of the way through gestation; they were removed from the uterus (but left connected to the maternal circulation), operated, and then returned to the uterus to await Caesarian delivery shortly before full term.

Both the infants deafferented and those deafferented/blinded on the day of birth developed the ability to use the forelimbs for postural support during sitting, for standing, and for ambulation (Taub et al., 1972b, 1972c, 1973, 1975c). Table 1 compares the time of appearance of different age-characteristic postures and movements in the experimental animals with their time of appearance in intact and blinded-only infant monkeys. As compared with normal infants, the emergence of these postures and movements was retarded by approximately 2 weeks in the deafferented infants and by approximately 3-1/2 weeks in the deafferented/blinded infants. By the age of 3 months, the movements of the proximal portion of the deafferented forelimbs appeared to be as good as that of animals deafferented in adolescence. However, all the infants exhibited a greater deficit in the use of the hands. Whether sighted or blinded, they displayed a tendency toward predominant placement on dorsal rather than palmar surfaces. The deafferented/ sighted infants were far less accurate in reaching for objects than adolescent deafferented animals, and these movements tended to be ballistic. Loose grasp, in which all four fingers were used in unison and the thumb did not participate, was seen on occasion, but it was used only in climbing and for support of weight on a ladder or swing--not for picking up objects. Thumb-forefinger prehension did not emerge in any of the animals. There is evidence, however, that this grave impairment in spontaneously developing movements of the hand may have been an artifact due to the need to keep the lower portions of the arms bandaged most of the day from an early age in order to prevent self-inflicted damage (Taub et al., 1973, 1975c).

Though most movements of the hand did not develop spontaneously, they could be brought into existence through the use of operant shaping techniques. These are training methods in which the difficulty of the required behavior is increased in small steps. Work was started with two animals at the age of 4 months, but with one animal was not initiated until 15 months. Even after extensive training, performance was not quite as good as can be observed following deafferentation in adolescence. However, within 30 sessions use of thumb and forefinger occurred on fully half the trials, and on approximately 20 percent of the trials there was true primitive pincer grasp involving independent movement of the thumb to the radial side of the forefinger.

Table 1. Age (in weeks) of appearance and comparative retardation of different types of posture and motor activities. The values are the mean ages of the first definite appearance of the behaviors (Taub et al., 1973).

Type of posture or motor activity	Deaff. only[a] (1)	Intact[b] (2)	Blind deaff.[a] (3)	Blind only (4)	Retardation (1) vs (3)
Visually guided reaching	1.1	0.4	--	--	--
Crouching, arms crossed	1.5	--[c]	2.5	--[c]	1.0
Sitting-crouching, arms uncrossed	1.8	0.6	4.0	0.4	2.2
Standing on all fours	2.4	0.6	3.5	0.7	1.1
First step	3.1	0.3	4.5	0.6	1.4
Sequential steps	3.3	0.7	4.5	0.7	1.2
Crude ambulation	3.6	2.0	5.5	2.0	1.9
Mean retardation	(1) vs (2) = 1.8		(3) vs (4) = 3.5		1.5

[a] Time of emergence of each of the behaviors was very consistent across animals in the group.

[b] Normal data derived from Hines (1942).

[c] This stage is exhibited only in deafferented animals.

The blinded/deafferented infants used the deafferented forelimbs well in coordination with other, intact portions of the body, but developed little independent use of these extremities. Independent movement of the limbs would be of relatively little value to an animal that could not determine the location of objects at a distance from itself, either through vision or through contact with the surfaces of the deafferented members. Indeed, if such movements were to develop spontaneously, there would be a tendency for them to be exhibited infrequently or even extinguished completely, due to lack of reward for their performance. However, it proved relatively easy to train these animals to make reaching, laterally differentiated pointing, crude grasping, and grasp-releasing movements through the use of shaping techniques (Taub et al., 1975c).

Berman and Berman (1973) have carried out forelimb deafferentation in term infants during the course of Caesarian delivery. These animals exhibited most of the behaviors that developed spontaneously in the animals described above, but the movements tended to emerge much later in life.

Since the rhesus neonate is capable of a fairly extensive variety of motor activities at birth, a great deal of motor development must occur during intrauterine life. Consequently, in order to completely resolve the issue of the role of somatic sensation and spinal reflexes in the ontogenetic development of behavior, it would be necessary to carry out deafferentation prenatally.

In an initial experiment 11 monkey fetuses were given either unilateral or bilateral forelimb deafferentation (C_1 or C_2 through T_4) two-thirds of the way through gestation (Taub et al., 1974), using a modification of techniques developed in other laboratories (Taub et al., 1975c). Five of these infants (45%) survived through Caesarian delivery and at least 5 months of infancy. All 5 animals were severely quadriparetic. Upon postmortem examination it was found that malformation of the vertebrae in the operative region had permitted the overlying muscle to compress and severely damage the spinal cord. In subsequent work, to prevent this type of injury, a prosthetic bridge (fabricated by Heyer-Schulte Corp., Goleta, California) was emplaced at time of fetal operation to substitute for the portions of the vertebrae removed by laminectomy (Taub et al., 1975a). A second device was designed by John Boretos to replace the originally implanted prosthesis after birth. Two animals treated in this fashion following prenatal left-unilateral forelimb deafferentation survived and are 6 and 8-1/2 months old, respectively, at the time of this writing. Motor function in the hindlimbs and the intact forelimb is apparently normal, indicating that the prostheses served their intended purpose in protecting the spinal cord. Consequently, these infants are valid subjects for study. Both animals currently use their deafferented extremity as well as did the monkeys deafferented at birth (see Figures 1-3).

MECHANISMS ENABLING THE PERFORMANCE OF PURPOSIVE MOVEMENT FOLLOWING DEAFFERENTATION

The experimental results described above pose some fundamental questions about the nature of the recovery of function after deafferentation. As has been reported, when deafferentation is carried out in the adolescent, there is initially a complete loss of the ability to use the limb. Indeed, the affected extremities are

Figure 1. The younger infant at 5-1/2 months supports its full weight solely by its deafferented forelimb while hanging from a swing.

Figure 2. The younger infant at 5-1/2 months employs the deafferented forelimb (identified by a white band) in good coordination with the other extremities during ambulation.

almost atonic and exhibit very little movement even of a random nature. Subsequently, there is a recovery of function, but the restitution process is gradual. After recovery is complete, rhesus monkeys are able to use a deafferented limb in a clumsy but effective way to perform the majority of functions of which intact animals are capable. Each postdeafferentation stage raises its own central enigma: (1) Why is there a loss in motor function following deafferentation: (2) Why is the recovery process gradual? (3) What are the mechanisms that enable a deafferented

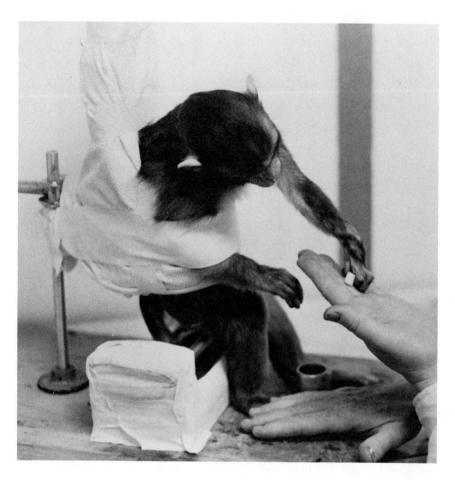

Figure 3. The older infant at 8 months exhibits use of the thumb and forefinger of its deafferented hand in prehension. This ability did not develop spontaneously, but had to be trained, as in the infants deafferented at birth. Note that the distal phalange of the forefinger has been lost as a result of previous injury. The intact limb is restrained to simplify training.

monkey deprived of spinal reflexes and somatosensory input to perform the movements that appear during the recovery process, not only when able to see its limbs but also when prevented from doing so? Still another question is the significance of the motor function that develops following deafferentation at or before birth.

The data presently available do not yet provide any definitive answers to these questions. They do, however, permit the formulation of hypotheses, but of differing degrees of inferential strength in each case.

The Role of Learning in the Recovery of Function

Because the recovery of function is gradual, it might be inferred that learning is an important factor. The process would presumably involve a slow, stepwise acquisition of new methods of accomplishing behavioral objectives to replace the basis on which behavior was previously elaborated. Indeed, recovery of function after different types of CNS damage is frequently referred to as a reeducation process. In this regard, it is relevant to point out that the full range of motor ability that one normally observed in the free situation develops spontaneously following bilateral forelimb deafferentation. No special training is required to generate its expression. This, of course, does not indicate that learning is not involved in the recovery process. The need to carry out routine functions can obviously be a powerful motivating force. In effect, the normal environment itself provides the contingencies of reinforcement for performing the maximal motor function of which a bilaterally deafferented animal is capable.

However, there is one piece of evidence that strongly militates against the assignment of an important role to learning in the recovery process. This is the swiftness with which unilaterally deafferented monkeys developed the ability to use the affected extremity when subjected to restraint of the intact, paired limb. These animals had never before been observed to use their deafferented limb in the free situation, but when motivation was increased sufficiently to force them to do so, they soon employed it at the level of motor sophistication appropriate to the time that had elapsed since operation (as indicated by the data from animals given bilateral forelimb deafferentation). If learning had been an important factor in the recovery process, then motor ability should have returned slowly. It is true that motor ability emerged in a progressive fashion over a period of 1 or 2 hours following straitjacket restraint of the unimpaired forelimb. The emotionalization produced by the restraining device was certainly a factor in this delay. Some learning may also be involved, but this would not account for the abridgement of the months-long process required before asymptotic motor function is attained after bilateral forelimb deafferentation.

What, then, is the significance of all the motor learning that clearly takes place in the conditioning experiments that have been carried out with deafferented animals? Do these studies not indicate that learning is involved in the recovery process? It is true that training is needed when an animal is required to learn new skills with a deafferented limb, as in the experiment where pointing at a visual target had to be carried out, sometimes even without view of the responding limb. The need for training becomes especially important when a task imposes special demands upon a deafferented animal as a result of its inability to perceive aspects of the environment involved--e.g., in such a task as climbing while blindfolded. In this case, the animal must develop strategies for correctly directing its fingers through the openings in a mesh that it can neither see nor feel. However, intact animals must also be trained when they are required to perform a new skill. And, indeed, it is in this area that one of the major deficits following deafferentation lies. Deafferented animals require considerably longer to acquire most new skills than do intact animals. This was certainly true in the pointing experiment described above and was also a consistent finding in almost all other training experiments.

Another factor that would seem to weigh against an important role for learning in the recovery process is the fact that there is no apparent transfer from the conditioning to the free situation. Following training in either the pointing or grasp experiments, or in securing food from dexterity boards, there was no appreciable improvement when the animals were returned to their home cages. What they had apparently learned was simply to apply the motor capacity they already possessed in a new way under novel conditions. It is, of course, possible that training can provide a basis for adding precision to the finest coordination that a rhesus monkey deafferented in adolescence can display in the free situation, but even this intuitively reasonable notion has at present no data to support it. It is also possible that training and practice may speed the recovery of some movements, but again, there is currently no clear evidence to indicate that this is the case. Learning is almost certainly involved in the nonuse of a deafferented limb (learned helplessness) in the free situation following unilateral deafferentation, as described in a previous section. However, the learning process here produces the opposite of an improvement in motor ability.

If learning is not a central feature, what are the factors that are primarily responsible for recovery of function following deafferentation? The mechanisms may well be related to those that cause the postdeafferentation loss of function in the first place. Rhizotomy, when carried out proximal to the dorsal root ganglion, leads to degeneration of the central root fibers and to the consequent disappearance of primary afferent terminals from their central connections. In addition, it is presumed that the elimination of somatosensory input results initially in a reduction within the CNS in the background level of excitation. The amount of incoming stimulation necessary to produce movement would thereby be greatly elevated. The loss of somatosensory input could also release a generalized inhibitory process. In any event, it is highly probable that the postdeafferentation loss of motor function is related to the spinal shock that follows spinal transection, with the same processes involved in both. Cross-species variations in the speed of recovery from the two procedures are certainly similar, requiring minutes or hours in amphibia, several weeks in rodents and carnivores, and months in monkeys.

Stavraky and coworkers (Drake and Stavraky, 1948; Teasdall and Stavraky, 1953, 1955; Stavraky, 1961) have shown that after a period of time deafferented spinal motoneurons become supersensitive to a variety of influences. In addition, the debris from the degenerating afferent terminals appear to serve as a stimulus for the sprouting of collaterals from intact axons in the vicinity (Liu and Chambers, 1958; Goldberger, 1974; Goldberger and Murray, 1974). Collateral sprouting would presumably lead to the reinnervation of vacated synaptic surfaces and therefore would probably reduce or eliminate neuronal supersensitivity. Collateral sprouting and supersensitivity, either in sequence or together, probably contribute importantly to the recovery of function following deafferentation in the monkey. Both processes are essentially neutral with respect to an organism's behavioral interactions with the environment. Consequently, they would proceed whether or not learning took place. Since both are progressive over time, a gradual return of motor ability would result, unless it was masked in some manner, as is the case following unilateral deafferentation.

Central Programs and Central Feedback

The foregoing discussion concerns the nature of the processes responsible for the initial loss of motor function after deafferentation and the gradual nature of the recovery of function. It tells little about the mechanisms which enable deafferented monkeys to perform specific movements in the absence of spinal reflexes and sensory input relating to position sense.

A possible explanation of some aspects of the phenomenon would be the existence of central motor programs. Presumably these could either be established through learning or be part of a species' genetic endowment. In either case, once the program was activated, the encoded behavior or sequence of behaviors would be run off without requiring the support of sensory feedback of any kind. Central programs for the control of various types of locomotion have been clearly demonstrated in a number of invertebrate species. Several papers in this volume provide relevant data. Central programs could also exist within the primate CNS. These could be located at the spinal level, and the work of Grillner (1973 and this volume) has recently focused considerable attention on this possibility. The research with deafferented monkeys strongly suggests that central programs could also exist at cephalic level. Many of the complex motor sequences in which these animals have been observed to engage could not conceivably be mediated only by spinal centers. Reaching toward visual objects, thumb-forefinger prehension and other manipulator activities must clearly be integrated at cephalic level, and in the primate much of the integration probably occurs in cortical structures.

It is reasonable to assume that this type of mechanism is responsible for much of the purposive movement observed in deafferented monkeys. However, it could not account for all the behavioral phenomena that have been observed. In particular, it could not explain how an animal could learn to make new uses of a deafferented limb while unable to see the limb, as has occurred in many operant or instrumental conditioning experiments. This type of learning involves the ability to repeat certain, specified movements consistently. In order for this to be accomplished, it is essential that an animal obtain information about the topography of previously executed movements. But how can this be accomplished with a limb that the animal can neither feel nor see? How does it know where the limb is, whether it has moved, and, if so, in what way? Since electrophysiological (Taub and Berman, 1968; Vaughan et al., 1970; Cohn et al., 1972; Wylie et al., 1973), behavioral, and anatomical (Taub, 1976, Section V-A) evidence indicates that peripheral pathways are not involved, then the required information must be provided by some central mechanism. One such mechanism would involve the existence of wholly central feedback pathways. These could provide an organism with information concerning the nature of movements without any contribution from peripheral receptors by indicating the pattern of the descending impulses that will ultimately produce these movements. A number of intracentral loop pathways with points of inflection at several different levels of the neuraxis have been demonstrated both anatomically and electrophysiologically; these could serve the purpose of central efferent monitoring. (See Taub and Berman, 1968, and Taub, 1976, for citations of the literature and further discussion of central mechanisms in the control of movement following deafferentation.)

Endogenous Motor Programs at Birth

The results with young monkeys indicated that no special training is necessary to enable the emergence of well-coordinated movements of the proximal forelimb musculature when deafferentation is carried out at or before birth. Use of the distal musculature, particularly the fingers and hands, was greatly impaired, and operant shaping was needed to bring such movements as thumb-forefinger prehension into existence. The necessity for bandaging the affected limbs of these animals, most of them from an early age, prevents conclusive interpretation of this result. There are indications in the data that, if bandaging could have been avoided, then greater use of the hands might have developed spontaneously. However, it may well be that some learning is important for the elaboration of manual dexterity.

There was no ambiguity with respect to the spontaneous development of such movements as ambulation, climbing, and reaching toward objects. An inference that can be drawn from the data is that the motor programs for these activities are already present within the primate CNS, at least as early as the end of the second trimester of prenatal life. The evidence that learning is not involved in the formation of the basic central templates for these patterns of coordination cannot yet be considered conclusive, but it is suggestive. The role of learning for these movements, then, would be to sharpen and refine the genetically determined motor endowment of an animal.

Acknowledgment

The author's work with adolescent deafferented monkeys before 1968 was supported by NIH Grant NB 3045 to Dr. A.J. Berman. The work with adolescent monkeys after 1968 and the neonatal deafferentation research were supported by NIH Grants MH 16954 and FR 5501RR05636. The studies involving prenatal deafferentation were supported initially by a grant from the W.T. Grant Foundation, currently by NIH Grant HD 08579, and throughout their duration by NIH Grant FR 5501RR05636. Early pilot work for the prenatal studies was supported by AFOSR Grant 1042. The prism adaptation research was made possible first by AFOSR Grant 1042 and subsequently by NIH Grant MH 18821. We thank Jean Swauger for her editorial assistance.

REFERENCES

Berman, A.J. and Berman, D., (1973) Fetal deafferentation: The ontogenesis of movement in the absence of peripheral sensory feedback. Exp. Neurol. 38, 170-176.

Bossom, J. and Ommaya, A.K., (1968) Visuo-motor adaptation (to prismatic transformation of the retinal image) in monkeys with bilateral dorsal rhizotomy. Brain. 91, 161-172.

Bossom, J., (1972) Time of recovery of voluntary movement following dorsal rhizotomy. Brain Res. 45, 247-250.

Bossom, J., (1974) Movement without proprioception. Brain Res. 71, 285-296.

Cohn, R., Jakniunas, A. and Taub, E., (1972) Summated cortical evoked response testing in the deafferented primate. Science. 178, 1113-1115.

Cole, J., (1952) Three tests for study of motor and sensory abilities in monkeys. J. Comp. Physiol. Psychol. 45, 226-230.

Denny-Brown, D., (1966) The Cerebral Control of Movements. Liverpool University Press, Liverpool.

Drake, G.D. and Stavraky, G.W., (1948) An extension of the "law of denervation" to efferent neurons. J. Neurophysiol. 11, 229-238.

Forssberg, H. and Grillner, S., (1973) The locomotion of the acute spinal cat injected with clonidine i.v. Brain Res. 50, 184-186.

Gilman, S., (1970) "The nature of cerebellar dyssynergia," In Modern Trends in Neurology, (Williams, D., ed.), Butterworths, London, (60-79).

Goldberger, M.E., (1974) "Recovery of movement after CNS lesions in monkeys," In Recovery of Function after Neural Lesions, (Stein, D., ed.), Academic Press, New York, (265-337).

Goldberger, M.E. and Murray, M., (1974) Restitution of function and collateral sprouting in the cat spinal cord: The deafferented animal. J. Comp. Neurol. 158, 37-54.

Gorska, T. and Jankowska, E., (1959) Instrumental conditioned reflexes of the deafferented limb in cats and rats. Bull. Acad. Pol. Sci. 7, 161-164.

Gorska, T. and Jankowska, E., (1961) The effects of deafferentation on instrumental (Type II) conditioned reflexes in dogs. Acta Biol. Exp. 21, 219-234.

Gorska, T., Jankowska, E. and Kozak, W., (1961) The effects of deafferentation on instrumental (Type II) cleaning reflex in cats. Acta Biol. Exp. 21, 207-217.

Goslow, G.E., Jr., Reinking, R.M. and Stuart, D.G., (1973) The cat step cycle: Hind limb joint angles and muscle lengths during unrestrained locomotion. J. Morphol. 141, 1-41.

Grillner, S., (1973a) "Locomotion in the spinal cat," In Control of Posture and Locomotion, (Stein, R.B., Pearson, K.G., Smith, R.S. and Redford, J.B., eds.), Plenum, New York, (515-535).

Grillner, S., (1973b) "On the spinal generation of locomotion," In Sensory Organization of Movements, (Batuev, A.S., ed.), Leningrad.

Grillner, S., (1975) Locomotion in vertebrates: Central mechanisms and reflex interaction. Physiol. Rev. 55, 247-304.

Herman, R., Cook, T., Cozzens, B. and Freedman, W., (1974) "Control of postural reactions in man: The initiation of gait," In Control of Posture and Locomotion, (Stein, R.B., Pearson, K.G., Smith, R.S. and Redford, J.B., eds.), Plenum, New York, (363-388).

Hines, M., (1942) The development and regression of reflexes, posture, and progression in the young macaque. Contr. Embryol. Carnegie Inst. Wash. 30, 153-209.

Jankowska, E., (1959) Instrumental scratch reflex of the deafferented limb in cats and rats. Acta Biol. Exp. 19, 233-242.

Jensen, G.D., (1961) The development of prehension in a macaque. J. Comp. Physiol. Psychol. 54, 11-12.

Knapp, H.D., Taub, E. and Berman, A.J., (1958) Effect of deafferentation on a conditioned avoidance response. Science. 128, 842-843.

Knapp, H.D., Taub, E. and Berman, A.J., (1963) Movements in monkeys with deafferented forelimbs. Exp. Neurol. 7, 305-315.

Konorski, J., (1962) Changing concepts concerning physiological mechanisms in animal motor behavior. Brain. 84, 227-294.

Lassek, A.M., (1953) Inactivation of voluntary motor function following rhizotomy. J. Neuropath. Exp. Neurol. 3, 83-87.

Liu, C. and Chambers, W.W., (1958) Intraspinal sprouting of dorsal root axons. AMA Arch. Neurol. Psychiat. 79, 46-61.

Liu, C.N. and Chambers, W.W., (1971) A study of cerebellar dyskinesis in the bilaterally deafferented forelimb of the monkey (Macaca mulatta and Macaca speciosa). Acta Neurobiol. Exp. 31, 263-289.

Mott, F.W. and Sherrington, C.S., (1895) Experiments upon the influence of sensory nerves upon movement and nutrition of the limbs. Proc. R. Soc. London. 57, 481-488.

Munk, H., (1909) Uber die Functionen von Hirn und Ruckenmark. Hirschwald, Berlin, (247-285).

Seligman, M.E.P. and Maier, S.F. (1967) Failure to escape traumatic shock. J. Exp. Psychol. 74, 1-9.

Seligman, M.E.P., Maier, S.F. and Solomon, R.L., eds., (1971) "Unpredictable and uncontrollable aversive events," In Aversive Conditioning and Learning, Academic Press, New York, (347-400).

Skipin, G.V., Ivanova, I.G., Kozlovskaya, I.B. and Vinnik, R.L., (1969) "Defense conditioning and avoidance," In A Handbook of Contemporary Soviet Psychology, (Cole, M. and Maltzman, E., eds.), Basic Books, New York, (785-810).

Sperry, R.W., (1945) The problem of central nervous reorganization after nerve regeneration and muscle transposition. Quart. Rev. Biol. 20, 311-369.

Stavraky, G.W., (1961) Supersensitivity Following Lesions of the Nervous System. University of Toronto Press, Toronto.

Stein, B.M. and Carpenter, M.W., (1965) Effects of dorsal rhizotomy upon subthalamic dyskinesis in the monkey. Arch. Neurol. 13, 567-583.

Stuart, D.G., Withey, T.P., Wetzel, M.C. and Goslow, G.E., Jr., (1974) "Time constraints for inter-limb co-ordination in the cat during unrestrained locomotion," In Control of Posture and Locomotion, (Stein, R.B., Pearson, K.G., Smith, R.S. and Redford, J.B., eds.), Plenum, New York, (537-560).

Taub, E. and Berman, A.J., (1963) Avoidance conditioning in the absence of relevant proprioceptive and exteroceptive feedback. J. Comp. Physiol. Psychol. 56, 1012-1016.

Taub, E., Bacon, R. and Berman, A.J., (1965) The acquisition of a trace-conditioned avoidance response after deafferentation of the responding limb. J. Comp. Physiol. Psychol. 58, 275-279.

Taub, E., Ellman, S.J. and Berman, A.J., (1966a) Deafferentation in monkeys: Effect on conditioned grasp response. Science. 151, 593-594.

Taub, E., Goldberg, I.A., Bossom, J. and Berman, A.J., (1966b) Deafferentation in monkeys: Adaptation to prismatic displacement of vision. Paper presented at meeting of Eastern Psychological Association, New York.

Taub, E., Teodoru, D., Ellman, S.J., Bloom, R.F. and Berman, A.J., (1966c) Deafferentation in monkeys: Extinction of avoidance responses, discrimination and discrimination reversal. Psychonom. Sci. 4, 323-324.

Taub, E., Schlossberg, S., Teodoru, D. and Berman, A.J., (1967) Deafferentation in monkeys and sensory prosthesis. Paper presented at meeting of Psychonomic Society, St. Louis.

Taub, E., (1968) "Prism compensation as a learning phenomenon: A phylogenetic perspective," In The Neuropsychology of Spatially Oriented Behavior, (Freedman, S.J., ed.), Dorsey Press, Homewood, Illinois, (77-106).

Taub, E. and Berman, A.J., (1968) "Movement and learning in the absence of sensory feedback," In The Neuropsychology of Spatially Oriented Behavior, (Freedman, S.J., ed.), Dorsey Press, Homewood, Illinois, (173-192).

Taub, E., Barro, G., Parker, B. and Gorska, T., (1972a) Utility of a limb following unilateral deafferentation in monkeys. Paper presented at meeting of the Neuroscience Society, Houston.

Taub, E., Perrella, P.N. and Barro, G., (1972b) Behavioral development in monkeys following bilateral forelimb deafferentation on the first day of life. Trans. Am. Neurol. Assn. 97, 101-104.

Taub, E., Perrella, P.N. and Barro, G., (1972c) Motor capacity following somatosensory deafferentation on first day of life in monkeys. Fed. Proc. 31, 821. (Abstract).

Taub, E., Perrella, P.N. and Barro, G., (1973) Behavioral development following forelimb deafferentation on day of birth in monkeys with and without blinding. Science. 181, 959-960.

Taub, E. and Goldberg, A., (1974) Use of sensory recombination and somatosensory deafferentation techniques in the investigation of sensory-motor integration. Perception. 3, 393-408.

Taub, E., Barro, G., Miller, E.A., Perrella, P.N., Jakniunas, A., Goldman, P.S., Petras, J.M., Darrow, C.C., II and Martin, D.P., (1974) Feasibility of spinal cord or brain surgery in fetal rhesus monkeys. Paper presented at meeting of Society for Neuroscience, St. Louis.

Taub, E., Barro, G., Heitmann, R., Grier, H.C., Boretos, J.W. and Cicmanec, J.L., (1975a) Behavioral development in monkey infants following forelimb deafferentation of exteriorized fetuses at the end of the second trimester of pregnancy. Neurosci. Abst. 1, 786.

Taub, E., Goldberg, I.A. and Taub, P.B., (1975b) Deafferentation in monkeys: Pointing at a target without visual feedback. Exp. Neurol. 46, 178-186.

Taub, E., Perrella, P.N., Miller, E.A. and Barro, G., (1975c) Diminution of early environmental control through perinatal and prenatal somatosensory deafferentation. Biol. Psychiat. 10, 609-626.

Taub, E., (1976) Movement in nonhuman primates deprived of somatosensory feedback. Exercise Sport Sci. Rev. 4, (In press).

Teasdall, R.D. and Stavraky, G.W., (1953) Responses of deafferented spinal neurons to cortico-spinal impulses. J. Neurophysiol. 16, 367-375.

Twitchell, T.E., (1954) Sensory factors in purposive movement. J. Neurophysiol. 17, 239-254.

Vaughan, H.G., Jr., Gross, E.G. and Bossom, J., (1970) Cortical motor potential in monkeys before and after upper limb-deafferentation. Exp. Neurol. 26, 253-262.

Vierck, C.J., Jr., (1976) "Proprioceptive deficits after dorsal column lesions in monkeys," In Somatosensory System, (Kornhuber, H.H., ed.), Georg Thieme, Stuttgart, Germany. (In press).

Wylie, R.M., Barro, G., Perrella, P.N., Weinberg, S.G. and Taub, E., (1973) An electrophysiological study of the spinal cords of deafferented monkeys. Paper presented at meeting of the Society for Neuroscience, San Diego, California.

Yamamoto, C., (1960) Role of the peripheral sensory pathway in performance of the conditioned leg flexion reflex. Tohuku J. Exp. Med. 72, 83-90.

THE DEVELOPMENT OF NEURAL CIRCUITS IN THE LIMB MOVING SEGMENTS OF THE SPINAL CORD

Lynn Landmesser

Biology Department, Yale University

New Haven, Connecticut 06520

The problem of how motoneurons in the vertebrate spinal cord establish appropriate peripheral and central connections during embryogenesis has been considered with emphasis on the chick. The sequence of synapse formation in the cord does not allow one to determine whether motoneurons are intrinsically specified, or specified by the muscles with which they synapse or by other mechanisms such as position in the cord. It was shown however that motoneurons grow out very selectively to the appropriate muscles from the start and that from the time that spontaneous movement begins, it is co-ordinated with reciprocal activation of antagonists, possibly mediated by reciprocal inhibitory connections. Various models for bringing about the observed connectivity are reviewed and experimental manipulations utilizing transplanted supernumerary limbs proposed that might allow one to distinguish between the different possibilities.

Introduction

During embryogenesis, the parts of the vertebrate spinal cord which eventually innervate the limbs develop regional differences, both anatomical and functional, presumably associated with their locomotor capacity. Although the thoracic segments of the spinal cord in both amphibians and birds are capable of innervating adjacent supernumerary limbs, these limbs never move in the characteristic motor pattern and eventually display severe muscle atrophy (Székely, 1963; Straznicky, 1963; Hollyday and Mendell, 1975).

There also appear to be regional peculiarities associated with the fore and hind limb. This is especially evident in the chick where hindlimbs innervated by transplanted brachial spinal cord perform flapping movements characteristic of wings, rather than the alternate stepping associated with the hindlimb (Narayanan and Hamburger, 1971).

What interactions occur during development to bring about the characteristic structure of the limb innervating cord, and how do motoneurons become synaptically connected both to appropriate muscles and to the other neurons that terminate on them? Unfortunately, the actual mechanisms involved in bringing about spinal cord connectivity remain largely unknown.

This chapter will summarize the classical and contemporary anatomical and behavioral studies dealing with this matter in order to identify problems and to suggest areas for further research. The focus will be on the chick spinal cord since its early development has been studied anatomically, behaviorally, and more recently electrophysiologically. In addition, it seems to possess the greatest potential for experimental manipulation in the early embryo. In addition, results obtained with chicks should be applicable to the developing mammalian spinal cord, which appears generally similar.

The potential relevance of the study of spinal cord development to the understanding of locomotion is twofold. The first, recognized by the early anatomists (Ramon y Cajal, 1929), is one of simplifying the system. In the early cord only motoneurons have differentiated and sent out axons. Other types of cells differentiate and synapse in a definable sequence. Since spontaneous motility begins very early, especially in the chick, it is possible to study functionally the development of motor capacity, and to determine how the addition of various anatomically defined elements modifies it. The application of electrophysiological techniques to this system should greatly extend the information already acquired with behavioral techniques. It should also be possible to destroy selected elements in the system early in development and to observe the effect on subsequent locomotor function (Hamburger et al., 1966; Oppenheim, 1975). Secondly, knowing the rules through which specific connectivity is achieved may help in understanding the functional organization of the cord. Since connectivity is modified more easily experimentally during development, it may be possible to determine whether motoneurons actually are identified by the muscle in which they terminate, by their geometric position in the cord, or by other factors.

Models for Bringing about Connectivity

Although there are numerous fine differences, the basic organization of the spinal cord is similar for most vertebrates. Motoneurons innervating specific muscles are grouped into columns in the spinal cord, with more cranial columns innervating proximal muscles, more caudal ones distal muscles. In broad outline then, the somatotopic representation of motoneurons is similar for the cat (Romanes, 1964), frog (Cruce, 1974) and chick (Landmesser and Morris, 1975). Thus motoneurons can be defined both by the muscles in which they terminate and by the position of their cell bodies in the cord. Which of these is of primary importance in bringing about specific connectivity needs to be determined.

In order to place the problems confronting the motoneuron in a broader context, several basic mechanisms by which specific connectivity may be accomplished will be reviewed. These include the following models:

(1) The motoneuron is, as the result of an early developmental interaction, specified with respect to its peripheral destination and must selectively grow to the appropriate target guided by chemical, mechanical or other clues. Alternatively, axon outgrowth may be random, with the ultimate death of motoneurons that fail to establish appropriate neuromuscular contacts. Synapse formation onto motoneurons in either case would result by selective recognition of the chemical identity of such cells, perhaps encoded in their surface membrane. (For a more extensive review of these concepts the reader is referred to Cotman and Banker, 1974.)

(2) The motoneuron is not early specified, but has its chemical identity impressed upon it by the muscle with which it synapses (myotypic specification). Synapses onto it again would result through a recognition process.

(3) The motoneuron does not possess a distinct biochemical identity. Rather, specific connections are brought about either by virtue of its position in the cord or by temporal sequences of innervation.

All of the above mechanisms are supported partially by developmental studies in other systems (Hughes and Prestige, 1967; Landmesser and Pilar, 1972, 1975; Lopresti et al., 1973; Crossland et al., 1974; Hollyday and Mendell, 1975) and the problem is to decide how they relate to the spinal cord motoneuron.

Early Development of the Chick Cord

The motoneurons arise by mitosis from the ependymal neuroepithelium and migrate into the mantle layer. Recent studies, (Hollyday, Unpublished observations) fix the birthdays of the brachial motoneurons between stages 15-22 of Hamburger and Hamilton, and of the lumbosacral motoneurons, stages 17-23 (roughly 2 - 2-1/2 days). The elongate cells, containing scanty cytoplasm, early send their axons out the ventral roots (Windle and Orr, 1934; Lyser, 1964) (Figure 1A).

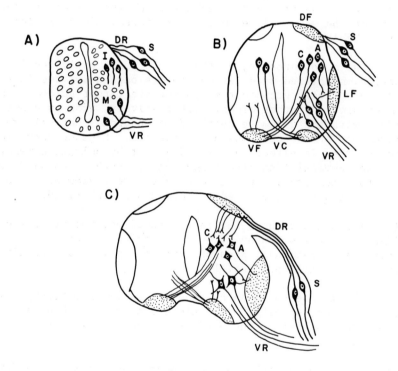

Figure 1. Early anatomical development of the chick spinal cord. (Drawn with modifications from Ramon y Cajal, 1929.) A) Early cord (3-1/2 - 4 days). B) Cord at 5 - 6 days when synapses between interneurons and motoneurons have formed. C) Cord after 8 days with cutaneous input onto interneurons. (A, association interneuron; C, commissural interneuron; DR, dorsal root; DF, dorsal funiculus; LF, lateral funiculus; M, motoneurons; S, sensory neurons; VC, ventral commissure; VF, ventral funiculus.

By 4-1/2 - 5 days there exists in the lumbosacral cord a prominent lateral motor column of more mature motoneurons, which contain more rough endoplasmic reticulum, and which are beginning to send out dendrites forming a motor neuropil (Windle and Orr, 1934; Hamburger, 1948; Lyser, 1964). At this time commissural and associational interneurons differentiate. The former, located more medially, send their axons across the ventral commissure to form the contralateral ventral funiculus. Here most appear to turn rostad, and to send collaterals into the motor neuropil. The associational interneurons, located more laterally, send their axons into the ipsilateral lateral funiculus where they contact the motoneuron dendrites (Windle and Orr, 1934) (Figure 1B). These primitive pathways may be the structural basis for the bilateral flexions of the axial musculature that begin to occur at this time.

The first synapses, axodendritic in nature, are seen with the EM in the lumbosacral cord at 5 days, and presumably derive from commissural and association interneurons (Foelix and Oppenheim, 1973; Oppenheim et al., 1975). At this time at least some of the motoneurons can be shown to have formed functional neuromuscular junctions, since electrical stimulation of spinal nerves results in limb movements that are blocked by dTC (Landmesser and Morris, 1975). (These events are outlined in Figure 2.)

At about this time electrical burst activity can be recorded with microelectrodes in the ventral horn, and can be shown to be synchronous with similar bursts recorded in the sciatic nerve (Provine, 1973). This neuronal discharge therefore appears to be the basis of the spontaneous limb motility occurring at this time (5 - 6 days). Many sensory neurons already have sent processes both peripherally into the limb and centrally to form the dorsal funiculus. However, at this time collaterals are not sent into the grey matter (Windle and Orr, 1934) (Figure 1B). Therefore, in the chick, initial limb movement is spontaneous, pre-reflexogenic, and occurs without any sensory input. The latter contention is supported by deafferentation experiments, which do not affect spontaneous motility assessed behaviorally until late in development (Hamburger et al., 1966; Narayanan and Mallow, 1974). Of course EMG recordings from such preparations might detect differences in the finer aspects of motility.

A short time later, collaterals from afferent fibers in the dorsal funiculus contact commissural and association interneurons and complete the first reflex arcs (Windle and Orr, 1934) (Figure 1C). This observation, based on silver stained preparations, needs to be confirmed with the electron microscope. Synchronous with this, at day 7.5 - 8, tactile stimulation results in the first cutaneous reflexes.

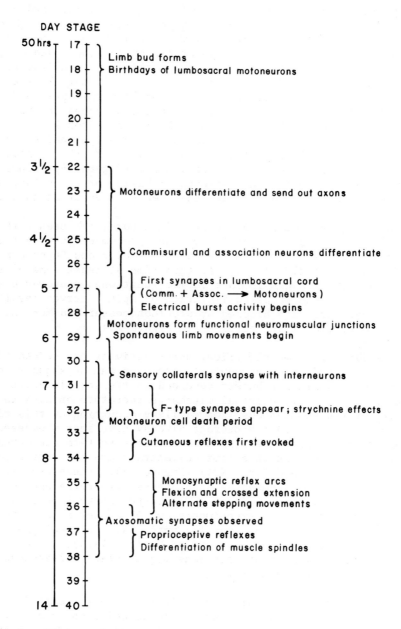

Figure 2. Timing of important events in the neurogenesis of the chick spinal cord, according to embryonic day and Stage of Hamburger and Hamilton. Note: the timing may not be precise as these results have been compiled from different investigators, and in some cases the embryonic stage was not recorded.

It is interesting that these first spontaneous movements appear to be entirely of spinal origin, with neither sensory (Hamburger et al., 1966) nor descending supraspinal (Oppenheim, 1975) inputs necessary. Both the behavioral observations and the recording of the electrical burst activity which sweeps through the cord in diffuse waves suggested a rather unco-ordinated movement of the whole embryo. However, more recent electrophysiological observations (Bekoff et al., 1975) indicate that activation of muscles within a single limb is co-ordinated. By recording electromyograms from the synergists, peroneus and gastrocnemius, and the antagonists, tibialis and gastrocnemius, they were able to show that at day 7, shortly after the inception of movement, the synergists were co-activated, while there was a phase lag between onset of activity in antagonists (see Figure 3).

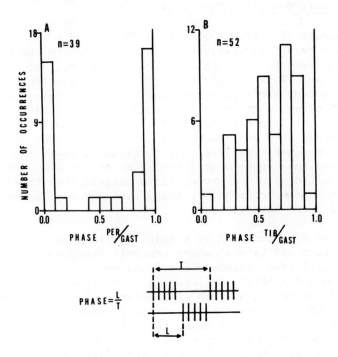

Figure 3. Histograms of the phase relationships of two pairs of ankle muscles during spontaneous activity at day 7. In A, it can be seen that the synergists, peroneus and gastrocnemius are mostly co-activated (are in phase) while in B the antagonists, tibialis and gastrocnemius can be seen to contract out of phase. Taken from Bekoff et al., 1975.

At the least these observations would suggest that flexor and extensor motoneurons are distinct by day 7, and are connected in a specific pattern. Since F-type, presumably inhibitory, synapses (Oppenheim et al., 1975) and sensitivity to strychnine (Oppenheim and Reitzel, 1975) develop about the same time, it is reasonable to infer that these early circuits contain both excitatory and inhibitory synapses. In summary then, there is co-ordination of the earliest limb movements, and the neural organization for it appears to be built into the cord prior to the completion of reflex circuits.

Somewhat later, day 9 - 12, monosynaptic reflex arcs can be demonstrated with silver stains, axosomatic synapses become common, and flexion and crossed extension as well as alternate stepping movements occur. At 10 - 12 days the first proprioceptive reflexes have been reported (Visintini and Levi-Montalcini, 1939).

Comparison with Mammalian Development

It appears that from the early work of Windle and Baxter (1936) and the more recent ultrastructural study of Vaughn and Grieshaber (1973) that development of the rat brachial spinal cord is essentially similar. Thus by embryonic day 13.5 motoneurons and sensory neurons have sent axons into the limb. Sensory neurons also have central processes in the dorsal funiculus, but collaterals do not penetrate as yet into the grey matter. However, association interneurons already have sent axons into the marginal zone adjacent to the ipsilateral motor nucleus, and commissural axons can be seen coursing into the contralateral medial motor nucleus. Thus as in the chick, silver-stained preparations suggest that interneurons synapse with motoneurons before being synapsed upon by sensory input (Windle and Baxter, 1936).

This was substantiated by a quantitative study of synapses in the association and lateral motor neuropil (Vaughn and Grieshaber, 1973). At 13.5 days synapses were found only in the lateral motor neuropil. One day later, collaterals from the dorsal funiculus could be seen penetrating into the neuropil of the association and commissural interneurons, and the first synapses could be observed ultrastructurally. By 15.5 days, the collaterals from the dorsal funiculus were numerous as were synapses, and cutaneous reflexes first could be elicited. Only much later (day 19, Tello, 1917; Zelena, 1957) do muscle spindles form and it is possible to demonstrate monosynaptic connections from dorsal root afferents onto motoneurons (Gilbert and Stelzner, 1975). Therefore, in the rat as in the chick, synapse formation occurs in a retrograde fashion, (i.e., motoneuron to muscle, interneuron to motoneuron, and sensory neuron to interneuron) and the first sensory input is exteroceptive, with proprioceptive connections forming later.

Similar events apparently occur both in the cat (Windle et al., 1933) and in humans (Windle and Fitzgerald, 1937). Bodian (1970), however, has proposed an orthograde sequence of synapse formation in the monkey based on a temporal relationship between initiation of synapse formation in the motor neuropil and reflex movements. It is possible that in the latter case synapse formation in the association neuropil could precede that in the motor neuropil, but by such a brief time that synapses in the motor neuropil appear to be related to the onset of reflexes. This problem might be resolved by a quantitative study of synapses similar to those done on the rat.

In any system consisting of several links, caution must be exercised in relating the function of the whole with the development of any one link. Each link, techniques permitting, should be investigated separately both with the EM and with electrophysiological techniques. During development many circuits may go undetected merely because they are not spontaneously active. For example, it has been clearly shown (Oppenheim, 1975) that supraspinal inputs have little effect on chick spontaneous motility until late in development. Yet silver-stained preparations of Windle and Orr (1934) indicate that some descending tracts may exist earlier. It is possible that such input may go undetected because it is electrically quiescent.

The basic and important observations made thus far which utilize behavior and anatomy would be greatly enhanced by an electrophysiological investigation into all components present at any developmental stage. Even if one were limited to extracellular recording due to size limitations, much valuable information could still be obtained. Such techniques could be expanded by the use of retrograde and orthograde labelling techniques (horseradish peroxidase and radioactive isotopes), which could be used to ascertain the cells that contribute synapses at any time (Oppenheim and Heaton, 1975).

In summary the basic structure of the cord with flexors and extensors reciprocally organized develops without sensory input. Thus cutaneous afferents grow into a partially organized cord, this being even more true for the muscle afferents which enter the cord considerably later. (The latter has been documented in the rat spinal cord for afferents making monosynaptic connections onto motoneurones, which are presumably from muscle spindles (Gilbert and Stelzner, 1975). It has not yet been possible to separate Type 1A and Type 11 afferents, nor to study the 1B afferents from Golgi tendon organs separately.)

The first of the possibilities for bringing about connectivity that is consistent with these observations is early specification (see Figure 4A). This would require that the specified motoneurons

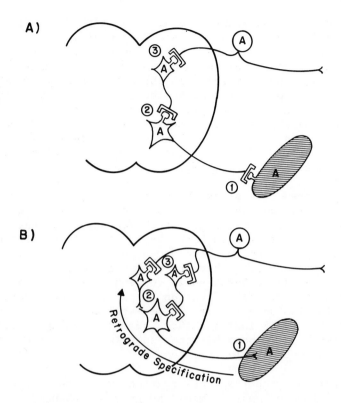

Figure 4. Possible pathways by which spinal cord neurons become specified. A) The motoneuron has an intrinsic biochemical identity (A) and must grow to and recognize muscle A of like identity. Interneuron and sensory neuron are also intrinsically specified and must recognize their respective targets. Numbers indicate temporal order in which synapses occur. B) Motoneuron is specified by muscle. Interneurons can either be specified in turn by motoneuron or can possess an intrinsic specificity. In both cases sensory neurons, forming synapses last are intrinsically specified and make selective connections with interneurons.

grow to and recognize the proper target. A recognition step (diagramed as a lock and key mechanism in Figure 4) also would be required for the interneuron to motor synapse and for the sensory neuron to interneuron synapse. Alternatively (Figure 4B), the motoneuron could be specified by its target muscle. This retrograde specification in turn could be impressed upon the interneurons so that the only recognition step required would be when the sensory neuron synapses onto either inter or motoneuron. The necessity of the latter follows from the fact that the sensory neurons already have sent peripheral processes to certain places in the skin, or to certain muscles before they form central synapses. Therefore they must form synapses in the cord in accordance with their peripheral termination. Another myotypic specification scheme would have the interneurons being intrinsically specified as well (Figure 4B).

It is apparent that the above questions are not resolvable by studying normal development, which at best is consistent with one or more of the above schemes. Some experimental manipulations designed to answer these questions will be presented after a consideration of how motoneurons establish proper peripheral synapses.

The Establishment of Connections between Motoneurons and Appropriate Peripheral Targets

Intrinsic specification, random outgrowth, and selective loss.
Since in the adult, motoneurons at a certain level in the spinal cord invariably synapse with certain muscles and not with others, one may consider the developmental mechanism that could bring about this correspondence. One possibility is that motoneurons are specified for a certain peripheral destination, are produced in excess numbers, and grow out somewhat randomly. Failure to synapse with the appropriate muscle results in motoneuron death. This mechanism would not require that axons be guided selectively into the limb, and it might explain the substantial death of chick spinal cord motoneurons that occurs at about the time peripheral synapses are first formed (Hamburger, 1975).

In other systems, (amphibian cord and chick ciliary ganglion) cell death has been shown to be correlated temporally with the development of peripheral synapses, to be enhanced greatly by prior removal of the peripheral target, and to remove motoneurons that have already sent out axons (Prestige, 1967; Prestige and Wilson, 1974; Landmesser and Pilar, 1974a and b, 1975).

The possibility that cell death played a major role in the development of specific connections was ruled out, at least in the chick spinal cord, by electrically stimulating the spinal nerves that contribute to the hindlimb innervation at different developmental stages and by recording either movement, contraction of specific muscles, or compound action potentials from specific muscle nerves (Landmesser and Morris, 1975). The segmental pattern of hindlimb innervation determined by these methods for late embryonic stages is shown in Figure 5. Exactly the same pattern, shown for two muscles in Figure 6, was found before (triangles) and after (filled circles) the period of cell death indicating that neurons at a certain position in the spinal cord selectively innervated certain muscles from the very beginning.

In these experiments the spinal nerve producing maximal tension was given a value of 100% and the contribution of other spinal nerves was expressed as a fraction of this value. This technique would detect only those motoneurons that had formed functional neuromuscular connections. It would not detect any axons that had grown down aberrant pathways but that had failed to form connections. Therefore experiments similar to the one described were

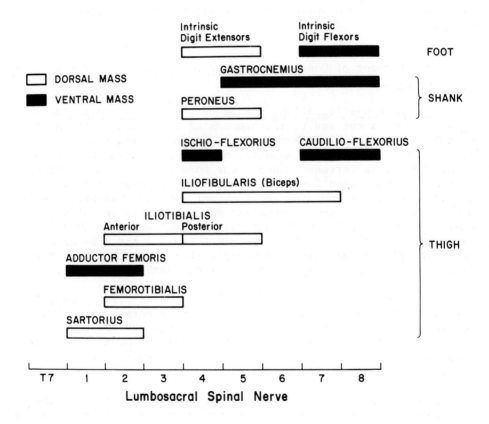

Figure 5. Segmental Innervation of Major Hind Limb Muscles in the Chick Embryo. This figure is a summary of results derived from movement observations, tension recording and action potential recording. Each bar represents the lumbosacral spinal nerves innervating a selected muscle. Open bars represent muscles derived from dorsal muscle mass, filled bars those from the ventral muscle mass. Lumbosacral spinal nerves 1 - 8 correspond to spinal nerves 23 - 30. Taken from Landmesser and Morris, 1975.

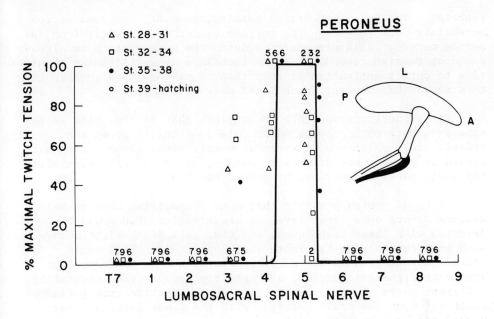

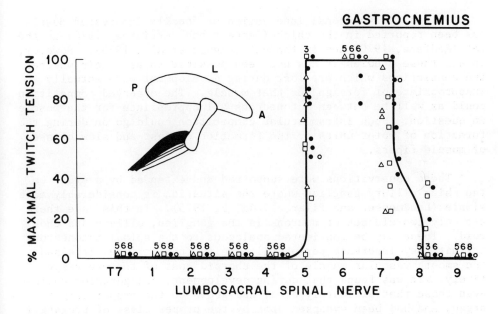

Figure 6. Twitch tension produced in the gastrocnemius and peroneus muscles by stimulating the various spinal nerves at different developmental stages. △, St. 28 - 31; □, St. 32 - 34; ●, St. 35 - 38; O, St. 39 to hatching. Numbers indicate number of different embryos sampled at each state.

repeated. Instead of recording tension, however, compound action potentials were recorded with suction electrodes directly from the muscle nerves. This procedure produced the same pattern as already found by tension recording. This technique is sufficiently sensitive to detect activation of less than 5% of the axons contributing to a nerve (Landmesser, Unpublished observations).

It is therefore possible to conclude that in the chick selective axon outgrowth plays a major role in bringing about the orderly adult pattern of motoneuron-muscle connections. Axon outgrowth is clearly not diffuse or random, nor does cell death alter the basic pattern described by these techniques.

It is of course possible that some competition between motoneurons occurs on a finer level of organization than would be detected with these techniques, and that cell death might remove such motoneurons that terminate inappropriately. This would include motoneurons in the same segment of the spinal cord that innervate different muscles, and possibly motoneurons innervating different parts of the same muscle. Thus selective axon guidance could set down the basic pattern, with the finer details being worked out by competitive interactions.

Initial polyneuronal innervation of focally innervated muscle has been reported in the chick (Bennett and Pettigrew, 1974) in the rat (Redfern, 1970) and in the cat (Bagust et al., 1973). However, in all these studies it has not been possible to ascertain whether the connections which are lost during development were actually inappropriate or foreign for that muscle. The observed competition could as well be between motoneurons all appropriate for the muscle in question. Such intramuscular competition could go on during the formation of motor units or the formation of fast and slow classes of muscle fibers.

These observations were supported and extended by a study on the chick ciliary ganglion where the situation is considerably more simple (Landmesser and Pilar, 1974a, b, 1975). In this case there are only two classes of neurons in the ganglion, ciliary and choroid. These may be considered analagous to the spinal motoneurons in that they innervate peripheral muscles, the striated iris and the smooth vascular muscle of the choroid coat of the eye respectively. It was found that during development, all ganglion cells, even those that ultimately died, had grown to the proper target organ, and had been synapsed upon by the proper class of preganglionic fibers (ganglion cell death is approximately 50%, similar to the amount of motoneuron death occurring in the chick spinal cord). Thus both peripheral synapse formation and synapse formation onto the ganglion cells was selective from the beginning. Oppenheim (Unpublished observations) has shown that, similar to the ciliary ganglion, motoneurons in the chick lumbosacral cord that eventually die, mature normally up to a point, send out dendrites, and are synapsed upon. It would be interesting to test the appropriateness of these synapses.

It should be pointed out, however, that in the amphibian (Hughes and Prestige, 1967) initial limb innervation assessed by electrical stimulation of spinal nerves appears somewhat random. The final specific pattern emerges by death of motoneurons that presumably have failed to form appropriate peripheral synapses. Thus real differences may exist between amphibians and higher vertebrates.

Timed outgrowth hypothesis. This hypothesis, based on the observed cranio-caudal differentiation of motoneurons together with a proximo-distal sequence of limb innervation, proposes that the most cranial motoneurons send axons into the limb first and synapse with the most proximal muscle. Axons from more caudally located motoneurons, entering the limb a short time later, find these muscles innervated and therefore grow further to synapse with progressively more distal muscle (Jacobson, 1970; Wyman, 1973). This hypothesis has the advantage of simplicity and neither requires selective axon guidance nor recognition of appropriate muscles. However, the bulk of available evidence on initial innervation of the chick hindlimb fails to support it.

Although there is a slight cranio-caudal difference in the time of motoneuron proliferation and differentiation (M. Hollyday, Unpublished data) both the crural and ischiadic plexuses form at approximately the same time (Fouvet, 1973) and most of the limb musculature is innervated functionally at the same time with no proximo-distal sequence being apparent (Landmesser and Morris, 1975). In fact, muscle primordia lying in close anatomical apposition, such as the sartorius and femorotibialis, or caudilioflexorius and ischioflexorius, are innervated by their appropriate spinal segments (Figure 7) from the outset and at the same time. Thus lumbosacral spinal nerve 1 always innervates the sartorius but never the femorotibialis, from as early as Stage 30 soon after movement can first be elicited. In addition, by consulting Figure 5, it can be seen that while a general cranio-caudal, proximo-distal relationship exists, it does not occur on a sufficiently fine level to explain the very selective innervation that was observed. For example, segments 4, 5, 7 and 8 all contribute to the innervation of the thigh, shank, and intrinsic foot musculature, yet each contributes to only certain muscles at each level.

The only muscles that appear to differentiate and to be innervated later are the intrinsic foot muscles. In the adult the muscles on the dorsal surface of the foot are innervated by spinal

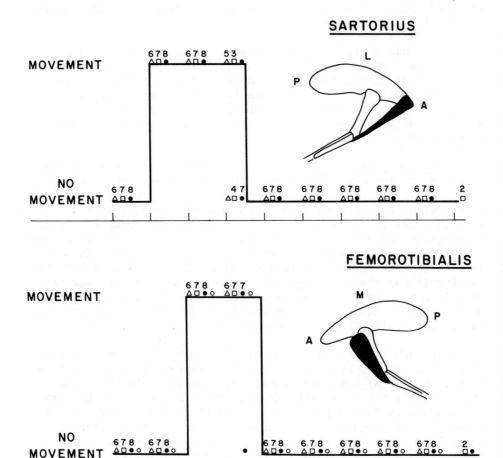

Figure 7. Observations of movements elicited in two anatomically adjacent muscles by stimulation of spinal nerves at different developmental stages. Symbols, same as for Figure 6. It can be seen that from St. 28, stimulation of lumbosacral spinal nerve 1 always activates the sartorius but never the femorotibialis.

nerves 4 and 5; those on the ventral surface by spinal nerves 7 and 8. Yet, even before these muscles differentiate and when the muscles in the thigh and shank are being innervated, the nerves that will innervate the foot muscles already are present and can be recorded from. It was found that only those spinal nerves that eventually will innervate these muscles contribute axons to the nerves at this stage. This strongly suggests that motoneurons in a certain place in the spinal cord are destined to grow to specific places in the limb, even before the muscles that they will innervate have differentiated. Thus growing axons do not appear to be attracted to specific muscles.

The timed outgrowth hypothesis also assumes that motoneurons can be prevented from synapsing with innervated muscle fibers, a situation that does seem to occur in the adult. There is evidence, however, that while one synapse can make much of a developing muscle fiber refractory to further innervation, multiple innervation of the endplate sites does occur (Bennett and Pettigrew, 1974). These additional synapses are lost some time later. Since it was found that muscles are innervated selectively from the beginning when multiple innervation pertains, any later tendency of a synapse to prevent hyper-innervation probably is not related to the formation of specific connections.

Selective outgrowth hypothesis. The evidence from the stimulation of spinal nerves (Landmesser and Morris, 1975) as well as the analysis of spontaneous muscle contractions (Bekoff et al., 1975) show that motoneurons lying in certain cranio-caudal positions in the spinal cord selectively grow to certain positions in the limb and there establish synapses with appropriate muscles from the start.

It was postulated that such neurons could be guided to specific geometric positions in the limb by some sort of gradient mechanism, and that once there they might synapse with any muscle fibers in that position (Landmesser and Morris, 1975). This model would not require that muscles possess specific biochemical identities nor that they be recognized by outgrowing nerve fibers. It would place the burden of insuring specific connections on the guided growth of axons. It is known from other studies that retinal ganglion cell axons grow to certain positions on the surface of the tectum (Jacobson, 1970; Crossland et al., 1974) and differences in adhesive interactions between ganglion cell and tectum have been postulated to explain this selective growth (Roth et al., 1973). Changes in substrate adhesivity have been shown to affect direction of axon growth in tissue culture (Letourneau, 1975), although it is not known what factors determine selective axon growth in vivo.

Alternatively some recognition between axon and muscle still might be required for synapse formation (i.e., growth to the right place might not be sufficient). Until more is known about the mechanisms bringing about specific connections, it seems wise to keep separate, at least conceptually, guided growth on the one hand and recognition and synapse formation on the other.

Experimental Manipulations of Normal Development

Unfortunately, the study of normal development is not capable of distinguishing between the above alternatives. Nor does it allow one to conclude anything about how appropriate synapses onto motoneurons are formed, and whether motoneurons are specified by their periphery, their position in the spinal cord or by some other combination of mechanisms.

In order to attempt to answer at least some of these questions a series of experimental manipulations chiefly utilizing supernumerary limbs has been begun (Morris, 1975 and Unpublished observations). The rationale is to transplant a supernumerary limb anterior to the normal limb. Since such limbs generally are innervated by adjacent spinal cord segments, this procedure allows motoneurons lying in a certain place in the spinal cord to grow into the supernumerary limb from a different position than is normal. For example (Figure 8) lumbosacral spinal nerve 1 that normally grows into the limb in the crural plexus and innervates the sartorius and adductor femoris, now finds itself entering the supernumerary limb from a caudal position. Does it innervate a part of the limb in accord with its new cranio-caudal position with respect to the limb, perhaps by growth down the ischiadic plexus and by innervation of the flexorius and gastrocnemius muscles? Or rather does it seek out the muscles (sartorius, adductor femoris) with which it normally would have synapsed?

In accord with earlier anatomical work (Hamburger, 1939; Narayanan, 1964) it was found that the limb influenced the anatomical nerve pattern so that a grossly normal plexus and nerve branching pattern resulted. Furthermore, electrical stimulation of the original lumbosacral spinal nerve 1 showed that not only did it grow to muscles in accord with its new position (flexorius, gastrocnemius) but also that it actually formed functional synapses with them. Motoneurons, then, were not attracted to nor did they form synapses with muscles which they normally would have innervated (Morris, 1975).

From this it is possible to conclude that motoneurons residing in a certain place in the spinal cord are not destined invariably to innervate certain muscles. In other words, they have not been specified rigidly in a way that prevents them from synapsing with something other than the appropriate muscle, even though they normally synapse very selectively. Several explanations are

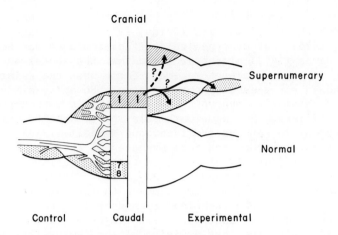

Figure 8. Supernumerary Limb Experiment. Left, control lumbosacral segment 1 innervates sartorius and adductor muscles (diagonal lines) while segments 7 and 8 innervate flexorius and gastrocnemius (stippling). On right, segment 1 grows into supernumerary limb from abnormal position. Rationale of experiment is to determine what muscles it innervates and whether synapses onto motoneurons in that segment are in accord with muscles innervated or position of motoneurons in cord.

possible for the above data. One is that the motorneurons in question have retained their intrinsic identity (i.e., are still sartorius motoneurons) and merely have grown to the wrong place in the supernumerary limb. This might imply that neurons entering the limb from a certain position invariably are guided to certain places in the limb by a hypothetical gradient mechanism. However, another possibility is that the motoneurons have been respecified by an early interaction with the supernumerary limb.

The hypothesis of myotypic specification has fallen into disrepute in recent years, because much of the supernumerary limb work on amphibians from whence the concept originally arose, can be explained by selective innervation (Grimm, 1971). This is because in the amphibian, motoneurons innervating any muscle are widely distributed in the cord. Also, with only three spinal nerves innervating the limb, any spinal nerve probably contains motoneurons for most of the limb muscles. Thus the experiments just proposed for chick embryos could not be interpreted unambiguously in the amphibian. In this respect the chick is probably uniquely well endowed since it has a precise segmental innervation pattern and, unlike mammals, is amenable to experimental modifications such as supernumerary limb and spinal cord transplants.

In nerve-cross experiments in non-embryonic stages, it has been shown both in salamanders (Grimm, 1971) and in cats (Mendell and Scott, 1975) that myotypic respecification does not occur. Davis and Davis (1973) also have shown that the neural network underlying lobster swimmeret movement forms when the swimmerets are removed at early developmental stages. However, almost no evidence pertaining to myotypic specification in vertebrate early development exists. Yet some work relating to specification of sensory neurons in Xenopus (Hollyday and Mendell, 1975) shows that thoracic sensory neurons can form synapses in the cord as if they had been specified as limb innervating neurons following an interaction with transplanted supernumerary limbs. Retinal ganglion cells are specified at about the time of their last division (Jacobson, 1970). Since technical considerations require that hind limb buds be transplanted between Stage 16 - 22, a time period that coincides with the last division of the lumbosacral motoneurons, the possibility that motoneurons are specified by the supernumerary limb seems plausible. This may not represent myotypic specification in the sense that motoneurons are specified by muscles to which they have grown, but rather by an early interaction with the limb bud as a whole before any axon outgrowth occurs.

In order to decide if the motoneurons supplying transplanted segments have been respecified, it will be necessary to assess in detail the synapses onto the motoneurons innervating the supernumerary limb, including cutaneous, proprioceptive and descending input. It should be possible to determine if these synapses are in accord with the original specificity of the motoneuron. This can be assessed by EMG recordings from muscles during spontaneous movement, during evoked reflexes and in the hatched bird during locomotion itself. Experiments of this design currently are being carried out (D.G. Morris, Unpublished observations) and it is hoped that the information obtained will aid in resolving these problems.

Since early spontaneous motility has been shown to be coordinated prior to the addition of sensory input (Bekoff et al., 1975) and since this probably results from organized networks of interneurons, with the above rationale it should be possible to determine some of the rules governing the connectivity of these interneurons. This could be assessed by recording electromyograms during spontaneous contractions in normal and supernumerary limbs at day 6 - 7 in a preparation similar to that depicted in Figure 8. Later in development, cutaneous, proprioceptive, and descending input could be activated either physiologically or electrically and the response of selected muscles in supernumerary and normal limbs determined. Of course in all these cases it will be necessary to determine by spinal nerve stimulation the actual innervation pattern of the supernumerary limb.

Somatotopic Matching Up

The spinal cord, as with many other neural structures, is organized in a somatotopic fashion (Romanes, 1964; Sterling and Kuypers, 1967; Burton and McFarlane, 1973; Brown and Fuchs, 1975). Thus far motoneurons have been considered as being identified with particular muscles. However, another factor which might be important in determining connectivity is merely position in the spinal cord itself (Sterling and Kuypers, 1967). Thus motoneurons at a particular locus could be postulated to have more similar connections than those located further away. It has been proposed (Wyman, 1973) that such connectivity based on topology might occur in the spinal cord, and further that such an organization would greatly simplify the developmental problem of specifying large numbers of connections with a limited amount of genetic information.

In considering this hypothesis it is important to distinguish between two possibilities. One is that the motoneuron to an individual muscle has no biochemical identity whatsoever, and that connections develop with it solely on the basis of its position. Another quite different and more plausible possibility is that early in development motoneurons are specified according to their position in the cord in a graded fashion as has been proposed for the topologically organized retino-tectal system (Jacobson, 1970). Marchase et al., (1975) have proposed a simple biochemical model by which the specificities of such cells could be set down, with correct matching up resulting from greater adhesive interactions between topologically corresponding points in the two systems (i.e., retina and optic tectum). Yet in this model the cells do possess a biochemical identity encoded in their surface membrane. However, instead of varying in a discontinuous fashion (i.e., separate species of muscles) the identity would vary continuously along any given axis as has been proposed by Wyman (1973).

Recent data obtained by mapping the connections that single 1A afferents make with homonymous and heteronomous motoneurons (Scott and Mendell, 1975) has indicated that separate species of motoneurons do apparently exist. Thus the soleus and medial gastrocnemius motoneuron pools overlap completely in position, yet still receive different 1A connections. A slight effect of position was detected but it was not sufficient to completely explain the difference in connectivity between the two motoneuron pools. These results clearly rule out position per se as a dominant feature in bringing about connectivity. They further suggest that specificity of 1A connections depends on inherent differences between motoneuron properties.

A more direct test of this later conclusion could be made in developing systems by disrupting the topology in a defined way to determine how the connectivities are affected. This may be possible to do in supernumerary limb preparations similar to those already described. In some cases it has been found that the same spinal cord segment (1 in the diagram of Figure 8) contributes to different muscles in the supernumerary and normal limb. Thus, motoneurons at this level make synapses with the sartorius in the normal limb and with the flexorius (semi-membranosis and semi-tendinosis) in the supernumerary. These muscles are distinct functionally and the motoneurons innervating them normally receive different connections (see Figure 5). Therefore, it should be possible to sample the input onto topologically adjacent motoneurons that innervate distinctly different muscles and to see whether their inputs are similar or different. This will allow one to decide whether they are synapsed upon according to the muscle they innervate or according to their position in the cord.

Conclusion

The basic sequence of synapse formation has been determined in the spinal cords of both birds and mammals. In the chick it has further been observed that motoneurons appear to be early specified with respect to their peripheral destination and that they selectively grow to certain positions in the limb. Both a random outgrowth of axons followed by death of improperly connective motoneurons as well as a timed sequence of innervation seem to be ruled out as possible mechanisms for bringing about the orderly pattern of peripheral innervation that is observed. The early co-ordination of muscles within the chick limb suggests that the circuits within the spinal cord responsible for these movements also develop rather selectively. These observations need to be made in more detail and to be extended to other vertebrates, including mammals. It would appear that a combination of electrophysiological techniques and of various embryonic manipulations is capable of greatly extending the currently sparse knowledge and may in fact be capable of resolving some major questions. Among these are the mechanism by which neurons become specified, by which they are selectively guided to the appropriate destinations, and by which they make synapses with the proper cells, both peripherally and centrally. It is hoped that this chapter, while incomplete and somewhat speculative, has helped in defining some of the basic problems and has identified areas where future research is needed.

Acknowledgment

I would like to thank Deborah Morris for allowing me to quote some of her unpublished results and for helpful discussions on the material presented, and to thank Drs. Bekoff, Hollyday, Hamburger,

Provine, Oppenheim, and Stein for letting me cite unpublished results. Thanks are also due to Mrs. F. Hunihan for typing the manuscript. Part of the research reported in this paper was supported by NIH grant NS 10666.

REFERENCES

Bagust, J., Lewis, D.M. and Westerman, R.A., (1973) Polyneuronal innervation of kitten skeletal muscle. J. Physiol. 229, 241-255.

Bekoff, A., Stein, P. and Hamburger, V., (1975) Coordinated motor output in the hindlimb of the 7 - day chick embryo. Proc. Nat. Acad. Sci. 72, 1245-1248.

Bennet, M.R. and Pettigrew, G., (1974) The formation of synapses in striated muscle during development. J. Physiol. 241, 515-545.

Bodian, D., (1970) "A model of synaptic and behavioral ontogeny," In The Neurosciences Second Study Program. (Schmitt, F.A., ed.), Rockefeller University Press, New York, (129-140).

Brown, P.B. and Fuchs, J.L., (1975) Somatotopic representation of hindlimb skin in the cat dorsal horn. J. Neurophys. 38, 1-9.

Burton, H. and McFarlane, J.J., (1973) The organization of the seventh lumbar spinal ganglion of the cat. J. Comp. Neurol. 149, 215-232.

Cotman, C.W. and Banker, G.A., (1974) The making of a synapse. Review of Neuroscience. 1, 1-57.

Crossland, W.J., Cowan, W.M., Rogers, L. and Kelly, J., (1974) The specification of the retino-tectal projection in the chick. J. Comp. Neurol. 155, 127-164.

Cruce, W., (1974) The anatomical organization of hindlimb motoneurons in the lumbar spinal cord of the frog, Rana cateshiana. J. Comp. Neurol. 153, 59-76.

Davis, W.J. and Davis, K.B., (1973) Ontogeny of a simple locomotor system: Role of the periphery in the development of central nervous circuitry. Amer. Zool. 13, 109-425.

Foelix, R.F. and Oppenheim, R.W., (1973) "Synaptogenesis in the avian embryo: Ultrastructural and possible behavioral correlates," In Behavioral Embryology. (Gottlieb, G., ed.), Academic Press, New York, (103-139).

Fouvet, B., (1973) Innervation et morphogenese de la patte chez l'embryon de poulet. Archiv. Anat. Microscop. et Morph. Exp. 62, 269-280.

Gilbert, M. and Stelzner, D.J., (1975) The pattern of supraspinal and dorsal root afferents connections in the lumbosacral spinal cord of the newborn rat. Neuroscience Abstracts. 1, 747.

Grimm, L.M., (1971) An evaluation of myotypic respecification in axolotls. J. Exp. Zool. 178, 479-496.

Hamburger, V., (1939) The development and innervation of transplanted limb primordia of chick embryos. J. Exp. Zool. 80, 347-385.

Hamburger, V., (1948) The mitotic patterns in the spinal cord of the chick embryo and their relation to histogenetic processes. J. Comp. Neurol. 88, 221-283.

Hamburger, V., Wenger, E. and Oppenheim, R., (1966) Motility in the chick embryo in the absence of sensory input. J. Exp. Zool. 162, 133-160.

Hamburger, V., (1975) Cell death in the development of the lateral motor column of the chick embryo. J. Comp. Neurol. 160, 535-546.

Hollyday, M. and Mendell, L., (1975) Area specific reflexes from normal and supernumerary hindlimbs of Xenopus laevis. J. Comp. Neurol. (In press).

Hughes, A. and Prestige, M.C., (1967) Development of behavior in the hindlimb of Xenopus laevis. J. Zool. Res. 152, 347-359.

Jacobson, M., (1970) Developmental Neurobiology. Holt, Rinehart and Winston, New York.

Landmesser, L. and Pilar, G., (1972) The onset and development of transmission in the chick ciliary ganglion. J. Physiol. 222, 691-713.

Landmesser, L. and Pilar, G., (1974) Synapse formation during embryogenesis on ganglion cells lacking a periphery. J. Physiol. 241, 715-736.

Landmesser, L. and Pilar, G., (1974) Synaptic transmission and cell death during normal ganglion development. J. Physiol. 241, 737-749.

Landmesser, L. and Morris, D.G., (1975) The development of functional innervation in the hindlimb of the chick embryo. J. Physiol. 249, 301-327.

Landmesser, L. and Pilar, G., (1975) Fate of ganglionic synapses and ganglion cell axons during normal and induced cell death. J. Cell Biol. (In press).

Letourneau, P.C., (1975) Possible roles for cell-to-substratum adhesion in neuronal morphogenesis. Cell-to-substratum adhesion and guidance of axonal elongation. Developmental Biology. 44, 77-92.

Lopresti, V., Macagno, E.R. and Levinthal, C., (1973) "Structure and development of neuronal connections," In Isogenic Organisms. P.N.A.S. 70, 443.

Lyser, K.M., (1964) Early differentiation of motor neuroblasts in the chick embryo as studied by electron microscopy. Dev. Biol. 10, 433-466.

Marchase, R.B., Barbera, A.J. and Roth, S., (1975) "A molecular approach to retinotectal specificity," In Cell Patterning. CIBA Foundation Symposium. 29, 315-341.

Mendell, L.M. and Scott, J.G., (1975) The effect of peripheral nerve cross-union on connections of single 1A fibers to motoneurons. Exp. Brain Res. 22, 221-234.

Morris, D., (1975) Development of motor innervation in supernumerary hindlimbs of chick embryos. Soc. for Neuroscience; 5th Ann. Meeting. Neuroscience Abstracts. 1, 753.

Narayanan, C.H., (1964) An experimental analysis of peripheral nerve pattern development in the chick. J. Exp. Zool. 156, 49-60.

Narayanan, C.H. and Hamburger, V., (1971) Motility in chick embryos with substitution of lumbosacral by brachial and brachial by lumbosacral spinal cord segments. J. Exp. Zool. 178, 415-432.

Narayanan, C.H. and Malloy, R.B., (1974) Deafferentation studes on motor activity in the chick. J. Exp. Zool. 189, 163-176.

Oppenheim, R.W., (1975) The role of supraspinal input in embryonic motility: A re-examination in the chick. J. Comp. Neurol. 160, 37-50.

Oppenheim, R., Chu-Wang, I.W. and Foelix, R.E., (1975) Some aspects of synaptogenesis in the spinal cord of the chick embryo: A quantitative electron microscopic study. J. Comp. Neurol. 161, 383-418.

Oppenheim, R. and Heaton, M.B., (1975) The retrograde transport of horseradish peroxidase from the developing limb of the chick embryo. Brain Res. (In press).

Oppenheim, R. and Reitzel, J., (1975) The ontogeny of behavioral sensitivity to strichnine in the chick embryo: evidence for the early onset of CNS inhibition. Brain Behavior and Evolution. (In press).

Prestige, M.C., (1967) The control of cell number in the lumbar ventral horns during the development of Xenopus laevis tadpoles. J. Embryol. Exp. Morph. 18, 359-387.

Prestige, M.C. and Wilson, M.A., (1974) A quantitative study of the growth and development of the ventral root in normal and experimental conditions. J. Embryol. Exp. Morph. 32, 819-833.

Provine, R., (1973) "Neurophysiological aspects of behavior development in the chick embryo," In Behavioral Embryology. (Gottlieb, G., ed.), Acad. Press, New York, (77-102).

Ramon y Cajal, S., (1929) Etudes sur la neurogenese de quelques vetebres. Madrid, 1929; translated by L. Guth. Studies on Vertebrate Neurogenesis. Thomas, Springfield, Ill. 1960.

Romanes, G.J., (1964) The motor pools of the spinal cord. Progress in Brain Res. 11, 93-119.

Redfern, P.A., (1970) Neuromuscular transmission in newborn rat. Neuroscience Abstracts. 1, 747.

Scott, J.G. and Mendell, L.M., (1975) "Specificity of projection of single triceps 1A afferent fibers to homonymous and heteronomous motoneurons in the cat," Soc. for Neuroscience 5th Ann. Meeting, Neuros. Abstracts. 1, 168.

Sterling, P. and Kuypers, H.G.J.M., (1962) Anatomical organization of the brachial spinal cord of the cat. II. The motoneuron plexus. Brain Res. 4, 16-32.

Sterling, P. and Kuypers, H.G.J.M., (1967) Anatomical organization of the brachial spinal cord of the cat. III. The propriospinal connections. Brain Res. 7, 419-443.

Straznicky, K., (1963) Function of heterotopic spinal cord segments investigated in the chick. Acta Biol. Hung. 14, 145-155.

Székely, G., (1963) Functional specificity of spinal cord segments in the control of limb movements. J. Embryol. Exp. Morph. 11, 431-444.

Tello, J.F., (1917) Genesis de las terminaciones nervioses motrices y sensitivas. Trab. Lab. Invest. Biol. Univ. Madr. 15, 101-199.

Vaughn, J.E. and Grieshaber, J.A., (1973) A morphological investigation of an early reflex pathway in developing a rat spinal cord. J. Comp. Neurol. 148, 177-210.

Visintini, F. and Levi-Montalcini, R., (1939) Relazione tra differenziazione structurale e funzionale dei centri e delle vie nervose nell' embrione di pollo. Schweiz Arch. Neurol. 44, 119-150.

Windle, W.F. and Griffen, A.M., (1931) Observations on embryonic and fetal movements of the cat. J. Comp. Neurol. 52, 149-189.

Windle, W.F., O'Donnell, J.E. and Glasshagle, E.E., (1933) The early development of spontaneous and reflex behavior in cat embryos and fetuses. Physiol. Zool. 6, 521-541.

Windle, W.F. and Orr, D.W., (1934) The development of behavior in chick embryos: Spinal cord structure correlated with early somatic motility. J. Comp. Neurol. 60, 287-307.

Windle, W.F. and Baxter, R.E., (1936) Development of reflex mechanisms in the spinal cord of albino rat embryos. J. Comp. Neurol. 63, 189-209.

Windle, W.F. and Fitzgerald, J.E., (1937) Development of the spinal reflex mechanisms in human embryos. J. Comp. Neurol. 67, 493-509.

Wyman, R.J., (1973) "Somatotopic connectivity or species recognition connectivity," In Control of Posture and Locomotion. (Stein, R.B. et al., eds.), Plenum Press, New York, (45-53).

Zelena, J., (1957) The morphogenetic influence of innervation on the ontogenetic development of muscle spindle. J. Embryol. Exp. Morph. 5, 283-292.

DEVELOPMENTAL ASPECTS OF LOCOMOTION

G. Székely

Department of Anatomy, Medical University

H-4012 Debrecen   Hungary

A few selected experiments studying the development of spinal cord segments for limb movement in amphibia are reviewed. Limb ablation experiments indicate that the histological differentiation of the spinal cord is determined in early embryonic age, but the maintenance of differentiated motoneurons depends on the periphery innervated. The spinal cord is determined also functionally in the sense that immediately after its closure the medullary tube at the limb level is endowed with the capacity to move a limb in a well defined rhythm and form. Thoracic spinal cord segments, if they are transferred in early embryonic ages, are capable of replacing the function of brachial segments. If the transplantation is done in increasingly older embryonic ages, limb movements fail to develop in a proximo-distal order. Experimental results that isolated brachial segments are able to control coordinated limb movements, suggest that the motor output pattern is programmed in the structure of the limb segments of the cord. In order to control the alternating coordination of a pair of limbs, the limb segments must be in contact with the medulla, or with a certain length of the thoracic spinal cord. Histological investigations suggest that the structure responsible for control of coordinated limb movements, is located in the central gray matter of the spinal cord.

Introduction

This chapter will discuss the contribution of developmental neurobiology to the study of neural mechanisms underlying locomotion. Since the amphibian embryo is the preferred object of experimental neuroembryology, the discussion will be confined to experiments performed on amphibian material. Studying the problems of locomotion in amphibians has several advantages. Amphibian locomotion depends on afferent and descending impulses to a lesser degree, and the central control of limb movements can be investigated under better conditions than in mammalian species. The amphibian embryo can be reared easily in vitro, which is an aid to transplantation and isolation experiments on the spinal cord and on the limb. This review will include an investigation both into that part of the spinal cord which moves limbs and into the development of the neural structure which controls coordinated limb movements. Transplantation experiments provide data about the functional capacity of the spinal cord segments for limb movement. Using histology, questions will be raised about which parts of the spinal cord are responsible for the generation of the motor pattern.

EMBRYONIC DEVELOPMENT OF THE SPINAL CORD SEGMENTS FOR LIMB MOVEMENTS

Histologic Differentiation

The only difference observed with present histologic techniques is that the limb innervating (brachial and lumbosacral) segments of the spinal cord have more cells than the non-limb innervating (thoracic) segments. This multitude of cells accounts for the enlarged gray matter and in particular for a large ventral horn at the level of the limb segments. As indicated by several studies on different embryos (Hamburger, 1958; Harris, 1965; Levi-Montalcini, 1975), the enlarged ventral horn develops from a continuous motor column as a result of cell proliferations and degeneration during growth. The cells in the limb segments of the cord have a lower rate of cell death than in other segments. Factors controlling cell proliferation and degeneration are unknown. Experiments indicate that this distinct feature of the cord at limb and non-limb levels largely is independent of external influences (Wenger, 1951; Székely and Szentágothai, 1962b; Straznicky, 1963).

The development of the ventral horn has been studied (Perri, 1956; Hughes, 1961, 1968; Prestige, 1967) in detail on anuran material. By counting the cells in the ventral horn of Xenopus larvae, Hughes (1961, 1968) found about five times more cells at early embryonic stages than at the time of metamorphosis. The rate of cell degeneration which is greater than that required for the overall reduction in cell numbers, indicates a turnover of cells in the ventral horn during development. Hughes (1968) calculated that for each mature cell which finally emerges, eight or nine others have degenerated. Ventral horn cells decline in a relatively short period; between the 12th and 32nd day of the 60-day larval life. The meaning of this extensive cell death and continued recruitment of new cells, is problematical. The beginning of this histogenetic degeneration coincides with the onset of first limb movements and therefore suggests that the neurons would reach a certain degree of differentiation before they would degenerate. This is supported also by the observation that limb ablation does not bring about the well known reduction of cell number before the 10th and 12th day of larval life. Limb amputation performed after this time, causes a large rapid loss of cells from which the ventral horn does not recover unless the limb regenerates. If the amputation is done in increasingly older larval ages, the extent of the rapid cell loss decreases. However, usually after an extended period of chromatolysis, an increased cell depletion reduces the ventral horn to less than half its normal size. On the basis of their behavior to limb amputation, Prestige (1967) distinguishes 3 phases in the development of ventral horn cells. In the first phase they are unaffected by removal of the periphery; in the second phase the cells die within 3-4 days of amputation (rapid loss), and in the third phase the cells are mature in appearance and die only after an extended period of chromatolysis following limb removal (late loss). He argues that the lack of a certain maintenance factor kills the differentiated neurons that survive histogenetic degeneration.

In limb ablation experiments, a phenomenon occurs that slows the rate of cell degeneration that normally appears in the form of histogenetic degeneration. This decline leads to an excess of cells on the amputated side for the duration between rapid and late cell loss. This transient increase in cells suggests that removal of the periphery does not influence cell proliferation, but does influence the delicate balance apparently existing between cell differentiation and histogenetic degeneration.

Tritiated thymidine autoradiography to study the proliferation and migration of differentiating ventral horn neurons, revealed that the histodifferentiation of the spinal cord takes place along an anteroposterior and a medioventral time axis (Prestige, 1973). The anterior part is advanced more than the posterior between levels of the spinal cord; and within any level, younger cells are lateral to older ones. These investigations also indicate that ventral horn cells originate only in a restricted region of the ventral part of the medullary tube. According to Prestige (1973), the result that the final division of presumtive ventral horn cells precedes the increase of their number by about 4 days cannot be regarded as support for cell turnover in the Xenopus ventral horn. This is at variance with Pollack's (1972) results. He observed labelled cells entering the already established motor column in Rana tadpoles.

These data upon microscopic investigation reveal that regional differences in the spinal cord are determined in early embryonic age. The maintenance of these differences following ventral horn differentiation depends on the periphery.

Functional Determination of the Spinal Cord

The early experiments of Detwiler (1920) on urodela larvae demonstrated that transplanted limbs innervated by the thoracic segments of the spinal cord do not exhibit coordinated movements. Nerve formation under such circumstances essentially is characteristic of limbs innervated from normal sources (Detwiler, 1920; Piatt, 1956). If, however, the supernumerary limb is transplanted close to the normal limb and receives innervation from the brachial plexus, coordinated stepping movements develop in the transplanted limb (Detwiler and Carpenter, 1929; Weiss, 1937). The findings indicate functional differences between limb and non-limb segments of the cord. Rogers' (1934) results that the brachial segments of the cord isolated in early embryonic life can control coordinated limb movements, suggest that the functional differences are determined at the premotile stage of embryonic life.

The problem of functional determination of limb innervating segments can be investigated further by transplanting brachial or lumbosacral segments into the place of thoracic segments (Figure 1), and by implanting supernumerary limbs at the same level. In urodela embryos these limbs would be innervated by the grafted cord segments, and limb movements in larvae and in metamorphosed animals are observed (Székely, 1963). The supernumerary limbs of these animals show regular coordinated stepping movements, in which the rhythm of movement is determined by the origin of the grafted segments. If the extra limbs are innervated by grafted brachial segments they will move in the rhythm of the normal forelimbs (Figure 2). If innervated by grafted lumbosacral segments, they will move in synchrony with the hindlimbs, regardless of whether fore or hindlimbs were transplanted. Spinal cord transplantations made at the earliest possible stages, that is, immediately after the closure of the medullary tube, give rise to the same results. This means that the functional capacity to move a limb is determined in the prospective limb segments of the cord. The rhythm of movement also is predetermined and does not evolve under external influences. This follows from the fact that the origin of the extra limbs, whether fore or hindlimbs, has no influence on the rhythm of movement. Furthermore if supernumerary limbs are grafted into the thoracic region of normal embryos, these limbs remain motionless. Transplantation of limb primordia can be made as early as at the neurula stage; the extra limbs normally develop but remain motionless if innervated only from thoracic segments.

DEVELOPMENTAL ASPECTS OF LOCOMOTION

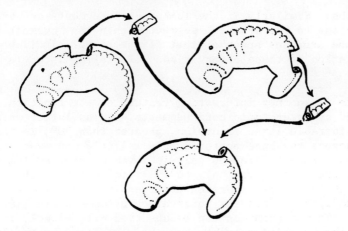

Figure 1. Schematic drawing illustrating the transplantation of the brachial and lumbosacral spinal cord segments in the place of the removed thoracic segments.

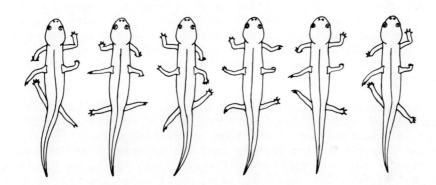

Figure 2. One complete locomotor cycle of a young metamorphosed newt (Pleurodeles Wialtlii), in which the midthoracic (8th, 9th and 10th) spinal cord segments are replaced by the brachial (3rd, 4th and 5th) segments. The supernumerary limbs (middle pair) are moving synchronously with the normal forelimbs on the same side. Redrawn from selected frames of a cinematographic film.

Minute observations of limb movements on the slow motion film leave some doubt about the predetermined character of the rhythm of movement. If the brachial segments are grafted into the place of the caudal thoracic segments, that is, close to the lumbosacral segments, then little delay can be observed in the movement of the extra limbs. Similarly if the lumbosacral segments are transplanted close to the brachial segments, then the movement of the extra limbs precedes somewhat that of the normal hindlimbs. An interpretation of these observations will follow later in this chapter.

There is another interesting feature of the function of limbs innervated by heterotopic cord segments. In an intact animal there is a considerable large excursion, greater than $90°$, in the elbow joint, whereas the knee performs only a little movement during locomotion. When a forelimb pair is moved by grafted lumbosacral segments, the elbow moves only to a limited extent. This corresponds with the small excursion observed in a normal knee joint. A forelimb with lumbosacral innervation assumes a hindlimb-type of movement. This phenomenon can be observed more clearly in chickens in which the brachial segments have been replaced by the lumbosacral segments on the third day of incubation (Straznicky, 1963). The wings do not perform wing-like fluttering movements; there is only an elevation and abduction of moderate excursion in the shoulder and a very slight elbow extension. These movements occur at the same time as the stepping movements of the leg. If the wing is replaced by a leg primordium in an otherwise normal chicken embryo, then the knee and ankle joints remain in an extended position. A movement similar to fluttering occurs in the hip joint at the same time as the movement of the normal wing on the other side. The wing produces a leg-type movement, and the leg a wing-type movement in these experiments.

These results indicate that not only the rhythm but also the character or movement (whether fore or hindlimb-type, leg or wing-type) is predetermined in the limb innervating spinal cord segments. This conclusion implies that the brachial segments and lumbosacral segments generate different output patterns for the movement of forelimbs and hindlimbs, or of wings and legs, respectively.

The Time Course of the Functional Determination

If the functional determination occurred gradually during development, then a study of the time course of determination in similar transplantation experiments performed at later stages of embryonic age might reveal more about the nature of the limb moving segments. Unfortunately, the limb segments suddenly become determined at the time of the closure of the medullary tube. It is known, however, from Detwiler's (1923) experiment that thoracic segments, if transferred in early embryonic stages, do replace the function of brachial segments. This possibility offers an indirect approach to the problem. The early thoracic segments may be considered as a "neutral" spinal cord. Then, it is possible to study

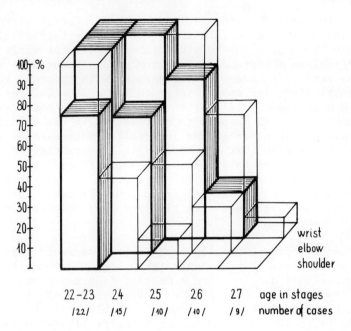

Figure 3. Stereogram illustrating the motility of the three principal joints (shoulder, elbow, wrist) of the forelimbs in animals in which the brachial spinal cord segments have been replaced by thoracic segments in older embryonic ages. Columns drawn with heavy and thin lines represent complete and defective movements, respectively. The vertical axis indicates the percentage of cases in which movements occurred. On the first horizontal axis the embryonic stages of operation and (in parenthesis) the number of observed cases are given. The second horizontal axis refers to the three forelimb joints. (From Straznicky and Székely, 1967, with permission of the Publishing House of Hungarian Academy of Sciences.)

their adaptation to move a limb when grafted into the place of brachial segments in increasingly older embryonic stages (Straznicky and Székely, 1967).

The developmental stages during which this experiment can be performed, encompass a time span of about two and a half days; between the 3rd and 5th day of development in salamander embryos. These developmental forms are labelled conventionally by numbers from 21 through 27, according to a scheme by Sato (1933). The medullary tube closes at stages 20-21 (3rd day), and the first myogen movements (that is, the contractions of the developing muscle fibers before they receive their motor innervations), appear at stage 31 (6th day). Normal limb movements develop if the transplantation occurs at the earliest stages. Embryos operated on in increasingly older ages fail to develop movements first in the shoulder, and then also in the elbow, and finally the whole limb remains motionless (Figure 3). In other words, functional adaptation of the grafted thoracic cord segments fails to occur in a proximo-distal direction with respect to the anatomical positions of the limb muscles. It is accepted generally that the proximo-distal sequence of limb muscles is represented in the spinal cord by the cranio-caudal order of the innervating motoneurons. This leads to an assumption that the histologic differentiation of the cord gradually fails to take place in the cranio-caudal direction, and the limb muscles gradually fail to receive the appropriate innervation in the proximo-distal direction. Fiber counts of the ventral roots and recordings from limb muscles, however, contradict this simple conclusion.

The brachial plexus largely emerges from segments 3 and 4 of the cord in the newt. Fiber counts taken from the ventral roots of these segments indicate that the number of fibers decreases as limb activity decreases, and this occurs approximately in the same measure throughout the brachial segments (Table 1). If coordinated limb movements require a given number of motor fibers to supply the corresponding muscle groups, then the two results, the proximo-distal worsening of limb movements and the equally decreasing number of fibers, suggest that limb muscles are not represented in a somatotopic order in the spinal cord. This, in fact, could be verified in subsequent experiments on localization of motoneurons innervating individual limb muscles (Székely and Czéh, 1967).

In animals which attain a certain size, myograms can be recorded simultaneously from eight limb muscles (4 shoulder girdle, 2 arm and 2 forearm) during ambulation (Figure 4). Intact animals produce a characteristic sequence of muscle activities in which extensive co-contraction of antagonistic pairs of muscles is conspicuous (Székely et al., 1969). The comparison of the muscle activity pattern with the movement of the limb clearly shows that simple visual observation of limb movements does not yield reliable information about the activities of individual muscles. Muscles of motionless joints usually show activities which may be stronger than those recorded from muscles of moving joints (for example in the shoulder in Figure 4B). Movement may be present in a particular joint (wrist in Figure 4D) even if one of its principal muscles (extensor digitorum) has undergone atrophy. The comparison of the muscle activity patterns obtained from operated and intact animals suggests that good motility of the different joints is associated with an activity pattern that resembles the normal pattern of muscle activities. This means that coordinated limb movements require a well defined output pattern which will secure the appropriate distribution of muscle strength for the movement of individual joints.

Table 1. Number of fibers in the ventral roots of grafted thoracic segments. The first rows indicate the number of fibers in ventral roots 3 and 4 in normal animals. (The 5th root is very small in newts and has not been used 'for fiber counts). The following rows show the data obtained from animals operated on at successively older embryonic stages. The bold-face figures are the sum of the average numbers of fibers in roots 3 and 4. The last column indicates the motility of the forelimbs: +, normal, ±, defective movements; −, no movement in the corresponding joints. (Reproduced from Székely, 1968, with permission of the Ciba Foundation, London.)

Stage of operation	Number of animals	Fibre counts in ventral roots		Average numbers		Motility		
		Root 3	Root 4	Root 3	Root 4	Shoulder	Elbow	Wrist
Normal embryos	1	415	470					
	2	468	496	435	431			
	3	471	351	**866**		+	+	+
	4	385	406					
23	1	354	281			±	+	+
	2	539	503	408	387	+	+	+
	3	419	405	**795**		±	+	+
	4	320	358			−	+	+
24	1	433	329			±	+	+
	2	292	318	343	343	−	±	+
	3	304	382	**686**		±	±	+
25	1	257	202			±	±	+
	2	231	168	221	171	−	±	+
	3	240	157	**392**		−	−	+
	4	274	164			−	±	+
	5	129	162			−	−	±
26	2	119	115			−	−	±
	3	69	81	105	77	−	−	−
	4	124	—*	**182**		−	−	+
	6	123	188			−	±	+
	7	92	—*			−	−	±

*Only root 3 was present.

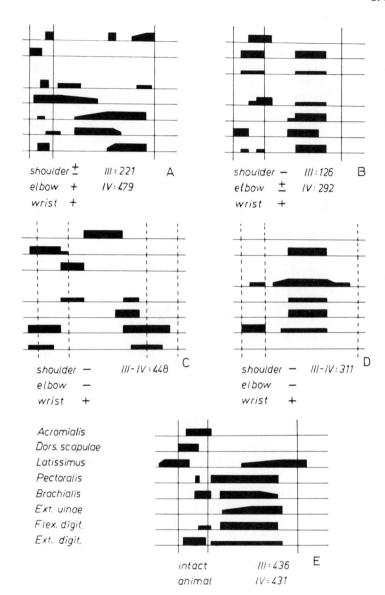

Figure 4. Schematic representation of myograms recorded from limbs innervated by transplanted thoracic spinal cord segments. The motility of individual joints is indicated with + (meaning normal movement), ± (defective movement) and − (lack of movement). The numbers after Roman numerals refer to fiber counts in the respective ventral root. Vertical lines separate the phase of protraction (narrow space) from the phase of retraction (wide space) of a stepping movement.   (Legend continued on next page.)

The data from these experiments are incomplete, although several implications result. Attempts to define the difference in differentiation between parts of the grafted thoracic cord (as to which parts move and which do not move the limb), may include using the number of motoneurons as a measure of differentiation. In fact, there is a decrease in the number of motor fibers which parallels the diminution of limb movements. However, there is no correlation between the movement of a particular joint and the activities of the respective muscle groups. It appears that the development of a neural mechanism which controls the appropriate order of motoneuron activities, is the significant factor in differentiation. In addition to the ventral horn, the intermediate gray matter is enlarged also in the limb segments of the spinal cord. It is conceivable that the innumerable small interneurons are organized into a delicate system of networks, and that this structure is actually the repository for a central program concerned with the control of coordinated limb movements. If these interneurons are present only in a reduced number, the thousands of intricate interconnections among one another and among motoneurons could not be established and part of the program would be absent from the structure. Movement is present in a particular joint if the motoneurons which innervate the muscles of this joint are appropriately triggered by the interneurons. This speculation is supported by histologic investigations described later in this chapter.

FUNCTIONAL CAPACITY OF THE SPINAL SEGMENTS

It is known from several experiments that limb deafferentation does not impair coordinated limb movements in amphibia (Weiss, 1936; Gray, 1939; Székely et al., 1969; Smith-Harcombe and Wyman, 1970). Transplanting the brachial section of the medullary tube together with the surrounding myotom tissue and limb primordia into the flank of a host salamander embryo fosters development of regular limb movements in the graft (Rogers, 1934). These results indicate that if either the limb is deprived of sensory innervation

---

Broken vertical lines in C and D mean that because of incomplete limb movements these phases could not be established clearly in these cases. Black stripes indicate muscle activities, the width of stripes indicates the height of the amplitude (strength of activity) in an arbitrary scale. The names of muscles refer to the corresponding lines for each myogram; the lack of horizontal lines indicates technical failure in recording. The thoracic spinal cord transplantation was made at stages 23, 24, 25 and 26 in A, B, C and D, respectively. E shows the myogram and fiber counts of an intact animal.

or the supraspinal control is eliminated, coordinated limb movements are maintained. Investigating what effect the absence of both the sensory and supraspinal controls has on limb movements, would be the same as asking whether the spinal structure generating the appropriate output requires a patterned input for its function.

The deplantation experiments devised by Weiss (1950) are suited to this type of investigation. The dorsal fin of salamander larvae consists of a gelatinous connective tissue with a rich blood supply, and serves as an excellent culture chamber for the survival and further differentiation of deplanted tissues (Figure 5). If a piece of spinal cord (without the sensory ganglia) and a limb are deplanted into a common tunnel made in the dorsal fin, neural connections develop between the two organs, and the limb shows a variety of movements (Székely and Szentágothai, 1962a; Székely and Czéh, 1971b). Since the fin does not contain any muscle tissue or motor fibers, all limb movements reflect deplanted spinal cord functioning. Regular stepping movements of the limb graft can be observed only if the spinal cord deplant derives from the brachial region of the donor larva (Figure 6). The movement appears either spontaneously, or is evoked by mechanical stimuli applied to the fin or to the limb. The movement of the grafted limb remains unaltered even when the host's spinal cord is destroyed. The muscle activity pattern recorded from such limbs is comparable to the normal pattern. If the spinal cord is split longitudinally so that only one half of it is deplanted, similar results are seen (Schrameck and Székely, Unpublished data). However, as expected, spinal cord deplants taken from the thoracic region initiate irregular muscle twitches which never transfer to coordinated limb movements.

Histologic studies indicate that structural integrity of the deplanted spinal cord is the prerequisite for the development of

Figure 5. Schematic drawing illustrating the deplantation of a piece of spinal cord and a limb into a common tunnel made in the dorsal fin of an axolotl larva.

DEVELOPMENTAL ASPECTS OF LOCOMOTION 747

Figure 6. Selected frames from a cinematographic film illustrating the step-like movement of the deplanted limb (arrow). The limb movement starts with an extension of the elbow, and continues into the stepping movement. At arrows with asterisk the contours of the limb fade into the background. (From Székely and Czéh, 1971b, with permission of the Publishing House of the Hungarian Academy of Sciences.)

stepping movements. A small fragment, or a structurally impaired spinal cord tissue could not move the limb better if it was innervated by thoracic segments. Serial sections made from the dorsal fin also show that this spinal cord deplant is not entirely without sensory supply. The dorsal fin is innervated richly by sensory fibers. Some of them course into the spinal cord deplant, leave and terminate in either the surrounding fin or limb graft. There are distinct sensitive spots on the limb and dorsal fin responsible for evoking stepping movements. This suggests the sensory fibers travelling through the cord deplant may establish connections with its neurons either by collateral sprouting or by ephasis-like appositions, and provide the cord with a sensory input. It is obvious this input cannot be patterned in the sense that it could determine the rhythm of output. It is more probable that the sensory impulses conveyed by the skin nerves of the dorsal fin do not provide more than a general background stimulation, and the small interneurons, by virtue of their structural organization, transform this stimulation into the appropriate output pattern.

COORDINATION OF LIMB PAIRS

If two limbs are grafted onto the dorsal fin in a common tunnel with the deplanted brachial section of the spinal cord (Czéh, Unpublished data), the two limbs move in exactly the same rhythm, similar to oar-strokes. This observation indicates that the existing structure in the spinal cord at the limb level can control stepping movements of individual limbs, but cannot control alternating movements of limb pairs. The conditions necessary for alternating movement in limb pairs can be studied in the following experimental arrangement (Brändle and Székely, 1973).

Various lengths of the body, including the prospective limb segments of the medullary tube, have been transplanted in the side of newt embryos. Except for neural connections, complete parabiosis develops between hosts and grafts. Therefore, limb movements that appear spontaneously or in response to mechanical stimulations in the graft are controlled solely by the graft's own nervous system. As indicated in Figure 7, the limbs perform the synchronous, oar-stroke movement, if the graft contains only brachial or only lumbosacral segments (groups 1, 3). If the medulla, or a part of the thoracic spinal cord also is included in the graft, the limbs show the regular alternating movement (groups 2, 4). Alternating limb movements develop in grafts in which a section of the thoracic cord, or another brachial cord, is transplanted forward to a set of brachial segments (groups 6, 7). In late larval life the distal end of the graft's neuraxis begins to deteriorate. In grafts which contain the medulla and the brachial segments, the deterioration causes alternating coordination to

DEVELOPMENTAL ASPECTS OF LOCOMOTION

Group	Recombination of neuraxis	No.	Limb coordination
1		13 (1)	Synchronous (Defective limb movements)
2	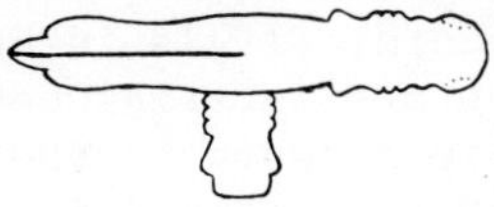	12 (3)	Alternating (Synchronous)
3	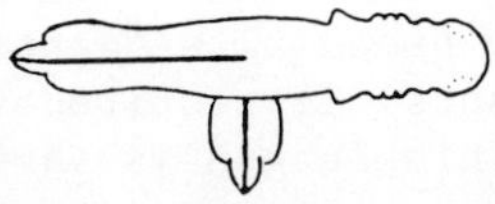	12 (7)	Synchronous (Defective limb movements)
4	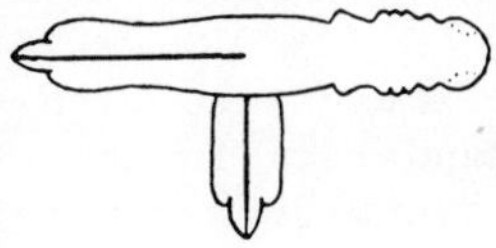	8	Alternating
5	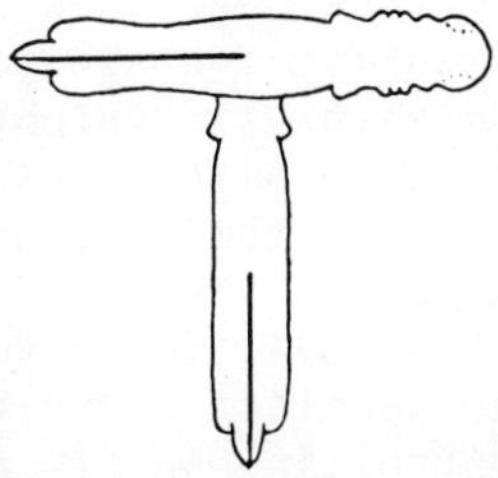	15	Forelimbs synchronous Hindlimbs alternating
6	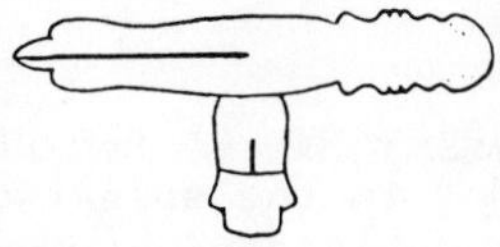	19 (1)	Alternating (Synchronous)
7	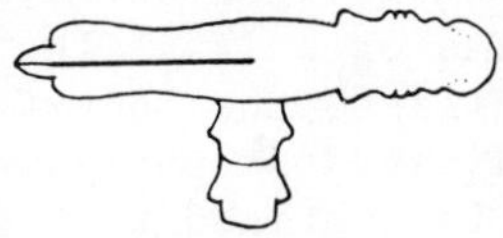	6	First limb pair synchronous Second limb pair alternating

Figure 7. Diagramatic illustration of the scheme of operation showing the composition of the graft's neuraxis and the movement of limbs in the different experimental groups. Numbers and description of limb movements in parenthesis refer to cases with faulty grafts. For further details see text. (From Brändle and Székely, 1973, with permission of S. Karger AG, Basel).

decline. Gradually, synchronous replaces alternating movement about a month after metamorphosis. Similar phenomena occur in group 6 animals following transection of the thoracic spinal cord forward to the brachial segments. If two or fewer thoracic segments remain in contact with the brachial segments, the limbs move synchronously after recovery from the operation.

The results indicate that control of alternating movements requires the limb segments to be elongated with the medulla, or with a few segments of the thoracic cord. It is interesting that the medulla and the thoracic segments are interchangeable in this respect, but these structures must lie forward to the limb segments. Another interesting aspect of the results is that a critical length of the spinal cord (medulla + brachial segments; more than 2 thoracic segments + brachial segments) is required for the control of alternating coordination. This length is about one third of the length of the medullo-spinal neuraxis.

The limb movements appear to be integrated with sinuous trunk movements during ambulation in salamanders. In this integration limb protraction is associated with trunk convexity; limb retraction with trunk concavity; while the trunk describes a sine wave in the diagonal pattern of limb movements. The sinuous movement of the trunk is actually the swimming movement, in which the bilateral axial musculature contracts in alternation. It is supposed that swimming is controlled by an endogenous rhythm generated automatically by a kind of oscillator mechanism in the spinal cord if the central excitation is kept at a certain level (Gray and Sand, 1936; von Holst, 1939; Roberts, 1969). The oscillation is maintained in short sections of the spinal cord controlling as few as 8 segments in dogfish (Grillner, 1974).

The experimental results are suggestive of the presence of two neural autochtonous mechanisms working in the spinal cord. One controls swimming and generates a travelling sine wave along the axis of the body when the animal swims or walks. The structure of this mechanism extends from the medulla to the end of the spinal cord. The second controls the coordinated stepping movement of individual limbs, and its structure is confined to the limb moving segments. The two structures may be interconnected, or the second may be part of the first in such a manner that trunk convexity facilitates protraction and trunk concavity promotes retraction in the limb moving structure. This speculation is supported by the observations described earlier in this chapter under functional determination of the cord. As mentioned a phase shift exists in the rhythm of supernumerary limbs innervated by brachial segments grafted into the place of lower thoracic segments and of those innervated by lumbosacral segments grafted into the mid-thoracic region. In the first case the convexity of the travelling sine wave reaches the grafted brachial cord later than does the normal one, and triggers protraction in the grafted limb later as compared to the normal forelimb. Likewise, trunk convexity occurs sooner in the region of the grafted lumbosacral cord than in the region of the normal one, and the grafted limb is protracted sooner than the normal hindlimb.

General Conclusions

Neuroembryological studies indicate that both the structure and the function of spinal cord segments responsible for limb movement are predetermined. They develop independently of external influences. The internal factors which influence the differentiation of this structure are unknown. This problem awaits clarification by further studies with the methodology of developmental mechanics.

Embryology also has revealed that rhythm and character (fore- or hindlimb, wing or leg) of movement is programmed in the structure. This is especially clear in the case of chicken extremities with heterotopic innervation, and the results suggest that the output pattern and the gross anatomical structure (freedom of motility) must match in order to obtain the whole complement of movement. This implies that quality rather than quantity of innervation is the significant factor in the control of coordinated limb movement. The notation obtained further support from myogram recordings from partially moving limbs innervated by transplanted thoracic segments. Some of the muscles showed an activity stronger than normal, but the joint in question remained motionless. Presumably, the firing sequence of motoneurons, regardless of their number, was not correct. This suggests the development of the structure which controls the time pattern of motoneuron activities was incomplete.

In another series of experiments recording muscle activities from partially innervated limbs of salamanders (Székely and Czéh, 1971a) it was found that one limb moving segment or a part of it generates a rhythmic output pattern that contains commands for both protraction and retraction activities of the limb. It has been proposed that this pattern is generated by numerous small, interlocking, cyclic networks. Admittedly this conclusion needs more direct evidence than found thus far. Nevertheless the loss of movement in partially moving limbs with adequate peripheral innervation can be explained in the following manner. A number of small cyclic networks failed to develop in grafted thoracic segments, and the rhythmicity required for the control of coordinated limb movement was absent from a part of the output pattern. It has been shown in model experiments (Székely, 1965; Kling and Székely, 1968) that cyclic networks are capable of converting a continuous input into a patterned rhythmic output. Neither does the spinal cord need any kind of patterned input to generate a patterned rhythmic output for the control of coordinated limb movements. The results of the dorsal fin experiments are interpreted in this sense.

The suggestion that the structure responsible for the generation of a rhythmic output pattern, lies topographically at the same level with, but hierarchically above, the motoneurons, is supported by recent histologic findings. The results derive from experiments in which the cobaltous chloride staining technique introduced by Pitman et al., (1972) has been adapted successfully to the frog's spinal cord (Székely, 1975; Székely and Gallyas, 1975). Co^{++}ions are taken up by axons and transported retrograde to the perikaryon, or anterograde to the terminals. Subsequent treatments precipitate the Co^{++}ions and intensify the Co precipitate, and the neuron is visible in its entirety.

In these specimens it can be demonstrated that the dendrites of motoneurons take three different directions in their courses (Figure 8). A group of dendrites turns laterally and arborizes in the lateral dendritic array. Their branches are in the way of descending fibers. It has been shown in electrophysiological experiments (Brookhart and Kubota, 1963) that motoneurons easily can be fired monosynaptically by descending volleys set up in the lateral funiculus. Another group of dendrites turns dorsally and forms the dorsal dendritic array at the border of the gray and white matter of the spinal cord. This group of dendrites, and only this, receives the synapses of dorsal root fibers. Also, motoneurons can be discharged monosynaptically through the activation of the dorsal dendrite array alone (Czéh and Székely, 1971; Czéh, 1972). The third group of dendrites turns dorsally and medially and constitutes the dorsomedial dendritic array which reaches into the central gray matter. These dendrites do not seem to be in direct contact with dorsal root fibers, but they are impinged upon presumably by the axon terminals of short axon interneurons of the spinal cord.

There are no data about the morphology and physiology of the synaptic contacts of the dorsomedial dendritic array. Possibly, the results of the dorsal fin deplantation experiments may help determine the functional significance of the synaptic relationship proposed here. In a spinal cord fragment isolated in the dorsal fin, the influence of descending fibers and of primary afferent fibers obviously is turned off, and the motoneurons are activated only by interneurons. The fact that coordinated limb movements can be observed under this experimental condition, suggests that the appropriate output pattern is processed by the networks of spinal interneurons and transmitted to the motoneurons, probably

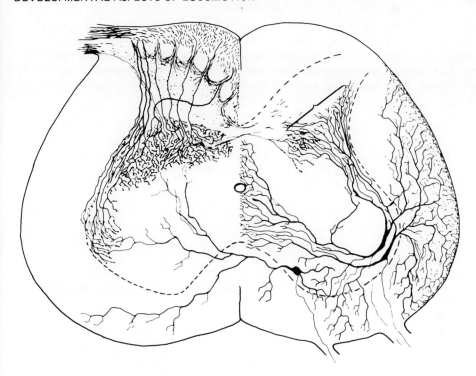

Figure 8. Schematic representation of the crossection of the frog's spinal cord showing the dendritic arborization of motoneurons (right side) and the ventral projection of dorsal root fibers (left side) as revealed by the cobaltous chloride staining technique. Of the three motoneurons illustrated the medial one with crossing dendrites innervates axial musculature; the lateral ones supply limb muscles. The dendrites of these latter neurons constitute the lateral, dorsal (indicated by a brace) and the dorsomedial dendritic arrays. On the left side, the lightly dotted area underneath the posterior funiculus represents the substantia gelatinosa. Deeper in the gray matter a triangular area receives dorsal root fibers. From this area a "tail" protrudes ventrally and supplies motoneuron somata. The comparison of the two sides indicate that dorsal root fibers can establish direct contacts only with the dorsal dendritic array. (From Székely, 1975, with permission of Brain Research, Amsterdam).

through the dorsomedial dendritic array. The central gray matter may be regarded then as the repository for the mechanism of movement. Neural information conveyed by descending and sensory fibers interacts with this spinal structure, both at the level of interneurons and motoneurons, and modifies its output pattern. This

information controls what may be called the mechanism of posture.
This enables the moving limb to adapt to the actual environmental
conditions which affect locomotion. In order to obtain support
for this speculation, more data on the structural organization of
the spinal interneurons are necessary.

REFERENCES

Brändle, K. and Székely, G., (1973) The control of alternating coordination of limb pairs in the newt (Triturus vulgaris). Brain Behav. Evol. 8, 366-385.

Brookhart, J.M. and Kubota, K., (1963) "Studies of the integrative function of the motor neurone," In Progress in Brain Research. (Moruzzi, G., ed.), Vol. I, Elsevier, Amsterdam, (38-61).

Czéh, G., (1972) The role of dendritic events in the initiation of monosynaptic spikes in the frog motoneurons. Brain Research. 39, 505-509.

Czéh, G. and Székely, G., (1971) Monosynaptic spike discharges initiated by dorsal root activation of spinal motoneurons of the frog. Acta Physiol. Hung. 39, 401-406.

Detwiler, S.R., (1920) Experiments on transplantation of limbs in Amblystoma. The formation of nerve plexuses, and the function of the limbs. J. Exp. Zool. 31, 117-169.

Detwiler, S.R., (1923) Experiments on the transplantation of the spinal cord in Amblysoma, and their bearing upon the stimuli involved in the differentiation of nerve cells. J. Exp. Zool. 37, 339-393.

Detwiler, S.R. and Carpenter, R.L., (1929) An experimental study of the mechanism of coordinated movements in heterotopic limbs. J. Comp. Neur. 47, 427-447.

Gray, J., (1939) Aspects of animal locomotion. Proc. R. Soc. B. 128, 28-61.

Gray, J. and Sand, A., (1936) The locomotory rhythm of the dogfish (Scillium canicula). J. Exp. Biol. 13, 200-209.

Grillner, S., (1974) On the generation of locomotion in the spinal dogfish. Exp. Brain Res. 20, 459-470.

Hamburger, V., (1958) Regression versus peripheral control of differentiation in motor hypoplasia. Am. J. Anat. 102, 365-410.

Harris, A.E., (1965) Differentiation and degeneration in the motor horn of the foetal mouse. Ph.D. thesis, University of Cambridge. (Cited in Prestige, 1967.)

Holst, E. von., (1939) Die relative Koordination als Phanomen und als Methode Zentralnervoser Funktionsanalyse. Ergebn. Physiol. 42, 228-306.

Hughes, A.F.W., (1961) Cell degeneration in the larval ventral horn of Xenopus laevis (Daudin). J. Embryol. exp. Morphol. 9, 269-284.

Hughes, A.F.W., (1968) Aspects of Neural Ontogeny. Logos. P., London.

Kling, U. and Székely, G., (1968) Simulation of rhythmic nervous activities, I. Function of networks with cyclic inhibitions. Kybernetik. 5, 89-103.

Levi-Montalcini, R., (1950) The origin and development of the visceral system in the spinal cord of the chick embryo. J. Morphol. 86, 253-283.

Perri, T., (1956) Trapianti enbrionali de midollo spinale e genesi dei corni motori in Bufo vulgaris, Atti Accad. naz. Lincei Rc. 20, 666-670.

Piatt, J., (1956) Studies on the problem of nerve pattern. I. Transplantation of the forelimb primordium to ectopic sites in Amblystoma. J. Exp. Zool. 131, 173-202.

Pitman, R.M., Tweedle, Ch. D. and Cohen, M.J., (1972) Branching of central neurons: Intracellular cobalt injection for light and electron microscopy. Science N.Y. 176, 412-414.

Pollack, E.D., (1972) Cell migration into the "established" lateral motor column in Rana pipiens larvae: I. Brachial spinal cord. J. Exp. Zool. 179, 183-190.

Prestige, M.C., (1967) The control of cell number in the lumbar ventral horns during the development of Xenopus laevis tadpoles. J. Embryol. exp. Morph. 18, 359-387.

Prestige, M.C., (1973) Gradients in time of origin of tadpole motoneurons. Brain Research. 59, 400-404.

Roberts, B.L., (1969) Spontaneous rhythms in the motoneurons of spinal dogfish (Scyliorhinus canicula). J. Mar. Biol. Ass. U.K. 49, 33-49.

Rogers, W.M., (1934) Heterotopic spinal cord grafts in salamander embryos. Proc. nat. Acad. Sci. Wash. 20, 247-249.

Sato, T., (1933) Uber die Determination des fetalen Augenspalts bei Triton taeniatus. Arch. Entw. Mech. Org. 128, 342-347.

Smith-Harcombe, E. and Wyman, R.J., (1970) Diagonal locomotion in deafferented toads. J. Exp. Biol. 53, 225-263.

Straznicky, K., (1963) Function of heterotopic spinal cord segments investigated in the chick. Acta Biol. Hung. 14, 143-155.

Székely, G. and Szentágothai, J., (1962a) Experiments with "model nervous systems". Acta Biol. Hung. 12, 253-269.

Székely, G. and Szentágothai, J., (1962b) Reflex and behavior patterns elicited from implanted supernumerary limbs in the chick. J. Embryol. exp. Morph. 10, 140-151.

Székely, G., (1963) Functional specificity of spinal cord segments in the control of limb movements. J. Embryol. exp. Morph. 11, 431-444.

Székely, G., (1965) Logical network for controlling limb movements in urodela. Acta Physiol. Hung. 27, 285-289.

Székely, G. and Czéh, G., (1967) Localization of motoneurons in the limb moving spinal cord segments of Amblystoma. Acta Physiol. Hung. 32, 3-18.

Székely, G., (1968) Development of limb movements: Embryological, physiological and model studies. Ciba Foundation Symposium on Growth of the Nervous System. (Wolstenholme, G.E.W. and O'Connor, M., eds.), Churchill Ltd., London, (77-93).

Székely, G., Czéh, G. and Voros, G., (1969) Activity pattern of limb muscles in freely moving normal and deafferented newts. Exp. Brain Res. 9, 55-62.

Székely, G. and Czéh, G., (1971a) Muscle activities of partially innervated limbs during locomotion in Ablystoma. Acta Physiol. Hung. 40, 269-286.

Székely, G. and Czéh, G., (1971b) Activity of spinal cord fragments and limbs deplanted in the dorsal fin of urodela larvae. Acta Physiol. Hung. 40, 303-312.

Székely, G., (1975) The morphology of motoneurons and dorsal root fibers in the frog's spinal cord. Brain Research. (In press).

Székely, G. and Gallyas, I., (1975) Intensification of cobaltous sulphide precipitated in the frog nervous system. Acta Biol. Hung. (In press).

Weiss, P., (1936) Study of motor coordination and tonus in deafferented limbs of amphibia. Amer. J. Physiol. 115, 461-475.

Weiss, P., (1937) Further experimental investigations on the phenomenon of homologous response in transplanted amphibian limbs. II. Nerve regeneration and innervation of the transplanted limbs. J. comp. Neurol. 66, 481-535.

Weiss, P., (1950) The deplantation of fragments of nervous system in amphibians. J. Exp. Zool. 113, 397-461.

Wenger, B.S., (1951) Determination of structural patterns in the spinal cord of chick embryo studied by transplantation between brachial and adjacent levels. J. Exp. Zool. 116, 123-164.

A KINETIC ANALYSIS OF THE TROT IN CATS

Gideon Ariel and Ruth Maulucci

Computer and Information Science, University of

Massachusetts, Amherst, Massachusetts

The biomechanics of locomotion in cats was investigated in 1938 by Manter. The present study was designed to supplement Manter's results, thereby obtaining further kinetic information useful in the construction of models of locomotion.

METHODS

The forces exerted by the intact cat during the trot were analyzed. The experiment consisted of two distinct recording sessions using the same five cats for each session. The data were gathered by employing a Kistler force plate. As the cat progressed across the force plate, the instantaneous vertical, horizontal, and lateral components of the force exerted by the cat were obtained. These force components were converted into analog electrical signals and displayed on a storage oscilloscope. Cinematographic data were obtained simultaneously in order to provide information as to which limbs were contributing to the forces. Data reduction was accomplished by digitizing the force components. The digitizer results were analyzed using a CRT display terminal interfaced with a CDC-Cyber-74 computer.

A typical trot sequence across the force plate consisted of the following stages:

(1) A single forelimb touched the plate.

(2) The forelimb was removed and the ipsilateral hindlimb was placed on the plate.

(3) The hindlimb was joined on the plate by the contralateral forelimb.

(4) This pair of diagonal limbs was removed and the remaining hindlimb was placed on the plate.

(5) The hindlimb was removed and the sequence was concluded.

RESULTS

Figure 1 is a typical graph of this trot sequence as displayed on the oscilloscope. The sweep is 100 msec/division. The bottom graph, F_z, is the vertical force (recorded at 10 newtons/division). Positive values for F_z indicate that the force is applied downward. The middle graph, F_y, is the horizontal force (recorded at 4 newtons/division). Positive values for F_y indicate that the force is applied in the direction opposite to that of the locomotion and negative values for F_y indicate that the force is applied in the direction of the locomotion. The top graph, F_x, is the lateral force (recorded at 2 newtons/division). Positive values for F_x indicate that the force is applied to the cat's right. The graphs are divided into three distinct intervals of activity. In the interval $[a_1,b_1]$, the graphs correspond to the action of the left forelimb on the plate; thus, $[a_1,b_1]$ is the stance phase of the left forelimb. This is followed by a brief return to the base lines as this forelimb is removed. In the interval $[a_2,b_2]$, the graphs correspond to the combined action of the left hindlimb followed by the right forelimb on the plate. It is noteworthy that the vertical force is remarkably smooth as the forelimb is added to the hindlimb on the plate. Next is another brief return to the base lines as these diagonal limbs are removed. Finally, in the interval $[a_3,b_3]$, the graphs correspond to the action of the right hindlimb on the plate; thus, $[a_3,b_3]$ is the stance phase of the right hindlimb.

With respect to the vertical force, the following may be noted.

$$\text{Max}(\{|F_z(t)| \mid a_1 \leq t \leq b_1\}) > \max(\{|F_z(t)| \mid a_3 \leq t \leq b_3\}),$$

i.e., the peak vertical force in the stance phase of the forelimb is significantly greater than the peak vertical force in the stance phase of the hindlimb. Also,

$$\left| \int_{a_1}^{b_1} F_z(t)dt \right| > \left| \int_{a_3}^{b_3} F_z(t)dt \right|,$$

i.e., the absolute impulse due to the vertical force during the stance phase of the forelimb is significantly greater than the

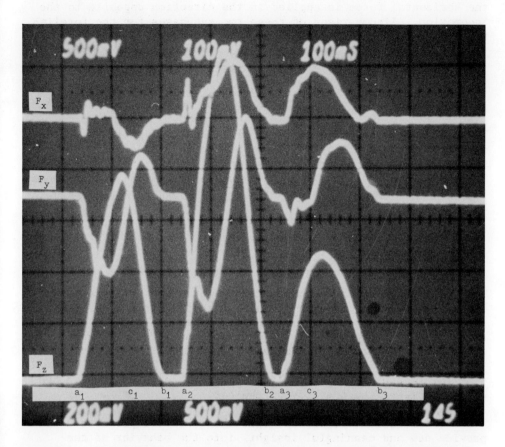

Figure 1. An oscilloscope display of the vertical, horizontal, and lateral components of the force exerted by the cat during the trot.

absolute impulse due to the vertical force during the stance phase of the hindlimb. In both the forelimb and the hindlimb, the peak vertical force increases as the speed of the cat increases, but the absolute impulse remains relatively constant over a wide range of trot speeds.

The horizontal force exhibits the following characteristics. There is a point $c_1 \epsilon [a_1, b_1]$ such that $F_y(t) < 0$ for $a_1 < t < c_1$ and $F_y(t) > 0$ for $c_1 < t < b_1$. This indicates that for the forelimb, the stance phase may be further subdivided into an initial impact phase $[a_1, c_1]$ in which the horizontal force is applied in the direction of the locomotion and a final propelling phase $[c_1, b_1]$ in which the horizontal force is applied in the direction opposite to the locomotion. A comparison of these two subphases for the forelimb reveals that

$$|c_1-a_1|>|b_1-c_1|, \max(\{|F_y(t)| \mid a_1 \leq t \leq c_1\}) > \max(\{|F_y(t)| \mid c_1 \leq t \leq b_1\}), \text{ and } \left|\int_{a_1}^{c_1} F_y(t)dt\right| > \left|\int_{c_1}^{b_1} F_y(t)dt\right|.$$

Thus, in the stance phase of the forelimb, the magnitudes of the duration, peak, and impulse of the horizontal force are respectively greater in the impact phase than in the propelling phase. Similarly, for the hindlimb, there is a point $c_3 \epsilon [a_3,b_3]$ such that $[a_3,c_3]$ is an impact phase and $[c_3,b_3]$ is a propelling phase. However, for the hindlimb,

$$|b_3-c_3|<|c_3-a_3|, \max(\{|F_y(t)| \mid c_3 \leq t \leq b_3\}) > \max(\{|F_y(t)| \mid a_3 \leq t \leq c_3\}),$$
$$\text{and } \left|\int_{c_3}^{b_3} F_y(t)dt\right| > \left|\int_{a_3}^{c_3} F_y(t)dt\right|.$$

Thus, in the stance phase of the hindlimb, the magnitudes of the duration, peak, and impulse of the horizontal force are respectively greater in the propelling phase than in the impact phase. In general, the impact phase tends to dominate the forelimb, whereas the propelling phase tends to dominate the hindlimb.

The lateral force, for all limbs, is chiefly applied in the direction away from the body of the cat.

In conclusion, it is felt that these kinetic parameters will provide new and meaningful insights into the behavior of the neural mechanisms of locomotion.

IPSILATERAL EXTENSOR REFLEXES AND CAT LOCOMOTION

J. Duysens and K.G. Pearson

Department of Physiology, University of Alberta

Edmonton, Alberta, Canada

Exteroceptive extensor reflexes such as the extensor thrust or the positive supporting reaction are obtained in cats by stimulating the skin of the foot sole. Since stimulation of the foot naturally occurs during locomotion, it was suggested that these extensor reflexes possibly play a role in the propulsion of the animal (Sherrington, 1906). This idea, however, was never proven and instead, evidence was obtained that reasonable normal walking is still possible after foot denervation (Sherrington, 1910).

In the present experiments, the role of extensor reflexes during locomotion was investigated using electrical stimulation of the skin and the skin nerves in thalamic cats showing spontaneous walking on a treadmill (for details of brain transection see Pearson and Duysens, this volume). Silver plate stimulating electrodes were attached to the pad and to the plantar surface of the foot while EMG electrodes were inserted bilaterally in the ankle extensors and flexors. Stimuli consisted of brief trains of 1 msec shocks at 60 Hz, evoking an extensor reflex at rest. During walking, stimulation of the hindlimb pad resulted in different reflex actions, depending on the phase of the step cycle during which the shocks were given. No extensor reflexes were seen when the hindlimb was in the swing phase at the moment of stimulation. However, when pulses were applied during the stance phase, a prolongation of the extensor activity was noted, with a concomitant delay in the onset of the next flexor activity (Figure 1A). Comparable results were obtained when the plantar surface of the foot was stimulated. As shown in Figure 1B plantar stimulation during the stance greatly prolonged and increased the ipsilateral extensor EMG activity, while delaying the onset of the contralateral extension.

To further evaluate the cutaneous nature of these reflex actions, a series of cats was used for direct stimulation of the sural nerve. These cats had one hindlimb denervated except for the ankle extensors and flexors. This denervated hindlimb was fixed to a frame over a treadmill on which the cat could walk with the remaining three free limbs. The ankle extensors and flexors of the fixed limb contracted alternatingly during the periods of walking. The sural nerve of the fixed limb was freed and mounted on stimulating electrodes. Stimulation of the sural nerve with low current strength evoked an extensor reflex in the ipsilateral triceps surae muscles when the animal was at rest. With the animal walking, the same stimuli applied during the contraction phase of

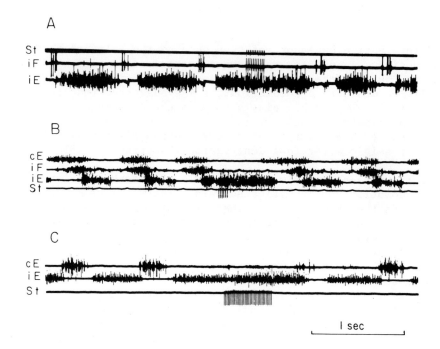

Figure 1. Prolongation of ipsilateral extensor (iE) activity due to cutaneous stimulation (St) during the stance phase. Stimulation of the (A) pad, (B) plantar surface of the foot or (C) sural nerve has an extensor reinforcing effect. Note the delay in the onset of the subsequent ipsilateral flexor (iF) and contralateral extensor (cE) burst.

the triceps surae resulted in a pronounced prolongation of the extensor activity (Figure 1C). Note that the prolongation of the extensor activity was accompanied by a delay in the onset of the contralateral extensor activity. No extensor reflex was observed when stimulation was given during the phase of flexor activity.

The data obtained with stimulation of the pad, plantar surface of the foot and sural nerve all indicate that reflex activation of extensors is easily evoked during the periods of rhythmic extensor activity seen in locomotion. Therefore it seems quite likely that such exteroceptive reflex actions are an integral part of the reflex control of stepping.

REFERENCES

Sherrington, C.S., (1906) The Integrative Action of the Nervous System. Yale University Press, New Haven, Connecticut.

Sherrington, C.S., (1910) Flexion-reflex of the limb, crossed extension reflex and reflex stepping and standing. J. Physiol. (Lond.) 40, 28-121.

LONG-TERM PERIPHERAL NERVE ACTIVITY DURING BEHAVIOR IN THE RABBIT

Joaquín Andres Hoffer[*] and William B. Marks[**]

[*]Department of Physiology, University of Alberta

Edmonton, Alberta, Canada

We report here the first successful attempts at recording from intact fine peripheral nerve filaments of mammals during normal unrestrained locomotion and posture. In 20 rabbits the nerve (200 μm thick, ~175 myelinated axons) supplying the tenuissimus muscle was isolated over ~15 mm with original connectivity and blood supply kept intact, and captured in two consecutive, longitudinally slit, insulating silastic tubes (300 μm x 6 mm) mounted onto a cuff that surrounded the sciatic nerve for support. Differential recording between contacts at the midpoint and ends of each tube allowed discrimination of multiunit nerve spike activity while potentials from neighboring muscles (EMG) were effectively screened out by the addition of shunts and shields.

Electrodes remained implanted for periods of up to 10 months. The largest spikes, initially 10-60 μV, reached 30 to 120 μV during the first week, but over the following weeks became considerably smaller. This was attributed to connective tissue growth inside the tubes, at first perhaps raising the impedance by replacing fluid, but later coating the recording contacts and wrapping tightly around the captured nerve, apparently causing compression injury to the largest fibers. In five rabbits (25%) nerve activity in tenuissimus was continuously present up to the day of the final acute experiment (5,6,18,70 and 136 d). In thirteen rabbits activity ceased by the second to twenty-fifth day, but in four of these neural activity was recorded again several weeks later, and this

---

[**]Laboratory of Neural Control, National Institutes of Health, Bethesda, Maryland 20014.

recovery was confirmed acutely 80 to 308 d post-implant, when the presence of function α-motoneurons that could cause the muscle to contract in response to ipsilateral foot squeeze, or in response to electrical stimulation through the recording contacts was demonstrated. Primary spindle afferents driveable by vibration were also present. Histology after 4-1/2 months in one rabbit demonstrated that 83% of the original fibers had survived, most of the loss occurring at the large-diameter end of the fiber spectrum.

Afferent and efferent fiber activity could sometimes be distinguished by differences in arrival time to the two consecutive tube electrodes. Activity in the nerve to tenuissimus, a functional knee flexor/abductor and hip retractor, and EMG activity from posterior biceps, a synergist of tenuissimus, were monitored during unrestrained hopping and stepping on a treadmill. Hip and knee joint angles were determined from videotaped records of the motion. Tenuissimus and biceps activity characteristically peaked twice during the step cycle (before footlift and during the first extension phase) with a distinct silent period lasting ≤ 100 ms between peaks, during the late flexion phase; activity was low during other phases. From differences in fiber activity in the proximal and distal electrodes, and from sampling only afferents in preparations where large efferents were blocked proximally to the electrode, we determined that afferent activity from tenuissimus usually began before, peaked together with, and continued after α-motoneuron discharge, suggesting γ-motoneuron coactivation. The first flexor burst was interpreted to effect the development of tension that caused swing. The second burst was interpreted to be of central origin, and to brake the fast-extending hindlimb in order to guide foot contact. Evidence of fusimotor coactivation was obtained also during postural changes, where spindle afferent activity from tenuissimus was sometimes tightly correlated with small EMG bursts in biceps. Functionally, tenuissimus was observed to behave very similarly to posterior biceps.

Our results have shown that it is possible to interface directly with the mammalian peripheral nervous system over extended periods of time, thus allowing the long-term study of patterns of peripheral activity during behavior, and, on the clinical side, suggesting a new option for the control of prostheses.

CHEMICAL LESIONING OF THE SPINAL NORADRENALINE PATHWAY: EFFECTS ON LOCOMOTION IN THE CAT

L.M. Jordan and J.D. Steeves

University of Manitoba, Faculty of Medicine, Department of Physiology, Winnipeg, Canada R3E 0W3

Recent evidence has accumulated which indicates that noradrenaline (NA) is important in the initiation of locomotion (Jankowska et al., 1967; Grillner, 1969; Forssberg and Grillner, 1973; Grillner and Shik, 1973). Grillner and Shik (1973) have proposed that stimulation of the mesencephalic locomotor region (MLR), described by Shik et al., (1966), leads to the activation of descending noradrenergic fibers and subsequent release of spinal mechanisms responsible for locomotion in the cat. Steeves et al., (1976), have presented evidence that direct activation of catecholamine-containing neurons in the nucleus locus coeruleus occurs during stimulation of the MLR, and other evidence suggests that locus coeruleus neurons in the cat project to the spinal cord (Kuypers and Maisky, 1975).

In order to determine whether spinal cord NA is essential for the initiation of locomotion by stimulation of the MLR, 6-hydroxydopamine (6-OHDA) was used to chemically lesion the NA fibers in the spinal cord. 6-OHDA was injected bilaterally at the thoraco-lumbar junction of pentobarbital anesthetized cats through 30 gauge needles inserted through the dorsal columns into the gray matter. Solutions of 2 µg/µl, 3 µg/µl or 4 µg/µl 6-OHDA were made in deoxygenated normal saline containing 0.1% ascorbic acid. The volume injected was 4, 5 or 6 µl, with 2, 2.5 or 3 µl, respectively, injected into each side of the cord. Total dose ranged from 8 µg to 24 µg. The rate of injection never exceeded 1 µl per minute and the injection needles were left in place for several minutes at the conclusion of the injection.

Following recovery from the anesthesia, the animals were maintained for a period of 7-10 days, during which time any changes in walking pattern during overground locomotion were noted. Although all the treated animals were able to walk following recovery from the anesthesia, they frequently exhibited an abnormality in locomotion with their hindlimbs. Some of the freely walking cats tended to hyperextend their hindlimbs at the termination of the stance phase, often to a degree which prevented forward progress. The hyperextension was clearly involuntary, since the animals would occasionally hop with the opposite hindlimb in an effort to continue their forward progression. Comparable effects were not seen in sham injected animals which received bilateral injections of 0.1% ascorbic acid in normal saline. Thus, 6-OHDA appears to alter hindlimb locomotor capability in freely moving animals. A detailed analysis of this phenomenon is now in progress.

Following the recovery period, the animals were prepared for MLR stimulation according to the method described by Steeves et al., (1976). Each halothane anesthetized cat was placed in a stereotaxic device, which was positioned over a treadmill, and decerebrated at the precollicular level. Upon termination of the anesthesia, the 6-OHDA treated animals were capable of supporting their weight when the treadmill was not in motion and exhibited decerebrate rigidity in both forelimbs and hindlimbs. Subsequently, monopolar stimulation of the MLR was employed in an effort to evoke locomotion.

In 7 animals previously treated with 6-OHDA it was possible to evoke locomotion which appeared indistinguishable from evoked locomotion in untreated mesencephalic animals. In 5 additional cats it was not possible to evoke coordinated locomotion by MLR stimulation, but in each case these animals were judged to be in poor condition following the decerebration on the basis of both poor recovery of reflexes and failure to develop decerebrate rigidity. As commonly observed in untreated cats, the 6-OHDA treated animals frequently commenced stepping movements with their hindlimbs prior to initiation of locomotor movements with the forelimbs. No difference in the threshold stimulus required to evoke locomotion in 6-OHDA treated versus untreated mesencephalic cats was observed.

Biochemical analysis of the NA and 5-hydroxytryptamine (5-HT) content of the cervical and lumbar enlargements were carried out according to the method described by Shellenberger and Gordon (1971). Control levels were determined in cervical and lumbar enlargements of untreated cats, and residual levels of NA in the lumbar cord were determined following degeneration of descending fibers in chronic spinal cats. Tables 1 and 2 summarize the effects of 6-OHDA on NA levels in the lumbar cord and show that the NA content was substantially reduced in 6-OHDA treated animals. Even the cats

Table 1. Effect of 6-OHDA Treatment on NA Levels
in Lumbar Cord

Group	Mean NA Levels (ng/g)
6-OHDA Treated (n=12)	35.6 (13.7 - 62.3)*
Chronic Spinal (n=3)	13.5 (5.2 - 21.3)
Control (Untreated) (n=5)	156.2 (118.0 - 235.0)

*Numbers in brackets indicate range of results obtained in each group.

Table 2. Mean % Depletion of NA in 6-OHDA
Treated Cats

Compared with Controls:	77.2% (60.1% - 91.2%)
Utilizing Residual NA Levels in Chronic Spinal Cats:*	86.5% (68.8% - 99.9%)

*Levels in 6-OHDA treated animals less residual NA levels in chronic spinal cats.

with maximum depletion of NA responded to MLR stimulation with coordinated locomotion which was indistinguishable from evoked locomotion in untreated mesencephalic cats. The mean depletion of NA in the cervical cord of 6-OHDA treated cats was 33.7%. 5-HT content of the spinal cord was unchanged.

The results of this study indicate that a marked reduction in NA content of the lumbar spinal cord does not alter evoked locomotion in the mesencephalic cat. This suggests that the descending NA pathway is not likely to be the sole means for activation of spinal locomotor systems from the MLR. It remains plausible that propriospinal influences arising in the cervical cord facilitate locomotion in the hindlimbs and mask the effects of reduced NA in the spinal cord, but this is rendered unlikely by the observation that hindlimb locomotion often commences prior to forelimb stepping. Alternatively, the NA depletion may not be sufficiently complete to prevent initiation of locomotion. The descending serotonergic pathway, which is intact after 6-OHDA treatment, may contribute to the initiation of locomotion, as suggested by Viala and Buser (1969) for rabbits, but Grillner (Personal communication) has shown that the serotonin precursor 5-hydroxytryptophan does not evoke stepping in the acute spinal cat. The hindlimb hyperextension observed in freely walking cats indicates that NA depletion alters locomotor capability, but these alterations are not evident during treadmill locomotion evoked by MLR stimulation in decerebrate cats. This suggests that there may be parallel pathways concerned with locomotor control or that certain features of overground locomotion cannot be demonstrated during locomotion on a treadmill.

REFERENCES

Forssberg, H. and Grillner, S., (1973) The locomotion of the acute spinal cat injected with clonidine i.v.. Brain Research. 50, 184-186.

Grillner, S., (1969) Supraspinal and segmental control of static and dynamic γ-motoneurones in the cat. Acta Physiol. Scand., Suppl. 327, 1-34.

Grillner, S. and Shik, M.L., (1973) On the descending control of the lumbosacral spinal cord from the "mesencephalic locomotor region." Acta Physiol. Scand. 87, 320-333.

Jankowska, E., Jukes, M.G.M., Lund, S. and Lundberg, A., (1967) The effect of DOPA on the spinal cord. 5. Reciprocal organization of pathways transmitting excitatory actions to alpha motoneurons of flexors and extensors. Acta Physiol. Scand. 70, 369-388.

Kuypers, H.G.J.M. and Maisky, V.A., (1975) Retrograde axonal transport of horseradish peroxidase from spinal cord to brainstem cell groups in the cat. Neurosci. Letters. 1, 9-14.

Shellenberger, M.K. and Gordon, J.H., (1971) A rapid, simplified procedure for simultaneous assay of norepinephrine, dopamine and 5-hydroxytryptamine from discrete brain areas. Anal. Biochem. 39, 356-372.

Shik, M.L., Severin, F.V. and Orlovsky, G.N., (1966) Control of walking and running by means of electrical stimulation of the midbrain. Biofizyka. 11, 659-66 (Engl. transl.).

Steeves, J.D., Jordan, L.M. and Lake, N., (1976) The close proximity of catecholamine containing cells to the "mesencephalic locomotor region" (MLR). Brain Research. (In press).

Viala, G. and Buser, P., (1969) The effects of DOPA and 5-HTP on
    rhythmic efferent discharges in hind limb nerves in the rabbit.
    Brain Research. 12, 437-443.

METAMORPHOSIS OF THE MOTOR SYSTEM DURING THE DEVELOPMENT OF MOTHS

A.E. Kammer* and M.B. Rheuben**

*Division of Biology, Kansas State University

Manhattan, Kansas

The basic motor patterns for locomotion in insects are well known, and in locust flight and cockroach walking there has been significant progress in understanding the central nervous mechanisms producing the motor output (Burrows; Fourtner; Pearson; this volume). A wealth of information about adult patterns provides the necessary background for investigations of the development of the motor system (Young, 1973; Altman, 1975; Kammer and Rheuben, 1976). Large Lepidoptera provide particularly favorable material for such investigations. This communication summarizes the results of studies on adults, larvae, and pupae of a hawkmoth, Manduca sexta.

During flight in moths, the power-producing thoracic muscles are excited during each wingbeat cycle by a brief burst of impulses, and antagonistic muscles are excited approximately in antiphase. During the pre-flight warm-up by which the thoracic temperature is elevated, the phase relationships are different but the cycle-time is similar to that of flight (Figure 1A, B; cf. Kammer, 1968). One of the main wing depressor muscles in flight is the mesothoracic dorsal longitudinal muscle. This muscle has the fine structure of an arthropod "fast" or phasic muscle: 6-8 thin filaments surrounding each thick filament, short sarcomeres, numerous mitochondria, and extensive sarcoplasmic reticulum, (Rheuben, 1975 and in preparation). [Insect muscles are divisible into phasic and tonic types (Cochrane et al., 1972), on grounds very dissimilar to those

---

**Department of Biology, Yale University, New Haven, Connecticut

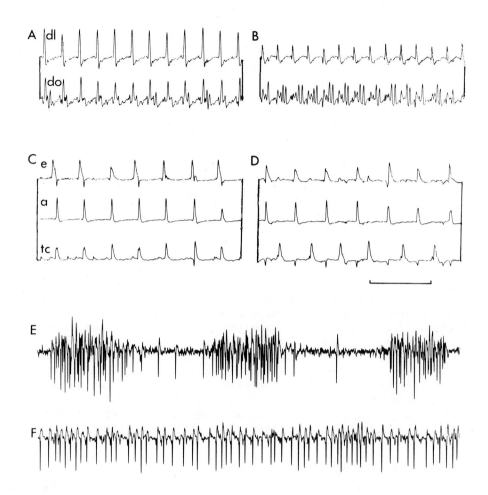

Figure 1. Motor patterns from different instars of Manduca secta, recorded extracellularly from mesothoracic muscles. A, Adult moth, pre-flight warm-up; the dorsal longitudinal wing depressor muscle (dl) and the dorsal oblique wing elevator (do) are excited synchronously. B, Adult moth, flight; the same units as in A are excited alternately. C, Pupa, warm-up pattern with longer cycle-time (a,e, two units of dorsal longitudinal muscle; tc, tergocoxal elevator muscle). D, Pupa, flight pattern, same units and approximately the same cycle-time as in C. E and F, Restrained caterpillar, unidentified dorsal muscles; similar patterns were seen in freely moving caterpillars. (Time mark: 200 msec for A-D, 1 sec. for E and F).

applied to vertebrates, since they all are multiterminally innervated and few have propagating, all-or-none action potentials.] In response to stimulation of the motor nerve, a fiber of the dorsal longitudinal muscle produces a large end-plate potential which gives rise to a large active response. The twitch is fast, reaching maximum tension in about 12 msec., at 35°C. At flight temperatures of 35-38°C the muscle follows stimuli at 40/sec. with only slight summation of tension. This flight muscle is innervated by five motor neurons, four of which have their cell bodies in the prothoracic ganglion, and one of which has a cell body in the contralateral side of the mesothoracic ganglion.

In the larval and pupal stages of Manduca there are five neurons with somata located in positions comparable to those of the adult (Casaday, 1975) and with axons supplying homologous dorsal muscles. Thus, the same motor neurons apparently persist from larva to adult, although the muscle fibers they supply are different in the two stages. In contrast to the adult "fast" muscle, the larval dorsal muscles have the fine structure of an insect "slow" or tonic muscle: 10-12 thin filaments surrounding each thick filament, relatively longer sarcomeres, few mitochondria, and less well-developed sarcoplasmic reticulum (Rheuben, 1975 and in preparation). A single contraction of a larval dorsal muscle is about 5 times slower, and the tetanus/twitch ratio is 6-7 times greater than that of the adult dorsal longitudinal muscle.

The precise roles of the dorsal muscles in caterpillars is not yet known, and only motor patterns observed by multiunit extracellular recordings can be reported at this time. The dorsal muscles can be activated tonically by a long train of single impulses (Figure 1F) or by long bursts at long but variable cycle-times (Figure 1E). Both patterns can occur during walking. The latter pattern seems to be associated with rhythmic turning (searching?) movements of the anterior end of the caterpillar, rather than with the peristaltic movements which are most obvious in abdominal segments. Many of the mesothoracic dorsal motor units are not used while the animal is feeding, an activity which involves the maintenance of posture and an occasional small step.

Comparison of adult and caterpillar thus shows that in these two different developmental stages the same motor neurons comprise the output pathway for different motor patterns and control two different types of muscle.

A caterpillar is converted into a moth during the three week pupal stage. Extracellular recordings from the mesothoracic muscles of pupae show that the adult motor patterns develop gradually. During the few days before ecdysis the developing moth, still enclosed within the pupal exoskeleton and almost immobile, produces motor patterns characteristic of adult flight and pre-flight shivering (Figure 1C, D). The pupal patterns, although similar to adult patterns in burst length and phase relationships, have a longer cycle-time (Kammer and Rheuben, 1975). It is probable that the long cycle-time reflects developmental processes in the central nervous system, rather than being a consequence of sensory feedback. Long cycle-times also occur during the development of the flight control system in Orthoptera (Kutsch, 1973; Altman, 1975) and in certain grooming movements of nymphal cockroaches (Reingold, 1975).

What developmental changes are responsible for the maturation of the central pattern-generator are not known, nor is it known what functions, if any, are served by the motor patterns of the pupa. Possibly patterned activity in the pupa is necessary for the maturation of the muscles or the development of the central motor-pattern generator.

Acknowledgment

This work was supported in part by a faculty research grant to A.E.K., in part by U.S.P.H.S. Grant 2R0INS 08996 to M.J. Cohen, and in part by a Research Fellowship of the Muscular Dystrophy Association of America to M.B.R.

REFERENCES

Altman, J.S., (1975) Changes in the flight motor pattern during the development of the Australian plague locust, Chortoicetes terminifera. J. Comp. Physiol. 97, 127-142.

Casaday, G.G., (1975) Neurodevelopment in a thoracic ganglion of a moth: identified motor neurons and microanatomical structure. Ph.D. Thesis, Cornell University.

Cochrane, D.G., Elder, H.Y. and Usherwood, P.N.R., (1972) Physiology and ultrastructure of phasic and tonic skeletal muscle fibres in the locust, Schistocerca gregaria. J. Cell Sci. 10, 419-441.

Kammer, A.E., (1968) Motor patterns during flight and warm-up in Lepidoptera. J. Exp. Biol. 48, 89-109.

Kammer, A.E. and Rheuben, M.B., (1976) Adult motor patterns produced by moth pupae during development. J. Exp. Biol. (In press).

Kutsch, W., (1973) The influence of age and culture on the wingbeat frequency of the migratory locust, Locusta migratoria. J. Insect Physiol. 19, 763-772.

Reingold, S., (1975) Developmental and short-term changes in behavioral outputs of the cockroach Periplaneta americana L. Ph.D. Thesis, Cornell University.

Rheuben, M.B., (1975) Change in structure from "slow" to "fast" of moth muscle fibers during development. Neurosci. Abs. 1, 778.

Young, D., ed., (1973) Developmental Neurobiology of Arthropods. Cambridge University Press, Cambridge, England.

CEREBELLAR NEURONAL FIRING PATTERNS IN THE INTACT AND UNRESTRAINED CAT DURING WALKING

James G. McElligott

Department of Pharmacology, Temple University School of Medicine, Philadelphia, Pennsylvania

The cerebellum plays a major role in the coordination of motor behavior. When there is damage to this area of the brain, a subject is still capable of motor activity but the smooth coordinated aspect of the behavior is no longer present. The object of this study was to investigate the firing patterns of cerebellar neurons in the awake, intact and unrestrained cat during a coordinated motor activity, namely, peg walking. Previous studies on decerebrate cats employed electrical stimulation in the 'locomotor region' of the mesencephalon to produce treadmill walking while cerebellar neuronal activity was recorded (Orlovsky, 1972a, b). These studies have indicated that the cerebellum plays a major role in locomotion and that Purkinje neurons and the nuclear cells modulate their activity during walking.

In order to investigate cerebellar involvement during locomotion, cats were trained to walk from one platform to another on a series of small pegs (surface area 4cm x 4cm) protruding above a shallow trough of water. These pegs were arranged in two rows about 3 meters long and each was equipped with a sensor to indicate forepaw placement. The animals could be trained to walk on these pegs in several days for food reinforcement. They would confine their walking to the pegs due to the presence of the water in the trough. The cat's walk required precise paw placements on the pegs and was self pacing. Peg positions could be varied and their particular arrangement would dictate the walking pattern that the animal used.

After an initial training period, the cats were surgically prepared under barbiturate anesthesia. Craniotomy was performed over the anterior lobe of the cerebellum and the exposed area was covered with a cylindrical housing that contained independent micro-manipulators and electrodes. This housing, an electrical connector plug, and restraining nuts were attached to the skull by means of dental acrylic. After a recovery period of a week, each cat was restrained via the head implant and a search was made for single unit activity of Purkinje neurons in the intermediate forelimb region of lobe V. When a cell was isolated, the animal was released from the restraint apparatus and the neural spike data was telemetered as the animal walked on the pegs (McElligott, 1973). Data was recorded on magnetic tape and subsequently analyzed on a PDP 11 computer.

In a normal cat before or after the operation it was an extremely rare occurrence to miss or to overstep a peg even when walking rapidly. In a number of initial surgical operations, an edema was inadvertently produced over the anterior cerebellar cortex that resulted in a motor deficit which lasted for about 3 weeks. The walking pattern of these animals was slower, more deliberate and characterized by over-stepping the peg (past pointing). The cat would then correct the error when it contacted the peg with the lower part of its forelimb. In addition, there was minimal disruption of equilibrium since these cats could walk on the pegs without losing their balance. Cell firing patterns in these animals were not investigated.

In the normal cats two classes of Purkinje neurons were found; those whose simple spike discharge pattern varied cyclically with the locomotor pattern and those that did not. An individual neuron in this former category was found to fire maximally at a constant phase with regard to ipsilateral forelimb placement for a specific gait. Different neurons fired maximally at different phases with regard to this placement. These results are in basic agreement with the studies of Orlovsky on the hindlimb of the decerebrate cat (Orlovsky, 1972b). Thach, (1968) has also reported similar results but for the forearm system of an awake monkey during conditioned wrist flexion and extension movements.

A further finding in our study was that when the pegs were rearranged to produce different gaits, there was a shift in the point of maximal firing with regard to ipsilateral forepaw placement. An example of this is given for 3 different gaits in Figure 1.

For the particular neuron in Figure 1, the peak of maximum firing during a normal walking pattern (i.e., equal extension of ipsilateral and contralateral forelimb) was about 300 msec after the placement of the ipsilateral limb (Figure 1A). When the animal used an asymmetrical gait (Figure 1B, ipsilateral forelimb

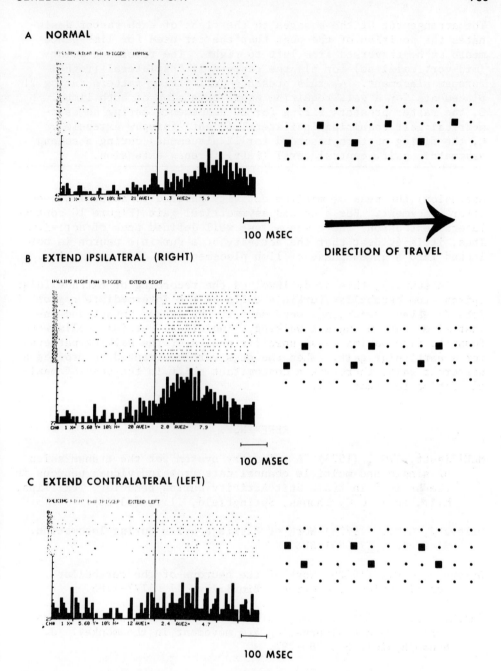

Figure 1. This figure illustrates the activity of one cerebellar Purkinje neuron in raster and cumulative raster form during 3 different locomotor gaits. (Legend continued on next page.)

The arrangement of the squares to the right of each raster designates the position of the pegs that the cat used for limb placements as he traversed from left to right. The reference point (mid-vertical line) for all the rasters is ipsilateral (right) forepaw placement. A, this presents cell firing pattern during 21 placements for a gait requiring equal extension of both limbs. B, the raster presents firing for 20 placements during an asymmetrical gait requiring ipsilateral (right) forepaw extension. C, the firing pattern recorded for 12 placements during a second asymmetric gait, contralateral (left) forepaw extension.

extension) the peak of maximum activity shifted to about 150 msec after placement. For a second asymmetrical gait (Figure 1C contralateral extension) there was a less well defined peak of activity. Thus, it is evident that the activity of a Purkinje neuron is not locked into a given phase of limb placement for different gaits.

In summary, this study involved the recording of extracellular spikes from cerebellar Purkinje cells of the intermediate region of lobe V. These recordings were made in the awake, intact and unrestrained cat during peg walking. Individual Purkinje cells were found to fire maximally at specific phases of the walking pattern for a particular gait. When the pegs were rearranged to produce a different gait, there was a concomitant shift in the peak of maximal firing for that cell.

REFERENCES

McElligott, J.G., (1973) "A telemetry system for the transmission of single and multiple channel data from individual neurons in the brain," In Brain Unit Activity During Behavior, (Phillips, M.I., ed.), C.C. Thomas, Springfield, Ill., (53-66).

Orlovsky, G.N., (1972a) Work of Purkinje cells during locomotion. Biofizika. 17, 891-896.

Orlovsky, G.N., (1972b) Work of the neurons of the cerebellar nuclei during locomotion. Biofizika. 17, 1177-1185.

Thach, W.T., (1968) Discharge of Purkinje and cerebellar neurons during rapidly alternating arm movement in the monkey. J. Neurophysiol. 31, 785-797.

PROPRIOCEPTION FROM NONSPIKING SENSORY CELLS IN A SWIMMING

BEHAVIOR OF THE SAND CRAB, EMERITA analoga

Dorothy Hayman Paul

Biology Department, University of Victoria

Victoria, B.C., Canada

Crabs of the genus Emerita swim by beating the uropods (the most posterior pair of segmental appendages). Two patterns of motor coordination of the uropod beat correlate with two kinds of behavior: "true" swimming and treading water. The essential difference between the two motor patterns is the timing of the power stroke in the interval between the beginning of successive return strokes (cycle). During swimming the latency of the power stroke is correlated positively with cycle duration (80-400 msec), so that the power-stroke phase is fairly constant at 0.5-0.6, whereas during treading water the latency is short (40-60 msec) and increases only slightly over a six fold increase in cycle duration (100-700 msec), (Paul, 1971a, and this study).

Proprioceptive input during the return-stroke movement of a uropod arises from stretching of a receptor strand along with the dendritic terminals of four nonspiking sensory cells insert (Paul, 1972). The reflexes mediated by these cells are negative; motor output to the return-stroke muscle is shut off and its peripheral inhibitor driven; concurrently the peripheral inhibitor to the power-stroke muscle is inhibited and the power-stroke excitor driven. The role of these resistance reflexes is unclear, especially as there is evidence for central neural generation of the motor pattern underlying uropod beating (Paul, 1971b).

The hypothesis which dictated these experiments is that the power stroke in treading water is driven by the negative feedback reflex mediated by the nonspiking cells which results from each return-stroke movement, the series of return strokes being centrally produced. This would explain the lack of correlation of power-stroke latency and cycle duration during treading water, for the timing of the power stroke would be determined by a peripheral (sensory) rather than a central mechanism. In contrast, the swimming motor pattern would be generated by the central nervous system, and independent of sensory feedback from the nonspiking cells. To test this hypothesis, comparison was made of electromyograms recorded from unrestrained Emerita during periods of uropod beating before and after the nonspiking cells had been ablated bilaterally.

All power-stroke latencies in post-ablation electromyograms were correlated positively with cycle duration so that the phase remained constant (0.5-0.7), typical of the swimming motor pattern; no small (0.2-0.4) phase values associated with treading water behavior were observed. Thus, without proprioceptive feedback from the nonspiking mechanoreceptors only the swimming motor pattern (and behavior) has been observed. Reflex drive from the nonspiking cells appears to be essential for generation of power strokes in the motor pattern underlying treading water.

Since in cycles of the same duration the power stroke occurs earlier in treading water than in swimming, the reafferent drive from the nonspiking mechanoreceptors must be circumvented during swimming in order to assure that a reflexly-driven power stroke does not occur too early for the swimming pattern of motor coordination. Uropod beating in Emerita is thought to be an evolutionary derivative of typical crustacean tailflipping (Paul, 1971a) and may be anticipated to use a control system at least partially homologous to that of crayfish (cf. Larimer, this volume). An integral part of the central command system in crayfish is inhibition (presynaptic) of reafference in order to protect habituating synapses from fatigue (Krasne and Bryan, 1973; Kennedy et al., 1974; Evoy, this volume; Larimer, this volume). Whether central suppression of reafference occurs in Emerita, in this case to allow a central motor pattern (swimming) to be expressed, or whether an alternative, central or peripheral, mechanism is responsible remains to be determined.

The swimming and treading water patterns actually represent the ends of a continuum in terms of both cycle duration and power stroke latency. Intermediate values of power stroke latency occur in cycles of less than 400 msec duration. They may be the compromise of differently-timed excitations impinging on the power-stroke motoneuron from center and periphery. The total repertoire of motor coordination underlying uropod beating in Emerita is thus thought to consist of (1) swimming, a centrally generated behavior, (2) treading water, a behavior in which proprioceptive feedback from nonspiking sensory cells plays a formative role and (3) a transitional behavior, which is the result of interactions between the mechanisms generating the first two.

Acknowledgment

Supported by a grant from the National Research Council of Canada to G.O. Mackie.

REFERENCES

Kennedy, D., Calabrese, R.L. and Wine, J.J., (1974) Presynaptic inhibition: Primary afferent depolarization in Crayfish neurons. Science. 186, 451-454.

Krasne, F.B. and Bryan, J.S., (1973) Habituation: regulation through presynaptic inhibition. Science. 182, 590-592.

Paul, D.H., (1971a) The swimming behavior of Emerita analoga (Crustacea, Anomura) I. Analysis of the uropod stroke. Z. vergl. Physiol. 75, 233-258.

Paul, D.H., (1971b) The swimming behavior of Emerita analoga (Crustacea, Anomura) III. Neuronal basis of uropod beating. Z. vergl. Physiol. 75, 286-302.

Paul, D.H., (1972) Decremental conduction over "Giant" afferent processes in an Arthropod. Science. 176, 680-682.

DISCHARGE PATTERNS OF MOTOR UNITS DURING CAT LOCOMOTION AND THEIR

RELATION TO MUSCLE PERFORMANCE

F.E. Zajac and J.L. Young

Electrical Engineering Department, University of

Maryland, College Park, Maryland   20742

Introduction

The objectives of these studies were to record and analyze the firing patterns of single motor units during cat locomotion and to correlate these patterns with those pulse trains which foster efficient tension production in single muscle units. We conclude that the discharge of motoneurones might be "matched" to the contractile properties of their innervated muscle fibers to produce quick development and efficient maintenance of tension.

Methods

Seven cats were prepared for the locomotion experiments similar to the methods described by Severin et al., (1967) except halothane was used rather than ether for anesthesia and decerebration was performed after all other major surgery. While the cats were recuperating from the anesthesia (about 2 hours), (i) the animals were mounted in an apparatus designed to accommodate a stereotaxic micromanipulator, vertebral column support, and a treadmill, (ii) electrodes were placed in hindlimb muscles, and (iii) the spinal cords were prepared for later ventral root dissection and recording from single units. Stimulation of the mesencephalic locomotory region induced walking and trotting (0.9 to 3.5 mph) at a speed controlled by the motor-driven treadmill. Single motor units were functionally isolated in the proximal ends of cut L7-S1 ventral root filaments. Units were presumed extensor if their discharges were in synchrony with the electromyographic (EMG) activity recorded by small copper wires inserted into the gastrocnemius since EMG activity from different hindlimb extensor muscles are mostly in-phase during cat locomotion (Engberg and Lundberg, 1969); otherwise units were presumed flexor.

Twenty cats were investigated for tension production in single medial gastrocnemius (MG) muscle units. The innervating motor fibers functionally isolated in the distal ends of cut ventral root filaments were stimulated with various pulse trains. The time integral of the tension response (tension area) resulting from each stimulus train was measured by a digital computer. The computer generated both trains with different sequences of interpulse intervals (test trains) and trains with the same sequence (standard trains). A test train was only generated if the tension area resulting from a periodically interjected standard train was within a prescribed tolerance (usually 5%) of a pre-determined value. Invariant responses resulting from these standard trains were used to verify that the unit's contractile mechanism was not undergoing slow changes in state such as those which occur during fatigue or during post-tetanic potentiation. Computer control of the experiment in this way ensured that different tension areas resulted from different pulse sequences in the test trains and thus from rapid rather than slow changes in the contractile mechanism's state.

Results

All flexor α-motoneurons (23 units) and 86% of the 50 extensor units began firing in most step cycles with a short interval (doublet) and continued to fire at a mean discharge frequency (Figure 1). Extensor and flexor motoneurons discharged with 2 to 21 and with 2 to 10 spikes per step cycle burst, respectively. Motoneuronal firing was not synchronized with brainstem stimulation in over 90% of the units. Discharge patterns of a given unit remained basically the same for all treadmill speeds, as evidenced by the existence of initial doublets in almost all bursts and by similar mean firing frequencies. However, the number of discharges per burst at faster speeds decreased in most extensor units (75%) and either increased or did not change in flexor units (88%).

For each of the 53 MG units studied in the other experiments (which included both type S and F units; Burke et al., 1973), the train which maximized tension area for a given number of stimuli (the optimal pulse train) was characterized by an initial doublet with a transition of no more than four intervals before the interpulse intervals reached steady-state. This result is consistent with the "catch-like" enhancement observed in type S units for repetitive trains preceded by a doublet (Burke et al., 1970). The steady-state interpulse interval was correlated with the unit's

DISCHARGE PATTERNS IN CAT

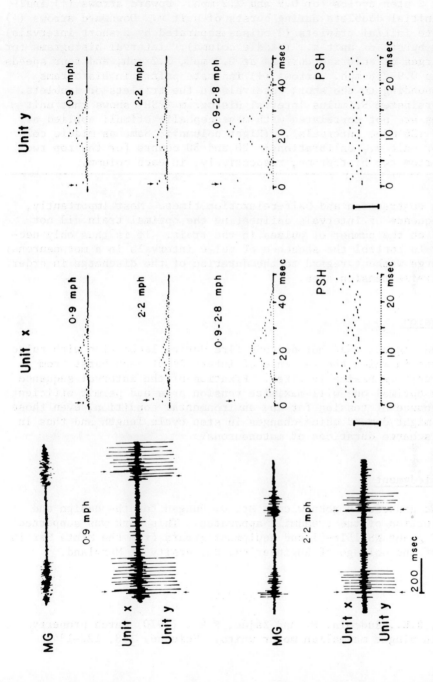

Figure 1. Discharge patterns of extensor motoneurons during locomotion. (Left column): EMG activity in MG muscle recorded simultaneously
(Legend continued on next page.)

with discharge from two extensor units x and y in the same filament during 2 step cycles for 0.9 and 2.2 mph. Upward arrows (↑) indicate initial doublets during bursts of unit y. Downward arrows (↓) indicate initial triplets (3 pulses separated by 2 short intervals) during bursts of unit x. (Middle column): Interval histograms for discharges of unit x occurring at 0.9 mph, 2.2 mph, and four speeds between 0.9-2.8 mph. Arrows (↓) indicate points in histograms corresponding to the short intervals in the triplets or doublets. Post-brainstem Stimulus Interval Histogram (PSH) shows that unit firings are not correlated with mesencephalic stimuli applied at 38 pps (26 msec intervals). (Right column): Same as middle column for unit y. Calibrations: 20 and 50 counts for the top two and bottom two histograms, respectively, in each column.

twitch contraction and half-relaxation times. Most importantly, the sequence of intervals delineating the optimal train did not depend on the number of pulses in the train. It is thus only necessary to control the sequence of pulse intervals in a motoneuronal discharge without regard to the duration of the discharge in order to maximize tension area.

Conclusion

We suggest that motoneurons fire during locomotion with pulse patterns in which the sequence of intervals of each burst from any given unit is basically fixed. Fixation of the interval sequence to the optimal one will maximize tension area and permit efficient maintenance of tension for any environmental condition, even those which might demand quick changes in step cycle length and thus in the discharge durations of motoneurons.

Acknowledgment

We gratefully acknowledge Mr. D. Dungan for the design and construction of the treadmill apparatus. This work was supported by NIH grant NS11518-01 and equipment grants from the Minta Martin Fund of the College of Engineering, University of Maryland.

REFERENCES

Burke, R.E., Rudomin, P. and Zajac, F.E., (1970) Catch property in single mammalian motor units. Science. 168, 122-124.

Burke, R.E., Levine, D.N., Tsairis, P. and Zajac III, F.E., (1973) Physiological types and histochemical profiles in motor units of the cat gastrocnemius. J. Physiol. (London). 234, 723-748.

Engberg, I. and Lundberg, A., (1969) An electromyographic analysis of muscular activity in the hindlimb of the cat during unrestrained locomotion. Acta Physiol. Scand. 75, 614-630.

Severin, F.V., Shik, M.L. and Orlovsky, G.N. (1967) Work of the muscles and single motoneurons during controlled locomotion. Biofizika. 12, 762-776 (Engl. transl.).

CONCLUSION

DR. PEARSON: What we want to do today is to try to assess some of the ideas that have come out of this conference, and also to fill in some of the gaps by talking about areas that were not fully discussed in the last few days. We have received a wealth of information so we must ask: what is all this information telling us about the nervous control of walking? I think the important thing in the panel discussion this morning is to focus on this question. Also it is my hope that following this panel discussion you will leave here with views about the current state of research in locomotion, the trends in research on locomotion and, some general concepts about how locomotory systems work. The latter point is most important since the development of good general concepts can aid and guide us in our research work. There is the danger, of course, if we get locked into these ideas and if we persist with outmoded models and so on, but nonetheless, I think their formulation is valuable.

One of the very exciting things for me at this conference has been the <u>integration of the invertebrate studies with the vertebrate studies.</u> Very often you go to conferences where there are papers on vertebrate and invertebrate work, and you find that those studying vertebrates talk at one time while those studying invertebrates talk at another. And this is certainly what happened in Edmonton two years ago (Control of Posture and Locomotion edited by R.B. Stein et al., Plenum Press). At that time we thought it was probably worthwhile to include some papers on invertebrate work so we put all the invertebrate papers in one morning session in a three day conference. This was an incredibly successful session, and after it, everybody was very excited because we realized that we did have a lot to contribute to the understanding of mammalian locomotion. I think what we have seen here at this conference is that vertebrate and invertebrate physiologists can learn from each other. The idea of the myotatic reflex was developed from work on mammalian systems but is now readily discussed in many invertebrate systems. On the other hand, the idea of central oscillators was given tremendous impetus by the studies on the invertebrate systems.

Over the next few years I anticipate we shall have more gatherings at which both invertebrate and vertebrate physiologists will talk together and thereby facilitate the development of concepts for the nervous control of locomotion.

Drs. Bizzi, Evarts, Davis and myself each will give a fifteen minute presentation on what we consider some of the critical issues developed from this conference. Each presentation will be followed by a discussion period.

INTRASEGMENTAL MECHANISMS FOR STEPPING

DR. PEARSON[x]: I wish to discuss a model for describing the nervous mechanisms controlling stepping in a single limb. I have only one slide to show, (see Figure 6 in Pearson and Duysens, this volume). Now there is a danger in putting up a scheme like this because it could obviously be wrong, but as I pointed out earlier, I think it is valuable to do this because we can then start formulating hypotheses and testing them. Now, on reviewing the literature, and in my own research, I have come to the conclusion that one of the most fruitful ways for regarding the central generation of the reciprocal bursting patterns of activity that we observe in walking systems is to postulate what I call a swing generator, this becomes active for a short period of time to produce bursts of activity mainly in flexor motoneurons, and to postulate that activity in this swing generator inhibits the extensor motoneurons. The rhythmicity resides in the swing generator and the reciprocity results from the inhibitory pathway going to the extensors switching off the maintained tonic activity produced by some central command as well as reflex inputs. Now, the data lead one to think that a system such as this is the most reasonable at the moment. Fourtner (this volume) has described what we think it is in the cockroach so I shall not repeat that. Grillner (this volume) in his talk on central patterning of locomotory rhythm in the cat pointed out an asymmetry in the motor pattern in deafferented preparations; namely, that the flexor bursts tend to be of fairly constant duration and that there is variation in the extensor burst durations. This is precisely what we see in the cockroach. Moreover this type of asymmetry seems to be a feature of a very large number of rhythmically active systems. Since I think a comparative approach is very useful, I shall talk briefly about some other systems just to show you that this scheme is certainly not unique to locomotory systems. The first is vestibular nystagmus. Recently Berthoz et al., (Brain Research. 71, 233, 1974) have postulated that within this system there is a fast phase generator, which is equivalent to what I am calling a swing generator. Another vertebrate system where we have

---

[x]K.G. Pearson, Department of Physiology, University of Alberta, Edmonton, Alberta, Canada

an asymmetry in the central organization of a central rhythm generator is the respiratory system. The respiratory rhythm seems to be produced by an inspiratory oscillator which is equivalent to the proposed swing generator for walking. In medullary inspiratory neurons, the inspiratory oscillator drives and concomitantly inhibits medullary expiratory interneurons. When we observe the invertebrate systems, we note a large number of examples that are organized on this principle of asymmetry. One of these was discussed here at this conference; namely the pyloric system in the stomatogastric ganglion (Selverston, this volume). In this system PD cells are spontaneous pacemakers, and these excite one group of muscles and inhibit motoneurons supplying other muscles. Here the primary pacemaker, the PD neurons, is equivalent to the swing generator in walking systems, and reciprocal patterning occurs because of inhibitory linkages from this pacemaker onto other neurons. Other examples of asymmetry in the rhythm generating systems are the snail masticatory system (Kater and Powell, J. Neurophysiol. 36, 142, 1973), and the ventilatory systems of crustaceans (Mendelson, Science. 171, 1170, 1971) and insects (Miller, Adv. Insec. Physiol. 3, 279, 1966).

Are we any further ahead now, having postulated a swing generator? This is an important question. It is all very well to propose the existence of swing generators but does that lead us on to be able to better describe the locomotory system? I think it does because a swing generator may be regarded as a decision making element; the decision being whether to step or not to step. For reasons of stability it is far more important that an animal ask the question: is it alright to step now, rather than ask is it alright to put my foot down now? By postulating a swing generator the major question now is, what controls the initiation of activity in this generator? There are probably two major inputs to the generator to control its activity apart from some sort of central command. The first is sensory input from peripheral receptors of the single limb. This gives information about the state of the limb (position and load). Provided the limb is in a certain position and unloaded the swing generator can be activated. Otherwise it is inhibited. The second is information from other segments signalling the state of the other limbs. This is important because one limb should only be lifted if the other limbs are in the appropriate positions for the maintenance of stability. It is probable that these intrasegmental inputs to the swing generator come from other swing generators and proprioceptive pathways originating in other limbs. Thus we now have the concept that the swing phase in any leg (produced by activity in the swing generator) will only be initiated when the lifting of that leg will not lead to instability. Thus conceptually we are further ahead by postulating a swing generator. But we may also be further ahead in experiments aimed at determining the mechanisms for generating the locomotory rhythm, since the rhythm generating system may be localized in a small region of the nervous system.

Finally, I want to discuss the function of the excitatory reflex input onto the extensor motoneurons, or the power producing motoneurons during the stance phase. Yesterday I called this a reinforcing reflex. A reinforcing reflex is a convenient way to talk about this, because essentially what happens is that when the extensors become active, there is sensory feedback during the stance that further excites and facilitates the activity in the extensors. This reflex has been identified in the cat and the cockroach (see Pearson and Duysens, this volume). Moreover if we do a comparative analysis all over again, we can see this same sort of reinforcing reflex in a large number of motor systems. Now we must ask ourselves the question: what is the function of the reinforcing reflex on extensor motoneurons? The obvious explanation is that it compensates for any unexpected increase in load. In other words, if a limb is extending and it runs up against some extra load for some reason, then additional feedback from peripheral receptors would further excite extensor motoneurons and there would be more force in the extensor muscles to overcome the load. But I am not sure force is the variable that is controlled in the walking system. The important parameter to be controlled in the walking system is the speed of extension. In other words, this reinforcing reflex functions to maintain the speed of muscular contraction. Now why is speed important in a walking animal? It is important because the movements of one limb must be coordinated with movements of the other limbs. In other words, an animal does not want one limb extending slowly just because that limb happens to run up against an external perturbation while the other three limbs continue extending at the same rate as they were before.

COORDINATION OF MOVEMENTS

DR. BIZZI[x]: In the course of this brief presentation on coordination of movements, I would like to consider two points. First, some aspect of motor coordination and second, the plasticity of eye-head motor coordination. To discuss these points, I will have to make reference to the motor coordinations that I have studied, namely the coordination between eyes and head. Therefore, I will briefly describe the main feature of eye-head coordination as observed in monkeys and man. The appearance of a target in space is followed by a saccadic eye movement which will bring the fovea to the image of the target. After a delay of between 20 and 40 milliseconds, the head turns in the same direction. Since the

---
[x]E. Bizzi, Department of Psychology, Massachusetts Institute of Technology, Cambridge, Massachusetts.

CONCLUSION

eyes have moved first, and with a higher velocity than the head, their lines of sight reach and fixate on the target while the head is still moving. Then for the duration of the head movement the eyes maintain their target fixation by performing a rotational movement that, by being counter to the movement of the head, allows the fovea to remain constantly on the target it has just acquired. This maneuver is termed compensatory eye movement.

To achieve the orderly sequence of eye-head movements, that is, to direct the eyes and the head toward the target and ultimately fixate the target with the fovea, the monkeys must make a number of computations. Initially, the animal must compute the angular distance between the initial lines of sight and the position of the target that is to be acquired. Although we do not have a clear understanding of how the angular distance is computed by the cortical and subcortical visual areas of the brain, we do know that a signal corresponding to that distance is translated into both the oculomotor system and the head-motor system at about the same time. In fact, the electrical activity recorded from eye muscle and neck-muscle fibers shows that motor commands are delivered almost synchronously to those muscles (Bizzi et al., Science, 1971).

The activation of eye and neck muscles leads not only to movements of the eye and head but also to the activation of a number of sensory receptors. These include neck-muscle spindles, neck-tendon organs, receptors located in the joints of vertebrae in the neck and receptors located in the vestibule of the inner ear. All will give rise directly or indirectly to nerve impulses whenever the head is turning.

I will now turn to the question of the role of these sensory receptors in modifying the ongoing eye and head motor programs. To begin with, I will examine the role of visual afferents. We investigated whether visual input is a factor in eye stabilization during spontaneous head turning by observing eye-head movements executed with and without a visible target light (see Figure 1).

In Figure 1(A), the saccadic eye movement, the compensatory eye movement and the head movement are shown. In addition, the sum of eye and head (Gaze) is indicated. Note the stability of the gaze during eye-head movements when the target light remained on during the movement. In Figure 1(B), the target light was switched off just prior to the eye movement. A comparison of compensatory eye movements recorded under these two conditions, namely A and B, indicated that ocular stabilization during active head turning in darkness is completely adequate within the limits of electrooculographic measurement. Thus, visual input is important in initiating these movements but once they are initiated, they will proceed without visual assistance.

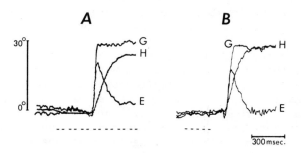

Figure 1. Eye-head coordination in light and darkness. Eye movements (E). Head movements (H). Gaze movements (G) represent the sum of E and H. A) Target light (dotted line) was kept on throughout the coordinated movement. B) Target light turned off prior to the initiation of eye-head movement. (From Dichgans et al., Exp. Brain Res. 18, 1973.)

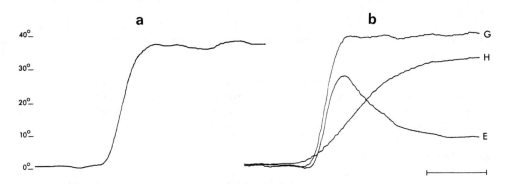

Figure 2. Comparison of eye saccades and gaze. A) Eye saccade to a suddenly appearing target with head fixed. B) Coordinated eye saccade (E) and head movement (H) to the same target with head free. The gaze movement (G) represents the sum of E and H. Note the remarkable similarity of eye saccade in (a) and gaze trajectory in (b) as well as reduced saccade amplitude in (b). Time calibration 100 msec. (From Morasso et al., Exp. Brain Res. 16, 1973.)

CONCLUSION

In contrast, the vestibular afferents play a crucial role in eye-head coordination. The role of these afferents during saccadic eye movements and compensatory eye movements can be demonstrated. I will begin with the effect of vestibular afferents upon the saccades which occur during head movements. The following observations will make the point clear.

If we compare the amplitude of the eye movement in the acquisition of a target when a monkey's head is allowed to turn and when it is held fixed, we observe that the amplitude of the eye movement is greater when the head is held fixed (see Figure 2).

We found that afferent, or incoming, signals arising from the vestibular apparatus are responsible for modulating the saccade amplitude (Figure 2B). Positive evidence of the crucial role of the vestibular afferent signals was demonstrated in monkeys by surgically interrupting the pathway linking the vestibular receptors to the vestibular nuclei. For several weeks after the operation (before the monkey had learned to compensate) the saccade amplitude during head-turning was identical with the saccade amplitude in the absence of head movement. This resulted in a remarkable overshooting of the target because the unmodulated eye movement was simply added to the head movement.

Having established that saccades are adjusted in scale by the vestibular activity initiated by the head movement, let us turn to the question of why this reflex-adaptive arrangement is more advantageous in the animal than one based on a centrally preprogrammed modification of saccadic parameters. Clearly, the reflex mode of organization greatly simplifies the task to the motor-programming systems required for eye-head coordination. The eye and head movements can be programmed independently, and by relying on vestibular reflexes that monitor the actual movement of the head, the resulting adjustment of eye movements will be able to compensate for all the unpredictable peripheral loads and resistances that might change the course of the centrally initiated (intended) head movement.

The modification of saccade characteristics is one aspect of the interaction of central programming and reflex activities. Although this interaction plays a decisive part in the process of target acquisition by a combined eye-head movement, the role of feedback from peripheral sensory organs (vestibular and neck afferents) extends beyond saccadic modulation to control and generation of compensatory eye movements.

Compensatory eye movements have been studied by several investigators. Although it is generally agreed that such eye movements are influenced by vestibular reflexes, it has been hypothesized that compensatory eye movements are initiated centrally and hence are not primarily dependent on feedback information. In our own recent work we have found no evidence for central initiation. We have been able to demonstrate, however, that compensatory eye movements result from the reflex action of the vestibular system. As a consequence of the head movement vestibular receptors are stimulated, and their activity induces a compensatory eye movement that enables the fovea to remain fixed in relation to a point in visual space while the head is rotating.

These findings indicate that the central command is only indirectly responsible for the compensatory eye movement insofar as it initiates the head movement. The head movement, in turn, provides by way of vestibular afferences the reflex excitation necessary for the compensatory eye movement to take place.

In summary, we are now in a position to outline a realistic scheme for how movements of the eye and head are coordinated when a visual target is being acquired. The sequence begins with the detection of a target somewhere in the visual field. Motor programs involving the head and the eyes are activated and respond by sending impulses to eye and neck muscles. This results in a saccadic eye movement and a head movement that activates vestibular receptors, which in turn generate a compensatory eye movement. The compensatory eye movement allows the fovea to remain fixed in relation to a point in visual space during head-turning. The fixation allows a second visual sampling, then a third and so on, with opportunities for correcting errors at each sampling.

If our hypothesized closed loop correctly describes the coordination of eye-head movements, it is clear that the role of the motor program stored in the central nervous system is simply to initiate, in an impulsive manner, movements of the eyes and head. Since there is no central programming of saccadic adjustment and of compensatory eye movement, it follows that the functional, or behavioral coordination of head and eyes is the joint result of a central initiation (following a stored program) modified by the crucial intervention of modulating signals triggered by receptors in the vestibule of the inner ear. This conclusion somewhat simplifies our views of the neural mechanisms underlying motor coordination. Contrary to common assumptions, we find no need to postulate a special central population of "executive" neurons with exclusive responsibility for coordinating the eyes and the head. In this way, one can suggest, by using Davis' terminology (see this volume), that the coordination is an emergent property of the circuit which includes the oculomotor system and the vestibular apparatus.

Having indicated the crucial role of vestibular afference for eye-head coordination, I will now briefly describe a set of experiments that were designed to investigate the plastic changes in the

CONCLUSION

central organization of the eye-head motor system that follow the elimination of vestibular input.

Our aim was not only to ascertain the degree of functional recovery of sensory-motor coordination but also to understand the mechanism underlying the recovery of the coordination of eye-head movements.

Our results have shown that a variety of several basically different mechanisms are developed and jointly brought into play. Among them is the development of new eye programs, including one that provides compensatory eye movements (Dichgans et al., Exp. Brain Res., 1973).

We have already seen that in the normal monkey compensatory eye movements are achieved by way of vestibular afferences. Two or more months after monkeys were surgically deprived of vestibular sensors, centrally programmed compensatory eye movements were found to contribute to ocular stability during active turning of the head. Thus the oculomotor system is capable of taking on, albeit in a crude and inadequate form, functions previously elicited by vestibular activity.

Another mechanism contributing to the remarkable recovery in the coordination of eye-head movements that occurs within the first two to three months following vestibulectomy entails a "recalibration" of saccadic eye movements with respect to visual input. As I have indicated, immediately after vestibulectomy the gaze of the monkey whose head is free to turn will badly overshoot a visual target because the saccade amplitude is no longer modulated by corrective signals from the vestibular apparatus. In other words, the eye movement appropriate for fixating the target when the head is held fixed is simply added to the head movement.

After two to three months, however, vestibulectomized monkeys learn to make saccades that are smaller than normal as they turn their head in the direction of a visual target, thereby reducing the tendency of the gaze to overshoot its mark. Hence the oculomotor system is recalibrated when the head is free to turn.

These two oculomotor functions are an important part of the mechanism underlying recovery of the coordination of eye-head movements. They provide a striking example of the remarkable plasticity of the central motor apparatus, a plasticity that comes into play whenever the organism is forced to compensate for a handicap or deficit imposed on it by events over which it has no control.

Acknowledgment

Research supported by NIH grant R01-NS09343.

REFERENCES

Bizzi, E., Kalil, R.E. and Tagliasco, V., (1971) Eye-head coordination in monkeys: evidence for centrally patterned organization. Science. <u>173</u>, 452-454.

Dichgans, J., Bizzi, E., Morasso, P. and Tagliasco, V., (1973) Mechanisms underlying recovery of eye-head coordination following bilateral labyrinthectomy in monkeys. Exp. Brain Res. <u>18</u>, 548-562.

Morasso, P., Bizzi, E. and Dichgans, J., (1973) Adjustment of saccade characteristics during head movements. Exp. Brain Res. <u>16</u>, 492-500.

CENTRAL ACTIVATION OF MOVEMENTS

DR. DAVIS[x]: One of the things that has been mentioned which has contributed to the success of this symposium is the integration of the type of thinking and the kinds of data that can be generated by people working on vastly different sorts of animals, cats and dogs on the one hand and worms, crayfish and snails on the other. If there is any one take-home lesson which I have derived from these very intensive three days of meetings, it is that <u>evolution is basically conservative</u>. It seems that once nature discovered the solution to a problem such as locomotion, then she applied this solution at all different levels in the animal kingdom. As a result, we can extract a number of general principles which I think apply equally to vertebrate and invertebrate locomotion. I am going to deal with six of these general issues: (1) redundancy in motor systems, (2) positive feedback, (3) command flexibility, (4) control flexibility, (5) command specialization, and (6) plasticity.

I think one thing that we could agree upon is that the nervous system is redundant. There seem to be many different units doing the same thing, and it appears that this generality applies also to locomotor systems. At first sight, redundancy in the nervous system might seem like a waste of precious central nervous space. On the asset side, however, redundancy provides several advantages. One is reliability. The more cells there are, the more likely it is that the given job is going to get done properly. In addition, there is the somewhat more esoteric advantage of increased resolution. When any system is redundant, be it a sensory system or a motor system, the control resolution is correspondingly higher. Finally, there is a possibility which was recalled to mind by the

---

[x]J. Davis, The Thimann Laboratories, University of California at Santa Cruz, Santa Cruz, California

lovely papers by Landmesser and Székely. In both of those papers it was emphasized that cell death is a natural part of the development of motor systems during ontogeny. Perhaps redundancy is the strategy which evolution has used to avoid the vulnerability of a motor system to the loss of strategic units during development. If there is redundancy, then in the process of development, while cells are necessarily dying, the system is not likely to lose any particular unit that is crucial to the output of the whole system. This is just one example of how developmental neurobiology is likely to provide insights into the operation of motor systems for us. Another point emerges from Landmesser's talk; she mentioned the fact that many spatial and temporal features of motor programs develop before the ingrowth of dorsal root afferents. There is hardly a more telling piece of evidence on behalf of the centralist theory of locomotion. Clearly, these motor patterns develop in the absence of sensory input, which must mean that within the spinal cord there exists machinery for generating patterns in the first place. I suspect we can think of many more examples of how developmental neurobiology is likely to help us understand adult motor systems, and perhaps we can discuss some of those.

Now, redundancy results in part from reciprocity. I think most of us would agree that, both in mammalian, higher primate, and in invertebrate motor systems, that there does exist reciprocity between the various elements of motor systems. And one of the consequences of reciprocity is positive feedback, the second of the general issues I wish to discuss. We have seen examples of reciprocal arrangement of neurons in both the stomatogastric ganglion of the lobster and also in the so-called "trigger" system of the mollusk Tritonia. These are two examples of the kind of reciprocity that can lead to positive feedback.

By positive feedback I mean self-reinforcement of the activity of a network which is caused by reciprocal connections within the network. Imagine, for example, delivering an input signal to a reciprocally-arranged network of two cells. Neuron A will excite neuron B which will excite A, etc. In other words, the network is at least in principle capable of sustained activity without sustained input, which is what we mean by reverberation. Reverberation, discussed as a possibility by Ramon y Cajal on morphological grounds, is one of the implications of positive feedback.

There is a most interesting potential consequence of reverberation in motor systems, relating to the distinction between command cells and trigger cells. As we have heard in this symposium, certain invertebrate neuroscientists believe that we should distinguish between command and trigger cells on the basis of whether the response is strictly concomitant with the stimulus (command) or outlasts the stimulus (trigger). Let us see whether, in view of reverberation, this is a realistic and useful distinction. Let us suppose that our simple, reciprocally connected network of two neurons is excited into activity by a neuron I will call, for purposes of neutrality, an "activating" neuron. Such a cell need be active for only a moment in order to evoke activity which may reverberate for many seconds within the affected motor network. If we call this activating cell a trigger cell, we may be identifying the neuron not according to its own properties, but rather according to the properties of the network that it drives. It is for this reason that we cannot presently draw a sensible distinction between trigger and command cells. We simply cannot distinguish the properties of the cells from the properties of the networks that they drive. We do not yet know what kinds of cells exist to activate motor networks. We do not know whether cells we have always called command cells in fact represent different interneural levels, perhaps arranged in a command "hierarchy." We do not know whether activating neurons fall into different categories, each having clearly distinct properties. So let us avoid stimulating but fruitless terminological disputes, pending the acquisition of the necessary data. Let us not argue whether we should call a cell a command or trigger, decision or executive, gating neuron or driver, godfather or grandmother cell. I propose that for the present we label not cells, but functions. There is no doubt that the functions of command and triggering are real and distinct. Our job is to find out whether these functions reflect properties of the activating cells, the activated network, or both. This is bound to be one of the more exciting areas of research on motor systems in the years to come.

Positive feedback, based on reciprocity, has yet another interesting consequence, and that is command flexibility, the third general issue I wish to discuss. Information theory holds that command flexibility serves the interest of efficiency, and that the property of command should be conferred according to the locus of the "most important information."

This same notion may be applied to motor networks, where the source of the most important information ought also to be the source from which the command is derived. It can be implemented by incorporating reciprocity into motor networks, with the attendant possibility of many different command loci. By this arrangement we have the command flexibility which information theorists tell us we should have.

Now on top of command flexibility there is yet another consequence of reciprocity and positive feedback, and that is control flexibility, the fourth general issue I wish to discuss. Whenever we deal with a network that exhibits reciprocal interconnections, we deal with a network in which there are many sites at which the network's activity can be modulated, and this leads to interesting and broad possibilities for functional gating. As to the practical application of such thinking, we saw in Rossignol's interesting paper that an afferent input from a cat's paw can supplement ongoing extensor <u>or</u> flexor discharge, depending upon when in the step cycle the stimulus is delivered. We may speculate that the cat locomotor network is a reciprocally connected one, with sensory inputs from the paw routed to flexor and extensor units alike, resulting in the demonstrated control flexibility. This is but one attempt at practical application of the kind of network thinking I am advocating. Perhaps some of the people here that study cat walking could comment on this possibility. Specifically, is cutaneous input from the dorsum of the foot routed broadly to both extensors and flexors? In this case the suppression of an agonist population's discharge would have to be accomplished by another reflex or by central motor program -- i.e., extensor units would have to inhibit flexors strongly to cancel the reflex excitation.

Finally, the last two issues that I would like to discuss in a very general way are the issues of command specialization and plasticity. Some of us work on motor systems because we want to know how we walk. Others of us, including myself, work on motor systems because they provide a good place to look for the cellular changes that underlie plasticity. I think that we are discussing issues in this symposium that have profound implications for thinking of plasticity. If, in fact, there are many different sites in a network at which the activity of that network can be modulated, then it follows that there are many different sites within a network at which the cellular changes that underlie plasticity could take place. Now, in theory at least, in a completely reciprocally coupled network, changes of the kind that underlie learning and memory could happen anyplace and everyplace. Lashley's unsuccessful search for the memory engram in a rat brain suggests the possibility that plasticity is distributed throughout the entire network which undergoes the plastic changes. This is a somewhat disconcerting possibility because motor networks have a lot of cells in them, and if we have to look everywhere for the changes in plasticity, it is conceivable that the changes in any single cell may be undetectable. And that of course would mean that the powerful methods of intercellular analysis of single neurons may not provide quite as much information about plasticity as we hoped.

I think there is a ray of hope, however. We talked about the fact that reciprocal interconnections within a network result in a loss of identity of individual elements within the network, but we also talked about the possibility -- and it has to be labeled a possibility because we don't have enough evidence to label it anything else -- that certain cells within the network have privileged access to information outside the network. It seems to me in fact possible that this is the way so-called command cells exercise their function of command. That is to say, they may have privileged access to information outside the network they drive. If this is the case, then even though the functional changes which underlie plasticity may be widely distributed, perhaps they will be concentrated in the very key cells whose function is to turn on the network by virtue of this unique access to external information. In short, my thinking on motor networks leads me to seek the cellular basis of plasticity in the higher order cells that function to activate motor networks -- the command neurons.

THE INTERACTION OF CENTRAL COMMANDS AND PERIPHERAL FEEDBACK IN PYRAMIDAL TRACT NEURONS (PTNs) OF THE MONKEY

DR. EVARTS[x]: Within recent years there have been a series of important advances in our understanding of spinal mechanisms subserving locomotion. These advances, led by workers in the Soviet Union and Sweden, have shown that segmental mechanisms alone can subserve rhythmical stepping in cats with brain-stem transections and/or limb deafferentation. The finding that such complex patterns can be mediated by spinal mechanisms has raised questions as to the roles of the cerebrum and of peripheral feedback in locomotion. A second example of spinal mediation of a process long thought to require the intact cerebral cortex is seen in the work by Forssberg, Grillner, and Sjöström, (1974) showing that tactile placing can occur in chronic spinal kittens.

In considering the implications of these findings as to the role of the cerebral cortex and sensory feedback in motor activity generally or locomotion in particular, it must be recognized that the capacity of a spinal or a deafferented animal to exhibit a given motor response when the cerebrum or the afferent input is absent does not exclude a role of the cerebrum or sensory input in mediating the motor response in the intact animal. Spinal and supraspinal centers normally function together and interact with continuous sensory input. It will be the purpose of this discussion to consider this interaction in control of movement in intact subjects.

The movement to be discussed was described by Hammond, (1956) in a paper entitled "The influence of prior instruction on an apparently involuntary neuromuscular response." Hammond, using human subjects, found that sudden extension of the elbow (which stretched the biceps muscle) elicited both a 20-msec latency tendon jerk (TJ) and a 50-msec latency later muscle response which was

---
[x]E. Evarts, Laboratory of Neurophysiology, National Institute of Mental Health, Bethesda, Maryland 20014.

CONCLUSION

present when the subject had been instructed to resist the extension but disappeared when the subject had been instructed not to resist. The consistency and relatively short (50-msec) latency of this intended EMG response in Hammond's subjects led him to suggest that it was a stretch reflex. He added, however, that "this must be reconciled with the fact that prior instructions to 'let go' can interfere so rapidly and effectively with the subject's response." Hammond accordingly suggested that the short-latency but intention-dependent EMG discharge was a spinal stretch reflex that had been preset by supraspinal activity determined by the prior instruction.

We, (Evarts and Tanji, 1974), have studied the central events associated with this phenomenon in monkeys trained to grasp a handle and to react to a sudden perturbation of the handle according to a prior instruction. A red light "instructed" the monkey to PULL the handle toward itself when the perturbation occurred and a green light meant that the monkey should PUSH the handle away when the perturbation occurred. There were two different directions of perturbation. One direction of perturbation (PER-PULL) opposed push, moving the handle toward the monkey. The other direction of perturbation (PER-PUSH) opposed pull, moving the handle away from the monkey. A given instruction (INS-PULL or INS-PUSH) called for a movement synergistic with the TJ for one direction of perturbation and antagonistic to this reflex for the other direction of perturbation. The instruction told the monkey how to move and the handle perturbation told the monkey when to move. The instruction itself did not induce any changes in EMG activity, muscular responses occurring only after the perturbation.

The interaction between instruction and perturbation for biceps muscle is illustrated in Figure 1. Maximum biceps activity occurred when PER-PUSH involving biceps stretch and eliciting a TJ was paired with INS-PULL, an instruction which had called for a movement involving biceps contraction. For this combination, the perturbation and prior instruction both called for biceps activity. Minimum biceps activity occurred when neither the perturbation nor the prior instruction called for biceps activity: a perturbation (PER-PULL) which shortened the biceps was paired with a prior instruction to PUSH. In this situation both the instruction and the perturbation called for triceps discharge and biceps quiescence.

For the two remaining pairings, the perturbation and prior instruction were antagonistic. For one case, the instruction (INS-PULL) called for biceps contraction but the segmental reflex effects of the perturbation (PER-PULL, involving biceps shortening) tended to silence biceps activity. Here the TJ was absent, and the intended response of biceps called for by the prior instruction had a longer latency than when the perturbation involved biceps stretch.

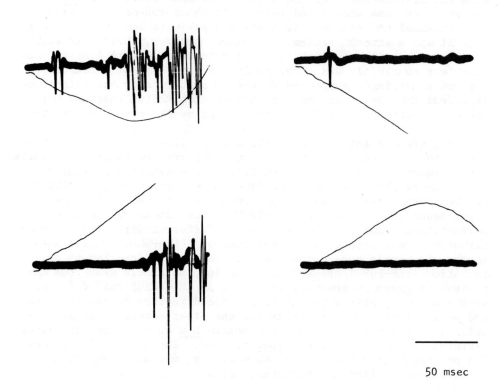

Figure 1. The four sets of traces show biceps activity with the four different instruction-perturbation combinations described in the text. Instructions to PULL or PUSH and perturbations toward or away from the monkey could be combined in four possible ways. Each pair of tracings shows biceps EMG activity and the output of a potentiometer coupled to the handle, with upward deflection of the potentiometer tracing indicating movement of the handle toward the monkey and downward deflection indicating movement of the handle away from the monkey. For the set of traces at upper left, the prior instruction was PULL, calling for biceps contraction, and the PUSH-perturbation (indicated by the potentiometer tracing) moved the handle away from the monkey, thereby stretching the biceps. At lower right, the instruction was PUSH, and the PULL-perturbation moved the handle toward the monkey, resulting in biceps shortening. For further explanation, see text. (Reprinted from Evarts and Tanji, 1974.)

CONCLUSION

The last of the four instruction-perturbation pairings involved a perturbation (PER-PUSH) which stretched biceps (tending to elicit biceps contraction) but the prior instruction was INS-PUSH, which did not call for biceps contraction. Here again there was antagonism between the perturbation and the instruction, and this antagonism resulted in a reduction in the TJ elicited by biceps stretch: the TJ elicited by biceps stretch was smaller when the stretch was coupled with a prior instruction calling for PUSH than when coupled with a prior instruction calling for PULL. It is thus apparent that the two different instructions give rise to differential presetting of spinal cord reflex mechanisms mediating the TJ.

Recordings of arm area motor cortex pyramidal tract neurons (PTNs) in these situations were aimed at determining the possible role of PTNs in 1) the presetting of spinal reflexes and 2) the mediation of the intended motor response, which though enhanced by stretch, was more dependent on a prior instruction calling for a movement involving activity of the muscle in question. Recordings were obtained from the side contralateral to the arm used by the monkey.

Responses to the Instruction

Recordings in pre- and postcentral sensorimotor cortex revealed instruction-induced changes of neuronal activity during the period intervening between the instruction and the perturbation-triggered movement. Effects of the instruction were differential depending on which of the two instructions was given, such differential responses to the instruction being detected in 61% of precentral PTNs, 44% of precentral non-PTNs, and 11% of postcentral neurons. The magnitude as well as the prevalence of the instruction effect was greater in precentral than in postcentral neurons.

Since motor cortex PTN axons end on alpha and gamma motoneurons and on interneurons of the spinal cord, changes of PTN activity with "intention" or "motor set" provide a mechanism for suprasegmental control and presetting of spinal cord reflex excitability specific to the nature of an impending movement. The modification of TJ by prior instruction, described by Hagbarth, (1967) may in part depend on this "presetting."

Responses to Perturbation

PTN discharge associated with movements triggered by perturbations to the arm has two components: 1) a relatively short-latency reflex component which depends on the direction of the perturbation to the arm, and 2) a longer-latency intended component which depends on the movement that the monkey has been instructed to perform. Figures 2 and 3 show these two components, and illustrate how an individual PTN is the site of both intended and reflex controls. While the first (reflex) component is input-related and the second (intended) component is output-related, interaction of input and output is seen for both components. Thus, a perturbation which is reflexly excitatory for the first (or input) component of PTN discharge speeds the occurrence of and intensifies the second (or output) component. Also, an intended movement which involves discharge of a given PTN enhances the response of the PTN to an excitatory perturbation. When a perturbation which is excitatory for a PTN triggers a movement which requires quiescence of the PTN (as in Figure 2, right), only the first (or input-related) component occurs. Conversely, when a movement involving discharge of the PTN is triggered by a perturbation which suppresses activity of the PTN, only the second (or output-related) component of PTN discharge occurs (Figure 3, right).

Analyses of the interactions between the intended movement and the perturbation triggering this movement were carried out with the aim of determining whether the changes of perturbation-evoked PTN discharge as a result of prior instruction might in part mediate the short-latency but intended motor response. It was found that the timing of the changes of this perturbation-evoked PTN activity was consistent with the hypothesis that PTN output is one of the factors mediating the intended response, i.e., that phase which is so critically dependent on the prior instruction. The expression "one of the factors" is to be emphasized, for even prior to the triggering perturbation and the PTN activity which it evokes, the instruction has already led to alterations of PTN (and non-PTN) discharge in motor cortex. Thus, in seeking to understand the mechanisms whereby a given peripheral event can trigger a given movement or its opposite depending on the prior instruction, one must bear in mind that the state of excitability of spinal cord reflex centers depends upon the "set" of "intention" of the subject. It follows that the presence or absence of the intended response depending on the prior instruction involves both segmental and suprasegmental loops.

These results show that both spinal and cerebral responses are modified by prior instruction, and it seems reasonable to conclude that both of these modifications underlie the difference in the intended response, i.e., the phase of muscular response which is responsible for the performance of the movement called for by the instruction. Hammond, (1956) first proposed that this phase was a spinal reflex which could be preset by nervous activity from the brain, but later (Hammond, 1961) suggested that the instruction-dependent phase of muscle activity might be mediated by "centres in

CONCLUSION

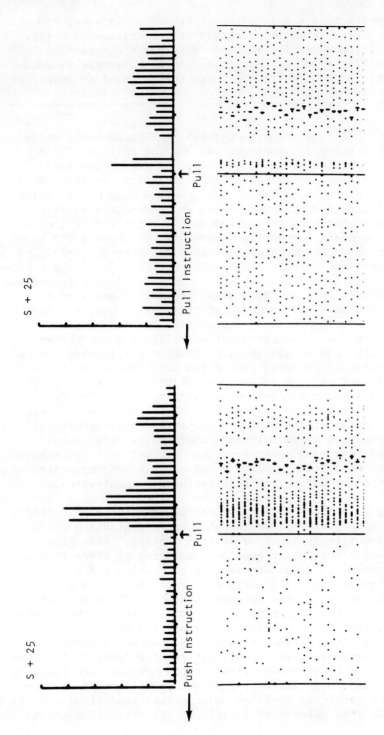

Figure 2. This PTN discharged with intended PUSH-movements and was reflexly activated by opposing PULL-perturbations. In the displays above, a PULL-perturbation was delivered at the center arrow, and activity is shown for 500 msec before and after the PULL-perturbation. In the display at the left, where the prior instruction was PUSH, reflex PTN discharge evoked by the perturbation merged with intended PTN discharge. At the right, however, where the intended PULL-movement involved PTN silence, the reflex PTN discharge (Legend continued on next page.)

evoked by the perturbation was quickly replaced by intended PTN silence. Here, then, is an illustration of convergence of reflex excitation and intended inhibition onto this motor cortex PTN. Histogram bin width is 20 msec, and the single heavy mark in each row of the raster shows the time at which the intended movement was completed.

---

spine or brain." We have observed changes of PTN discharge which vary depending on the prior instruction, and the ability of motor cortex PTN output to preset spinal reflexes is well documented. Granted that PTNs function in presetting spinal cord reflexes as a result of prior instruction, is it also possible that motor cortex neurons participate in a high speed loop which actually <u>mediates</u> the later phase of perturbation-evoked muscle activity? Evidence for such cerebral mediation has been provided both by our PTN recordings and by the work of Marsden, Merton and Morton, (1973). These authors obtained latency measurements compatible with a cortical pathway for the intended response in man. But is the response spinal or cerebral? This question might persist if one were to accept the assumption that the muscular response in question must be exclusively one or exclusively the other. Perhaps we should reject this assumption. The perturbation triggers motor cortex PTNs into activity soon enough to allow them to participate in controlling motoneuronal discharge during the intended phase of muscular discharge, and, of course, segmental inputs are reaching these same motoneurons simultaneously and are also participating in this control. The answer to the question "spinal or cerebral?" is therefore "both." The fact that motor cortex participates in control does not prove failure of spinal participation, or vice versa. Indeed, it seems that many motor functions are multiply represented and that control of these functions involves simultaneous action of all levels of the nervous system's hierarchical organization.

In closing, I would like to consider these results in relation to some of the concepts of hierarchical motor organization which have been prominent in discussions at this meeting. This hierarchical organization has been viewed as consisting of three levels: 1) a command system, 2) a pattern generator to which the command system sends signals, and 3) a peripheral apparatus which gives rise to movement and generates sensory feedback. There is sometimes a tendency to think that if an element is part of the command system, then it should <u>not</u> be part of the pattern generator. Along the same lines, it is sometimes thought that command elements responsible for central programs should <u>not</u> receive feedback from the periphery. We know that command elements can operate when feedback has been eliminated, but this fact does not prove that the command elements are not receiving feedback under normal conditions. Davis has already made this point most lucidly in the discussion which

CONCLUSION

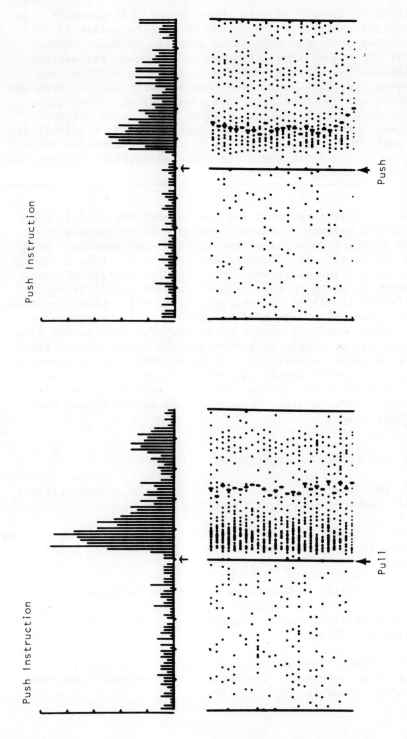

Figure 3. This PTN was reflexly excited by PULL-perturbations (left) and reflexly inhibited by PUSH-perturbations (right). It discharged with intended PUSH-movements, which were called for by PUSH-instructions in the displays at both left and right. At the left, reflex excitation due to the PULL-perturbation merges with
(Legend continued on next page.)

intended discharge associated with the intended PUSH-movement. At the right, reflex inhibition due to the PUSH-perturbation is followed by intended discharge associated with the intended PUSH-movement. At the right, then, is shown a case where reflex inhibition and intended excitation converge on the same motor cortex PTN. Histogram bin width is 10 msec, and activity is displayed for 500 msec before and 500 msec after the perturbation. The single heavy mark in each row of the raster shows the time at which the intended movement was completed. This mark occurs more quickly at the right, where the PUSH-perturbation assists the intended PUSH-movement, than at the left, where the PULL-perturbation opposes the intended PUSH-movement.

has preceded my present remarks, and has pointed out that in a normally functioning network, command elements will be quickly impinged on by feedback from the movements that they command. PTNs have several properties of command elements. First, they discharge prior to volitional movements even after deafferentation of the body part which is to be moved. Second, the axons of PTNs project to many different levels of the neuraxis (cerebral cortex, thalamus, pontine nuclei, dorsal column nuclei, spinal cord) whose coordinated functioning underlies normal movement. For purposes of discussion, then, let us assume that PTNs are the motor command elements of the cerebral cortex. If this be so, then the existence of rapid feedback onto PTNs during intended movement in the monkey provides an example of what Davis proposed: peripheral sensory inputs rapidly modify central commands in the intact animal with intact feedback loops.

REFERENCES

Evarts, E.V. and Tanji, J., (1974) Gating of motor cortex reflexes by prior instruction. Brain Res. 71, 479-494.

Forssberg, H., Grillner, S. and Sjöström, A., (1974) Tactile placing reactions in chronic spinal kittens. Acta Physiol. Scand. 92, 114-120.

Hagbarth, K.-E., (1967) "EMG studies of stretch reflexes in man," In Recent Advances in Clinical Neurophysiology, Electroenceph. Clin. Neurophysiol. Suppl. 25, (Widen, L., ed.), Amsterdam, Elsevier, (74-79).

Hammond, P.H., (1956) The influence of prior instruction to the subject on an apparently involuntary neuromuscular response. J. Physiol. (Lond.), 132, 17P-18P.

Hammond, P.H., (1961) "An experimental study of servo action in human muscular control," In Proc. Third Internatl. Conf. on Medical Electronics, Springfield, Ill., Charles C. Thomas, (190-199).

Marsden, C.D., Merton, P.A. and Morton, H.B., (1973) Latency measurements compatible with a cortical pathway for the stretch reflex in man. J. Physiol. (Lond.), 230, 58P-59P.

SUBJECT INDEX

Adaptation, 137-180, 265-292
Ankle, 13-50, 65-76, 99-136,
　　439-464, 647-674,
　　763-766, 789-794

Brain Stem, 351-376
　　(See Descending
　　Modulation, Rubospinal,
　　Reticulospinal,
　　Cerebellum)
　in cat, 587-616
Burst Generation, 796-798
　in lobster, 377-400
　in mollusk, 327-350

Cat, 99-136, 181-202, 439-464,
　　489-518, 519-538
　decerebrate (and mesence-
　　phalic preparation),
　　181-202, 439-464,
　　489-518, 519-538,
　　769-774, 789-794
　normal, 99-136, 489-518,
　　519-538
　spinal, 439-464, 489-518,
　　587-616, 647-674
　thalamic, 519-538, 587-616,
　　763-766
Central Program
　in cat, 439-464, 489-518,
　　587-616
　in crustacea, 137-180,
　　617-646
　in development, 735-758
　in fish, 181-202
　in primate, 675-706

Cerebellum, 351-376
　pathways, 351-376, 587-616
　unit recordings, 781-784
Characteristics of Gait
　　(See Step Cycle)
　interlimb, 51-64, 137-180,
　　203-236, 489-518
　single limb, 13-50, 65-76,
　　99-136, 401-418, 439-464,
　　519-538
　whole limb, 99-136
Chemical Mediation, 351-376
　in cat, IV Dopa, 439-464
　in cat, 6-hydroxydopamine,
　　769-774
Chick, 707-734
Cockroach, 401-418, 519-538,
　　561-586
Command Neurons (See Neurons)
Coordination, 798-804
　　(See Interlimb)
Crab, 785-788
Crayfish, 293-326, 465-488
Crustacea, 137-180, 293-326,
　　617-646

Deafferentation
　in cat, 439-464, 587-616
　in cockroach, 401-418
　in lobster, 377-400
　in primates, 675-706
Descending Modulation, 351-376
　activation of noradrenergic
　　pathways, 181-202, 439-464,
　　587-616, 769-774

Descending Modulation (Cont'd)
  extrapyramidal tracts, 351-376, 587-616
  in cat, 351-376, 587-616
  in lobster, 377-400
  on ascending pathways, 351-376, 587-616
Development, 707-734, 735-758, 775-780

Efferents
  in cat, 439-464, 587-616
EMG
  in cat, 439-464, 489-518, 519-538, 587-616, 647-674, 763-766, 781-784
  in crab, 785-788
  in fish, 181-202, 647-674
  in man, 13-50, 51-64, 65-76
  in rabbit, 767-768
Energy, 1-12, 13-50, 77-98
Eye, 798-804

Facultative Behavior, 99-136, 137-180
Fish, 181-202, 647-674
Footfall Patterns, 203-236
Fusimotor Activity, 351-376
  in cat, coactivation, 439-464, 587-616

Gaits (See Characteristics, Step Cycle)
  galloping, 99-136, 203-236, 489-518
  pacing, 99-136, 203-236, 489-518
  slow, 13-50, 51-64, 65-76, 77-98
  symmetry, 203-236
  trotting, 99-136, 203-236, 489-518
Galloping, 99-136, 203-236, 489-518
Ganglia
  metacerebral, 265-292
  metathoracic, 401-418

Ganglia (Cont'd)
  stomatogastric, 377-400
  thoracic, 419-438
Hip
  afferents, 439-464, 519-538
  amplitude, 13-50, 51-64, 489-518
  height, 99-136
  phase relations, 13-50, 51-64, 489-518

Initiation of Gait, 65-76
  (See also Transition)
Interlimb Coordination, 351-376
  in cat, 489-518
  in man, 51-64
  in tetrapods, 203-236
Interlimb Phase
  in cat, 489-518
  in crayfish, 465-488
  in crustacea, 137-180
  in man, 51-64
Interneurons (See Neurons)

Kinematics
  in cat, 99-136, 401-418, 438-464, 489-518, 519-538, 796-798
  in man, 13-50, 51-64, 65-76, 77-98
  in tetrapods, 203-236
Kinetics
  effect of loading, 13-50, 519-538
  ground reaction forces, 13-50, 65-76, 759-763
  joint forces, 13-50
  muscle forces, 13-50, 439-464, 519-538, 789-794
  physical properties, 1-12
Knee
  in cat, 439-464, 489-518, 647-674, 767-768
  in man, 13-50

Load Compensation, 13-50, 65-76, 137-180, 439-464
Loading Reflexes, 137-180, 519-538

INDEX

Lobster, 377-400
Locust, 419-438

Man, 1-12, 13-50, 51-64,
    65-76, 77-98
Mollusk, 327-350
Momentum, 13-50, 539-560
Moth, 775-780
Motoneuron (See Neurons)
Motor Units, 789-794
  contractile properties of,
    789-794
Muscle
  "catch and throw" properties,
    539-560
  filtering properties, 13-50
  force, 1-12, 13-50, 77-98,
    439-464, 519-538,
    539-560
  stiffness, 13-50, 65-76,
    439-464
Muscle Spindle
  primaries of, 767-768

Neck Receptors, 539-560
Neuronal Noise, 13-50
Neurons, Classes of
  command neurons, 265-292,
    293-326, 327-350,
    351-376
  in crayfish, 293-326
  in crustacea, 293-326
  in mollusk, 327-350
  in pleurobranchaea,
    265-292
  properties of, 265-292,
    293-326, 327-350,
    804-808
  sensory feedback on,
    265-292, 293-326,
    804-808
  coordinating interneurons,
    265-292, 351-376, 465-488
  in crayfish, 465-488
  motoneurons, 265-292, 293-326,
    351-376, 419-438
  in cat, 439-464, 587-616
  in moth, 775-780
  in rabbit, 767-768

Neurons, Classes of (Cont'd)
  oscillator and rhythm generator
    interneurons, 265-292,
    293-326, 401-418, 419-438
  in cockroach (Interneuron I),
    401-418
  in locust, 419-438
Neurons, Concepts/Organization
  theory of "least interaction,"
    13-50
  theories of network properties,
    265-292, 327-350, 351-376
Noradrenergic Pathways (See
  Descending Modulation)

Ontogeny (See Development)

Pacing, 99-136, 203-236, 489-518
Pattern Generation
  in cat, 351-376, 439-464,
    489-518, 519-538, 647-674
  in cockroach, 401-418, 519-538
  in crustacea, 617-646
  in fish, 181-202
  in lobster, 377-400
  in locust flight, 419-438
  in man, 13-50, 65-76
  in mollusk, 327-350
  in moths, 775-780
  in primates, 675-706
Phase Coupling, 465-488
  alternate, in-phase, 51-64,
    489-518
  reflexes, 561-586
  switching, 51-64, 489-518,
    647-674
  transition, 51-64, 65-76,
    489-518
Phase "Reversal", 647-674
Pleurobranchaea, 265-292
Plasticity, 137-180, 265-292,
  804-808
Primates, 675-706
Proprioception
  and learning, 137-180
  in crab, 785-788
  in crustacea, 617-646
Propriospinal Pathways, 489-518

Pyramidal Tract, 351-376
  neurons, 808-817

Rabbits, 767-768
Redundancy, 804-808
Reflex Control
  in cat, 519-538
  in cockroach, 519-538
  in crustacea, 617-646
Reflex Effects, 519-538,
    561-586, 647-674,
    796-798
  and intended movement,
    808-817
  extensor behavior, 763-766
  flexor behavior, cats,
    519-538, 587-616
  on pattern generator
    in crab, 785-788
    in locusts, 419-438
  on phase relations, in
    cockroach, 561-586
  phase dependency, cats,
    647-674
Reticulospinal Pathways,
    351-376, 587-616
Robot, 237-264
Rubospinal System, 351-376,
    587-616

Scapula, 489-518
Shoulder, 51-64, 489-518
Spinal Generator
  in cat, 439-464, 489-518,
    587-616
  models of, 439-464, 519-538
Stance Phase (*See* Kinematics)
Statistical Treatment, 13-50
Step Cycle (*See* Kinematics,
    Kinetics, Transition)
  ground reaction model,
    13-50
  Philippson model, 13-50,
    99-136, 351-376

Stretch Reflexes, 808-817
Swimming, 181-202, 465-488,
    647-674
Swing Phase (*See* Kinematics
Switching
  central-peripheral organization,
    137-180, 785-788
  phase coupling, 51-64, 489-518,
    647-674
Synapse Formation, 707-734

Tetrapods
  ground reaction measures,
    203-236
Trajectory of Limb, 99-136
Transition in Gait Cycle
  in cat, 519-538
  in man, 65-76
Trotting, 99-136, 203-236,
    351-376, 489-518, 759-763

Velocity
  joint displacement, 13-50,
    51-64, 65-76, 439-464,
    489-518
  muscle lengthening, 13-50,
    65-76, 439-464, 519-538
  step cycle, 13-50, 51-64,
    65-76, 351-376
Vestibular, 65-76, 351-376,
    539-560
  head-eye control, 798-804
  initiation of gait, 65-76
  receptors, 539-560
  vestibulospinal pathways, 65-76,
    351-376

Work
  in man, 1-12, 77-98